Horticulture and Plant Breeding

Horticulture and Plant Breeding

Editor: Clive Koelling

Callisto Reference
www.callistoreference.com

Callisto Reference,
118-35 Queens Blvd., Suite 400,
Forest Hills, NY 11375, USA

Visit us on the World Wide Web at:
www.callistoreference.com

ISBN: 978-1-63239-906-9 (Hardback)

The publisher's policy is to use permanent paper from mills that operate a sustainable forestry policy. Furthermore, the publisher ensures that the text paper and cover boards used have met acceptable environmental accreditation standards.

Printed in China.

Cataloging-in-Publication Data

Horticulture and plant breeding / edited by Clive Koelling.
p. cm.
Includes bibliographical references and index.
ISBN 978-1-63239-906-9
1. Horticulture. 2. Plant breeding. I. Koelling, Clive.
SB318 .H67 2018
635--dc23

Table of Contents

Preface

Horticulture is the study of plant production techniques that are aimed at intensive cultivation, high-yielding variety crops and plant propagation. Plant breeding concentrates on the growing of plants that have favorable characteristics for longevity and improved yields. Horticulture and plant breeding are important sectors of the food industry. Modern plant breeding creates transgenic plants through genetic modification. High standards are kept in selecting the right species for genetic modification. This book includes some of the vital pieces of work being conducted across the world, on various topics related to plant breeding and horticulture. The various studies that are constantly contributing towards advancing technologies and evolution of this field are examined in detail. Students, researchers, experts and all associated with this field will benefit alike from this book.

This book is a result of research of several months to collate the most relevant data in the field.

When I was approached with the idea of this book and the proposal to edit it, I was overwhelmed. It gave me an opportunity to reach out to all those who share a common interest with me in this field. I had 3 main parameters for editing this text:

1. Accuracy – The data and information provided in this book should be up-to-date and valuable to the readers.

2. Structure – The data must be presented in a structured format for easy understanding and better grasping of the readers.

3. Universal Approach – This book not only targets students but also experts and innovators in the field, thus my aim was to present topics which are of use to all.

Thus, it took me a couple of months to finish the editing of this book.

I would like to make a special mention of my publisher who considered me worthy of this opportunity and also supported me throughout the editing process. I would also like to thank the editing team at the back-end who extended their help whenever required.

Editor

Variation in Broccoli Cultivar Phytochemical Content under Organic and Conventional Management Systems: Implications in Breeding for Nutrition

Erica N. C. Renaud[1]*, Edith T. Lammerts van Bueren[1], James R. Myers[2], Maria João Paulo[3], Fred A. van Eeuwijk[3], Ning Zhu[4], John A. Juvik[4]

1 Wageningen UR Plant Breeding, Plant Sciences Group, Wageningen University, Wageningen, The Netherlands, **2** Department of Horticulture, Oregon State University, Corvallis, Oregon, United States of America, **3** Biometris, Plant Sciences Group, Wageningen University, Wageningen, The Netherlands, **4** Department of Crop Sciences, University of Illinois, Urbana, Illinois, United States of America

Abstract

Organic agriculture requires cultivars that can adapt to organic crop management systems without the use of synthetic pesticides as well as genotypes with improved nutritional value. The aim of this study encompassing 16 experiments was to compare 23 broccoli cultivars for the content of phytochemicals associated with health promotion grown under organic and conventional management in spring and fall plantings in two broccoli growing regions in the US (Oregon and Maine). The phytochemicals quantified included: glucosinolates (glucoraphanin, glucobrassicin, neoglucobrassin), tocopherols (δ-, γ-, α-tocopherol) and carotenoids (lutein, zeaxanthin, β-carotene). For glucoraphanin (17.5%) and lutein (13%), genotype was the major source of total variation; for glucobrassicin, region (36%) and the interaction of location and season (27.5%); and for neoglucobrassicin, both genotype (36.8%) and its interactions (34.4%) with season were important. For δ- and γ-tocopherols, season played the largest role in the total variation followed by location and genotype; for total carotenoids, genotype (8.41–13.03%) was the largest source of variation and its interactions with location and season. Overall, phytochemicals were not significantly influenced by management system. We observed that the cultivars with the highest concentrations of glucoraphanin had the lowest for glucobrassicin and neoglucobrassicin. The genotypes with high concentrations of glucobrassicin and neoglucobrassicin were the same cultivars and were early maturing F_1 hybrids. Cultivars highest in tocopherols and carotenoids were open pollinated or early maturing F_1 hybrids. We identified distinct locations and seasons where phytochemical performance was higher for each compound. Correlations among horticulture traits and phytochemicals demonstrated that glucoraphanin was negatively correlated with the carotenoids and the carotenoids were correlated with one another. Little or no association between phytochemical concentration and date of cultivar release was observed, suggesting that modern breeding has not negatively influenced the level of tested compounds. We found no significant differences among cultivars from different seed companies.

Editor: Hany A. El-Shemy, Cairo University, Egypt

Funding: Funding was provided by Seeds of Change (www.seedsofchange.com) and monetary funding from Wageningen University, University of Illinois and Oregon State University. The funder had no role in the study design, data collection and analysis, decision to publish, or preparation of the manuscript.

Competing Interests: The principle researcher, Erica Renaud was employed by Seeds of Change and later Vitalis Organic Seeds, the organic division of Enza Zaden, a vegetable seed company.

* Email: E.Renaud@enzazaden.com

Introduction

Organic food consumption is in part driven by consumer perception that organic foods are more nutritious and simultaneously less potentially harmful to human health [1–2]. Studies, such as Smith-Sprangler et al. [3], have concluded that there is little evidence for differences in health benefits between organic and conventional products, but other studies have indicated that organic vegetables and fruits contain higher concentrations of certain plant phytochemicals associated with health promotion than those produced conventionally [4–8]. A number of these compounds are produced by plants in response to environmental stress or pathogen infection, providing a potential explanation of why concentrations of these compounds might be higher in plants grown in organic systems without application of pesticides [9]. In addition, higher phytochemical levels may be due to the effects that different fertilization practices have on plant metabolism. Synthetic fertilizers used in conventional agriculture are more readily available to plants than organic fertilizers [10]. Nutrients derived from organic fertilizers need to be mineralized, and the availability of these nutrients depends on soil moisture, temperature and level of activity of soil organisms [11]. Conventional systems seek to maximize yields, resulting in a relative decrease of plant phytochemicals and secondary metabolites [12–15]. Correspondingly, compounds such as phenolics, flavonoids, and indolyl glucosinolates may be induced by biotic or abiotic stress [16–17].

Broccoli is an abundant source of nutrients, including provitamin A (β-carotene), vitamin C (ascorbate), and vitamin E (tocopherol) [18]. It is also a source of phytochemicals associated with health benefits and these include glucosinolates, carotenoids, tocopherols, and flavonoids [19–21]. Verhoeven et al. [22], Keck and Finley [23] and Here and Büchler [24], reported that diets rich in broccoli reduce cancer incidence in humans. Strong associations between consumption level and disease risk reduction exists for glucosinolates (anti-cancer), tocopherols (cardiovascular), and the carotenoids (eye-health) [25].

Sulfur containing glucosinolates are found in the tissues of many species of the *Brassicaceae* family. When glucosinolates are consumed, they are hydrolyzed into isothiocyanates (ITC) and other products that up-regulate genes associated with carcinogen detoxification and elimination. Aliphatic glucoraphanin (up to 50% of total glucosinolates) and the indolylic glucosinolates, glucobrassicin and neoglucobrassicin are abundant in broccoli florets [20,19,26]. Glucoraphanin is hydrolyzed either by the endogenous plant enzyme myrosinase [27–28] or by gut microbes to produce sulforaphane, an ITC. The indole glucosinolates are tryptophan-derived in a similar but alternate biosynthetic pathway [29]. The health promoting effects of the indolyl glucosinolates are attributed to indole-3-carbinol, a hydrolysis product of glucobrassicin, N-methoxyindole-3-carbinol and neoascorbigen, hydrolysis products from neoglucobrassicin, and the catabolic products derived from alkyl glucosinolates. Clinical studies have shown that the glucosinolate hydrolysis products reduce the incidence of certain forms of cancer (e.g., prostate, intestinal, liver, lung, breast, bladder) [30–35]. The lipophilic phytonutrients found in broccoli include the carotenoids lutein, zeaxanthin, β-carotene, and tocopherols (forms of vitamin E) [36–37]. In addition to their role as vitamins, these compounds are powerful antioxidants [38–39]. Consumption of vegetables high in tocopherols and carotenoids has decreased the incidence of certain forms of cancer [40]. Lutein and zeaxanthin protect against development of cataracts and age-related macular degeneration [41]. Tocopherols have also been associated with reduced risk of cardiovascular disease by preventing oxidative modification of low-density lipoproteins in blood vessels [42].

The genetic potential for high nutrient content has long been a concern of the organic industry in order to meet the expectations of organic consumers. This has often been manifested by questioning whether modern elite cultivars may have lower levels of nutritional content than older open pollinated cultivars. Indirect evidence supporting this argument comes from Davis et al. [43], who compared USDA nutrient content data for 43 garden crops released between 1950 and 1999. Statistically significant decreases were noted for six nutrients (protein, calcium, potassium, iron, riboflavin, and ascorbic acid), with declines ranging from 6% for protein to 38% for riboflavin. Crop varieties in 1950 had been bred to be adapted to specific regions and a relatively low input agriculture system, but contemporary cultivars are selected for yield, disease resistance, broad adaptation to high input agriculture systems, and for increased 'shipability' and shelf life. Traka et al. [44] recommend breeding with greater genetic diversity when the goal is enhanced phytochemical content by exploiting wild crop relatives. The genotype is important in determining the level of nutrients in a crop cultivar [45–47]. What is unclear, however, is whether the nutritional content of a cultivar is associated with certain genotypic categorization, e.g. old versus modern, open pollinated versus F_1 hybrid cultivars. In addition, there is no clear differentiation as to what extent nutritional content in a crop is determined by genotypic or by field management factors or by the interaction of both. Some studies comparing performance of genotypes in organic and conventional production systems have shown that for certain agronomic traits, cultivars perform differently between the two production systems (e.g. for winter wheat: Murphy et al. [48], Baresel et al. [49]; for lentils: Vlachostergios et al. [50]; for maize: Goldstein et al.[51]), while others have shown no differences in ranking performance (for maize: Lorenzana and Bernardo [52]; for onions: Osman et al. [53]; for cereals: Przystalski et al.[54]). The results of these studies have profound implications for organic cultivar selection and breeding strategies and raise questions as to the need for cultivars to be bred with broad adaptability or specific adaptation for the requirements of regional organic production and for designing breeding programs that optimize phytochemicals in an adapted management system.

Previous studies comparing organically versus conventionally grown broccoli for nutritional quality have been 'market basket' (off-the-shelf) studies [55–56]. Harker [57] explained that the limitation of market basket studies is that they either have purchased the products from the store shelf and cannot relate differences to specific growing conditions or that the number of cultivars is too small to generalize the results. While other studies have compared cultivars from one production season time period to another, knowledge of the actual cultivar and production system (soil quality, temperature, rainfall) was not available [58,43]. The concentrations and form of health-promoting nutrients in *Brassica* vegetables have been reported to vary significantly due to (1) genotype (cultivar and genotypic class) [59,20,26,60,21,37,61,44], (2) environmental conditions such as season [62–67], light [19], max/min temperature, irrigation [68–69], (3) genotype by environment interactions [19,70–71]; (4) management system including soil fertility [72–73], organic versus conventional [13,74–75], days to harvest [63–64], and (5) post-harvest management [76–77]. Identifying specific growing conditions and genotypes that produce cultivars with varying phytochemical content and putative disease-prevention activity could offer value-added commercial opportunities to the seed and food industry.

In addition to research conducted on how broccoli genotypes, management system and environment interact for horticultural traits [78], we address in this paper the question of how do genotypes, management system and environment interact to determine the nutritional contributions of broccoli to the human diet. We studied the relative importance and interaction among genotypes (cultivars, genotypic classes) and environment {management system [M: organic (O) or conventional (C)], season (S, a combination of year and season within year, i.e., fall 2006, spring 2007, fall 2007, spring 2008), location (E)} in a set of 23 broccoli cultivars for floret glucosinolate, tocopherol and carotenoid concentrations grown under organic and conventional production systems in two contrasting broccoli production regions of the US: Oregon and Maine. Specifically we addressed the following questions: (1) what is the impact of organic management system compared to the environmental factors including climatic region, season and their interactions [Genotype (G) x Environment (E) x Management System (M)]? (2) is there a significant difference in phytochemical content between different genotypes and genotypic classes (old and modern cultivars; open pollinated and F_1 hybrid cultivars; early and late maturing cultivars; and between different commercial seed sources)? (3) what is the best selection environment for a broccoli breeding program for enhanced phytochemical content?

Materials & Methods

Plant Material and Field Trial Locations

Twenty-three broccoli cultivars including open pollinated (OP) cultivars, inbred lines, and F_1 hybrids were included in field trials (**Table 1**). Cultivars were grown in a randomized complete block design with three replicates in Maine (ME)-Monmouth (Latitude 44.2386°N, Longitude 70.0356°W); and Oregon (OR)-Corvallis (Latitude 44.5647°N, Longitude123.2608°W)] with each location including organically (O) and conventionally (C) managed treatments. Plots contained 36 plants, planted in three rows of 12 plants at 46 cm equidistant spacing within and between rows. The 2006 trials had only 18 of the 23 entries, and the Oregon 2006 trial had only two replicates at the organic location. Field trials were conducted for three consecutive years with one production cycle in Fall 2006, two production cycles in Spring and Fall 2007 and one production cycle in Spring 2008. The primary management differences between the organic and conventional field trial sites are outlined in **Table S1 in File S1**, which describes the production system, soils, fertility applications, the applied supplemental irrigation, and weather conditions for the area of study. Further details of the field design are reported in Renaud et al. [78].

Field Data Collection

As plots approached maturity they were evaluated three times a week for field quality and broccoli heads that had reached commercial market maturity (approximately 10 to 12 cm in diameter for most of the cultivars while retaining firmness). Field quality traits evaluated on a 1 to 9 ordinal scale included head color, bead size, and bead uniformity. Average head weight was determined by taking the mean of the five individual heads per plot. Head diameter averaged for five heads at harvest maturity from each plot. Maturity was based on days to harvest from transplanting date. Detailed procedures and horticulture trait performance data are reported in Renaud et al. [78].

Broccoli Floret Samples and glucosinolate, tocopherol, and carotenoid analysis

In order to analyse nutritional compounds of the broccoli heads, the following procedure was followed: As plots approached maturity, five broccoli head tissue samples were harvested fresh from each subplot at each trial location and were composited into a single sample per replication. The samples were frozen at −20°C and shipped in a frozen state to the University of Illinois, Urbana-Champaign where they were freeze-dried and assessed for nutritional phytochemicals. Each sample was analyzed for the glucosinolates (glucoraphanin, glucobrassicin and neoglucobrassicin), carotenoids (β-carotene, lutein, and zeaxanthin), and tocopherols (δ-, γ-, α- tocopherol) by high-performance liquid chromatography (HPLC) analysis using analytical protocols described in Brown et al. [19] for glucosinolates, and Ibrahim and Juvik [37] for tocopherols and carotenoids. Glucosinolates in lyophilized floret tissue samples were extracted and analysed by HPLC using a reverse phase C18 column. Three hundred mg samples of broccoli floret tissue were weighed out for extraction and the HPLC quantification of the tocopherols and carotenoids.

Statistical Analysis

Various linear mixed models were used for the analysis of trait variation. We followed the same methodology as described in Renaud et al. [78], which was comparable to the approach followed by Lorenzana and Bernardo [52]. For fitting the linear

Table 1. Overview of commercially available broccoli cultivars, showing origin, main characteristics, included in paired organic - conventional field trials 2006–2008.

Cultivar	Abbreviation	Origin	Cultivar Type[a]	Date of Market Entry	Maturity Classification[b]
Arcadia	ARC	Sakata	F_1	1985	L
B1 10	B11	Rogers	F_1	1988	M
Batavia	BAT	Bejo	F_1	2001	M
Beaumont	BEA	Bejo	F_1	2003	L
Belstar	BEL	Bejo	F_1	1997	L
Diplomat	DIP	Sakata	F_1	2004	L
Early Green	EGR	Seeds of Change	OP	1985	E
Everest	EVE	Rogers	F_1	1988	E
Fiesta	FIE	Bejo	F_1	1992	L
Green Goliath	GRG	Burpee	F_1	1981	M
Green Magic	GRM	Sakata	F_1	2003	M
Gypsy	GYP	Sakata	F_1	2004	M
Imperial	IMP	Sakata	F_1	2005	L
Marathon	MAR	Sakata	F_1	1985	L
Maximo	MAX	Sakata	F_1	2004	L
Nutribud	NUT	Seeds of Change	OP	1990	E
OSU OP	OSU	Jim Myers, OSU	OP	2005	E
Packman	PAC	Petoseed	F_1	1983	E
Patriot	PAT	Sakata	F_1	1991	M
Patron	PAN	Sakata	F_1	2000	M
Premium Crop	PRC	Takii	F_1	1975	E
USVL 048	U48	Mark Farnham, USVL	Inbred	not released	L
USVL 093	U93	Mark Farnham, USVL	Inbred	not released	M

[a]Cultivar Type: F_1: hybrid; OP: Open Pollinated; Inbred.
[b]Maturity Classification: E: Early; M: Mid; L: Late.

mixed models, GenStat 15 (VSNi, 2012) was used. The models followed the set-up:

$$y = E + R(E) + G + G \times E + e.$$

Here y is the phytochemical response. Term E represents the environment in a very general sense, it includes all main effects and interactions of Season (S), Location (L) and Management (M). For analyses per location, the terms involving L were dropped. Similarly, for analyses regarding a specific management regime, the terms involving M were dropped. Term $R(E)$ is the effect of replicate within environment, and there were two or three replicates in individual trials. G and $G{\times}E$ are genotype and genotype by environment interaction effects, respectively. Finally e is a residual.

Variance components were reported as coefficients of variation, i.e., $CV = 100\sqrt{V}/x$ with V the variance corresponding to specific effects and x the trait mean. Repeatability was calculated from the variance components in its most general form as $H^2 = V_G/(V_G\ V_{GL}/nL + + V_{GS}/nS + V_{GM}/nM + V_{GLS}/(nL.nS) + V_{GLM}/(nL.nM) + V_{GSM}/(nS.nM) + V_{GLSM}/(nL.nS.nM) + V_e/(nL.nS.nM.nR))$, where the variance components correspond to the terms in the mixed model above. The terms nL, nS, nM and nR stand for the number of locations (2: Maine and Oregon), number of 'seasons' (4: Fall 2006, Spring 2007, Fall 2007, Spring 2008), management (2; organic and conventional), and replicates (2 or 3).

Genotypic means were calculated by taking genotypic main effects fixed instead of random in the mixed models above. Pairwise comparisons between genotypic means were performed using GenStat procedure *VMCOMPARISON*. Correlations on the basis of genotypic means were referred to as genetic correlations. Genotypic stabilities under organic and conventional conditions were calculated as the variance for individual genotypes across all trials in the system.

To assess the feasibility of selection for organic conditions (the target environment) under conventional conditions, we calculated the ratio of correlated response (for organic conditions using conventional conditions), CR, to direct response (for organic conditions in organic conditions), DR, as the product of the genetic correlation between organic and conventional systems (r_G) and the ratio of the roots of conventional and organic repeatabilities (H_C and H_O respectively): $CR/DR = r_G\ H_G/H_O$. A ratio smaller than 1 indicates that selection is better done directly under organic conditions when the aim is indeed to improve the performance in organic conditions.

Results

Comparison of phytochemicals means over the environments

Glucosinolates. Across all trials, glucoraphanin levels were comparable between locations and seasons but were more variable at the individual location and season trial analysis level (**Table 2**). Glucoraphanin, glucobrassicin and neoglucobrassicin levels were comparable between organic and conventional treatments. Comparisons of organic versus conventional by location and season for the glucosinolate phytochemicals are presented in **Figure S1**. Comparable levels of glucosinolates were observed in the organic - conventional comparisons within locations and seasons.

Tocopherols. Across trials compared regionally, Oregon had higher levels of all three tocopherols compared to Maine (**Table 2, Figure S2**). The tocopherols δ- and γ- were higher in Fall compared to Spring, but not so for α-tocopherol (**Figure S2**). Organic and conventional levels for all tocopherol concentrations were in the same range and not significantly different. When the three tocopherols were analysed by organic versus conventional within location and season, there were no clear significant differences in management system across the season and location combinations (**Table 2, Figure S2**).

Carotenoids. Overall, Oregon had higher levels of lutein and β-carotene compared to Maine (**Table 2, Figure S3**) and comparative levels of zeaxanthin (**Table 2, Figure S3**). Spring produced higher levels of all carotenoids compared to Fall levels in contrast to the glucosinolates and the δ- & γ- tocopherol concentrations. There were no significant differences between organic and conventional for any carotenoid measured. When carotenoids were analysed by management system within location and season, β-carotene showed significantly lower levels in Maine in the Fall compared to other location and season combinations (**Figure S3**).

Partitioning of variance components

Glucosinolates. For glucoraphanin across all trials in both regions, Genotype (G) main effect accounted for the largest proportion of variance, followed by G×L×S interaction (**Table 3**). There was no Management (M) main effect, but M contributed to the three (L×S×M and G×S×M) and four-way interactions (G×L×S×M). In contrast to glucoraphanin, Location (L) had the largest effect for glucobrassicin and neoglucobrassicin across all trials in both regions, followed by the L×S interactions. For neoglucobrassicin the S and G main effect was more important than for glucobrassicin. When trials were further partitioned by location, a G and S main effect was apparent for neoglucobrassicin in both locations; for glucobrassicin the S main effects was only apparent in Oregon and not in Maine (**Table S2 in File S1**). There was M main effect for glucobrassicin and neoglucobrassicin, but not for glucoraphanin, and no G×M interaction for all glucosinolates.

Tocopherols. For δ- and γ-tocopherol across all trials in both regions, the Season (S) main effect accounted for the largest proportion of variance (**Table 3**). In contrast the proportion of the variation associated with S for α-tocopherol across all trials was minor. For all three tocopherols there was minor to no M effect, but a large L main effect, being the greatest for γ-tocopherol. The G main effect showed a similar pattern to L.

Carotenoids. For all three carotenoids across all trials in both regions, the G main effect described a significant component of total variance and was of largest influence for lutein (**Table 3**). The S main effect played an important role for zeaxanthin, and to a lesser extent for lutein but not for β-carotene. For all three carotenoids the L effect was minor, but the L×S interaction for β-carotene was relatively large and mostly associated with Maine (**Table S2 in File S1**). There was no M main effect; only for β-carotene was there a small effect of the G×M interaction (mainly driven by Maine).

Repeatability, genetic correlation and ratio of correlated response to direct response

Organic versus conventional. In the present study, we were able to estimate the proportion of the genotypic variance relative to phenotypic variance, but because we did not have a genetically structured breeding population, we apply the term repeatability rather than broad sense heritability. Of the phytochemicals studied, repeatabilities for concentrations of seven of the nine were comparable or higher in organic compared to conventional systems (**Table 4**). Only for glucobrassicin and δ-tocopherol was repeatability under organic conditions lower than under conven-

Table 2. Trait means[1] of phytochemicals of 23 broccoli cultivars grown across four pair combinations of location (Maine/Oregon), season (Fall/Spring) two-years combined and management system (Conventional/Organic), 2006–2008.

	Maine					Oregon				
	Fall		Spring			Fall		Spring		
	2006–2007 Combined		2007–2008 Combined		Mean	2006–2007 Combined		2007–2008 Combined		Mean
	C	O	C	O		C	O	C	O	
Glucoraphanin	5.31 e	3.77 bc	3.56 b	4.06 c	4.18	3.46 b	3.03 a	4.64 d	4.51 d	3.91
Glucobrassicin	1.06 b	0.90 a	1.45 c	1.33 c	1.19	5.14 f	5.51 g	2.24 d	2.70 e	3.90
Neoglucobrassicin	0.46 a	0.40 a	2.16 c	1.85 b	1.22	2.34 c	3.20 d	4.32 e	5.10 f	3.74
δ-Tocopherol	2.34 c	2.77 d	1.91 b	1.70 a	2.18	3.53 e	3.66 e	1.91 b	2.24 c	2.83
γ-Tocopherol	4.67 c	4.40 c	2.63 a	2.98 b	3.67	8.48 d	8.73 d	3.31 b	3.22 b	5.94
α-Tocopherol	25.83 a	27.33 a	38.61 b	40.51 bc	33.07	43.04 c	43.20 c	40.52 bc	42.25 c	42.25
Lutein	11.49 a	12.47 a	15.53 b	15.93 b	13.85	15.91 b	16.04 b	16.48 b	17.81 c	16.56
Zeaxanthin	0.81 a	0.83 ab	0.87 ab	0.88 b	0.85	0.83 ab	0.84 ab	1.02 c	1.02 c	0.93
β-Carotene	12.98 a	13.25 a	28.73 c	29.71 c	21.16	29.10 c	30.10 c	25.16 b	25.80 b	27.54

[1]Values in the table are means. Means of the same letter in the same row are not significantly different at the $P<0.05$ level.

Table 3. Partitioning of variance components (%) presented as coefficients of variation for phytochemicals of 23 broccoli cultivars grown across eight pair combinations of location (Maine/Oregon), season (Fall/Spring) and management system (Conventional/Organic), 2006–2008.

	Location (L)	Season (S)	Management (M)	L×S	L×M	S×M	L×S×M	L×S×M×R Rep (R)	Genotype (G)	G×L	G×S	G×M	G×L×S	G×L×M	G×S×M	G×L×S×M	Residual
Glucoraphanin	0.01	5.45	0.00	7.20	0.01	0.00	11.86	1.56	17.45	0.01	0.01	0.00	15.97	0.01	7.49	12.62	11.94
Glucobrassicin	36.00	0.00	0.00	27.51	5.58	4.18	1.34	1.86	9.42	7.77	0.01	0.00	13.84	0.00	0.00	10.63	10.91
Neoglucobrassicin	36.81	13.51	0.00	34.36	8.47	6.84	4.50	4.76	15.16	6.24	0.01	0.00	16.40	0.00	0.01	13.40	15.80
δ-Tocopherol	6.83	35.22	0.43	7.87	0.01	0.01	3.39	0.01	5.57	5.65	6.01	0.00	13.65	0.01	0.00	12.21	12.74
γ-Tocopherol	12.02	19.09	0.01	12.11	0.01	3.64	4.47	2.12	13.79	4.85	15.82	0.00	12.95	0.00	0.01	11.08	12.03
α-Tocopherol	6.73	0.01	0.28	10.20	0.01	0.01	0.01	1.29	2.79	3.42	0.01	0.97	10.07	0.00	0.00	8.18	8.92
Lutein	3.71	4.91	0.00	7.52	0.01	2.70	1.85	1.40	13.03	4.14	0.01	0.01	10.76	0.73	0.01	9.21	9.95
Zeaxanthin	1.97	11.55	0.01	3.99	0.01	0.00	0.01	0.00	8.44	3.18	0.01	1.05	6.91	0.01	0.83	8.52	11.36
β-Carotene	4.61	0.00	0.00	17.84	0.01	0.70	0.01	0.71	8.41	4.45	0.00	2.31	11.32	4.63	0.65	12.83	10.99

tional. In the analyses δ- and α-tocopherol had relatively low repeatabilities. The highest repeatabilities were for glucoraphanin (0.82–0.84), neoglucobrassicin (0.75–0.76), γ-tocopherol (0.72–0.75), lutein (0.83–0.85) and zeaxanthin (0.76–0.77). Genetic correlations were high between organic and conventional for the glucosinolates, γ-tocopherol and lutein (0.84–0.95), while δ-tocopherol, α-tocopherol, zeaxanthin and β-carotene were lower (0.63–0.77). The ratio of the correlated response to direct response for selection in the organic system was less than 1.0 for all traits.

By location and season. For the glucosinolates, glucoraphanin and glucobrassicin repeatability at each location, season and treatment trial were comparable and generally high (0.83–0.97) between organic and conventional trials, while no clear trend for neoglucobrassicin repeatabilities was observed between organic and conventional aside from being much lower than glucoraphanin and glucobrassicin (**Table S3 in File S1**). For γ- and α-tocopherol, repeatabilities were comparable between organic and conventional, while for δ-tocopherol repeatabilities were comparable between systems or higher in conventional except for one paired trial. For the carotenoids, repeatabilities were comparable or higher in organic for all paired trials, while for lutein in seven of the eight paired trials organic was comparable or greater than conventional. Repeatabilities for zeaxanthin concentrations were comparable for six of the eight paired trials.

Comparison of cultivar ranking for phytochemical concentration and stability across trials

To determine trends in cultivars with both the highest concentration of phytochemical groups most stable across locations, seasons and production systems, phytochemical concentrations were plotted against stability per genotype across trials. A group of cultivars were identified as both highest in concentration and most stable and are indicated in the highlighted 'red circle' per phytochemical (**Figure 1A–I**). For glucoraphanin, the same group of cultivars had both the highest concentrations and were the most stable across production systems (**Figure 1A; Table S4 in File S1**). While for glucobrassicin, a different set of cultivars had the highest concentrations across production systems (**Figure 1B; Table S5 in File S1**). Overall stability of all cultivars across production system was less related to cultivar mean concentrations for glucobrassicin than for glucoraphanin. None of the cultivars with the highest concentration for neoglucobrassicin were in the top quartile for stability across trials; all cultivars with the highest neoglucobrassicin content were in the bottom half for stability (**Figure 1C; Table S6 in File S1**). Some but not all cultivars that had the highest concentrations of α-tocopherol were among the top group for δ- and/or γ-tocopherol. There was no relationship between δ-tocopherol concentrations and stability, but both γ- and α- tocopherols had higher concentrations associated with greater stability (**Figure 1D–F; Tables S7–9 in File S1**). Open pollinated and early maturing cultivars had the highest and most stable concentrations for all carotenoids. (**Figure 1G–I; Tables S10–12 in File S1**).

Comparison of phytochemical concentration by genotype classification

The open pollinated and F_1 hybrid cultivars were compared across trials for each phytochemical analysed (**Figure 2A**). The levels of glucoraphanin in F_1 hybrids tended to be higher than the open pollinated cultivars. But the inverse trend was observed for glucobrassicin, which was supported by the ranking and stability analysis where the F_1 hybrids showed higher levels and more stability across trials than the open pollinated cultivars for

Table 4. Repeatabilities, genetic correlation and ratio of correlated response to direct response for broccoli phytochemicals comparing organic versus conventional management systems over all trial season/location combinations, 2006–2008.

	Repeatability (H)			
	C	**O**	r_A[a]	CR_{org}/R_{org}[b]
Glucoraphanin	0.84	0.82	0.84	0.83
Glucobrassicin	0.70	0.64	0.88	0.84
Neoglucobrassicin	0.75	0.76	0.94	0.94
δ-Tocopherol	0.50	0.42	0.73	0.66
γ-Tocopherol	0.75	0.72	0.95	0.93
α-Tocopherol	0.23	0.35	0.61	0.76
Lutein	0.83	0.85	0.93	0.94
Zeaxanthin	0.76	0.77	0.77	0.78
β-Carotene	0.62	0.72	0.63	0.68

[a]Average genetic correlation between conventional and organic production systems across locations.
[b]Ratio of correlated response to direct response.

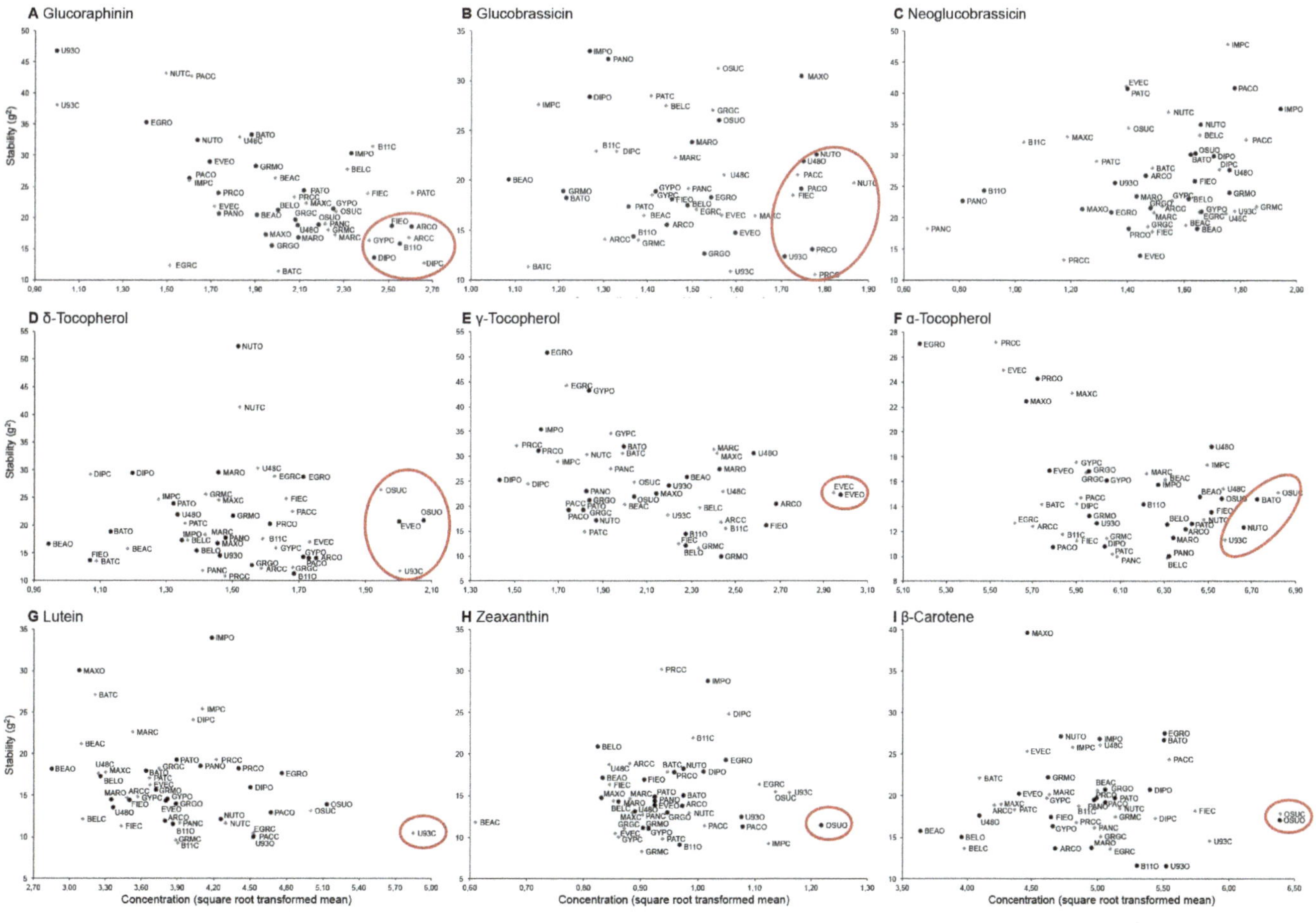

Figure 1. Broccoli cultivar stability from trials conducted in two locations over four seasons with two management systems plotted against phytochemical content. A. Glucoraphanin, B. Glucobrassicin, C. Neoglucobrassicin, D. δ-tocopherol, E. γ -tocopherol, F. α-tocopherol, G. Lutein, H. Zeaxanthin, I. β-carotene. See Table 1 for cultivar name abbreviations. The C or O at the end of the cultivar abbreviation indicates conventional or organic management system, respectively.

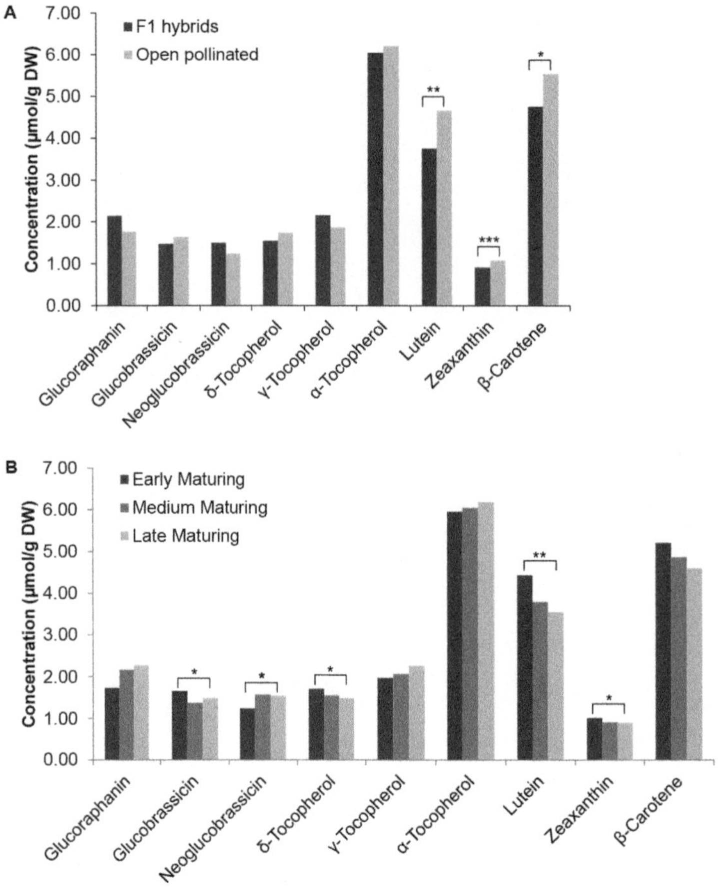

Figure 2. Mean phytochemical content of broccoli genotypic classes. A. Mean phytochemical content of broccoli F_1 hybrids versus open pollinated cultivars, and B. Mean phytochemical content of early, mid- and late-maturing cultivars grown across all trials at two locations (Maine and Oregon), in two seasons (Fall and Spring) and in two management systems (Conventional and Organic) and conventional management systems. See Table 1 for key to cultivar F1 hybrid versus open pollinated classification and maturity classification. Significance (* = $P<0.05$, ** = $P<0.01$, *** = $P<0.001$).

glucoraphanin. The reverse was observed for glucobrassicin. For the carotenoids, the open pollinated cultivars had a significantly higher mean value of lutein and zeaxanthin and tended to be higher for β-carotene compared to the F_1 hybrids.

Based on the results of our field trials, the 23 cultivars of broccoli were grouped into three distinct maturity classes: Early (55–63 days); Mid (64–71 days); and Late (72–80 days) and analysed for the effect of the maturity class on phytochemical content (**Figure 2B**). For glucoraphanin, late maturing cultivars had significantly higher content levels, while for the carotenoids, early maturing cultivars tended to have higher concentrations and were significantly higher for lutein.

When cultivar performance between genetic material originating from two primary broccoli breeding companies was compared for phytochemical content there were no significant differences with the exception of lutein, where company 1's cultivars had significantly higher concentrations than those of company 2 (**data not shown**).

A negative correlation between the date of release and levels of glucobrassicin ($R^2 = 0.21$; $p = 0.03$) (**Figure 3**) was observed, but no significant correlations for any other phytochemical were seen when 21 cultivars (the total set minus the two inbred lines) were analysed by their date of commercial release (1975–2005).

Correlation analysis among phytochemicals and horticulture traits

Phytochemical correlation across trials. Correlation among phytochemicals indicated that glucoraphanin was significantly negatively correlated to glucobrassicin (**Table 5**). Correlations between the glucosinolates and the tocopherols were not significant. Glucoraphanin and neoglucobrassicin were negatively correlated to all carotenoids but only lutein and glucoraphanin were statistically significant. Glucobrassicin demonstrated a positive trend with all carotenoids. No statistically significant correlations were observed within tocopherols. Δ-tocopherol was positively correlated, while γ-tocopherol was negatively correlated to all carotenoids. There were no significant correlations for α-tocopherol with carotenoids. All carotenoids were highly positively correlated with one another.

Phytochemical correlation to horticulture traits across trials. A correlation analysis was conducted for six horticulture traits, derived from the field study component of this research, Renaud et al. [78], and the nine phytochemicals across trials. The results indicated that greater head weight and head diameter were significantly positively correlated with glucoraphanin and negatively correlated with glucobrassicin, δ-tocopherol and the carotenoids. Increasing days to maturity was positively correlated with glucoraphanin, and negatively correlated to carotenoids.

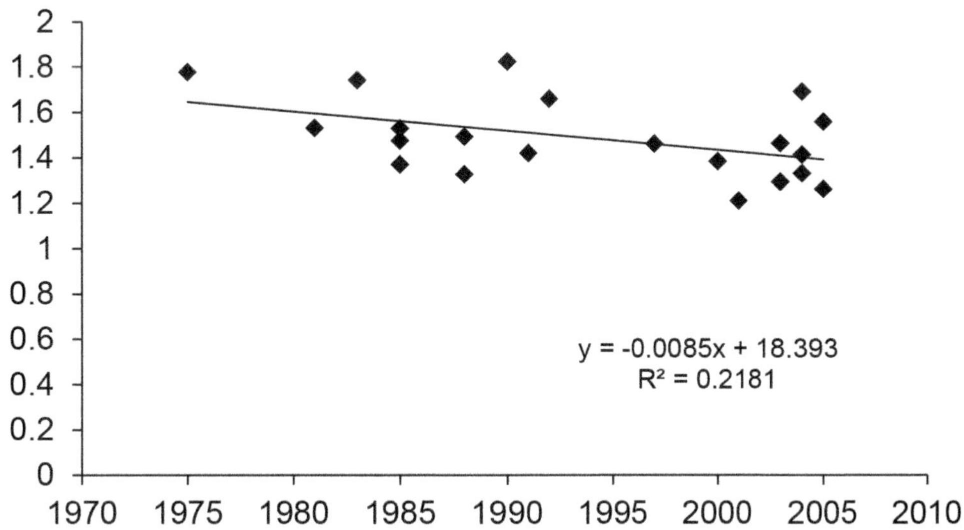

Figure 3. Regression of broccoli floret glucobrassicin concentrations on date of cultivar release for 23 cultivars grown across all trials in two locations (Maine and Oregon), in two seasons (Fall and Spring), in two management systems (Conventional and Organic), 2006–2008.

Head color was significantly correlated with δ-tocopherol and the carotenoids, but not with glucosinolates or γ- and α-tocopherol. Bead size and bead uniformity were positively correlated with glucoraphanin, neoglucobrassicin and γ-tocopherol and negatively correlated with glucobrassicin and the carotenoids.

Principal component biplot analysis: correlation between phytochemicals and cultivars by production system

In the principal component analysis the first PC axis accounted for similar amounts of the total variation in both conventional and organic production systems (43.5% vs. 39.6%). The second PC axis showed a similar trend with 17.02% for conventional and 16.93% for organic (**Figures 4A and 4B**). The first two PC axes together accounted for 60.53% and 56.57% of total variation for conventional and organic, respectively. The PCA biplot analysis supported our findings that carotenoids were highly associated across systems, while tocopherols were highly associated in conventional, but not in organic (tocopherols demonstrated the largest shift between production systems). Glucoraphanin and neoglucobrassicin were associated with one another, but not with glucobrassicin across production systems. Glucoraphanin was associated with α-tocopherol in organic, but not in conventional treatments. Glucobrassicin was associated with δ- and α-tocopherol in conventional, but not in organic treatments. δ-tocopherol had a higher association with the carotenoids in organic than conventional. The biplots show response of both cultivars and phytochemical traits to environment. Those cultivars close to the origin reveal little about the relationship of cultivars and trait vectors, whereas those located near the extremes of trait vectors are those with the highest (or lowest) values for those traits.

Discussion

Impact of organic management system compared to environmental factors on phytochemical content

Few studies have specifically compared the levels of health promoting compounds in *Brassica* vegetable species grown under organic and conventional production systems [13,74–75]. To our knowledge, this investigation is the most comprehensive study with the broadest range of phytochemical compounds (9) and a diverse set of broccoli cultivars (23) over regions (2), and management systems (2), with Fall and Spring season trials (2 each). In this study organic versus conventional management systems contributed the smallest source of variation compared to genotype, region and season. Within the phytochemicals studied individual compound concentrations responded differently. All compounds showed genetic variation, but also a substantial proportion of variance components were accounted for by high level interactions (**Table 3; Table S2 in File S1**). While M main effect was generally small, it had a substantial contribution in three- and four-way interactions. In particular, many G×L×S×M interactions were large relative to other variance components. This indicates that for the phytochemicals, M did have an influence on G, but that there were no consistent patterns across locations and seasons that would have shown up as significant G×M. Rather in each season and location, the paired organic and conventional environments differed significantly from one another but each situation was unique. In contrast to many comparisons between organic and conventional production systems [79], it should be noted that in our trials, yields averaged over the years did not differ significantly between the organic and conventional management systems [78].

Among the nine compounds, glucoraphanin was the most strongly influenced by genotype followed by lutein: supporting the findings of several other broccoli studies where variation in concentrations for glucoraphanin [19,70,65–66] and lutein [21,37] was primarily due to genotype. For γ-tocopherol, genotype was a large source of variation, but this compound was equally influenced by location and season (also found by Ibrahim and Juvik [37]). For glucobrassicin and neoglucobrassicin the location was the largest source of variation, but also L×S interaction was very influential, particularly for neoglucobrassicin, which is supported by Kushad et al. [20] and Schonhof et al. [26]. Jasmonic acid, a signal transduction compound in plants, is up-regulated under conditions of plant stress, wounding, and herbivory. Increased endogenous levels or exogenous application of this compound (or methyl jasmonate) increases biosynthesis and transport of neoglucobrassicin to broccoli florets. This up-

Table 5. Correlations coefficients (r) for six horticultural traits and nine phytochemicals, calculated using data standardized across trials.

	Head Weight	Head Diameter	Maturity	Head Color	Bead Size	Bead Uniformity	Glucoraphanin	Glucobrassicin	Neoglucobrassicin	δ-tocopherol	γ-tocopherol	α-tocopherol	Lutein	Zeaxanthin	β-Carotene
Head Weight															
Head Diameter	0.81														
Maturity															
Head Color															
Bead Size	0.63		0.69												
Bead Uniformity	0.49	0.48													
Glucoraphanin	0.47	0.44	0.43		0.63	0.51									
Glucobrassicin	−0.54	−0.50			−0.56	−0.64	−0.51								
Neoglucobrassicin			0.58		0.48										
δ-Tocopherol	−0.55			0.49											
γ-Tocopherol					0.43										
α-Tocopherol															
Lutein	−0.65		−0.70	0.56	−0.69		−0.41			0.55	−0.54				
Zeaxanthin	−0.68	−0.43	−0.62	0.49	−0.64					0.60	−0.42		0.95		
β-Carotene	−0.53		−0.54	0.59	−0.48					0.50	−0.43		0.90	0.90	

Correlation results include means from 23 cultivars, across eight pair combinations of location (Maine/Oregon), season (Fall/Spring) and management system (Conventional/Organic), 2006–2008[a].
[a]For empty cells, r is not significantly different from zero ($P<0.05$).

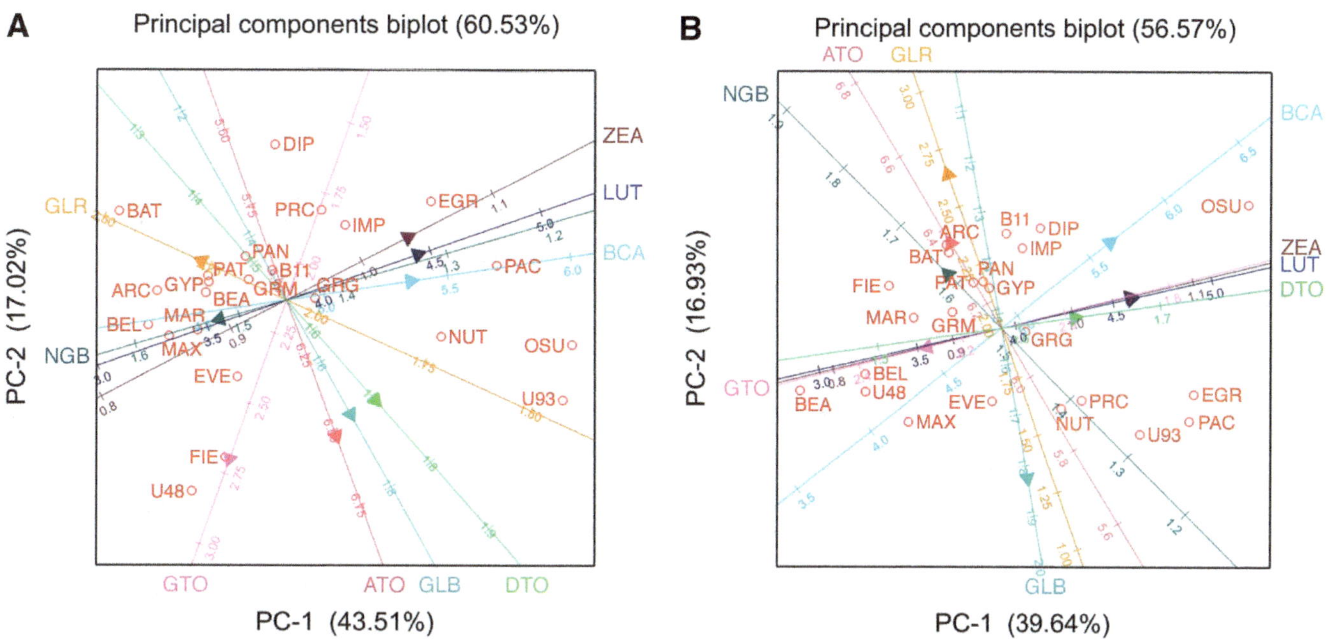

Figure 4. Principal components biplot of phytochemicals (vectors) and 23 cultivars (circles) grown in four seasons in Oregon and Maine. A. Biplot for conventional production, B. Biplot for organic production. See Table 1 for cultivar name abbreviations. Trait abbreviations: GLR: Glucoraphanin; GLB: Glucobrassicin; NGB: Neoglucobrassicin; DTO: δ-tocopherol; GTO: γ-tocopherol; ATO: α-tocopherol; LUT: Lutein; ZEA: Zeaxanthin; BCA: β-Carotene.

regulation was not observed for glucobrassicin biosynthesis [17]. This could explain why neoglucobrassicin was primarily under the control of Location and L×S interaction in our study. Season was the largest variance component for δ-tocopherol and zeaxanthin, which contrasts with the work of Ibrahim and Juvik [37] who found genotype had the largest influence on these compounds, followed by genotype by environment interaction although this study was constrained by the fact that the experiment was conducted in only one location over two growing seasons. For the other compounds such as α-tocopherol and β-carotene, L×S and the G×L×S interactions were most important.

Overall we found high genetic correlations between glucosinolates in organic and conventional trials. When trial locations were analysed separately, M main effect was present for glucobrassicin and neoglucobrassicin. The mean concentrations of glucobrassicin and neoglucobrassicin in broccoli from Oregon organic trials had higher concentrations compared to Oregon conventional trials, while Maine trials were comparable between management systems (**Table 2, Figure S1**). These results can be explained by the larger environment effect on glucobrassicin and genotype by environment effect on neoglucobrassicin found in the variance component analysis indicating sensitivity of these compounds to abiotic and/or biotic stresses. Our location specific findings are supported by those of Meyer and Adam [13] who performed a comparative study of the glucosinolate content of store bought organic and conventional broccoli and determined that the indolyl glucosinolates, glucobrassicin and neoglucobrassicin were significantly higher in the organically grown versus the conventionally grown. Evaluation of 10 broccoli genotypes over two years by Brown et al. [19] further supports our findings and those of Rosa and Rodrigues [62], Vallejo et al. [64], and Farnham et al. [70], that variation in concentration for glucoraphanin was primarily due to genetic variation, while differences in glucobrassicin was due to environmental variation (e.g. season, temperature) and genotype by environment interaction. The significantly higher levels of glucobrassicin in Oregon in the Fall harvested trials compared to Maine could be attributed to the higher maximum temperatures and GDD in Oregon compared to Maine (**Table S1 in File S1**).

Compared to glucosinolates, there is substantially less research on the genotype by environment interaction of tocopherol and carotenoid phytochemical groups in broccoli, and no specific studies exploring the influence of organic production system. In our study, minor management system effect at the overall trial analysis level was observed for the tocopherols and for carotenoids, there was management system effect only for lutein in Oregon Spring trials. Picchi et al. [75] also did not find differences in levels of carotenoids in cauliflower in organic versus conventional systems. In the tocopherols, there were no significant differences in location, but for δ- and γ- tocopherol concentration levels were higher in the fall compared to the spring, while for α-tocopherol, concentration levels were higher in the spring compared to the fall. For the carotenoids, there were no significant location differences, however there was a seasonal trend that all carotenoids were higher in spring compared to fall. Ibrahim and Juvik [39] found significant environmental variation among 24 broccoli cultivars for carotenoids and tocopherols which they attributed to the stressful production environments. Factors explaining the genotype and genotype by environment interaction components of variation in the carotenoids and tocopherols could be clarified by the fact that environmental stimuli are both up- and down-regulating genes associated with carotenoid and tocopherol biosynthesis. There is evidence in the literature that there are coordinated responses of the carotenoid and tocopherol antioxidants *in vivo*. There was a reduction in rape seed (*Brassica napus*) tocopherol content in response to increased carotenoid levels due to over expression of the enzyme phytoene synthase [80]. This response could explain the negative correlation between γ-tocopherol concentration and the carotenoids observed in our trials.

Differences in phytochemical content between different genotypes and genotypic classes

The partitioning of variance indicated that genotype was an important source of variation for all glucosinolates. The cultivar ranking and rank correlation analysis demonstrated that there was a pattern in genotype content of glucosinolates where cultivars with the highest concentrations of glucoraphanin had the lowest levels for glucobrassicin (**Figures S1**). In our trials, the range in glucoraphanin concentrations across cultivars was (1.15–7.02 µmol/g DW, **Table S4 in File S1**), while glucobrassicin was 1.46–3.89 µmol/g DW, **Table S5 in File S1**). Several of the cultivars with the highest concentrations of neoglucobrassicin were those that had the highest concentrations of glucobrassicin. Range in neoglucobrassicin concentrations across cultivars was 0.68–4.54 µmol/g DW, **Table S6 in File S1**). In earlier studies, glucosinolate concentrations in broccoli have shown dramatic variation among different genotypes. Rosa and Rodriguez [62] studied total glucosinolate levels in eleven cultivars of broccoli and found ranges from 15.2–59.3 µmol/g DW. Among 50 accessions of broccoli Kushad et al. [20] found glucoraphanin content ranges from 0.8–22 µmol/g DW with a mean concentration of 7.1 µmol/g DW, while Wang et al. [61] found glucoraphanin content of five commercial hybrids and 143 parent materials ranging from 1.57–5.95 µmol/g for the hybrids and 0.06–24.17 µmol/g in inbred lines and Charron et al. [65] found ranges from 6.4–14.9 µmol/g DW. While the means in our study are somewhat lower, they are within the range of other studies.

A genotype effect was observed for tocopherols, but predominantly for γ-tocopherol. The PCA biplots (**Figure 4AB**) and the correlation analysis (**Table 5**) demonstrated the high positive correlations between δ-tocopherol, α-tocopherol and the carotenoids (α-tocopherol and β-carotene were also highly correlated in the Kushad et al. [20] study). The cultivar relationship to different phytochemicals was represented in the biplots as well as in the cultivar content and stability analysis (**Figure 1**). Many cultivars with the highest concentrations in the tocopherols and carotenoids were open pollinated cultivars, inbreds and early maturing, older F_1 hybrids. Many of this same group were also relatively high in glucobrassicin concentrations. Kurilich et al. [38] found that carotenoid and tocopherol concentrations among 50 broccoli lines were highly variable and primarily genotype dependent. Specifically, levels of β-carotene ranged from 0.4–2.4 mg/100 g FW. Ibrahim and Juvik [37] also found broad ranges for total carotenoid and tocopherol concentrations among 24 genotypes ranging from 55–154 µg/g DW and 35–99 µg/g DW, respectively. Farnham and Kopsell [21] studied the carotenoid levels of nine double haploid lines of broccoli. Similar to our findings, lutein was the most abundant carotenoid in broccoli ranging from 65.3–139.6 µg/g DM. The sources of variation for lutein were predominantly genotype, followed by environment and G×E interaction, which also supports our findings. No genotypic differences were found for β-carotene in Farnham and Kopsell [21], which is in contrast to our findings. Overall, they found that most of the carotenoids measured were positively and highly correlated to one another as was observed in our study (**Table 5**). Kopsell et al. [81] found lutein levels in kale of 4.8–13.4 mg/100 g FW where the primary variance components for both lutein and β-carotene were also genotype and season.

Our research aimed also to address the question whether the phytochemical content of broccoli cultivars is associated with certain genotypic classes, e.g. open pollinated vs. F_1 hybrids; older vs. newer cultivar releases; and between commercial sources. Broccoli is typically a cross-pollinated, self-incompatible crop species and cultivars are either open pollinated and composed of heterogeneous genetically segregating individuals, or F_1 hybrids produced by crossing of two homozygous inbred lines, resulting in homogeneous populations of heterozygous individuals. In the 1960's virtually all broccoli grown was derived from OPs. By the 1990's almost all commercial cultivars were hybrids [82].

In our trials with 18 F_1 hybrids (released between 1975–2005) and 3 open pollinated cultivars (released from 1985–2005), we found several interesting trends related to genotype and genotypic class performance as it related to the three groups of phytochemicals. When analysing F_1 hybrid and open pollinated cultivars, they also demonstrated different performance patterns depending upon the individual phytochemical or group of compounds analysed. When cultivars were ranked for content and stability per phytochemical, there were distinct trends for certain compounds such as late maturing, F_1 hybrids outperforming early maturing F_1 hybrids and open pollinated cultivars for glucoraphanin, while the inverse was found for glucobrassicin and all carotenoids studied. This analysis was further supported by the PCA biplots that showed a strong relationship for select cultivars to certain phytochemicals or groups of phytochemicals such as 'OSU OP' to the carotenoids. When the full set of cultivars was divided into F_1 hybrid and open pollinated groups and the means compared by phytochemical, the results further supported the individual cultivar analysis where F_1 hybrids had higher mean values for glucoraphanin than the open pollinated cultivars (**Figure 2A**). Clear cultivar performance differences were identified where early maturing versus late maturing cultivars performed differently depending upon the phytochemical (**Figure 2B**). We also found that late maturing cultivars had higher concentrations for glucoraphanin than early maturing lines (and the inverse for glucobrassicin and the carotenoids). Picchi et al. [75] studied the quantity of glucosinolates of an early and late maturing cultivar of cauliflower grown in one conventional and three organic production systems, and found a significantly higher level of glucoraphanin in the later maturing cultivar compared to the early maturing cultivar in the organic production system. Another interesting trend was that cultivars with higher concentration levels for those phytochemicals whose expression is heavily influenced by environmental factors were not necessarily the most stable across trial environments; as was the case with neoglucobrassicin, δ- and γ-tocopherol in our study. For traits where genotype played a more significant role in contributing to variation, cultivars with a higher concentration level tended to also be those that were most stable across environments as was seen for lutein and glucoraphanin concentrations.

No significant differences were found for cultivar performance in phytochemical concentrations between genetic materials originating from two distinct commercial sources, with the exception of lutein (data not shown). When the full set of broccoli cultivars were analyzed for a correlation between date of release and mean level of phytochemical content across trials, no significant correlation was found with the exception of a negative trend for glucobrassicin (**Figure 3**). Our data does not support the idea that modern breeding for high yield performance and disease resistance necessarily leads to a trade-off in level of phytochemicals. Previous reports examining the relationship between year of release and performance had focussed on wheat vitamin and mineral content [83–85], and mineral content in broccoli [86–87]. However these authors did not study phytochemical content and their results were equivocal on the question on an innate biological trade-off between increased yield and nutritional content.

Not many studies have included two or more groups of phytochemicals. In our study with three phytochemical groups we found that phytochemicals demonstrating a negative correlation with one another (e.g. glucoraphanin with the carotenoids), showed an inverse cultivar response: e.g. cultivars with highest concentrations of glucoraphanin were the lowest in the carotenoids and vice versa. When both horticultural traits and phytochemicals were analysed for their phenotypic correlation, head weight was significantly and positively correlated with glucoraphanin and negatively correlated with δ- and α-tocopherol and the carotenoids. Farnham and Kopsell [21] explained that negative correlations may occur as a result of increased biomass accumulation in a certain genotype that is not accompanied by increased carotenoid production, effectively lowering the carotenoid concentration in the immature broccoli florets when pigments are expressed. Comparatively, head color was highly correlated to the carotenoids and negatively correlated to the glucosinolates overall. The cultivar 'OSU OP' was explicitly bred for a dark green stem and head color, not only for a darker green dome surface but also for a dark green interior color between the florets of the dome and in the stem (personal communication, Jim Myers 2013). 'OSU OP' was the highest in overall carotenoid concentrations across trials as it is known that carotenoids are correlated with chlorophyll concentrations and the intensity of green pigmentation [88].

Perspectives on breeding broccoli for enhanced phytochemical content specifically for organic agriculture

Our study included predominantly broccoli cultivars selected for broad adaptability in conventional production systems and not purposely bred for high phytochemical content nor for adaptation to organic agriculture. What we can conclude from our data is that there has been little change in levels of several phytochemicals over three decades of breeding. This may indicate genetic variation for phytochemicals is limited in elite germplasm, or it may be the result of the lack of selection tools for these traits. This may be changing with recent efforts to introgress high glucoraphanin from *B. villosa* to produce the high-glucoraphanin F_1 cultivar 'Beneforté' [89–90,44]. The seed industry needs to exploit known sources of variation in the genus *Brassica* to enhance levels of other health-promoting phytochemicals and to broaden the genetic diversity of commercial broccoli germplasm. Our finding of a strong correlation between dark green color and high carotenoid levels provides breeders with a simple and efficient means of increasing carotenoids. The three groups of phytochemicals studied contribute to health promotion in different ways. As these groups are related to different metabolic pathways selecting for one compound does not necessarily inadvertently improve the other compounds, and may even result in negative correlation as we have seen in our data between glucoraphanin and the carotenoids. Although these compounds belong to different metabolic pathways, their production may be coordinated through regulatory feedback loops, or the structural and/or regulatory genes controlling these pathways may be genetically linked.

Designing a breeding program for broccoli high in glucosinolates would require the following considerations generated from our research: (1) Glucoraphanin is a highly genetically determined compound with minor location and season main effects but with substantial G×L×S interaction. (2) Comparatively, glucobrassicin and neoglucobrassicin are more impacted by location and season and L×S interaction with highest glucobrassicin concentrations and largest range in our Oregon Fall trials and neoglucobrassicin highest in Oregon Spring trials. (3) Cultivar performance for glucoraphanin and glucobrassicin and neoglucobrassicin was negatively correlated indicating that there may be a trade-off between glucoraphanin on the one hand, and glucobrassicin and neoglucobrassicin on the other hand. (4) Selection for glucoraphanin without consideration of horticultural traits would probably result in larger headed and later maturing cultivars. Conversely, selection for smaller headed, early maturing cultivars would favor glucobrassicin and neoglucobrassicin at the expense of glucoraphanin.

A breeding program for broccoli for high tocopherol content would require: (1) Overall the tocopherols were more season, location and L×S dependent and had lower overall repeatabilities compared to the glucosinolates. In a structured genetic population where additive genetic variance could be partitioned, narrow sense heritability would likely be low, and increasing tocopherol content would best be conducted with breeding methods suited to low heritability traits; (2) δ- and γ-tocopherols were both season dependent and fall grown broccoli had higher concentrations of these compounds across trials and a wider range of content levels, whereas levels of α-tocopherol were higher in spring but the range was comparable under both seasons. Thus, fall would be the preferred environment for breeding for these compounds; (3) there were no significant differences for location for δ- or γ-tocopherol, but the average levels of α-tocopherol levels were significantly higher in Oregon than Maine, suggesting greater potential for genetic gain in the Oregon environment.

If the goal is to design a breeding program for broccoli enhancing the levels of carotenoids it would require the following considerations: (1) For all three carotenoids studied, genotypic variation, particularly for lutein, was relatively more important than location and season; (2) however, zeaxanthin exhibited a large S (spring) and L×S interaction. For both β-carotene and lutein, spring grown broccoli had significantly higher levels than fall produced. Thus, selection for carotenoids would probably be more effective in spring than in fall; (3) early maturing and small headed cultivars had higher levels of carotenoids. Since most of the carotenoids are associated with the outer surfaces of the inflorescence, smaller broccoli heads with a greater surface area to volume ratio should show higher concentrations of these compounds; (4) because carotenoids have high G main effect good germplasm sources as indicated in **Figure 1** have high concentrations of carotenoids and demonstrated stability across environments. As all three carotenoids are highly correlated with one another, selecting for one should effectively select for all; (5) selection for darker green colour more widely distributed throughout the tissues of the head should allow the breeder to relatively efficiently increase carotenoid content in broccoli.

In closing, we want to address the question of selecting in an organic or a conventional environment. The argument commonly used to support selecting in productive environments is that heritabilities are higher compared to resource poor environments [91–92]. Organic is often considered a low-external input environment, resulting on average in 20% less yield compared to conventional production [79]. Nevertheless, in our trials repeatabilities for some phytochemicals were higher or comparable to conventional (**Table 3**). Narrow sense heritabilities would be expected to be significantly lower. For those traits where repeatabilities were higher or comparable, direct selection under organic systems could enhance selection gain. In all cases, the ratio of correlated response to direct response was less than one suggesting that direct selection would allow more rapid progress than correlated selection. Our data on phytochemicals did not show a wider range of levels under organic conditions as we found for horticultural traits in the same trials [78], however, in several cases, repeatabilities in organic production were higher than in conventional.

To maximize efficiency in a breeding program, commercial breeders may seek to combine breeding for both conventional and organic markets, and a combination of strategies can be proposed. Some studies that utilized highly heritable (agronomic) traits, where cultivar yield performance ranked similarly between organic and conventional management systems and which had high genetic correlations, suggested that early breeding be conducted under conventional conditions, with the caveat that advanced breeding lines be tested under organic conditions for less heritable traits (e.g. Löschenberger et al. [93]; Lorenzano and Bernardo, 2008) [52]. In studies where cultivar yield performance differed between management systems and there were significant differences in cultivar ranking, and in some cases low genetic correlations for lower heritability traits (e.g. Kirk et al. [94]; Murphy et al. [48]), these studies recommended that cultivars intended for organic agriculture be selected only under organic conditions. In our study of phytochemicals, we would recommend for organic purposes selection under organic conditions for the compounds where genetic correlations between organic and conventional were moderate.

Supporting Information

Figure S1 Comparison of broccoli cultivars for glucosinolates (μmol/g DW) grown across all trials in two locations (Maine and Oregon), in two seasons (Fall and Spring), in two management systems (Conventional and Organic), and at the individual trial level, 2006–2008. A. Glucoraphanin, B. Glucobrassicin, C. Neoglucobrassicin.

Figure S2 Comparison of broccoli cultivars for tocopherols (μmol/g DW) grown across all trials in two locations (Maine and Oregon), in two seasons (Fall and Spring), in two management systems (Conventional and Organic), and at the individual trial level, 2006–2008. A. δ-tocopherol, B. γ-tocopherol, C. α-tocopherol.

Figure S3 Comparison of broccoli cultivars for carotenoids (μmol/g DW) grown across all trials in two locations (Maine and Oregon), in two seasons (Fall and Spring), in two management systems (Conventional and Organic), and at the individual trial level, 2006–2008. A. Lutein, B. Zeaxanthin, and C. β-carotene.

File S1 Supporting tables. Table S1. Description of agronomic and environmental factors of the trial locations with paired organically and conventionally managed fields, 2006–2008. **Table S2.** Partitioning (%) of variance components for various traits of 23 broccoli cultivars grown across four pair combinations in Maine, season (Fall/Spring) and management system (Conventional/Organic), 2006–2008. Variance components reported as coefficients of variation. **Table S3.** Repeatability for broccoli for phytochemicals and per trial of 23 broccoli cultivars grown across eight pair combinations of location (Maine/Oregon), season (Fall/Spring) and management system (Conventional/Organic), 2006–2008. **Table S4.** Glucoraphanin level (μmol/g DW) of 23 cultivars grown under conventional (C) and organic (O) conditions in two locations (Maine and Oregon) in two seasons (Fall and Spring) from 2006–2008. **Table S5.** Glucobrassicin level (μmol/g DW) of 23 cultivars grown under conventional (C) and organic (O) conditions in two locations (Maine and Oregon) in two seasons (Fall and Spring) from 2006–2008. **Table S6.** Neoglucobrassicin level (μmol/g DW) of 23 cultivars grown under conventional (C) and organic (O) conditions in two locations (Maine and Oregon) in two seasons (Fall and Spring) from 2006–2008. **Table S7.** δ-tocopherol level (μmol/g DW) of 23 cultivars grown under organic (O) and conventional (C) conditions in two locations (Maine and Oregon) in two seasons (Fall and Spring) from 2006–2008. **Table S8.** γ-tocopherol level (μmol/g DW) of 23 cultivars grown under conventional (C) and organic (O) conditions in two locations (Maine and Oregon) in two seasons (Fall and Spring) from 2006–2008. **Table S9.** α-tocopherol level (μmol/g DW) of 23 cultivars grown under conventional (C) and organic (O) conditions in two locations (Maine and Oregon) in two seasons (Fall and Spring) from 2006–2008. **Table S10.** Lutein level (μmol/g DW) of 23 cultivars grown under conventional (C) and organic (O) conditions in two locations (Maine and Oregon) in two seasons (Fall and Spring) from 2006–2008. **Table S11.** Zeaxanthin level (μmol/g DW) of 23 cultivars grown under conventional (C) and organic (O) and conditions in two locations (Maine and Oregon) in two seasons (Fall and Spring) from 2006–2008. **Table S12.** β-carotene level (μmol/g DW) of 23 cultivars grown under conventional (C) and organic (O) conditions in two locations (Maine and Oregon) in two seasons (Fall and Spring) from 2006–2008.

Acknowledgments

For support in understanding the genotype selection and providing select elite inbred lines for the study, the authors thank Dr. Mark Farnham from the USDA, Charleston, South Carolina. For the Oregon trials, the authors wish to thank the organic growers Jolene Jebbia and John Eveland at Gathering Together Farm for providing the location and support for the organic broccoli trials. Deborah Kean, Faculty Research Assistant at the Oregon State University Research Station, and the students Hank Keogh, Shawna Zimmerman, Miles Barrett, and Jennifer Fielder provided support in data collection of the field trials. For the Maine trials, the authors wish to thank the students Heather Bryant, Chris Hillard and Greg Koller for support in data collection of the field trials. We also thank the University of Maine Highmoor Farm Superintendent, Dr. David Handley University of Maine Cooperative Extension. For the soil analysis, we thank Dr. Michelle Wander from the University of Illinois, Urbana and her students. At Wageningen University, we thank Paul Keizers and Dr. Chris Maliepaard for support with the statistical analysis. We thank Carl Jones for his valuable input on iterations of this research paper and Ric Gaudet for support in data organization.

Author Contributions

Conceived and designed the experiments: ER ELvB JM MP FvE NZ JJ. Performed the experiments: ER JM NZ JJ. Analyzed the data: ER ELvB JM MP FE JJ. Contributed reagents/materials/analysis tools: ER JM MP FvE NZ JJ. Wrote the paper: ER ELvB JM JJ FvE.

References

1. Saba A, Messina F (2003) Attitudes towards organic foods and risk/benefit perception associated with pesticides. Food Quality and Preferences 14: 637–645.
2. Stolz H, Stolze M, Hamm U, Janssen M, Ruto E (2011) Consumer attitudes towards organic versus conventional food with specific quality attributes. Netherlands Journal of Agricultural Science 58: 67–72.
3. Smith-Spangler C, Brandeau ML, Hunter GE, Bavinger JC, Pearson M, et al (2012) Are organic foods safer or healthier than conventional alternatives? A systematic review. Annals of Internal Medicine 157: 348–366.
4. Asami DK, Hong YJ, Barrett DM, Mitchell AE (2003) Comparison of the total phenolic and ascorbic acid content of freeze-dried and air-dried marionberry, strawberry, and corn grown using conventional, organic and sustainable agricultural practices. Journal of Agriculture & Food Chemistry 51(5): 1237–1271.

5. Chassy AW, Bui L, Renaud ENC, Mitchell AE (2006) Three-year comparison of the content of antioxidant microconstituents and several quality characteristics in organic and conventionally managed tomatoes and bell peppers. Journal of Agriculture and Food Chemistry 54: 8244–8252.
6. Brandt K, Leifert C, Sanderson R, Seal CJ (2011) Agroecosystem management and nutritional quality of plant foods: The case of organic fruits and vegetables. Critical Reviews in Plant Sciences 30: 177–197.
7. Hunter D, Foster M, McArthur JO, Ojha R, Petocz P, et al (2011) Evaluation of the micronutrient composition of plant foods produced by organic and conventional methods. Critical Reviews in Food Science and Nutrition 51: 571–582.
8. Koh E, Charoenprasert S, Mitchell AE (2012) Effect of organic and conventional cropping systems on ascorbic acid, vitamin C, flavonoids, nitrate, and oxalate in 27 varieties of spinach (*Spinacia oleracea* L.) Journal of Agriculture and Food Chemistry. 60(12): 3144–3150.
9. Crozier A, Clifford MN, Ashihara H (eds) (2006) Plant secondary metabolites: occurrence, structure and role in the human diet. Blackwell Publishing Ltd.
10. Bourn D, Prescott J (2002) A comparison of the nutritional value, sensory qualities and food safety of organically and conventionally produced foods. Critical Review Food Science Nutrition 42(1): 1–34.
11. Mäder P, Fliessbach A, Dubois D, Gunst L, Fried P, et al (2002) Soil fertility and biodiversity in organic farming. Science 296: 1694–1697.
12. Martinez-Bellesta MC, Lopez-Perez L, Hernandez M, Lopez-Berenguer C, Fernandez-Garcia N, et al (2008) Agricultural practices for enhanced human health. Phytochemical Review 7: 251–260.
13. Meyer M, Adam ST (2008) Comparison of glucosinolate levels in commercial broccoli and red cabbage from conventional and ecological farming. European Food Research Technology 226: 1429–1437.
14. Mozafar A (1993) Nitrogen fertilizers and the amount of vitamins in plants: a review. Journal of Plant Nutrition 16: 2479–2506.
15. Zhao X, Carey EE, Wang W, Rajashekar CB (2006) Does organic production enhance phytochemical content of fruit and vegetables? Current knowledge and prospects for research. HortTechnology 16(3): 449–456.
16. Dixon RA, Paiva NL (1995) Stress-induced phenylpropanoid metabolism. The Plant Cell 7: 1085–1097.
17. Kim HS, Juvik JA (2011) Effect of selenium fertilization and methyl jasmonate treatment on glucosinolate accumulation in broccoli floret. Journal of the American Society for Horticulture Science 136(4): 239–246.
18. USDA National Nutrient Database, 2011. http://ndb.nal.usda.gov/(last visited 14 April 2014).
19. Brown AF, Yousef GG, Jeffrey EH, Klein BP, Wallig MA, et al (2002) Glucosinolate profiles in broccoli: variation in levels and implications in breeding for cancer chemoprotection. Journal of American Society Horticulture Science 127(5): 807–813.
20. Kushad MM, Brown AF, Kurilich AC, Juvik JA, Klein BP, et al (1999) Variations of glucosinolates in vegetable crops of *Brassica oleracea*. Journal of Agriculture and Food Chemistry 47: 1541–1548.
21. Farnham MW, Kopsell DA (2009) Importance of genotype on carotenoid and chlorophyll levels in broccoli heads. HortScience 44(5): 1248–1253.
22. Verhoeven DT, Goldbohm PA, van Poppel GA, Verhagen H, van den Brandt PA (1996) Epidemiological studies on *Brassica* vegetables and cancer risk. Biomarkers Prevention 5: 733–748.
23. Keck A-S, Finley JW (2004) Cruciferous vegetables: cancer protective mechanisms of glucosinolate hydrolysis products and selenium. Integrative Cancer Therapies 3(1): 5–12.
24. Here I, Büchler MW (2010) Dietary constituents of broccoli and other cruciferous vegetables: implications for prevention and therapy of cancer. Cancer Treatment Review 36: 377–383.
25. Higdon JV, Delage B, Williams DE, Dashwood RH (2007) Cruciferous vegetables and human cancer risk: epidemiologic evidence and mechanistic basis. Pharmacology Research 55(3): 224–236.
26. Schonhof I, Krumbein A, Bruchner B (2004) Genotypic effects on glucosinolates and sensory properties of broccoli and cauliflower. Nahrung/Food 48(1): 25–33.
27. Fenwick GR, Heaney RK, Mulllin WJ (1983) Glucosinolates and their breakdown products in food and food plants. Critical Reviews in Food Science and Nutrition 18(2): 123–201.
28. Juge N, Mithen RF, Traka M (2007) Molecular basis for chemoprevention by sulforaphane: a comprehensive review. Cellular and Molecular Life Sciences 64: 1105–1127.
29. Mithen RF, Dekker M, Verkerk R, Rabot S, Johnson IT (2000) The nutritional significance, biosynthesis and bioavailability of glucosinolates in human foods. Journal of the Science of Food and Agriculture 80: 967–984.
30. Wang LI, Giovannucci EL, Hunter D, Neuberg D, Su L, et al (2004) Dietary intake of cruciferous vegetables, glutathione S-transferase (GST) polymorphisms and lung cancer risk in a Caucasian population. Cancer Causes Control 15: 977–985.
31. Hsu CC, Chow WH, Boffetta P, Moore L, Zaridze D, et al (2007) Dietary risk factors for kidney cancer in eastern and central Europe. American Journal of Epidemiology 166: 62–70.
32. Kirsh VA, Peters U, Mayne ST, Subar AF, Chatterjee N, et al (2007) Prospective study of fruit and vegetable intake and risk of prostate cancer. Journal of the National Cancer Institute 99: 1200–1209.
33. Lam TK, Ruczinski I, Helzlsouer KJ, Shugart YY, Caulfield LE, et al (2010) Cruciferous vegetable intake and lung cancer risk: a nested case-control study matched on cigarette smoking. Cancer Epidemiology, Biomarkers and Prevention 19: 2534–2540.
34. Bosetti C, Filomeno M, Riso P, Polesel J, Levi F, et al (2012) Cruciferous vegetables and cancer risk in a network of case-control studies. Annals of Oncology 23: 2198–2203.
35. Wu QJ, Yang Y, Vogtmann E, Wang J, Han LH, et al (2012) Cruciferous vegetables intake and the risk of colorectal cancer: a meta-analysis of observational studies. Annals of Oncology 00: 1–9. doi:10.1093/annonc/mds601.
36. Kopsell DA, Kopsell DE (2006) Accumulation and bioavailability of dietary carotenoids in vegetable crops. Trends in Plant Science 11(10): 499–507.
37. Ibrahim KE, Juvik JA (2009) Feasibility for improving phytonutrient content in vegetable crops using conventional breeding strategies: case study with carotenoids and tocopherols in sweet corn and broccoli. Journal of Agriculture and Food Chemistry 57: 4636–4644.
38. Kurilich AC, Tsau GJ, Brown A, Howard L, Klein BP, et al (1999) Carotene, tocopherol, and ascorbate contents in subspecies of *Brassica oleracea*. Journal of Agriculture and Food Chemistry 47: 1576–1581.
39. Kurilich AC, Juvik JA (1999) Quantification of carotenoid and tocopherol antioxidants in *Zea mays*. Journal of Agriculture and Food Chemistry 47: 1948–1955.
40. Mayne ST (1996) βeta carotene, carotenoids, and disease prevention in humans. The FASEB Journal. 10(7): 690–701.
41. Krinsky NI, Landdrum JT, Bone RA (2003) Biologic mechanisms of the protective role of lutein and zeazanthin in the eye. Annual Review Nutrition. 23: 171–201.
42. Kritchevsky SB (1999) β-Carotene, carotenoids and the prevention of coronary heart disease. Journal of Nutrition 129: 5–8.
43. Davis D, Epp MD, Riordan HD (2004) Changes in USDA food composition data for 43 garden crops, 1950 to 1999. Journal of the American College of Nutrition 23(6): 1–14.
44. Traka MH, Saha S, Huseby S, Kopriva S, Walley PG, et al (2013). Genetic regulation of glucoraphanin accumulation in Beneforte broccoli. New Phytologist, doi: 10.1111/nph.12232.
45. Munger HM (1979) The potential of breeding fruits and vegetables for human nutrition. HortScience 14(3): 247–250.
46. Welch RM, Graham RD (2004) Breeding for micronutrients in staple food crops from a human nutrition perspective. Journal of Experimental Botany 55(396): 353–364.
47. Troxell Alrich H, Salandanan K, Kendall P, Bunning M, Stonaker F, et al (2010) Cultivar choice provides options for local production of organic and conventionally produced tomatoes with higher quality and antioxidant content. Journal of Science Food and Agriculture 90: 2548–2555.
48. Murphy KM, Campbell KG, Lyon SR, Jones SS (2007) Evidence of varietal adaptation to organic farming systems. Field Crops Research 102: 172–177.
49. Baresel JP, Zimmermann G, Reents HJ (2008) Effects of genotype and environment on N uptake and N partition in organically grown winter wheat (*Triticum aestivum* L.) in Germany. Euphytica 163: 347–354.
50. Vlachostergios DN, Roupakias DG (2008) Response to conventional and organic environment of thirty-six lentil (*Lens culinaris* Medik.) varieties. Euphytica 163: 449–457.
51. Goldstein W, Schmidt W, Burger H, Messmer M, Pollak LM, et al (2012) Maize: Breeding and field testing for organic farmers. In: Lammerts van Bueren ET, Myers, JR editors. Organic crop breeding. West Sussex, UK, Wiley-Blackwell, John Wiley & Sons, Inc.
52. Lorenzana RE, Bernardo R (2008) Genetic correlation between corn performance in organic and conventional production systems. Crop Science 48(3): 903–910.
53. Osman AM, Almekinders CJM, Struik PC, Lammerts van Bueren, ET (2008) Can conventional breeding programmes provide onion varieties that are suitable for organic farming in the Netherlands? Euphytica 163: 511–522.
54. Przystalski M, Osman AM, Thiemt EM, Rolland B, Ericson L, et al (2008) Do cereal varieties rank differently in organic and non-organic cropping systems? Euphytica 163: 417–435.
55. Wunderlich SM, Feldmand C, Kane S, Hazhin T (2008) Nutritional quality of organic, conventional, and seasonally grown broccoli using vitamin C as a marker. International Journal of Food Sciences and Nutrition 59(1): 34–45.
56. Koh E, Wimalasin KMS, Chassy AW, Mitchell AE (2009) Content of ascorbic acid, quercetin, kaempferol and total phenolics in commercial broccoli. Journal of Food Composition and Analysis. 22(7–8): 637–643.
57. Harker FR (2004) Organic food claims cannot be substantiated through testing of samples intercepted in the marketplace: a horticulturist's opinion. Food Quality and Preference 15: 91–99.
58. Benbrook C (2009) The impact of yield on nutritional quality: lessons from organic farming. HortScience 44(1): 12–14.
59. Carlson DG, Daxenbichler ME, VanEtten CH, Kwolek WF, Williams PH (1987) Glucosinolates in crucifer vegetables: broccoli, brussels sprouts, cauliflower, collards, kales, mustard greens, and kohlrabi. Journal of the American Society for Horticulture Science 112(1): 173–178.
60. Farnham MW, Stephenson KK, Fahey JW (2005) Glucoraphanin level in broccoli seed is largely determined by genotype. HortScience 40(1): 50–53.
61. Wang J, Gu H, Yu H, Zhao Z, Sheng X, Zhang X (2012) Genotypic variation of glucosinolates in broccoli (*Brassica oleracea* var. italica) florets from China. Food Chemistry 133: 735–741.

62. Rosa EAS, Rodrigues AS (2001) Total and individual glucosinolate content in 11 broccoli cultivars grown in early and late seasons. HortScience 36(1): 56–59.
63. Vallejo F, Tomas-Barberán FA, García-Viguera C (2003a) Effect of climatic and sulphur fertilisation conditions, on phenolic compounds and vitamin C, in the inflorescences of eight broccoli cultivars. European Food Research Technology 216: 395–401.
64. Vallejo F, Tomas-Barberán FA, Gonzalez Benavente-Garcia A, García-Viguera C (2003b) Total and individual glucosinolate contents in inflorescences of eight broccoli cultivars grown under various climatic and fertilisation conditions. Journal of the Science of Food and Agriculture 83: 307–313.
65. Charron CS, Saxton A, Sams CE (2005a) Relationship of climate and genotype to seasonal variation in the glucosinolate – myrosinase system. I. Glucosinolate content in ten cultivars of *Brassica oleracea* grown in fall and spring seasons. Journal of the Science of Food and Agriculture 85: 671–681.
66. Charron CS, Saxton A, Sams CE (2005b) Relationship of climate and genotype to seasonal variation in the glucosinolate – myrosinase system. II. Myrosinase activity in ten cultivars of *Brassica oleracea* grown in fall and spring seasons. Journal of the Science of Food and Agriculture 85: 682–690.
67. Aires A, Fernandes C, Carvalho R, Bennett RN, Saavedra MJ, et al (2011) Seasonal effects on bioactive compounds and antioxidant capacity of six economically important brassica vegetables. Molecules 16: 6816–6832.
68. Pek Z, Daood H, Nagne MG, Berki M, Tothne MM, et al (2012) Yield and phytochemical compounds of broccoli as affected by temperature, irrigation, and foliar sulphur supplementation. HortScience 47(11): 1646–1652.
69. Schonhof I, Blandenburg D, Müller S, Krumbein A (2007) Sulfur and nitrogen supply influence growth, product appearance, and glucosinolate concentration of broccoli. Journal of Plant Nutrition and Soil Science 170: 65–72.
70. Farnham MW, Wilson PE, Stephenson KK, Fahey JW (2004) Genetic and environmental effects on glucosinolate content and chemoprotective potency of broccoli. Plant Breeding 123: 60–65.
71. Björkman M, Klingen I, Birch ANE, Bones AM, Bruce TJA, et al (2011) Phytochemicals of Brassicaceae in plant protection and human health – influences of climate, environment and agronomic practice. Phytochemistry 72: 538–556.
72. Robbins RJ, Keck AS, Banuelos G, Finley JW (2005) Cultivation conditions and selenium fertilization alter the phenolic profile, glucosinolate, and sulforaphane content of broccoli. Journal of Medicinal Food 8(2): 204–214.
73. Xu C, Guo R, Yan H, Yuan J, Sun B, et al (2010) Effect of nitrogen fertililization on ascorbic acid, glucoraphanin content and quinone reductase activity in broccoli floret and stem. Journal of Food, Agriculture and Environment 8(1): 179–184.
74. Naguib AM, El-Baz FK, Salama ZA, Hanna HAEB, Ali HF, et al (2012) Enhancement of phenolics, flavonoids and glucosinolates of broccoli (*Brassica oleracea*, var. Italica) as antioxidants in response to organic and bio-organic fertilizers. Journal of the Saudi Society of Agricultural Sciences 11: 135–142
75. Picchi V, Migliori C, Scalzo RL, Campanelli G, Ferrari V, et al (2012) Phytochemical content in organic and conventionally grown Italian cauliflower. Food Chemistry 130: 501–509.
76. Hansen M, Moller P, Sorensen H, Cantwell de Trejo M (1995) Glucosinolates in broccoli stored under controlled atmosphere. Journal of the American Society for Horticulture Science 120(6): 1069–1074.
77. Tiwari U, Cummins E (2013) Factors influencing levels of phytochemicals in selected fruit and vegetables during pre- and post-harvest food processing operations. Food Research International 50: 497–506.
78. Renaud ENC, Lammerts van Bueren ET, Paulo MJ, van Eeuwijk FA, Juvik JA, et al(2014) Broccoli cultivar performance under organic and conventional management systems and implications for crop improvement. Crop Science DOI: 10.2135/cropsci201.
79. De Ponti T, Rijk B, van Ittersum MK, (2012) The crop yield gap between organic and conventional agriculture. Agricultural Systems 108: 1–9.
80. Shewmaker CK, Sheehy JA, Daley M, Colburn S, Ke DY (1999) Seed-specific overexpression of phytoene synthase: increase in carotenoids and other metabolic effects. Plant Journal 20: 401–412.
81. Kopsell DA, Kopsell DE, Lefsrud MG, Curran-Celentano J, Dukach LE (2004) Variation in lutein, β-carotene, and chlorophyll concentrations among *Brassica oleracea* cultigens and seasons. HortScience 39(2): 361–364.
82. Hale AL, Farnham MW, Ndambe Nzaramba M, Kimbeng CA (2007) Heterosis for horticultural traits in broccoli. Theoretical Applied Genetics 115(3): 351–60.
83. Murphy KM, Reeves PG, Jones SS (2008) Relationship between yield and mineral nutrient concentrations in historical and modern spring wheat cultivars. Euphytica 163: 381–390.
84. Hussain A, Larsson H, Kuktaite R, Johansson E (2010) Mineral composition of organically grown wheat genotypes: contribution to daily minerals intake. International Journal of Environmental Research and Public Health 7: 3443–3456.
85. Jones H, Clarke S, Haigh Z, Pearce H, Wolfe M (2010) The effect of the year of wheat variety release on productivity and stability of performance on two organic and two non-organic farms. Journal of Agricultural Science 148: 303–317.
86. Farnham MW, Keinath AP, Grusak MA (2011) Mineral concentration of broccoli florets in relation to year of cultivar release. Crop Science 51: 2721–2727.
87. Troxell-Alrich H, Kendall P, Bunning M, Stonaker F, Kulen O, et al (2011) Environmental temperatures influence antioxidant properties and mineral content in broccoli cultivars grown organically and conventionally. Journal of AgroCrop Science 2(2): 1–10.
88. Khoo H-E, Prasad KN, Kong K-W, Jiang Y, Ismail A (2011) Carotenoids and their isomers: color pigments in fruits and vegetables. Molecules 16(2): 1710–1738.
89. Faulkner K, Mithen R, Williamson G (1998) Selective increase of the potential anticarcinogen 4-methylsulphinylbutyl glucosinolate in broccoli. Carcinogenis 19(4): 605–609.
90. Mithen R, Faulkner K, Magrath R, Rose P, Williamson G, Marquez J (2003) Development of isothiocyanate-enriched broccoli, and its enhanced ability to induce phase 2 detoxification enzymes in mammalian cells. Theoretical Applied Genetics 106: 727–734.
91. Ceccarelli S (1996) Adaptation to low/high input cultivation. Euphytica 92: 203–214.
92. Ceccarelli S (1994) Specific adaptation and breeding for marginal conditions. Euphytica 77: 205–219.
93. Löschenberger F, Fleck A, Grausgruber H, Hetzendorfer H, Hof G, et al (2008) Breeding for organic agriculture: the example of winter wheat in Austria. Euphytica 163: 469–480.
94. Kirk AP, Fox SL, Entz MH (2012) Comparison of organic and conventional selection environments for spring wheat. Plant Breeding 131: 687–694.

2

Genetic Diversity and Population Structure of Tetraploid Wheats (*Triticum turgidum* L.) Estimated by SSR, DArT and Pedigree Data

Giovanni Laidò[1], Giacomo Mangini[2], Francesca Taranto[2], Agata Gadaleta[2], Antonio Blanco[2], Luigi Cattivelli[1¤], Daniela Marone[1], Anna M. Mastrangelo[1], Roberto Papa[1], Pasquale De Vita[1]*

1 Consiglio per la Ricerca e la sperimentazione in Agricoltura, Cereal Research Centre, Foggia, Italy, **2** Department of Soil, Plant, and Food Sciences, Section of Genetics and Plant Breeding, University of Bari, Via Amendola, Bari, Italy

Abstract

Levels of genetic diversity and population genetic structure of a collection of 230 accessions of seven tetraploid *Triticum turgidum* L. subspecies were investigated using six morphological, nine seed storage protein loci, 26 SSRs and 970 DArT markers. The genetic diversity of the morphological traits and seed storage proteins was always lower in the durum wheat compared to the wild and domesticated emmer. Using Bayesian clustering (K = 2), both of the sets of molecular markers distinguished the durum wheat cultivars from the other tetraploid subspecies, and two distinct subgroups were detected within the durum wheat subspecies, which is in agreement with their origin and year of release. The genetic diversity of morphological traits and seed storage proteins was always lower in the improved durum cultivars registered after 1990, than in the intermediate and older ones. This marked effect on diversity was not observed for molecular markers, where there was only a weak reduction. At K >2, the SSR markers showed a greater degree of resolution than for DArT, with their identification of a greater number of groups within each subspecies. Analysis of DArT marker differentiation between the wheat subspecies indicated outlier loci that are potentially linked to genes controlling some important agronomic traits. Among the 211 loci identified under selection, 109 markers were recently mapped, and some of these markers were clustered into specific regions on chromosome arms 2BL, 3BS and 4AL, where several genes/quantitative trait loci (QTLs) are involved in the domestication of tetraploid wheats, such as the tenacious glumes (*Tg*) and brittle rachis (*Br*) characteristics. On the basis of these results, it can be assumed that the population structure of the tetraploid wheat collection partially reflects the evolutionary history of *Triticum turgidum* L. subspecies and the genetic potential of landraces and wild accessions for the detection of unexplored alleles.

Editor: Massimo Labra, University of Milano Bicocca, Italy

Funding: This works has been funded by the research project Esplora, Biomassval, RGV-FAO and Sicerme2 from the Ministry of Agricultural, Food and Forestry Policies (Italy). The funders had no role in study design, data collection and analysis, decision to publish, or preparation of the manuscript.

Competing Interests: The authors have declared that no competing interests exist.

* E-mail: pasquale.devita@entecra.it

¤ Current address: Consiglio per la Ricerca e la sperimentazione in Agricoltura, Genomics Research Centre, Fiorenzuola d'Arda (PC), Italy

Introduction

It has been widely reported that the genetic diversity of the elite germplasm for the major crops has suffered an overall reduction with time in comparison to the wild forms, especially for self-pollinating cereals such as durum wheat. This has arisen primarily as a consequence of the combined effects of domestication processes, recurrent use of adapted germplasm, and adoption of breeding schemes that do not favour wide genetic recombination [1–5]. Durum wheat (*Triticum turgidum* ssp. *durum*) is the only tetraploid species with significant agricultural importance, because of its use for human consumption (e.g., pasta, cous cous, bread, bulgur). Since modern varieties of durum wheat are expected to have very strict quality parameters (i.e., protein content, gluten quality, semolina colour), the use of wide crosses with unadapted germplasm in breeding programmes is very limited and selection intensity is very high. Through drift and selection, a large amount of genetic diversity has been lost, which has thus reduced the potential for wheat improvement in modern agricultural systems [6–9]. In this context, landraces, wild forms and other related wild species can have crucial roles in breeding programmes [10–12] because of their wide variability in terms of phenological, morphological, abiotic, biotic and quality traits. The evaluation of the level and structure of genetic diversity in tetraploid wheats is a prerequisite for plant breeding and genetic resource conservation programmes.

The level and structure of genetic diversity can be estimated by different approaches, which include the use of pedigrees [13], biochemical and morphological markers [14–15], and molecular markers [9,16–18]. Pedigree analysis has been extensively used to define the genetic diversity in barley [17,19], maize [20], rice [21], bread wheat [22] and durum wheat [23–24]. Unfortunately, the degree of genetic relatedness among genotypes based on pedigree information can sometimes be erroneous or incomplete, and it does not necessarily reflect the underlying genetics [23]. The accuracy of a coefficient of parentage depends on the availability of reliable and detailed pedigree data. In addition, these calculations do not take into account the effects of selection, mutation, and genetic drift, and require several simplifying assumptions that are generally not met.

In contrast, DNA markers allow the assessment of relatedness directly at the DNA level by estimation of the proportion of alleles that are identical by state, with the underlying assumption of a strict relationship between identity by state and identity by descent. In this context, the extent of the information that they can provide might depend on the nature and number of markers (e.g., level of homoplasy, mutation rate), the genome coverage and distribution, and the population under investigation [25,16].

Different molecular markers can be used for genome analysis, and many of these have been applied with success for genetic mapping [26–27], phylogenetic relationships [28], comparative genomic studies, and diversity studies [9,29]. Many studies have demonstrated the effectiveness of simple sequence repeat (SSR) markers for the characterisation of germplasm collections, because of their ease of use, high polymorphism, locus specificity, and codominance [9,23,30]. Diversity array technology (DArT) markers are also widely used for genome analysis; these offer deep genome coverage and high effectiveness, as many genotypes can be screened in a time-effective and cost-effective manner [9,31–32]. More recently, the availability of the sequences of 2,000 wheat DArT clones (http://www.diversityarrays.com/sequences) has provided functional meaning to these markers, opening a number of applications, such as collinearity studies, fine mapping of loci of interest, and identification of candidate genes in association mapping [33].

Tetraploid wheats are genetically and morphologically diverse and their evolution under domestication has not been fully elucidated [34–35]. Almost all of the studies conducted to date have considered the subspecies of the tetraploid wheat (*Triticum turgidum* L.) separately for the analysis of genetic diversity. So far, ssp. *durum* [23–24], ssp. *dicoccum* [36–37], ssp. *polonicum* [38] and ssp. *dicoccoides* [39] have only rarely been analysed together [40–41]. In this context, the objectives of our study were to: (i) characterise the population structure and level of genetic diversity in a collection of 230 accessions of seven tetraploid *Triticum turgidum* L. subspecies using morphological traits, biochemical markers, SSR and DArT markers; (ii) assess the correspondence between similarity matrices based on these different types of traits; and (iii) consider the potential of the DArT markers to detect genomic regions that have been potentially subjected to selection events. The study was also focused on analysis of temporal diversity changes in durum wheat cultivars released during the recent breeding period.

Materials and Methods

Plant Material

The tetraploid wheat (*T. turgidum* L., 2 n = 4x = 28; AABB genome) collection was classified according to van Slageren [42] and MacKe [43], who considered all forms as subspecies of *T. turgidum* (Table 1). This consists of 230 accessions classified into seven subspecies: ssp. *durum* (128), ssp. *turanicum* (20), ssp. *turgidum* (19), ssp. *polonicum* (20), ssp. *carthlicum* (12), ssp. *dicoccum* (19) and ssp. *dicoccoides* (12).

The durum wheat accessions (96 are mainly elite cultivar) are representative of the Italian durum breeding programmes over the last 100 years. A further subset of these accessions (32) came from the most important durum production areas (Table 1). For the purpose of the study, the durum collection was subdivided into three groups according to the year in which each cultivar was released. Group 1 comprises the 'old' genotypes that were selected from indigenous and exotic landraces, and/or varieties selected from crosses involving landraces (from 1915 to 1973); group 2 comprises the 'intermediate' genotypes that were selected from crosses between CIMMYT breeding lines and group 1 materials (from 1974 to 1989); and group 3 comprises the 'modern' genotypes selected after the 1990s. Twenty plants of each accession were sown at Valenzano (Bari, Italy), and a single plant representing the prevalent biotype of the accession was selected and grown to maturity for self-seeding. Seed stocks can be obtained upon request from the corresponding author.

Morphological and Biochemical Analysis

The accessions were evaluated according to six morphological traits recorded following the descriptors defined by the International Plant Genetic Resources Institute (http://www.cgiar.org/ipgri). In particular, during the growing season of 2008–2009, outer glume colour, awn colour, awned/awnless form, and spike and culm glaucousness were recorded, while the trait of naked/hulled kernel was evaluated at harvest time. The collection was investigated for glutenin alleles located at five loci: *Glu-A1* and *Glu-B1* for the high-molecular-weight subunits (HMW-GS), and *Glu-A3, Glu-B3* and *Glu-B2* for the low-molecular-weight subunits (LMW-GS). The glutenins were extracted from flour samples (50 mg) according to Laemmli [44]. Electrophoresis was performed in an SE 600 Ruby Hoefer vertical electrophoresis unit with stacking and running gel concentrations of 3% and 10% acrylamide, respectively. The identification of HMW-GS was based on the classification of Branlard *et al.* [45]. The new subunits and alleles were designed according to Li *et al.* [46], McIntosh *et al.* [47] and Riefolo *et al.* [14]. The LMW-GS were classified according to Nieto-Taladriz *et al.* [48]. Moreover, the collection was assessed for gliadin alleles at four loci: *Gli-A1, Gli-B1, Gli-A2* and *Gli-B2*. Gliadins were extracted from single seeds and fractioned using acid polyacrylamide gel electrophoresis (pH 3.1), according to Lafiandra and Kasarda [49], with their classification according to Riefolo *et al.* [14], Boggini *et al.* [50] and Aguiriano *et al.* [51].

DNA Extraction and SSR and DArT Analyses

Leaf tissue of the plants that represented the prevalent biotype of each accession was used for DNA extraction, using the protocol described by Sharp *et al.* [52]. The wheat collection was genotyped with 26 SSR markers, which were selected based on published map data [53–55] according to the following criteria: locus-specific amplification, low complexity, robust amplification, and good genome coverage (nearly one marker per chromosome arm). PCR amplification was carried out in 15 µl volumes containing 2 µl DNA ($\approx$ 80 ng), 1.5 µl 10× PCR buffer (EuroClone), 0.4 µM of each microsatellite primer (the forward primers were fluorescently labelled), 1.5 mM $MgCl_2$ (EuroClone), 0.2 mM dNTP mixture (Fermentas), and 1 U Taq DNA-polymerase (EuroClone). The PCR was carried out as follows: 95°C for 3 min, followed by 35 cycles of 94°C for 30 s, the specific T_a for each primer for 30 s, 72°C for 1 min, with a final extension at 72°C for 10 min. The PCR products were detected by capillary electrophoresis using an ABI PRISM 3130xl analyser, and analysed using GeneMapper version 4.0 genotyping software. The internal molecular-weight standard was 500-ROX (Life Technologies).

Genotyping with DArT markers was performed by Triticarte Pty. Ltd. (Canberra, Australia; http://www.triticarte.com.au), a whole-genome profiling service laboratory, as described by Akbari *et al.* [31]. For both the SSR and DArT markers, alleles that

Table 1. List of accessions of the *T. turgidum* subspecies included in the wheat collection.

Taxonomic classification	Accession	Country	Taxonomic classification	Accession	Country
T. turgidum	Cappelli	Italy	*T. turgidum*	K cer	Egypt
ssp. *durum*	Aziziah	Italy	ssp. *turanicum*	Cltr-11390	United State
	Russello	Italy		PI 68287	Azerbaijan
	Timilia	Italy, Sicily		PI 113393	Iraq
	Taganrog	Italy		PI 167481	Turkey, Denizli
	Capeiti-8	Italy		PI 191599	Marocco, Rabat-Sale
	Grifoni	Italy		PI 192641	Morocco
	Kyperounda	Marocco		PI 254206	Iran
	Langdon	United States		PI 278350	Italy
	Hymera	Italy		PI 290530	Hungary, Pest
	Trinakria	Italy		PI 306665	France, Herault
	Appulo	Italy		PI 576854	Turkey, Diyarbakir
	Belfuggito	Italy		PI 623656	Iran, West Azerbaijan
	Lambro	Italy		PI 624429	Iran, Bakhtaran
	Creso	Italy		PI 127106	Afghanistan, Faryab
	Isa	Italy		PI 67343	Australia, Victoria
	Mexicali 75	Mexico		PI 192658	Morocco
	Mida	Italy		PI 184526	Portugal
	Polesine	Italy		PI 352514	Azerbaijan
	Valgerardo	Italy		PI 362067	Romania, Brasov
	Valnova	Italy	*T. turgidum*	PI 56263	Portugal, Lisboa
	Tito	Italy	ssp. *turgidum*	PI 134946	Portugal, Lisboa
	Sansone	Italy		PI 157983	Italy, Sicily
	Karel	Italy		PI 157985	Italy, Sicily
	Produra	United States		PI 173503	Turkey, Artvin
	Valforte	Italy		PI 185723	Portugal, Leira
	Berillo	Italy		PI 191104	Spain
	Appio	Italy		PI 191145	Spain, Baleares
	Athena	Italy		PI 191203	Spain
	Latino	Italy		PI 286075	Poland
	Messapia	Italy		PI 221423	Portugal
	Arcangelo	Italy		PI 352544	Switzerland, Vaud
	Lloyd	United States		PI 290522	Germany
	Altar84	Mexico		PI 290526	Hungary, Pest
	Duilio	Italy		PI 341391	Turkey, Burdur
	Primadur	France		PI 352538	United Kingdom
	Quadruro	Italy		PI 352541	Germany
	Tresor	Italy		PI 352542	France
	Adamello	Italy		PI 352543	France
	Grazia	Italy	*T. turgidum*	PI 266846	United Kingdom, England
	Ambral	France	ssp. *polonicum*	PI 278647	United Kingdom, England
	Amedeo	Italy		PI 289606	United Kingdom, England
	Brindur	France		PI 330554	United Kingdom, England
	Neodur	France		PI 330555	United Kingdom, England
	Agridur	France		PI 349051	Georgia
	Antas	Italy		PI 352487	Germany, Saxony-Anhalt
	Plinio	Italy		PI 352488	Italy
	Simeto	Italy		PI 352489	Cyprus
	Fenix	Italy		PI 361757	Denmark

Table 1. Cont.

Taxonomic classification	Accession	Country	Taxonomic classification	Accession	Country
	Ofanto	Italy		PI 366117	Egypt, Sinai
	Enduro	Italy		PI 387479	Ethiopia
	Cirillo	Italy		PI 566593	United States
	Cosmodur	France		PI 208911	Iraq
	Dauno	Italy		PI 210845	Iran
	Doral	France		PI 223171	Jordan
	Exeldur	France		PI 272564	Hungary, Pest
	Fauno	Italy		PI 286547	Ecuador
	Gianni	Italy		PI 290512	Portugal
	Granizo	Spain		PI 306549	Romania
	Parsifal	France	*T. turgidum*	Citr 7665	Russian Federation
	Zenit	Italy	ssp. *carthlicum*	PI 70738	Iraq
	Italo	Italy		PI 94755	Georgia
	Kronos	United States		PI 115816	Georgia
	Ceedur	France		PI 283888	Iran
	Arcobaleno	Italy/Spain		PI 341800	Russian Federation, Dagest
	Ares	Italy		PI 499972	Georgia
	Colosseo	Italy		PI 532501	Former Soviet Union
	Fortore	Italy		PI 572849	Georgia
	Platani	Italy		PI 573182	Turkey, Kars
	Preco	Italy		PI 585017	Georgia
	Saadi	France		PI 585018	Georgia
	Bronte	Italy	*T. turgidum*	Farvento	Italy
	Ciccio	Italy	ssp. *dicoccum*	Lucanica	Italy
	Durfort	France		Molise selezione Colli	Italy
	Iride	Italy		ISC Foggia 152	Iran
	Nefer	France		ISC Foggia 159	Morocco
	Rusticano	Italy		ISC Foggia 161	United Kingdom
	San Carlo	Italy		ISC Foggia 171	Ethiopia
	Svevo	Italy		ISC Foggia 175	Hungary
	Vitromax	Italy/Spain		MG 5350	Ethiopia
	Varano	Italy		MG 4387	United Kingdom
	AC-Navigator	Canada		MG 5416/1	Iran
	Baio	Italy		MG 5471/1	Spain
	Cannizzo	Italy		MG 5473	Spain
	Claudio	Italy		MG 15516/1	Syria
	Martino	Italy		MG 5344/1	Ethiopia
	Provenzal	Italy		MG 5293/1	Italy
	Giotto	Italy		MG 5323	n.a.
	Meridiano	Italy		MG 3521	n.a.
	Orobel	Italy		MG 5300/1	n.a.
	Quadrato	Italy	*T. turgidum*	PI 346783	Hungary, Pest
	Vesuvio	Italy	ssp. *dicoccoides*	PI 343446	Israel
	Avispa	Italy		PI 481539	Israel
	Fiore	Italy		PI 352323	Asia Minor
	Tiziana	Italy		PI 352324	Lebanon
	Duetto	Italy		PI 355459	Armenia
	Dylan	Italy		PI 470944	Syria, Al Qunaytirah
	Grecale	Italy		PI 470945	Syria, Al Qunaytirah

Table 1. Cont.

Taxonomic classification	Accession	Country	Taxonomic classification	Accession	Country
	Normanno	Italy		MG 4343	n.a.
	Virgilio	France		MG 4328/61	n.a.
	Ancomarzio	Italy		MG 5444/235	n.a.
	Casanova	Italy		MG 4330/66	n.a.
	Chiara	Italy			
	Latinur	France			
	Vendetta	Italy			
	L092	United States			
	L252	United States			
	Maestrale	Italy			
	Orfeo	Italy			
	S99B34	United States			
	Saragolla	Italy			
	Ariosto	Italy			
	Arnacoris	Italy			
	Canyon	Italy			
	Imhotep	Italy			
	PR22D89	Italy			
	Strongfield	Canada			
	Alemanno	Italy			
	Ciclope	Italy			
	K26	Italy			
	UC1113	Canada			
	Neolatino	Italy			
	5-BIL42	Italy			
	PC32	Italy			
	Barcarol	Italy			
	Pedroso	Spain			
	Sharm 5	Syria			
	West Bread 881	United States			

n.a. not available. Cltr and PI number indicate the accession number in USDA National Small Grains Collection, Aberdeen, Idaho, USA. MG number indicate the accession number in CNR Institute of Plants Genetics, Bari, Italy. ISC Foggia number indicate the accession number in CRA-CER Cereal Research Centre, Foggia, Italy.

occurred at a low frequency ($p<0.05$) were excluded from the analysis.

Co-ancestry Analysis

The pedigree records, as reported in Table S1, were obtained from the literature, web-based pedigree databases (http://genbank.vurv.cz/wheat/pedigree), and personal communication with the breeders. For some of the genotypes, the pedigree data were not available and/or consistent, and therefore a subset of 116 varieties for which the pedigree data could be traced back several generations was defined (see Table S1). The coefficient of co-ancestry between the genotypes, or the 'Kinship coefficient', was calculated using Winkin2, according to Tinker and Mather [56]. The coefficient of co-ancestry was assumed to be zero ($f=0$) in the absence of any degree of kinship, and one ($f=1$) when the maximum degree of kinship was observed. A kinship coefficient matrix was obtained and transformed into a distance matrix (using $d=1-r$), which was used for the multivariate analysis.

Mantel test genetic similarity (GS) matrices for the SSR and DArT markers, and the distance matrix obtained from the kinship values, were used for pair-wise comparisons to determine the degrees of association between pairs of matrices, using the ARLEQUIN software, version 3.5 [57]. The normalised Mantel statistic Z [58] was used to determine the level of association between the matrices. The association was considered significant if $r \geq 0.50$, $p<0.01$.

Genetic Diversity, Genetic Structure and Population Differentiation Analysis

The genetic diversity (morphological traits, seed storage protein loci, and molecular markers, separately) within each subspecies was estimated by calculation of the number of observed alleles (na) and the unbiased estimator of gene diversity (H_E [59]) as:

Table 2. Alleles number and genetic diversity of morphological traits and seed protein storage loci for each subspecies included in the wheat collection.

	T. turgidum						
	ssp. *durum*	ssp. *turanicum*	ssp. *polonicum.*	ssp. *turgidum*	ssp. *carthlicum*	ssp. *dicoccum*	ssp. *dicoccoides*
Sample size	128	20	20	19	12	18	12
Glume colour	3	1	1	3	3	2	2
Awn colour	3	3	3	2	3	2	3
Glaucosness culm	2	2	1	1	1	1	2
Glaucosness spike	2	2	1	2	2	2	2
Awnedness	1	1	1	1	1	1	1
Naked/hulled kernel	1	1	1	1	1	1	1
n_a morphological traits	12	10	8	10	11	9	11
H_E morphological traits	0.09	0.15	0.07	0.23	0.28	0.07	0.33
Glu-A1	4	1	4	6	1	4	4
Glu-B1	10	5	2	7	4	11	5
Glu-B2	2	2	2	2	2	2	2
Glu-A3	3	3	4	4	2	5	3
Glu-B3	2	2	2	3	1	4	3
n_a *Glu-*	21	13	14	22	10	26	17
H_E *Glu-*	0.22	0.43	0.55	0.67	0.32	0.73	0.67
Gli-A1	4	1	2	2	2	2	5
Gli-B1	21	6	7	13	7	9	9
Gli-A2	3	2	2	2	2	2	2
Gli-B2	3	2	1	2	1	2	2
n_a *Gli-*	31	11	12	19	12	15	18
H_E *Gli-*	0.39	0.42	0.42	0.44	0.44	0.54	0.65
n_a total	64	34	34	51	33	50	46
H_E mean	0.23	0.33	0.35	0.45	0.35	0.45	0.55

Glu- glutenin locus; *Gli-* gliadin locus; n_a: number of alleles; H_E: genetic diversity.

$$HE=\left[\frac{n}{(n-1)}\right]*\left[1-\sum_{j=1}^{r}\sum_{i=1}^{k}\frac{pij}{r}\right]$$

where *pij* is the frequency of the ith variants at the jth locus, and *r* is the number of loci.

The population structure was determined on 230 tetraploid wheat accessions. The molecular data (26 SSR and 970 DArT markers) were processed using the STRUCTURE program, version 2.2 (http://pritch.bsd.uchicago.edu/structure.html) [60]. The number of sub-groups (K) was estimated by 20 independent runs for each K (from 2 to 20) applying the admixture model, with allele frequencies uncorrelated for SSR markers and correlated for DArT markers, 100,000 Markov Chain Monte Carlo (MCMC) repetitions, and a 100,000 burn-in period. The means of the log-likelihood estimates for each K were calculated. The true K was determined using both an estimate of the posterior probability of the data for a given K (as proposed by Pritchard *et al.*, [60]), and the Evanno ΔK [61]. Successively, separate analyses for the

Table 3. Genetic diversity for SSR and DArT markers for each subspecies included in the wheat collection.

		T. turgidum						
		ssp. *durum*	ssp. *turanicum*	ssp. *polonicum*	ssp. *turgidum*	ssp. *carthlicum*	ssp. *dicoccum*	ssp. *dicoccoides*
	Sample size	128	20	20	19	12	18	12
SSRs	H_E mean	0.60	0.58	0.58	0.62	0.56	0.66	0.70
DArTs	H_E mean	0.33	0.29	0.28	0.28	0.22	0.29	0.30

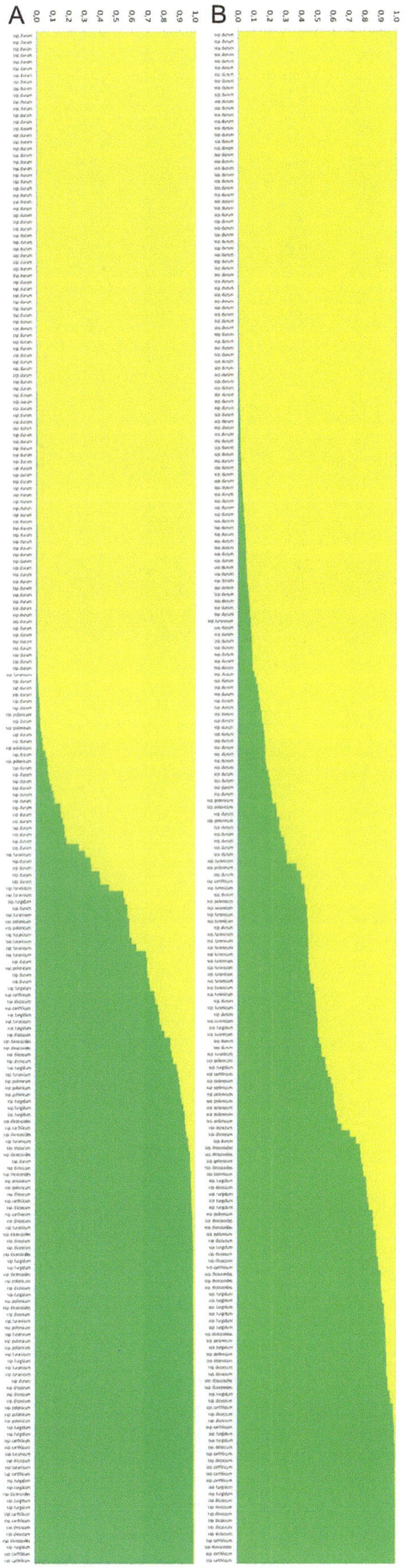

Figure 1. Population structure of the 230 accessions using SSR and DArT markers, as estimated using the model-based Bayesian algorithm implemented in the STRUCTURE programme (K = 2).

genetic diversity structure were performed on each sub-group, with the same parameters previously set. A genotype was considered to belong to a group if its membership coefficient was ≥0.50 [30]. For the population structure, we used the Spearman's rank correlation coefficient (ρ), to correlate the year of release of the durum cultivars and the taxonomic classification of the genetic structure. Dendrograms constructed with the neighbour-joining (NJ) algorithm from shared-allele distances were also used to analyse the genetic structure of the collection, which was implemented in the PAST software, version 2.1 [62].

The population differentiation was assessed by analysis of molecular variance (AMOVA) using the ARLEQUIN software, version 3.5 [57]. The significance levels for the variance components were estimated using 16,000 permutations. Loci under selection were identified by the *Fst*-outlier detection method, using 100,000 simulations. Based on the *Fst* values that fell outside the 99% confidence interval, candidate loci identified under positive selection were used for further analysis.

Results

Genetic Diversity for Morphological and Biochemical Traits

The wheat collection was evaluated for six morphological traits and nine seed-storage-protein loci (Table 2). Together with ssp. *turgidum*, ssp. *dicoccum* and ssp. *dicoccoides*, the ssp. *durum* showed a higher allele number than detected in ssp. *turanicum*, ssp. *polonicum* and ssp. *carthlicum*. The H_E of ssp. *durum* was the lowest (0.23), while ssp. *dicoccoides* showed the highest H_E (0.55).

The number of states of morphological traits ranged from 8 to 12. As expected, the awed/awned-less and naked/hulled kernel traits were monomorphic in each *T. turgidum* subspecies. The variability of the durum cultivars ($H_E = 0.09$) was lower than the ssp. *dicoccoides* accessions ($H_E = 0.33$).

The greatest allele number of glutenin loci was found in ssp. *dicoccum* (26 alleles), with the lowest in ssp. *carthlicum* (10 alleles). Three out of five glutenin loci analysed were polymorphic in each subspecies, with the exception of *Glu-A1*, monomorphic in ssp. *turanicum* and ssp. *carthlicum*, and *Glu-B2* monomorphic in ssp. *carthlicum*. *Glu-B1* was the most polymorphic locus (2–11 alleles). The H_Es in the durum cultivars were low in comparison to the ssp. *dicoccum* and ssp. *dicoccoides*.

For gliadins, the allele number of loci ranged from 11 to 31. *Gli-A1* was monomorphic in ssp. *turanicum*, as well as the *Gli-B2* locus in ssp. *polonicum* and ssp. *carthlicum*. *Gli-B1* was the most polymorphic one (6–21 alleles).

Genetic Diversity for SSR and DArT Markers

Twenty-six SSR loci broadly distributed over the genome and 970 DArT markers were used. The chromosomal position and number of alleles detected for each SSR are detailed in Table S2. A total of 436 alleles were detected, which ranged from 133 (ssp. *carthlicum*) to 211 (ssp. *durum*), while the number of alleles per locus varied from one (*BQ170801* and *BJ274952* for ssp. *carthlicum*) to 18 (*Xwmc606* for ssp. *durum*), with a mean of 16.8 alleles per locus. All of the genomic (g)SSRs were polymorphic in the seven subspecies, while the EST-SSRs were monomorphic in ssp. *carthlicum*. The

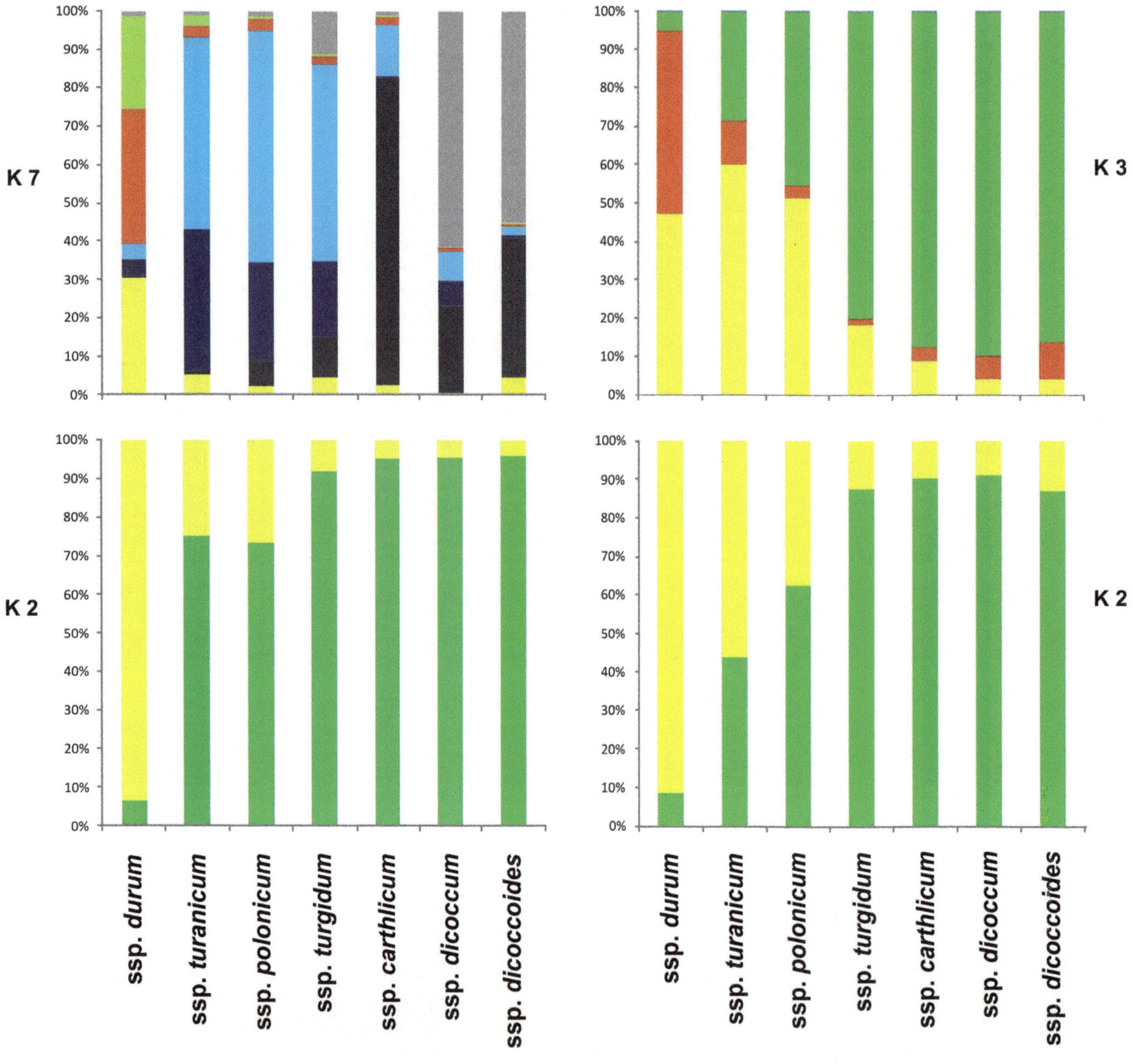

Figure 2. Population structure of each subspecies of wheat according to the taxonomical classification, using 26 SSR and 970 DArT markers.

largest allele number was detected for *Xwmc606* (6–18 alleles), while the lowest was found for *BJ274952* (1–3 alleles).

When considering the genetic diversity (H_E) among the subspecies, the differences were less evident (Table 3). Indeed, using SSRs, the H_E values ranged from 0.56 (ssp. *carthlicum*) to 0.70 (ssp. *dicoccoides*), and when computed for the DArT markers, from 0.22 (ssp. *carthlicum*) to 0.33 (ssp. *durum*), with a mean value of 0.28.

Population Structure of the Wheat Collection

The population structure was analysed using a Bayesian approach on 230 wheat accessions implemented in the STRUCTURE software. Following the method of Evanno *et al.* [61], the ΔK were plotted against the K numbers of the sub-groups. The maximum ΔK occurred at K = 2 and K = 7 for the SSR markers, and at K = 2 and K = 3 for the DArT markers. When considering K = 2, the collection was split in two sub-groups (group 1, group 2) containing 129 and 101 accessions based on SSRs data, respectively, and 147 and 83 accessions based on DArT data, respectively (Figure 1). In both cases, the structures assigned all of the durum accessions to the same group (group 1), with the exception of four cultivars (Timilia, Belfuggito, Lambro, Russello). Furthermore, based on the analysis carried out with only SSRs, Ceedur and Kyperounda were assigned to group 2, together with all of the other subspecies. In particular, all of the durum genotypes were assigned to cluster 1, with a q1 mean membership of 0.96 and 0.88 for SSRs and DArT, respectively. The remaining genotypes were assigned to group 2, with a q2 mean of 0.91 and 0.85 for SSRs and DArT, respectively, with the exception of some genotypes that showed high levels of admixture. Again, using DArTs, 31 genotypes showed q values lower than 0.6, while with

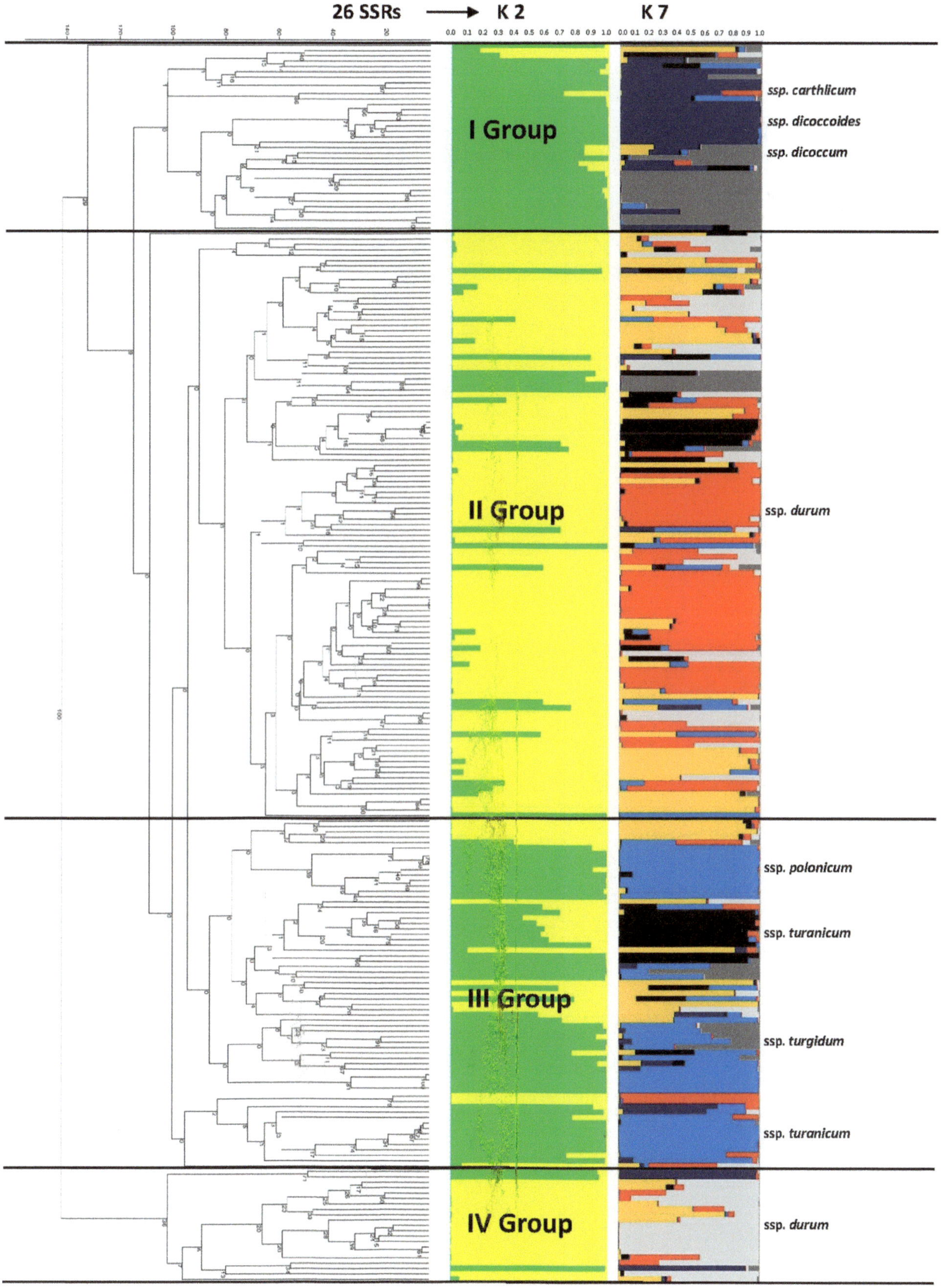

Figure 3. Population structure and dendrogram of the 230 accessions using SSR markers. The numbers on nodes are bootstrap probabilities estimated by permutation tests with 1000 replications.

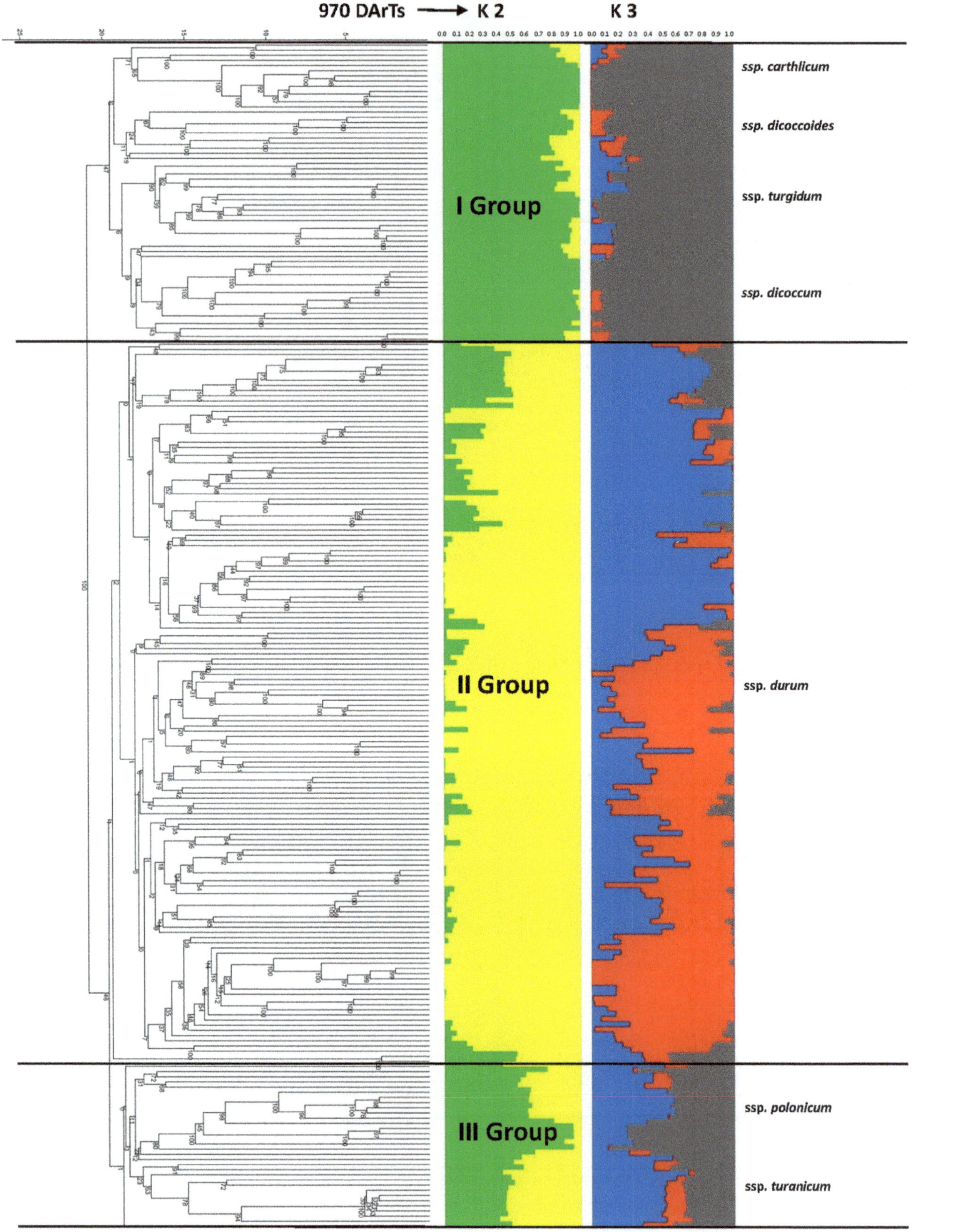

Figure 4. Population structure and dendrogram of the 230 accessions using DArT markers. The numbers on nodes are bootstrap probabilities estimated by permutation tests with 1000 replications.

SSRs, q values lower than 0.6 were only recorded for 10 genotypes. The difference in the sizes of these two groups was because when using DArT markers, most of the ssp. *turanicum* accessions (17 of 20) were assigned to the same group of ssp. *durum* with a greater number of admixed genotypes. Performing the analysis using DArT, at K = 3, the ssp. *turanicum* accessions were separated from the ssp. *durum*, while the remaining accessions were grouped together (data not shown).

Considering the mean q for each subspecies, at K = 2 (Figure 2), the data show the stratification for both of the sets of molecular markers. In particular, both sets distinguished the ssp. *durum* from the other subspecies, while the structure of the ssp. *turgidum* was relatively similar to other subspecies, such as ssp. *carthlicum*, ssp. *dicoccum* and ssp. *dicoccoides*. At K >2, the SSR markers showed a greater degree of resolution than the DArT, with their identification of a greater number of groups within each subspecies. In this case, with the exception of ssp. *durum*, the structure of ssp. *turgidum*, ssp. *turanicum*, and ssp. *polonicum* appeared to be more homogeneous and recognisable from other subspecies. At the same time, at K = 7, ssp. *carthlicum* showed a genetic structure significantly different from the other subspecies analysed, appearing more similar to ssp. *dicoccum* and ssp. *dicoccoides* using the DArT markers (K = 3).

To further refine the relationships between the tetraploid wheat accessions, cluster analyses were performed for each of the two marker types, and the general topology of these trees was compared with the Bayesian model structure shown in Figures 3 and 4. The genotypes were clustered into four and three large groups, corresponding to their botanical classification for the SSR and DArT markers, respectively. In particular, the dendrogram generated using the SSR matrices of similarity (Figure 3) assembled the ssp. *durum* genotypes into two main groups (groups II and IV) with 93 and 17 varieties, respectively, although a number of genotypes were admixed. The remaining accessions were clustered in the other two groups (groups I and III). The greater part of the accessions of ssp. *dicoccum* (14), ssp. *carthlicum* (10) and ssp. *dicoccoides* (8) belonged to group I, while group III comprised mainly the accessions of ssp. *polonicum* (13), ssp. *turanicum* (15) and ssp. *turgidum* (16), and some admixed accessions of ssp. *durum*, which were generally of unknown origin (i.e., Ceedur) or were derived from introgression with wild and domesticated wheat (Lambro and Belfuggito). Similar data are reported in Figure 4 using the similarity matrix generated by the DArT markers. In this case, three distinct groups were identified, although with lower mean genetic distances. The accessions of ssp. *dicoccum* (17), ssp. *dicoccoides* (12), ssp. *turgidum* (17), ssp. *carthlicum* (10) and 2 admixed accessions of ssp. *polonicum* were in group I. All ssp. *durum* accessions were in group II, together with 8 accessions of ssp. *turanicum*, 5 of ssp. *polonicum*, and 1 of ssp. *turgidum*. The largest part of ssp. *turanicum* (11) were in group III, together with a few accessions of ssp. *carthlicum* (2), ssp. *dicoccum* (2) and ssp. *turgidum* (1), and three old Italian durum varieties (Aziziah, Russello, Timilia).

To explore the relationships revealed using different types of data, the similarities between distance matrices based on the SSR, DArT, morphological and biochemical traits were measured using pair-wise comparisons among the genetic distance matrices with normalised Mantel statistics. The highest correlation was observed between the SSR and DArT matrices ($r = 0.66$, $p \leq 0.001$), which indicates that each set of markers yielded highly similar estimates of genetic distances between genotypes. The lowest values of correlation were recorded between morphological and biochemical matrices ($r = 0.22$, $p \leq 0.001$) and between morphological and DArT matrices ($r = 0.28$, $p \leq 0.001$).

Divergence Analysis Among the Wheat Subspecies

The effects of selection pressure on the evolution of the cultivated forms of tetraploid wheat were determined by the fixation index (Fst). From comparisons of the alleles among genotypes of the seven subspecies and measurements of the level of differentiation for each DArT marker, the genetic changes selected over the course of the wheat breeding were assessed. Furthermore, knowing the nucleotide sequences of most DArT markers, it is possible to suggest candidate genes for the loci that are under positive selection between subspecies. On a total of 970 DArT markers that were polymorphic across the tetraploid collection, a subset of 590 DArT markers were positioned on a consensus map developed in durum wheat by Marone *et al.* [33].

The analysis of Fst on a locus-by-locus basis provided no statistical cut-off for the identification of loci that might be under positive selection. Therefore, we used an outlier detection method implemented in the ARLEQUIN software. Across the subspecies, a total of 211 outlier loci were identified as under positive selection, and of these, 109 (51.7%) were mapped in the consensus map [54].

The markers under positive selection were spread across all chromosomes, with 42% and 58% of the markers located on genomes A and B, respectively (Table S3). In some cases, the markers under positive selection were spread over large chromosome regions, at great distances from each other, although regions in which a number of outlier markers were grouped together in a few cM were also identified on chromosomes 1B, 2B, 3B, 4A, 4B, 6A, 6B and 7B (Figure 5). Particularly interesting was the case of chromosome 4A, where three regions putatively under selection were found: 33.9–40.9 cM (three DArT markers), 89.9–104.2 cM and 118.5–129.6 cM (7 DArT markers). Among the 211 loci, NBS-LRR, protein kinase, peroxidase, putative cellulose synthase, and transposable element related sequences were identified as under selection between subspecies. These genes might be selected as having an important role in plant responses to biotic and abiotic stress, or, more simply, are in genetic linkage with the locus subjected to selection during the domestication process.

Genetic Distance and Relationships between Durum Cultivars

The relationships between the durum cultivars were first established by analysing the genetic structure of 128 genotypes (Figure 6). The analysis carried out with both sets of markers (SSR and DArT) indicated that the maximum ΔK occurred at K = 2. The durum cultivars were split in two sub-groups according to their origin and year of release (Table S1). Indeed, a strong correlation was observed between the structure results (the q1 mean) and durum cultivars according to year of release ($r = 0.39$, 0.38, $p \leq 0.001$, using SSRs and DArT, respectively). The first sub-group included most of the historic Italian varieties that were largely established during the first three quarters of the last century, while the second sub-group was primarily represented by modern cultivars released after 1990, with some exceptions.

The temporal trend of diversity in the Italian durum collection is reported in Table 4, and to avoid the strong bias for different sampling sizes, we analysed the means of the genetic variation (H_E) while combining the old and intermediate breeding groups in comparison with modern cultivars. One hundred and twenty-two out of 128 durum cultivars were grouped at K = 2 (SSR): 49 and 73 genotypes for the group before 1989 (old and intermediate) and after 1990 (modern), respectively. The number of alleles detected across the two groups were relatively similar for all of the traits, with the exception of the SSR markers, for which a similar level of diversity was found among the different groups (Table 4).

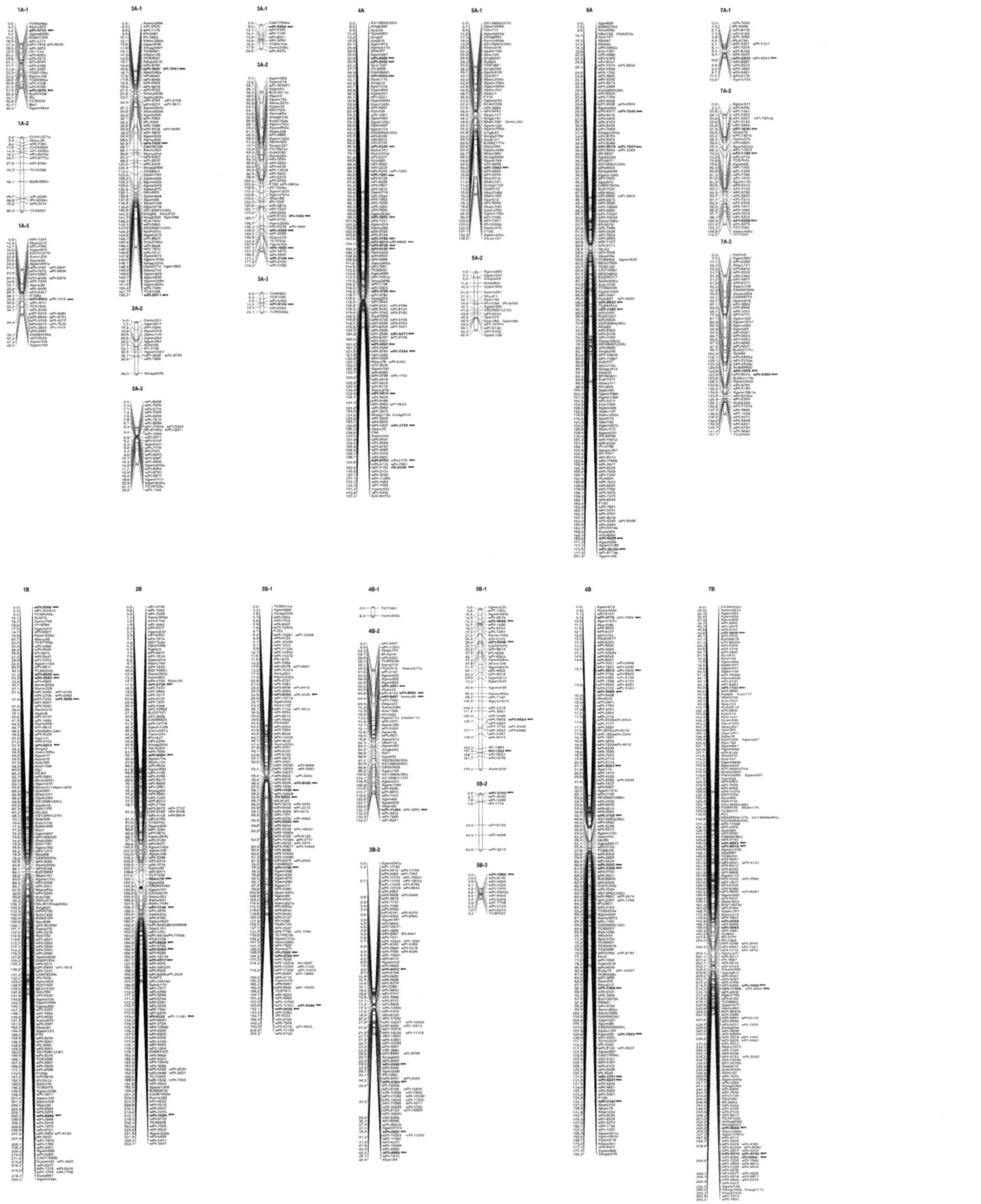

Figure 5. DArT markers under selection positioned (bold type and arrow) on a durum wheat consensus map (Marone et al. 2012a).

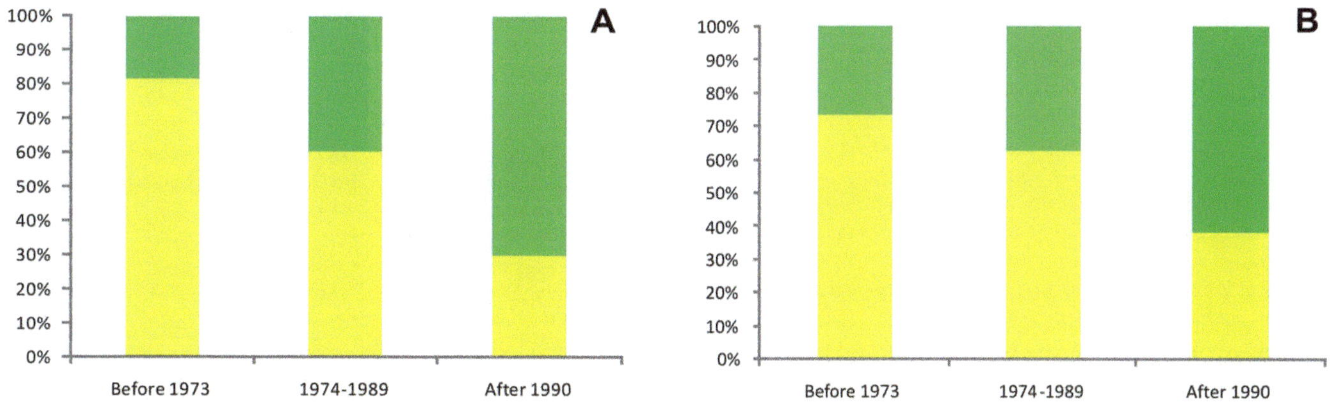

Figure 6. Population structure of the sub-group of durum cultivars inside the wheat collection with K = 2, using 26 SSR (A) and 970 DArT markers (B). The population structure of durum cultivars was grouped according to year of release.

The differentiation of the two groups was also confirmed by the analysis carried out to estimate the presence of private alleles in each group. Considering a threshold of 5%, to eliminate the rare alleles, 3 SSR markers (Xgwm408, Xgwm537, Xgwm299) were polymorphic in the first group of old and intermediate durum wheat varieties, and monomorphic in the second group, while 5 markers, as 4 SSR (Xgwm1084, Xgwm1093, Xgwm1017, Xwmc606) and 1 DArT (wPt-4142), were polymorphic in the modern durum wheat group, and monomorphic in the first group released before 1989.

Using morphological and biochemical traits (glutenin and gliadin), there was a loss of genetic diversity (H_E). Indeed, the 'old and intermediate' group showed $H_E = 0.19$, 0.29 and 0.48 for morphological, glutenin and gliadin traits, respectively, while the modern genotypes showed lower H_E, at 0.07, 0.19 and 0.34 for the same traits, respectively. For the SSR markers, a two-fold higher mean H_E was recorded with respect to those of the morphological and storage protein, and no trend was observed between the two groups ($H_E = 0.59$, 0.60, respectively). The same behaviour was seen for DArT, with the H_E mean for the old and intermediate groups (0.33) higher that for the modern (0.32) cultivars. The data obtained with the SSR and DArT marker technologies were in agreement.

A subset of 116 durum cultivars selected on the basis of the pedigree information available was used to investigate the relationships between the genetic distances estimated with the SSR, DArT and pedigree data. For the pedigree data, a matrix of genetic distances was derived from the kinship coefficients established on the basis of the pedigree for each genotype. The genetic distances ranged from 0.04 to 0.96 for SSRs, from 0.003 to 0.49 for DArT, and from 0.0 to 1.0 for pedigree, with means of 0.64, 0.33 and 0.89, respectively. The lowest genetic distance (0.003) was observed between three pairs of genotypes by DArT markers (Dylan vs. Fenix, Latino vs. Maestrale, Belfuggito vs. Lambro), while Colosseo and Doral showed the lowest genetic distance (0.04) using the SSR markers. The highest genetic distances (1.0) were observed in the pedigree matrix between different pairs of genotypes with no shared ancestors, and they skewed the distribution towards higher values. The majority of the kinship coefficients were between 0.0 (unrelated) and 1.0 (identical). Of the 13,340 pairwise comparisons, 4,670 were 0.0 (unrelated). Examples of varieties that showed a high number of zero coefficients (>110 pairwise comparisons) were Aziziah (113), Russello (115), Timilia (115), Tangarog (115), Kyperonda (115), Gianni (115) Cosmodur (115), Italo (115), Durfort (115), Provenzal (115) Sharm 5 (115), Athena (115), Tresor (115), Quadruro (115), Plinio (114), Doral (115), Parsifal (115), Nefer (115) and Virgilio (115). Several cultivars showed identical kinship, including Platani and Ciccio (0.185), Exeldur, Brindur and Neodur (0.092), Lambro and Belfuggito (0.072), and Vendetta, Simeto and Fortore (0.191).

As performed on the entire collection, the similarities between the distance matrices based on the SSRs, DArT and kinship coefficient were measured by comparing the three genetic distance matrices using the Mantel test. The highest correlation was observed between the SSR and DArT matrices ($r = 0.48$, $p \leq 0.001$), which indicates that each set of markers yielded highly similar estimates of genetic distances between genotypes. Lower but significant positive correlations were obtained between the DArT and pedigree matrices ($r = 0.21$, $p \leq 0.001$) and between the SSR and pedigree matrices ($r = 0.23$, $p \leq 0.001$). Of the three methods used, the pedigree analysis provided the lowest resolution.

Table 4. Alleles number and genetic diversity across the two genetic groups of *T. turgidum* ssp. *durum* grouped according to years of release.

T. turgidum	Years of release	Sample size	Morphological traits		Glutenin protein		Gliadin protein		SSR markers	DArT markers
			n_a	H_E	n_a	H_E	n_a	H_E	H_E	H_E
ssp. *durum*	Old and intermediate	49	11	0.19	15	0.29	19	0.48	0.59	0.33
	Modern	73	10	0.07	17	0.19	17	0.34	0.60	0.32

Discussion

The accurate description of population structure became extremely important with the advent of association genetics [63]. Indeed, when not accounted for sufficiently, population structure can lead to spurious detection of associations between markers and phenotypes of interest [64]. So, understanding the level and structure of the genetic diversity of a crop is a prerequisite for the conservation and efficient use of the available germplasm for plant breeding.

Several tools have been used to evaluate genetic diversity in domesticated and wild wheat, with those most widely used being morpho-agronomic traits [30,65], seed storage proteins [14,46,66], and molecular markers [4,9,23]. In the present study, we combined the analysis of information observed for phenotypic traits (morphological and biochemical markers) with molecular marker (SSRs and DArTs) and pedigree data. The combination of different types of information might be very important for the effects of different evolutionary forces on the structure of the genetic diversity to be disentangled, and also to highlight the evolutionary history of crop germplasm. For instance, while molecular-marker neutrality can be assumed, morphological and biochemical traits are more likely to be affected by selection. Moreover, among molecular markers, differences in describing the diversity of a population might be associated with different mutation rates [67–68]. Finally, pedigree information is also an important element to assess the structure of the diversity of modern cultivars.

Structure of the Whole Collection

The genetic diversity of the morphological traits and seed storage proteins was always lower in the cultivated durum group compared to the wild and domesticated emmer, while ssp. *turanicum*, ssp. *polonicum* and ssp. *carthlicum* were intermediate. In contrast, differences in the levels of molecular diversity for SSR and DArT markers were less strong, which suggests that selection for a well-defined phenotype was very strong during the domestication and development of durum cultivars. The ssp. *dicoccum* and ssp. *dicoccoides* were the ones that had the highest genetic variability for seed storage proteins and SSRs, which confirms that the unadapted germplasm (i.e., the wild and landrace germplasm) represents a powerful source of genes for the improvement of durum wheat [40]. Very low genetic diversity was observed in the ssp. *carthlicum* for all of the markers. This indicated the restricted genetic basis of these wheats, in agreement with Carmona *et al.* [66] and Riefolo *et al.* [14], which is probably associated to a more restricted area of geographical origin [69]. Moreover the molecular analysis showed ssp. *carthlicum* as a very distinct group from the other free-threshing tetraploid wheats, such as ssp. *durum*, ssp. *turgidum*, ssp. *turanicum* and ssp. *polonicum*. As it looks very much like common wheat (*T. aestivum* ssp. *vulgare*), ssp. *carthlicum* was initially classified as a hexaploid species [70]. A striking feature of ssp. *carthlicum* is the awned glume, so all of the spikelets show four awns. Later, due to its resistance to mildew and rust and a chromosome number with 2 n = 28, ssp. *carthlicum* was recognised as a tetraploid species [71].

When considering K = 2 using both SSRs and DArTs, the data were effective in discriminating about 95% of the *durum* genotypes from a second group that included all of the other accessions (122 out 128 durum cultivars were grouped together). Nevertheless the *durum* cultivars Belfuggito, Timilia, Ceedur, Kyperounda, Lambro and Russello were clustered with the other accessions. The presence in the second group of Kyperounda, Timilia and Russello might be explained as being derived from local landraces, while Belfuggito and Lambro have ssp. *dicoccoides* in their pedigree. This suggested that the cultivars Ceedur and Kyperounda could also be derived by hybridisation between an improved durum variety and a genotype extracted from a wild or landrace population.

According to Oliveira et al. [35] the structure of the dendrograms, obtained from SSR and DArT distances, strengthen the further taxonomic *T. turgidum* spp. subdivision proposed by MacKey [43], which classifies *durum*, *turanicum*, *turgidum* and *polonicum* as convarieties of the subspecies *T. turgidum* spp. *turgidum* (e.g., *T. turgidum* spp. *turgidum* conv. *durum*), and distinguishes the wild and domesticated emmer from the naked wheat gene pool.

Moreover, when some hexaploid wheat genotypes were added to the cluster analysis (data not shown), the spp. *carthlicum* accessions were more similar to common wheat than to ssp. *dicoccoides*, which confirms the uncertain origin of this subspecies [69–70], and suggests gene flow from hexaploid to tetraploid wheat [72–73].

Divergence Analysis Across Subspecies during Wheat Evolution

Tetraploid wheat has undergone intensive selection for certain desirable characteristics during domestication and the subsequent breeding process, such as high and stable yields. A Fst-outlier method was used to identify which loci might be under positive selection, and therefore might be linked to regions of the genome that are responsible for the phenotypic variation present in the germplasm analysed.

We identified 211 candidate loci under positive selection based on Fst values that fall outside of the 99% confidence interval established for the distribution [57]. These loci might be directly under selection, but they are more likely to mark regions of the genome that have been selected during evolution. Among the 211 loci identified under selection, 109 markers were recently mapped on the durum wheat consensus map [54]. Some of these markers were clustered into specific regions on chromosome arms 2BL, 3BS and 4AL, where several genes/QTLs involved in the domestication of tetraploid wheat are located, such as the tenacious glumes (*Tg*) [74–78] and brittle rachis (*Br*) [77,79] characteristics. The regions identified from the markers under selection, *wPt-6122* and *wPt-7004*, which were mapped on chromosome bin 2BL4 (0.50–0.89) [54], corresponded at the *Tg2* QTL reported by Peleg *et al.* [78]. Additional QTLs that affect threshability have been detected on chromosome 4AL [78]. In particular, there are two *Tg* QTLs on chromosome bin 4AL13 (0.59–0.66) that are linked at the markers *wPt6515*, *wPt7558* and *Xgwm610* [78]; these coincided with the same region as the *wPt8489*, *wPt5455*, *wPt6303*, *wPt1091* and *wPt6330* markers under selection that were mapped on the durum wheat consensus map [54]. Finally, the markers *wPt8686*, *wPt1349*, *rPt5853* and *wPt1159* that were identified in the present study as being under selection are in the same chromosome bin 3BS8 (0.78–1.00), to where the dominant gene *Br-A3* that affects the brittle rachis in tetraploid wheat was mapped by Nalam *et al.* [79].

These data suggest that the use of objective approaches to identify outliers will reveal portions of the genome that are under selection and that might represent candidates for further functional analyses to identify the loci underlying the phenotypic differences between these varieties.

Population Structure of Durum Wheat Cultivars

The population structure of the durum accessions identified two major groups associated with the year of release of the cultivar. Moreover, the genetic diversity of morphological traits and seed storage proteins was always lower in the modern durum cultivars registered after 1990, than in the intermediate and old ones. This marked effect on diversity was not observed for molecular markers, where there was only a weak reduction. This suggests that the reduction in diversity is likely to be due to selection for few adaptive traits (i.e., glaucousness) and quality characters (i.e., glutenin and gliadin subunits) that strongly correlate with grain yield and technological properties of gluten. Similar results were observed in durum wheat in terms of yield components and yellow pigment content [65,80]. The absence of a parallel reduction in the genetic diversity can be explained by two non-alternative hypotheses: (i) the effect of drift has not been very strong during modern breeding; and (ii) new germplasm from different gene pools has been introduced in the genetic background of the durum cultivars. Considering this first hypothesis, a possible explanation arises as the durum diversity was already significantly reduced at the very beginning of the breeding, while a second explanation might be associated with the introduction of new genotypes from CYMMIT or from countries such as France, Spain and USA. The second hypothesis is supported by data obtained by [24], and it can also explain the association between the population structure and the 'ages' of the different varieties.

Correlation between Diversity and Distance Estimates

The average genetic distances between the pairs of durum accessions that were obtained for molecular markers differed markedly. With DArT markers, a low mean genetic distance was obtained (0.33; range 0.0–0.49), whereas the pedigree information yielded a mean distance of 0.89 (range 0.0–1.0). SSRs produced a range of pairwise distances between 0.00 and 0.96, with a mean of 0.63. The low distances seen between these lines using DArT markers probably reflects the limited number of alleles (presence vs. absence) compared to SSRs. Thus, for any pair of accessions, a large proportion of the markers carries the same allele, even between very different genotypes. In contrast, using SSRs, the greater number of alleles increases the probability that two genotypes differ for any given marker. At the other extreme, pedigree information leads to a highly biased distribution of genetic distances (data not shown), with a large number of pairs showing a distance of 1, thus indicating that there are no common ancestors in their pedigrees. This bias in the distribution of distances based on the pedigree information also impacts heavily on the correlation between the genetic distances.

As reported above, the correlations between pedigree *versus* the SSR and DArT matrices were significant, but low (0.23 and 0.21, respectively), although in many previous studies, even weaker correlations (0.10–0.25) have been reported between genetic distance and similarities based on pedigree and molecular markers, such as single nucleotide polymorphism [25], random amplification of polymorphic DNA [81], restriction fragment length polymorphism [82] and amplified fragment length polymorphism [83].

The low level of genome similarities obtained from parentage analysis might be due to the distinct names that are often given to parents, which actually trace back to common progenitors and incomplete or uninformative pedigree records, especially if encoded varieties are included [83]. Nineteen of 116 durum genotypes appear to be unique, with no commonality with other genotypes. In addition to this, various assumptions that are made in the calculation of kinship coefficients based on pedigree can introduce inaccuracies [13,84–85]. These include: (i) the equal parental contributions to progeny; (ii) the absence of selection pressure or genetic drift; and (iii) the absence of relatedness of parents with unknown pedigree. So, the genetic similarity estimates based on molecular marker data are expected to be more accurate, as any polymorphism is a direct outcome of variations at the DNA level.

Most of the durum cultivars included in the present study were selected in Italy. The pedigree information indicated that many of these cultivars derived from a few durum lines. In particular, the Cappelli genotype can be considered as the true founder of the germplasm of the cultivated durum wheats. Moreover, Creso, Valnova and Valforte can be considered as the main founders of the modern durum cultivars. Indeed, these three genotypes, which are known as the first generation of the modern CIMMITY-related materials, are historically relevant in that they introduced the innovative semi-dwarf CIMMITY materials into the Italian durum germplasm [24]. This suggests that most of the varieties introduced and the advanced breeding lines developed by crossing exotic materials (introduced by CIMMYT, or from Mexico, USA, France), or the genotypes derived from exotic materials followed by selection of superior genotypes, make the gene pool smaller for all of the wheat cultivars. Therefore, there is the need to incorporate new variability into the existing wheat germplasm to address the new challenges, like climate change and food security.

Our data also demonstrate that dendrograms obtained with the two types of marker data are highly congruent, and that the durum cultivars were clearly divided into two major groups that reflect their origins and year of release. With few exceptions, such a clear division was also documented in a small collection of 28 durum cultivars by Maccaferri *et al.* [23] using SSR markers, and Zhang *et al.* [86] using DArT markers. This is, however, quite different from the situation reported by Rostoks *et al.* [87] among European barley cultivars, where the habitus (winter *vs.* spring) was found to be the primary determinant of the population structure. This suggests that for the most part of the last century, Italian breeders used predominantly Italian genetic material in their breeding programmes, and only in the last decades did they perform crosses using genetic material that came from other countries.

Conclusions

In conclusion, our data initially confirm that both sets of SSR and DArT markers provide an accurate picture of the population structure within tetraploid wheat collections, which is information of critical importance for the design of association analyses. Probably a subsequent increase in the number of accessions will allow us to better understand the influence of the geographical area of origin on the evolutionary behaviour of tetraploid wheats. The present study also suggests the genetic potential of the landraces and wild accessions for the detection of unexplored alleles. Overall, the panel of genotypes investigated in the present study represents a strategic platform for the study of traits related to evolution and domestication of tetraploid subspecies, and for association mapping studies. The information obtained from this collection of genotypes will help in the selection of parents to develop high-yield durum wheat lines in breeding programmes, and to determine the potential of this panel of varieties for association mapping in subsequent studies.

Supporting Information

Table S1 Year of release, country and pedigree information of the 128 durum wheat accessions assembled in the wheat collection.

Table S2 Alleles number and genetic diversity for each SSR marker for each subspecies included in the wheat collection.

Table S3 DArT markers under selection, chromosome position and putative function.

Acknowledgments

The authors wish to thank N. Tinker and B. Baum (Agriculture and Agri-Food Canada) for providing the Winkin2 software. We would like also to thank Dr. E. Bellucci (Università Politecnica delle Marche, Ancona, Italy) for helping in the statistical analysis, Dr G. Petruzzino for helping in the pedigree data analysis, and Dr. Christopher Berrie for scientific English language editorial assistance.

Author Contributions

Conceived and designed the experiments: AB RP GL PD. Performed the experiments: GL GM FT DM. Analyzed the data: GL GM RP PD. Contributed reagents/materials/analysis tools: GL GM DM FT AM AG LC. Wrote the paper: GL GM PD. Read, critically edited, and approved the final manuscript: GL GM FT AG AB LC DM AM RP PD.

References

1. Autrique E, Nachit MM, Monneveux P, Tanksley SD, Sorrells ME (1996) Genetic diversity in durum wheat based on RFLPs morphophysiological traits and coefficient of parentage. Crop Sci 36: 735–742.
2. Thuillet AC, Bataillon T, Poirier S, Santoni S, David JL (2005) Estimation of long-term effective population sizes through the history of durum wheat using microsatellite data. Genetics 169: 1589–1599.
3. Fu YB, Peterson GW, Clarke JM, Richards KW, Somers DJ (2005) Allelic reduction and genetic shift in the Canadian red hard spring wheat germplasm released from 1886 to 2004. Theor Appl Genet 110: 1505–1516.
4. Fu YB, Peterson GW, Yu JK, Gao LJJ, Richards KW (2006) Impact of plant breeding on genetic diversity of the Canadian hard red spring wheat germplasm as revealed by EST-derived SSR markers. Theor Appl Genet 112: 1239–1247.
5. Haudry A, Cenci A, Ravel C, Bataillon T, Brunel D, et al. (2007) Grinding up wheat: a massive loss of nucleotide diversity since domestication. Mol Biol Evol 24: 1506–1517.
6. Allard RW (1996) Genetic basis of the evolution of adaptedness in plants. Euphytica 92: 1–11.
7. Hoisington D, Khairallah M, Reeves T, Ribaut JM, Skovmand B, et al. (1999) Plant genetic resources: what can they contribute toward increased crop productivity? Proc Natl Acad Sci USA 96: 5937–5943.
8. Donini P, Law JR, Koebner RMD, Reeves JC, Cooke RJ (2000) Temporal trends in the diversity of UK wheat. Theor Appl Genet 100: 912–917.
9. Ruiz M, Giraldo P, Royo C, Villegas D, Aranzana MJ, et al. (2012) Diversity and genetic structure of a collection of Spanish durum wheat landraces. Crop Sci 52: 2262–2275.
10. Moragues M, Zarco-Hernández J, Moralejo MA, Royo C (2006) Genetic diversity of glutenin protein subunits composition in durum wheat landraces [*Triticum turgidum* ssp.*turgidum* convar. *durum* (Desf.) MacKey] from the Mediterranean basin. Genet Resour Crop Evol 53: 993–1002.
11. Moragues M, Moralejo MA, Sorrells ME, Royo C (2007) Dispersal of durum wheat landraces across the Mediterranean basin assessed by AFLPs and microsatellites. Genet Resour Crop Evol 54: 1133–1144.
12. Peleg Z, Saranga Y, Suprunova T, Ronin Y, Röder M, et al. (2008) High-density genetic map of durum wheat x wild emmer wheat based on SSR and DArT markers. Theor Appl Genet 117: 103–115.
13. Cox TS, Lookhart GL, Walker DE, Harrell LG, Albers LD, et al. (1985) Genetic relationships among hard red winter wheat cultivars as evaluated by pedigree analysis and gliadins polyacrylamide gel electrophoretic patterns. Crop Sci 25: 1058–1063.
14. Riefolo C, Ficco DBM, Cattivelli L, De Vita P (2011) Genetic diversity of gluten proteins in *T. turgidum* L. Cereal Res Commun 39: 405–414.
15. Damania AB, Porceddu E, Jackson MT (1983) A rapid method for the evaluation of variation in germplasm collections of cereals using polyacrylamide gel electrophoresis.Euphytica 32: 877–883.
16. Soleimani VD, Baum BR, Johnson DA (2007) Analysis of genetic diversity in barley cultivars reveals incongruence between S-SAP, SNP and pedigree data. Genet Resour Crop Evol 54: 83–97.
17. Mikel MA, Kolb F(2008) Genetic diversity of contemporary North American barley. Crop Sci 48: 1399–1407.
18. Martos A, Royo C, Rharrabti Y, Garcia Del Moral LF (2005) Using AFLPs to determine phylogenetic relationships and genetic erosion in durum wheat cultivars released in Italy and Spain throughout the 20th century. Field Crops Res 91: 107–116.
19. Martin JM, Blake TK, Hockett EA (1991) Diversity among North American spring barley cultivars based on coefficients of parentage. Crop Sci 31: 1131–1137.
20. Van Inghelandt D, Melchinger AE, Lebreton C, Stich B (2010) Population structure and genetic diversity in a commercial maize breeding program assessed with SSR and SNP markers. Theor Appl Genet 120: 1289–1299.
21. Jin L, Lu Y, Xiao P, Suan M, Corke H, et al. (2010) Genetic diversity and population structure of a diverse set of rice germplasm for association mapping. Theor Appl Genet 121: 475–487.
22. Fu YB, Somers DJ (2009) Genome-wide reduction of genetic diversity in wheat breeding. Crop Sci 49: 61–168.
23. Maccaferri M, Sanguineti MC, Donini P, Tuberosa R (2003) Microsatellite analysis reveals a progressive widening of the genetic basis in the elite durum wheat germplasm. Theor Appl Genet 107: 783–797.
24. Maccaferri M, Sanguineti MC, Noli E, Tuberosa R (2005) Population structure and long-range linkage disequilibrium in a durum wheat elite collection. Mol Breed 15: 271–289.
25. Soleimani VD, Baum BR, Johnson DA (2002)AFLP and pedigree based genetic diversity estimates in modern cultivars of durum wheat (*Triticum turgidum* L. Subsp. *durum*). Theor Appl Genet 104: 350–357.
26. Blanco A, Bellomo MP, Cenci A, De Giovanni C, D'Ovidio R, et al. (1998) A genetic linkare map of durum wheat. Theor Appl Genet 97: 721–728.
27. Gadaleta A, Giancaspro A, Giove SL, Zacheo S, Mangini G, et al. (2009) Genetic and physical mapping of new EST-derived SSRs on the A and B genome chromosomes of wheat.Theor Appl Genet 118: 1015–1025.
28. Medini M, Hamza S, Rebai A, Baum M (2005) Analysis of genetic diversity in Tunisian durum wheat cultivars and related wild species by SSR and AFLP markers. Genet Resour Crop Evol 52: 21–31.
29. Altintas S, Toklu F, Kafkas S, Kilian B, Brandolini A, et al. (2008) Estimating genetic diversity in durum and bread wheat cultivars from Turkey using AFLP and SAMPL markers. Plant Breed 127: 9–14.
30. Royo C, Maccaferri M, Álvaro F, Moragues M, Sanguineti MC, et al. (2010) Understanding the relationships between genetic and phenotypic structures of a collection of elite durum wheat accessions. Field Crops Res 119: 91–105.
31. Akbari M, Wenzl P, Caig V, Carling J, Xia L, et al. (2006) Diversity arrays technology (DArT) for high-throughput profiling of the hexaploid wheat genome. Theor Appl Genet 113: 1409–1420.
32. Zhang LY, Liu DC, Guo XL, Yang WL, Sun JZ, et al. (2011) Investigation of genetic diversity and population structure of common wheat cultivars in northern China using DArT markers.BMC Genetics 12: 42.
33. Marone D, Panio G, Ficco DBM, Russo MA, De Vita P, et al. (2012b) Characterization of wheat DArT markers: genetic and functional features. Mol Genet Genomics 287: 741–53.
34. Matsuoka Y (2011) Evolution of polyploid *Triticum* wheats under cultivation: the role of domestication, natural hybridization and allopolyploid speciation in their diversification. Plant Cell Physiol 52: 750–764.
35. Oliveira HR, Campana MG, Jones H, Hunt HV, Leigh F, et al. (2012) Tetraploid Wheat Landraces in the Mediterranean Basin: Taxonomy, Evolution and Genetic Diversity. PLoS ONE 7(5): e37063.
36. De Vita P, Riefolo C, Codianni P, Cattivelli L, Fares C (2006) Agronomic and qualitative traits of *T. turgidum* ssp. *dicoccum* genotypes cultivated in Italy. Euphytica, 50: 195–205.
37. Dedkova OS, Badaeva ED, Amosova AV, Martynov SP, Ruanet VV, et al. (2009) Diversity and the origin of the European population of *Triticum dicoccum* (Schrank) Schuebl. As revealed by chromosome analysis. Russ J Genet 45: 1082–1091.
38. Pan D, Hong L, Wei L, Peng-Fei Q, Yu-Ming W, et al. (2007) Genetic diversity of storage proteins in *Triticum polonicum* L. J Plant Sci 2: 416–424.
39. Ozbek O, Millet E, Anikster Y, Arslan O, Feldman M (2007) Spatio-temporal genetic variation in population of wild emmer wheat, *Triticum turgidum* ssp. *dicoccoides*, as revealed by AFLP analysis. Theor Appl Genet 115: 19–26.
40. Li W, Zhang DF, Wei YM, Yan ZH, Zheng YL (2006) Genetic diversity of *Triticum turgidum* L. based on microsatellite markers. Russ J Genet 42: 311–316.
41. Sissons MJ, Batey IL (2003) Protein and starch properties of some tetraploid wheats. Cereal Chem 80: 468–475.
42. van Slageren MW (1994) Wild wheats: a monograph of *Aegilops* Land *Amblyopyrum* Eig (Poaceae). Wageningen Agriculture University Paper 7: 513 pp.
43. MacKey J (1988) A plant breeder's perspective on taxonomy of cultivated plants. Biol Zbl 107: 369–379.
44. Laemmli UK (1970) Cleavage of structural proteins during the assembly of the head of bacteriofage T4.Nature 227: 680–685.
45. Branlard G, Autran JC, Monneveux P (1989) High molecular weight subunit in durum wheat (*T. durum*). Theor Appl Genet 78: 353–358.
46. Li QY, Yan YM, Wang AL, An XL, Zhang YZ, et al. (2006) Detection of HMW glutenin subunit variations among 205 cultivated emmer accessions (*Triticum turgidum* ssp. *dicoccum*). Plant Breed 125: 120–124.
47. McIntosh RA, Devos KM, Dubcovsky J, Rogers WJ, Morris CF, et al. (2006) Catalogue of gene symbols for wheat: 2006 Supplement.
48. Nieto-Taladriz MT, Ruiz M, Martinez MC, Vazquez JF, Carrello JM (1997) Variation and classification of B low-molecular-weight glutenin subunit alleles in durum wheat. Theor Appl Genet 95: 1155–1160.

49. Lafiandra D, Kasarda DD (1985) One- and two-dimensional (two-pH) polyacrylamide gel electrophoresis in a single gel: separation of wheat proteins. Cereal Chemistry 62: 314–319.
50. Boggini G, Tusa P, Pogna NE (1995) Breadmaking quality of durum wheat genotypes with some novel glutenin subunit composition. J Cereal Sci 22: 105–113.
51. Aguiriano E, Ruiz M, Fité MR, Carrillo JM (2008) Genetic variation for glutenin and gliadins associated with quality in durum wheat (*Triticum turgidum* L. ssp. *turgidum*) landraces from Spain. Span J Agric Res 6: 599–609.
52. Sharp PJ, Kreis M, Shewry PR, Gale MD (1988) Location of β-amylase sequences in wheat and its relatives.Theor Appl Genet 75: 286–290.
53. Marone D, Del Olmo AI, Laido G, Sillero JC, Emeran AA, et al. (2009) Genetic analysis of durable resistance against leaf rust in durum wheat.Mol Breed 24: 25–39.
54. Marone D, Laido' G, Gadaleta A, Colasuonno P, Ficco DBM, et al. (2012a) A high-density consensus map of A and B wheat genomes.Theor Appl Genet 125: 1619–1638.
55. Russo MA, Ficco DBM, Marone D, De Vita P, Vallega V, et al. (2012) A major QTL for resistance to soil-borne cereal mosaic virus derived from an old Italian durum wheat cultivar. J Plant Interact 7: 290–300.
56. Tinker NA, Mather DE (1993) KIN: Software for computing kinship coefficients. J Hered 84: 238.
57. Excoffier L, Laval G, Schneider S (2005) Arlequin (version 3.0): an integrated software package for population genetics data analysis. Evolutionary Bioinformatics Online 1: 47–50.
58. Mantel N (1967)The detection of disease clustering and a generalized regression approach. Cancer Res 27: 209–220.
59. Nei M (1973) Analysis of gene diversity in subdivided populations. Proc Natl Acad Sci USA 70: 3321–3323.
60. Pritchard JK, Stephens M, Donnelly P (2000) Inference of population structure using multilocus genotype data. Genetics 155: 945–959.
61. Evanno G, Regnaut S, Goudet J (2005) Detecting the number of clusters of individuals using the software STRUCTURE: A simulation study. Mol Ecol 14: 2611–2620.
62. Hammer Ø, Harper DAT, Ryan PD (2001) PAST: paleontological statistics software package for education and data analysis. Palaeontol Electron 4: http://palaeo-electronica.org/2001 1/past/issue1 01.htm.
63. Nordborg M, Weigel D (2008) Next-generation genetics in plants. Nature 456: 720–723.
64. Thornsberry JM, Goodman MM, Doebley J, Kresovich S, Nielsonet D, et al. (2001) Dwarf8 polymorphisms associate with variation in flowering time. Nat Genet 28: 286–289.
65. De Vita P, Li Destri Nicosia O, Nigro F, Platani C, et al. (2007) Breeding progress in morpho-physiological, agronomical and qualitative traits of durum wheat cultivars released in Italy during the 20th century. European J Agronomy 26: 39–53.
66. Carmona S, Alvarez JB, Caballero L (2010)Genetic diversity for morphological traits and seed storage proteins in Spanish rivet wheat. Biologia Plantarum54: 69–75.
67. Rossi M, Bitocchi E, Bellucci E, Nanni L, Rau D, et al. (2009) Linkage disequilibrium and population structure in wild and domesticated populations of *Phaseolus vulgaris L.* Evol Applic 2: 504–522.
68. Bitocchi E, Nanni L, Bellucci E, Rossi M, Giardini A, et al. (2012) Mesoamerican origin of the common bean (*Phaseolus vulgaris* L.) is revealed by sequence data. Proc Natl Acad Sci USA 109: 5148–5149.
69. Kuckuck H (1979) On the origin of *Triticum carthlicum* Neyski (= *Triticum persicum* Vav.). Wheat Information Service50: 1–5.
70. Bushuk W, Kerber ER (1978) The role of *Triticum carthlicum* in the origin of bread wheat based on gliadin electrophorograms. Can J Plant Sci 58: 1019–1024.
71. Schiemann E (1948) Weizen, Roggen, Gerste. Systematik, Geschichte und Verwendung (Verlag von Gustav Fischer, Jena), 1–102.
72. Dvorak J,Akhunov ED,Akhunov AR,Deal KR,Luo MC (2006) Molecular characterization of a diagnostic DNA marker for domesticated tetraploid wheat provides evidence for gene flow from wild tetraploid wheat to hexaploid wheat. Mol Biol Evol 23: 1386–1396.
73. Wang S, Li X, Wang K, Wang X, Li S, et al. (2011) Phylogenetic analysis of C, M, N, and U genomes and their relationships with *Triticum* and other related genomes as revealed by LMW-GS genes at Glu-3loci. Genome 54: 273–284.
74. Simonetti MC, Bellomo MP, Laghetti G, Perrino P, Simeone R, et al. (1999) Quantitative trait loci affecting free-threshing habit in tetraploid wheats. Genet Res Crop Evol 46: 267–277.
75. Sood S, Kuraparthy V, Bai G, Gill BS (2009) The major threshability genes soft glume (sog) and tenacious glume (Tg), of diploid and polyploid wheat, trace their origin to independent mutations at non-orthologous loci. Theor Appl Genet 119: 341–351.
76. Jantasuriyarat C, Vales MI, Watson CJW, Riera-Lizarazu O (2004) Identification and mapping of genetic loci affecting free-threshing habit and spike compactness in wheat (*Triticum aestivum* L.). Theor Appl Genet 108: 261–273.
77. Peng JH, Sun D, Nevo E (2011) Domestication evolution, genetics and genomics in wheat. Mol Breed 28: 281–301.
78. Peleg Z, Fahima T, Korol AB, Abbo S, Saranga Y (2011) Genetic analysis of wheat domestication and evolution under domestication. J Exp Bot 62: 5051–5061.
79. Nalam V, Vales M, Watson C, Kianian S, Riera-Lizarazu O (2006) Map-based analysis of genes affecting the brittle rachis character in tetraploid wheat (*Triticum turgidum* L.).Theor Appl Genet 112: 373–381.
80. Di Gesù AM, Platani C, Cattivelli L, Mangini G, Blanco A (2009) Genetic variability in yellow pigment components in cultivated and wild tetraploid wheats. J Cereal Sci 50: 210–218.
81. Manninen O, Nissila E (1997) Genetic diversity among Finnish six-rowed barley cultivars based on pedigree information and DNA markers. Hereditas 126: 87–93.
82. Graner A, Ludwig WF, Melchinger AE (1994) Relationships among European barley germplasm. II. Comparison of RFLP and pedigree data. Crop Sci 34: 1199–1205.
83. Ellis RP, McNicol JW, Baird E, Booth A, Lawrence P, et al. (1997) The use of AFLPs to examine genetic relatedness in barley Mol Breed 3: 359–369.
84. Souza E, Sorrells M (1989) Pedigree analysis of North American oat cultivars released from 1951 to 1985. Crop Sci 29: 595–601.
85. Cox TS, Murphy JP (1990)The effect of parental divergence on F_2 heterosis in winter wheat crosses. Theor Appl Genet 79: 241–250.
86. Zhang LY, Marchand S, Tinker NA, Belzile F (2009) Population structure and linkage disequilibrium in barley assessed by DArT markers. Theor Appl Genet 119: 43–52.
87. Rostoks N, Mudie S, Cardle L, Russell J, Ramsay L, et al. (2005) Genome-wide SNP discovery and linkage analysis in barley based on genes responsive to abiotic stress. Mol Genet Genomics 274: 515–527.

3

A Comparison of Screening Methods to Identify Waterlogging Tolerance in the Field in *Brassica napus* L. during Plant Ontogeny

Xiling Zou, Chengwei Hu, Liu Zeng, Yong Cheng, Mingyue Xu, Xuekun Zhang*

Key Laboratory of Oil Crop Biology and Genetic Improvement of the Ministry of Agriculture, Oil Crops Research Institute of the Chinese Academy of Agricultural Sciences, Wuhan, Hubei, China

Abstract

Waterlogging tolerance is typically evaluated at a specific development stage, with an implicit assumption that differences in waterlogging tolerance expressed in these systems will result in improved yield performance in fields. It is necessary to examine these criteria in fields. In the present study, three experiments were conducted to screen waterlogging tolerance in 25 rapeseed (*Brassica napus* L.) varieties at different developmental stages, such as seedling establishment stage and seedling stage at controlled environment, and maturity stage in the fields. The assessments for physiological parameters at three growth stages suggest that there were difference of waterlogging tolerance at all the development stages, providing an important basis for further development of breeding more tolerant materials. The results indicated that flash waterlogging restricts plant growth and growth is still restored after removal of the stress. Correlation analysis between waterlogging tolerance coefficient (WTC) of yield and other traits revealed that there was consistency in waterlogging tolerance of the genotypes until maturity, and good tolerance at seedling establishment stage and seedling stage can guarantee tolerance in later stages. The waterlogging-tolerant plants could be selected using some specific traits at any stage, and selections would be more effective at the seedling establishment stage. Thus, our study provides a method for screening waterlogging tolerance, which would enable the suitable basis for initial selection of a large number of germplasm or breeding populations for waterlogging tolerance and help for verifying their potential utility in crop-improvement.

Editor: Ji-Hong Liu, Key Laboratory of Horticultural Plant Biology (MOE), China

Funding: This work is supported by National Science Foundation of China (31171589), (31201244), Hi-Tech Research and Development Program of China (2011AA10A104), Special Fund for Agroscientific Research in the Public Interest (2101203032), China Postdoctoral Science Foundation (2013M530081) and Hubei agricultural science and Technology Innovation Center. The funders had no role in study design, data collection and analysis, decision to publish, or preparation of the manuscript.

Competing Interests: The authors have declared that no competing interests exist.

* E-mail: seedcq@263.net

Introduction

Waterlogging is an important abiotic determinant for crop growth [1]. A major feature of waterlogging is depletion of oxygen [2], which results in changes in the metabolism and morphology of plants. The formation of aerenchyma is an important adaptive response of waterlogged plants [3,4]. During waterlogging, plant growth is either inhibited or enhanced based on the types of species and survival strategy [5]. Photosynthesis is also inhibited due to stomatal closure, leaf chlorophyll degradation or early leaf senescence [6,7]. Moreover, waterlogging usually causes major or even total yield loss for crops [8,9].

Yield loss under waterlogging can be reduced by controlling the drainage in fields. However, this process is labor, time and cost intensive, which makes the waterlogging stress inevitable. To study the genetic mechanism underlying natural variation in waterlogging tolerance of species is an important step towards the waterlogging management. Before looking for an answer to the question "Why are tolerant genotypes tolerant", the key issue is to develop a selection criterion to answer the question "Which genotypes are tolerant?".

Previous studies have outlined methods that depend on traits that help in screening plants for waterlogging tolerance. When natural variation of tolerance among Arabidopsis accessions was screened, the median lethal times of 10-leaf seedlings subjected to complete submergence in the dark was estimated as a measure of tolerance [10]. Another study about phenotypic plasticity in response to flooded conditions in Arabidopsis measured three sets of traits, including vegetative, architectural and reproductive [11]. Leaf color, plant height, and growth rate were evaluated during the screening of 99 accessions of perennial ryegrass under submergence and recovery conditions [12]. Waterlogging injury scores were used for the identification of QTLs underlying waterlogging tolerance in soybean [13]. Qiu performed QTL mapping associated with waterlogging tolerance during seedling stage in maize using traits of root length, root dry weight, plant height, shoot dry weight, and total dry weight [14], and similar work were carried out in wheat [8]. In rice, Sub1A is mapped based cloned from rice by evaluating leaf death on individual plant after one week of recovery followed completely submerged [15], and SNORKEL1 and SNORKEL2 were cloned by using the trait of early elongation ability adapting to deep water [16]. Most of these studies were performed with seedlings growing in green house during treatments, and the traits of seedling were evaluated, with the implicit assumption that differences in waterlogging tolerance expressed in these systems would result in improved yield performance in the field. However, it is necessary to compare screening methods and criterions under controlled conditions to yield loss in the fields under waterlogging.

Rapeseed is the second largest oil crop in the world. In China, the largest planting country in the world, 80% of rapeseed is planted along the Yangtze River. In this area, rapeseed is planted in paddy fields as a rotation crop just after rice [17]. Owing to the lack of labor required for transplanting, direct seeding of rapeseed is increasingly being practiced by farmers, which can result in severe waterlogging. Heavy rainfall during this season further compounds the problem. Waterlogging always results in yield loss in rapeseed. However, there have been only a handful of studies on screening of waterlogging tolerance in rapeseed.

Therefore, the present study was conducted to compare the efficacy of various methods at different growth stages under artificial environments and in fields based on traits associated with waterlogging tolerance. We address the following questions: (1) is there genetic variation for yield loss among genotypes after being exposed to waterlogging in fields? (2) Is there plasticity and variation for waterlogging tolerance-related traits at the seedling establishment and seedling stage under controlled conditions? (3) Is it possible to screen at seedling establishment and seedling stages to identify differences in waterlogging tolerance in the fields?

Materials and Methods

Twenty-five varieties were selected for this study, including ZY821, YG2009, HYZ9, ZS9, ZS7, ZS12, Y108, 9548, 9508, 9981, B108, 21937, H0102, 95_3_12, Y208, JHTG, JHY201, GH01, HYZ62, HYZ6, FS9, HS3, NY1, and QY8 (supplied by Oil Crops Research Institute of the Chinese Academy of Agricultural Sciences).

Experiment 1: screening at maturity stage in the field

The field trials were conducted in two different environments: Xishui, and Wuxue, Hubei in China, which is managed by local people. The owners provided full access permissions to the study sites. This study did not involve specimen collection. The field studies did not involve any threatened or endangered species. No vertebrates were involved in this study.

Planting was conducted using a randomized complete-block design with two replicates in each environment. Seeds were directly sown in the fields at the end of September. Each plot was 13.2 m^2 (6.6 m×2 m). Plants were thinned to 450 individuals for one replicate and the plant density was about 33–37 plants m^{-2}. The field management followed standard agricultural practice. Waterlogging treatment was imposed at the seedling stage with 4–6 leaves. All the fields of treatment plots were filled with water to 1 cm above the soil surface, which was maintained for 15 d.

The plants were harvested at the beginning of the following May. In each replicate, 10 representative individuals from the middle of each plot were harvested at physiological maturity. The traits evaluated included plant height (*Ph*), branch height (*Br*), primary branches (*Pbr*), main shoot length (*Msl*), siliques on the main shoot (*Sqms*), siliques on branches (*Sqb*), siliques on a plant (*Sqp*), seeds in a siliques (*Ssq*) and root length. Primary and secondary branches were the number of branches arising from the main shoot and primary branches, respectively. Main shoot length was measured from the base of the last primary branch to the tip of the main shoot. Siliques on the main shoot and siliques on a plant were the number of well-filled, normal siliques in the main shoot and the whole plant, respectively. Seeds in a silique were counted as the number of well-developed seeds in a silique. Phenotyping was done from the ten selected plants. Moreover, the number of all the plants in each plot (*Sn*, seedlings number) and the yield in each plot were also evaluated.

Waterlogging tolerance coefficient (WTC) was used to measure waterlogging tolerance. WTC of samples was calculated using the following formula:

$$\text{Survival rate } (Sr) = \text{Plant numbers of treatment}/450$$

$$\text{Trait of treatment} =$$
$$\text{Mean value of the evaluated 10 plants} \times \text{Survival rate}$$

$$\text{WTC} = \text{Trait of treatment}/\text{Mean value of control}$$

Experiment 2: screening at the seedling establishment stage

Healthy seeds of each variety were selected and were placed on moist filter paper. Seeds were grown at 25°C (light:dark, 16:8 h) until their radicals were about 2–5 mm. Uniform seeds were selected and divided into two groups. One group was cultured with a normal water supply as the control and the other was followed by treatment. Thirty normal germinated seeds were treated for 12 h in a sealed tube completely filled with distilled water to avoid oxygen exchange with the air. Following removal from anaerobic chambers, seeds were sown in dishes filled with moist fine vermiculite and were cultured for 5 d.

All the seedlings of each variety per replicate were used for measuring of root length (*Rl*), shoot length (*Sl*), full length (*Fl*) and full weight (*Fw*) under control and waterlogged conditions.

WTC of samples was calculated using the following formula:

$$\text{WTC} =$$
$$\text{Mean value of treated seedling}/\text{Mean value of control seedlings}$$

Experiment 3: screening at the seedling stage in the green house

Uniform-sized seeds of each variety were sown in pots and the seedlings were thinned to ten per pot after 5 d. The experiment was conducted under the same growth conditions as described in experiment 2.

Each variety included control and waterlogging treatment. The experimental design was a split-plot design with three replications. All pots of the treatment were filled with water to 2–3 cm above the soil surface at the seedling stage with three leaves, which was maintained for 5 d. After waterlogging, the treated seedlings of treatment were returned to normal conditions to continue growing. Three days later, the data for traits from both treatment and control were recorded from these seedlings, including root length (*Rl*), shoot length (*Sl*), full length (*Fl*), root weight (*Rw*), shoot weight (*Sw*), weight and full weight (*Fw*).

WTC of samples was calculated using the following formula:

WTC=

Mean value of treated seedlings/Mean value of control seedlings

Statistical analysis

The means of all observations were calculated using Microsoft Excel. The relative analysis and the linear relativity were performed using SAS software [18].

Results

The effect on yield related traits by waterlogging

To study the effect of waterlogging on yield, the seedlings were waterlogged in fields and yield related traits were compared to those of control. Under normal condition, all the evaluated traits except of branch height, main shoot length and root length were significantly related with yield (Table 1), while all the evaluated traits were significantly related with yield under waterlogging condition (Table 2). Besides, there was significantly correlation between WTC of any two traits under waterlogging (Table 3).

Waterlogging caused an obvious decrease in yield (Table S1). The yield of each of 12 varieties tested under waterlogging was less than 20% of that under the control, while only 3 varieties under waterlogging had more than 75% of the yield in control. The yields of the majority of varieties (18/25) were lower by fifty percent of that of control plants.

Our results (Table S1) showed that compared with normal conditions all of the traits evaluated showed a decrease in response to waterlogging. Waterlogging resulted in plant mortality in most of the varieties. Four varieties (9.3%) had less than 20% of plants compared to the control, and 12 varieties (48%) had fewer than half plants of the control after waterlogging. Only 6 varieties (24%) had all of the plants alive under waterlogged conditions. All other traits were also decreased in response to waterlogging. Plant heights of 19 varieties (76%) were decreased more than 50% after waterlogging. Twenty varieties (80%) had only fewer than half of the branch height, primary branches, total branches and siliques of the control. Sixteen varieties (64%) had fewer than half of the siliques on the main shoot of the control. There were fewer than half the siliques on branches for 22 varieties (88%) after waterlogging. Sixteen varieties (64%) had only half of the seeds in a silique of the control. Waterlogging also caused inhibition of root growth. The length of the roots for 17 varieties (68%) after waterlogging was less than half of the control. The main shoot length of 21 (84%) varieties was less than half of that of the control.

A wide range of variation was observed for all the traits among the different varieties under waterlogging. The WTC of yield ranged from 6.8% in 9981 to 91.3% in ZS9 (Table S1). The variety "B108" had only 62 plants left after waterlogging, accounting for 13.8% of the control, while there were six varieties which had 450 plants as many as the control. WTC of *ph*, *bh* and *msl* ranged from 8.4% to 79.8%, 8.2% to 88.3% and 8.9 to 79.3%, respectively. WTC of *pbr* and *tbr* ranged from 9.1% to 85.9%, and 7.9% to 86.1%, respectively. WTC of *sqms*, *sqb*, and *sqp* varied from 8.9% to 94.4%, 5.2% to 74.8%, and 6.6% to 82.8%. 9981 had only 9.8% of root length of the control while HYZ9 had 92.7%.

The effect on seedling establishment by waterlogging

To compare waterlogging tolerance of the 25 rapeseed varieties at the seedling establishment stage, the germinated seeds were subjected to normal conditions and hypoxia conditions. The root length, shoot length, and full weight were examined to evaluate waterlogging tolerance.

Under control conditions, there was a significant positive correlation between root length and full length as well as between shoot length and full weight. After waterlogging treatment was applied, there was a significant positive correlation between root length and shoot length, root length and full length, and shoot length and full length (Table 4).

Besides, waterlogging decreased all of the four traits (Table S2). The average root length under waterlogging was only 33.8% of that of the control, while the average of shoot length, full length, and full weight under waterlogging was 67.9%, 40.7% and 71.2% of that under control, respectively. However, the changes in these parameters varied with varieties. ZS9 had the highest WTC of root length (61.6%), while that of Y208 was only 4.3%, and 76% of varieties (19/25) had less than half of root length under waterlogging when compared to that under the control. WTC of shoot length ranged from 40.0% to 99.9%. The largest full length rate was 65.9%, and the smallest one was only 17.9%. ZY821 represented the largest WTC of full weight, 93.5%, while that of Y208 was decreased to 44.5% by waterlogging.

Table 1. Correlation coefficients (*r*) between pairs of yield related traits under control.

	Ph	Bh	Msl	Pb	Tb	Sqms	Sqb	Sqp	Ssq	Rl	Yield
Ph	1	0.701***	0.212	0.530**	0.481*	0.471*	0.562**	0.5750**	−0.159	−0.158	0.486**
Bh		1	−0.173	0.157	−0.010	0.077	0.080	0.137	0.101	0.066	0.326
Msl			1	−0.159	0.011	0.298	0.049	0.042	−0.243	0.077	−0.159
Pb				1	0.915***	0.437*	0.868***	0.761***	−0.003	−0.358	0.382
Tb					1	0.4960*	0.891***	0.681***	−0.221	−0.388	0.315
Sqms						1	0.707***	0.729***	−0.223	−0.224	0.552**
Sqb							1	0.914***	−0.088	−0.367	0.485*
Sqp								1	0.042	−0.283	0.557**
Ssq									1	0.005	0.145
Rl										1	−0.209
Yield											1

Table 2. Correlation coefficients (*r*) between pairs of yield related traits after waterlogging.

	Ph	Bh	Msl	Pb	Tb	Sqms	Sqb	Sqp	Ssq	Rl	Yield	Sn
Ph	1	0.943***	0.932***	0.500***	0.897***	0.986***	0.760***	0.922***	0.940***	0.938***	0.853***	0.879***
Bh		1	0.939***	0.934***	0.882***	0.938***	0.785***	0.913***	0.795***	0.849***	0.804***	0.823***
Msl			1	0.853***	0.795***	0.926***	0.690***	0.852***	0.806***	0.879***	0.767***	0.796***
Pb				1	0.974***	0.919***	0.903***	0.971***	0.826***	0.826***	0.817***	0.819***
Tb					1	0.871***	0.879***	0.934***	0.807***	0.801***	0.798***	0.781***
Sqms						1	0.757***	0.927***	0.937***	0.936***	0.867***	0.868***
Sqb							1	0.946***	0.615**	0.598***	0.739***	0.610***
Sqp								1	0.814***	0.802***	0.854***	0.778***
Ssq									1	0.953***	0.828***	0.813***
Rl										1	0.772***	0.816***
Yield											1	0.600**
Sn												1

The effect on the seedlings by waterlogging

To compare waterlogging tolerance among the 25 rapeseed varieties at the seedling stage, the seedlings were waterlogged, and root length, shoot length, root weight, and shoot weight of the seedlings were evaluated after recovery.

Under normal conditions, correlation analysis among the traits indicated that there was significant positive correlation between root length and full length, and shoot length and full length (Table 5). A similar correlation was observed for root weight and full weight, and shoot weight and full weight. A significant positive correlation was also found for shoot length and full weight. After waterlogging, our analysis revealed that the full length was significantly related with root length, shoot length and root weight, while full weight was not significantly related with any of the other traits (Table 6 and 7).

Waterlogging inhibited the growth of seedlings to some extent (Table S3). Except for WTC of root weight, the average WTC of other traits ranged from 80% to 90%. The average WTC of root weight for all of the 25 varieties was 72.7%. WTC of root length, shoot length, and full length ranged from 66.0%~99.9%, 68.4%~99.6%, and 69.8%~98.4%, respectively. WTC of root weight, shoot weight, and full weight ranged from 43.8%~96.6%, 70.0%~99.5%, and 65.1%~98.9%, respectively.

Relationship between yield and other traits under waterlogging

Possible correlations were analyzed after waterlogging stress between WTC of yield and other traits. Significant correlations were found between WTC of yield and other 17 evaluated traits, including root length ($P<0.001$), full weight ($P<0.05$), full length ($P<0.01$) at seed germination stage, shoot length ($P<0.05$), root length ($P<0.01$) and full length ($P<0.01$) at seedling stage, and all the evaluated traits when maturity in fields. By contrast, the WTC of shoot length at the seedling establishment stage was not significantly related to the WTC of yield in the field, while there were no significant correlations between WTC of those weight related traits at the seedling stage and WTC of yield in the field.

Table 3. Correlation coefficients (*r*) between pairs of WTC of yield related traits.

	Ph	Bh	Msl	Pb	Tb	Sqms	Sqb	Sqp	Ssq	Rl	Sr	Yield
Ph	1	0.926***	0.946***	0.904***	0.858***	0.982***	0.777***	0.897***	0.924***	0.960***	0.871***	0.871***
Bh		1	0.949***	0.846***	0.772***	0.902***	0.715***	0.819***	0.754***	0.860***	0.811***	0.729***
Msl			1	0.810***	0.744***	0.901***	0.704***	0.815***	0.815***	0.911***	0.793***	0.773***
Pb				1	0.970***	0.930***	0.908***	0.947***	0.819***	0.845***	0.792***	0.855***
Tb					1	0.878***	0.849***	0.895***	0.836***	0.821***	0.748***	0.850***
Sqms						1	0.834***	0.940***	0.905***	0.934***	0.843***	0.893***
Sqb							1	0.962***	0.657***	0.678***	0.580**	0.859***
Sqp								1	0.800***	0.820***	0.706***	0.933***
Ssq									1	0.929***	0.833***	0.868***
Rl										1	0.840***	0.818***
Sr											1	0.620***
Yield												1

*** Significant at 0.05, 0.01 and 0.001 levels, respectively (n = 23).

Table 4. Correlation coefficients (*r*) between pairs of traits of seedlings at seedling establishment stage under control and waterlogging.

Control					Treatment				Waterlogging Tolerance Index			
	RL	SL	FL	FW	RL	SL	FL	FW	RL	SL	FL	FW
RL	1	−0.15	0.91***	−0.05	1	0.50**	0.96***	0.35	1	0.38	0.99***	0.78**
SL		1	0.27	0.40*		1	0.72***	0.23		1	0.45*	0.51**
FL			1	0.12			1	0.35			1	0.78***
FW				1				1				1

*, **, and *** Significant at 0.05, 0.01 and 0.001 levels, respectively (n = 23).

Waterlogging tolerance evaluation of 25 rapeseed varieties

The ranking of waterlogging tolerance for the 25 genotypes was done on the basis of their performance in WTC of those traits at the seedling establishment and seedling stages, which were significantly correlated with the WTC of yield. The genotypes were ranked as 1–25 with '1' being the most tolerant and '25' being as susceptible based upon their abilities to withstand waterlogging conditions. The ranks were pooled to calculate an overall ranking of the genotypes on the basis of these traits (Table S4).

Screening at the seedling establishment stage identified four waterlogging-tolerant varieties, ZS9, ZY821, HYZ9, and YG2009, of which ZS9, YG2009, and HYZ9 were also identified as high-tolerant varieties in the field. HYZ62, H0102, HYZ6, HYZ9 and YG 2009 were identified to be waterlogging-tolerant at the seedling stage, and YG2009 was identified as the tolerant one in the field. Although some of the varieties, which were evaluated to be tolerant at the seedling establishment stage and seedling stage, were not in the top list of tolerant ones on yield, most of these varieties were above the media tolerance of all the varieties.

Discussion

There was significant difference of yield loss in rapeseed under waterlogging stress

The first step of breeding waterlogging tolerant rapeseed is to identify difference of tolerance in species. As a crop, yield is the most important trait for breeding. Waterlogging tolerance for rapeseed means less yield loss under waterlogging. In this study, we screened 25 varieties of rapeseed in fields for waterlogging tolerance.

As expected, waterlogging decreased the yield in all of the varieties of rapeseed. As shown in Table S1, most of these varieties were very sensitive to waterlogging, and yield loss reached more than fifty percent. According to the breeding and research history, problem of waterlogging did not attract a lot of attention before, and breeders have not focused on waterlogging tolerance in breeding. Moreover, there is little research on waterlogging stress in rapeseed. Breeding experiments are generally carried out under normal condition. Although the yield is excellent under normal condition, it declines significantly when the plants are subjected to waterlogging.

Although waterlogging caused yield loss in all of these 25 varieties, considerable natural variation was observed among rapeseed varieties. There was a significant decrease in yield of non-tolerant genotypes. However, the tolerant genotypes had relatively less reduction in the yield (Table S1). The most tolerant one, YG2009, showed only a 10% reduction in yield compared to that of the control. Our results suggested that substantial genetic variations existed in waterlogging tolerance in rapeseed. This difference opens up the possibility to study the underlying mechanisms and genes responsible for these variations.

In our study, waterlogging treatment not only caused yield loss in these varieties, but also led to the death of plants for most of these varieties as shown in Table S1. In a previous study in Arabidopsis, flooding over all the growth stage did not lead to the death of any plants [11]. In another study in Arabidopsis, media lethal times were estimated as a measure of tolerance only when Arabidopsis accessions were subjected to complete submergence [10]. In our study, the seedlings were recovered after being waterlogged for 14 days. This system led to considerable lethal rate for these varieties, revealing that rapeseed is very sensitive to waterlogging and seedling stage is a sensitive stage during plant

Table 5. Correlation coefficients (*r*) between pairs of traits of seedlings at seedling stage under control.

	RL	SL	FL	RW	SW	FL
RL	1	0.04	0.84***	0.55***	−0.18	−0.03
SL		1	0.57***	0.18	0.61***	0.58***
FL			1	0.55***	0.18	0.29
RW				1	0.36	0.55***
SW					1	0.98***
FL						1

*** Significant at 0.05, 0.01 and 0.001 levels, respectively (n = 23).

Table 6. Correlation coefficients (*r*) between pairs of traits of seedlings at seedling stage under waterlogging.

	RL	SL	FL	RW	SW	FL
RL	1	−0.05	0.73***	0.54***	−0.04	0.09
SL		1	0.65***	0.27	0.62**	0.59**
FL			1	0.59**	0.39	0.47*
RW				1	0.53**	0.69***
SW					1	0.98***
FL						1

*, **, and *** Significant at 0.05, 0.01 and 0.001 levels, respectively (n = 23).

Table 7. Correlation coefficients (*r*) between WTC of pairs of traits of seedlings at seedling stage after waterlogging.

	RL	SL	FL	RW	SW	FL
RL	1	0.34	0.95***	0.29	0.01	0.15
SL		1	0.60**	−0.18	−0.26	−0.22
FL			1	0.18	−0.07	0.07
RW				1	0.66***	0.88***
SW					1	0.93***
FL						1

, and * Significant at 0.01 and 0.001 levels, respectively (n = 23).

ontogeny. After the plants encountered waterlogging at the seedling stage, these rapeseed varieties also showed a large diversity in survival ability. This is one reason different varieties had different yield loss after the stress, and it is also revealed by the results that the survival rate was significantly correlated with the yield index.

Besides the differences in yield loss and survival rates among the 25 varieties, other morphological traits also decreased. Generally, the growth of these varieties were all inhibited after waterlogging in our study. This was also found in Arabidopsis, maize, soybean, *etc* and plants showed an inhibition of growth under waterlogging or submergence [10,11,19]. Taking plant height as an example in this study, 19 out of 25 varieties after waterlogging had only half of the plant height compared to that of control (Table S1), suggesting that waterlogging in rapeseed inhibits growth instead of promotion of growth which is revealed in some species, such as *Rumex palustris* [20] or deepwater rice [21]. The inhibition also displayed on other morphological traits (Figure 1). All of the varieties showed a decrease in root length when exposed to waterlogging stress. The aerial structures were also affected by the stress. Under waterlogging, the plant height was decreased, leading to a lower position of branch height, fewer branches, and shorter main shoot length. Certainly, the smaller plants after the stress had fewer siliques on the main shoot and siliques on a plant. Besides, seed number in siliques (*Ssq*) was also decreased as a result of waterlogging. According to these factors above, yield under waterlogging was decreased compared to that of the control. It seems the most important problem is decreased oxygen concentration in the soil in the case of excess water. This primarily affects the root system and, as a consequence, the above-ground growth. Under flash waterlogging, the plant restricts its growth and growth is still inhibited compared to the control even after removal of water. The stress eventually decreases yield for rapeseed. Similar to yield, the morphological traits of roots and aerial structure varied among the 25 varieties of rapeseed. Our results revealed that WTC of these traits were all significantly correlated with that of yield, indicating the tolerant one would have a strong ability to keep growth potential under waterlogging and would recover better from the stress (Figure 1). Our data further suggests a consistency between the selected morphological traits and yield in waterlogging tolerance of these genotypes until maturity. Waterlogging-tolerant plants could be evaluated by any of these morphological traits at maturity.

Roots are the most sensitive part to waterlogging at the seedling establishment stage

Seedling establishment of rapeseed seeds in paddy fields that are either waterlogged or submerged is one instance in which rapeseed is exposed to hypoxia or even anoxia. Waterlogging which can impose stress on seedlings and has been shown to reduce crop yield [22,23]. However, waterlogging poses a significant problem to seeds prior to emergence. We exposed the 25 varieties of rapeseed to waterlogging during the seedling establishment stage and evaluated the morphological traits after the recovery. Seedling establishment stage is a sensitive stage to waterlogging. Seeds of a vast majority of higher plant species fail to germinate under waterlogging stress and the germination rates decline dramatically, including soybean, pea (*Pisum sativum*), wheat, maize, and rice [24,25]. Fatty seeds are more sensitive to hypoxia than starchy seeds because starchy ones can maintain a high energy metabolism even under the stress [26]. Rapeseed bears fatty seeds, therefore, waterlogging can cause low germination rates or even failure to germinate in several varieties. Although, the stress lasted for only 12 h in our study, it still caused extensive damage to the rate of germination.

The results also revealed that the length and weight of rapeseed seedlings decreased significantly under waterlogging during the seedling establishment stage. In some varieties of rice, the final length of coleoptiles can exceed the length of aerobic coleoptiles when germinated without oxygen [27]. Under the stress, rice seedlings elongate the coleoptile in an attempt to breach the surface of the water and thus allow oxygen transport to underwater tissues. On the other hand, in other varieties of rice and other species, waterlogging inhibits the growth of seedlings at the seedling establishment stage [28,29]. Our study in rapeseed corroborated these results as the growth of rapeseed seedlings was inhibited under waterlogging. Besides, the tolerant varieties mean strong ability of seed germination and seedling growth under waterlogging and the inhibition was shown to a much lower extent in tolerant ones.

Although the seedlings of some varieties had shoots and leaves like a normal plant, they did not have radicles as a result of waterlogging (Figure 2). Instead, some adventitious roots were generated, which is an adaptive mechanism under waterlogging [30,31]. In this seedling establishment stage, the root tips are very sensitive to hypoxia. The quick cell division and expansion is indeed oxygen requiring in the meristem tissues [32]. Meristems are the first target of stress and waterlogging stimulates root tip death.

Waterlogging affected the length of seedlings at the seedling stage

Most of waterlogging-tolerance screening experiments were carried out at the seedling stage. In our study, the results showed that waterlogging led to the inhibition of growth for both roots and shoots for rapeseed seedlings (Table S3). Under waterlogging, most plants, such as Arabidopsis [10,11,33], barley [34], maize [14], soybean [35,36], *etc*, produced smaller roots because of the anoxic conditions of the soil . This inhibition of growth of seedlings was because of a passive response to severe stress rather than an adaptive response, although Sub1A controls submergence tolerance through inhibition of growth under the stress in rice [37]. Waterlogging tolerance could be evaluated using the waterlogging tolerance coefficient (WTC = waterlogging treatment of each trait/control of each trait). In maize, Qiu performed QTL mapping associated with waterlogging tolerance using WTC during the seedling stage [38].

Figure 1. The effect of waterlogging on morphological traits of GH01 and ZS9 in field conditions. A: GH01 control and treated plants; B: GH01 control and treated roots; C: ZS9 control and treated roots; D: ZS9 control and treated plants.

Similar to that in the fields or at seedling establishment stage, variations of tolerance were found for the six morphological traits at the seedling stage. Although the seedlings were waterlogged for 5 days, the extent of inhibition of growth at seedling stage was not as pronounced as that at the seedling establishment stage. The results revealed that the WTC of the root weight had the largest range of variation among all varieties. There can be attributed to two reasons: 1. WTC of root length varied among the 25 varieties under waterlogging; 2. adventitious roots were generated under waterlogging stress. Although the number of adventitious roots was not evaluated in our study, there should be differences in different varieties, leading to the difference in root weight.

WTC of six traits at the other two development stages were significantly related to the WTC of yield

There are several kinds of systems for screening plants for waterlogging tolerance. Most of the previous studies dealt with tolerance at a specific growth-stage. In our study, three systems of screening for rapeseed over the ontogeny were performed. The growth of rapeseed was inhibited by waterlogging irrespective of the developmental stage. Moreover, the inhibition of growth led to the loss of yield in fields. Based on this idea, correlation between WTC of the yield and traits from two other development stages were analyzed. The results revealed that the WTC of six traits at the other two development stages were significantly related with the WTC of yield.

At the seedling establishment stage, the roots were very sensitive to waterlogging as described above. As expected, there was significant correlation between WTC of root length at the seedling establishment stage and yield in the field. The length of shoots was decreased by waterlogging, and the extent of decrease varied in different varieties (Table S2). However, the growth of shoots was not as sensitive as that of roots at the seedling establishment stage, not leading to correlation between WTC of this trait and yield. Radicle protrusion is the first stage in the emergence of rapeseed seedlings and the roots are generally much longer than shoots at this stage. Although the variation of shoot length was not correlated with yield, the full length and full weight of seedlings was significantly correlated with yield in fields for the roots were very sensitive to waterlogging stress.

At the seedling stage, it is interesting to note that the WTC of all the three traits of length were significantly correlated with the WTC of yield, while the WTC of the other three traits related with weight were not correlated with WTC of yield. Although the growth of main roots was inhibited by waterlogging, which led to

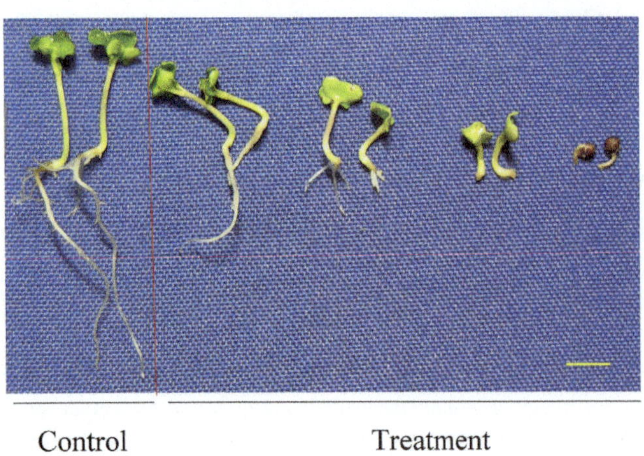

Figure 2. The effect of waterlogging on seedlings of 9981 at the seedling establishment stage. Yellow bars = 1.0 cm.

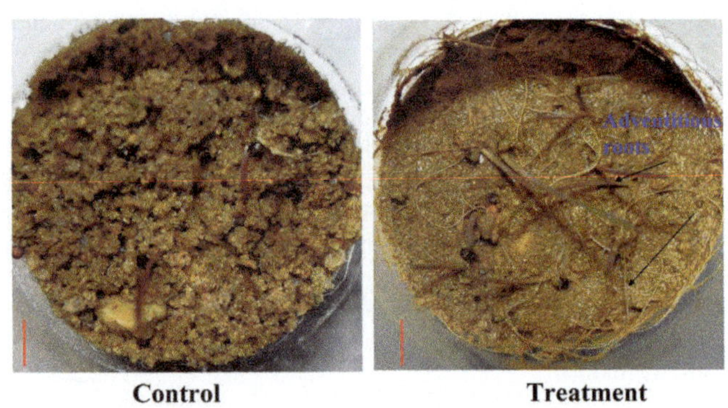

Figure 3. The adventitious roots of ZS12 induced by waterlogging at seedling stage. Red bars = 1.0 cm.

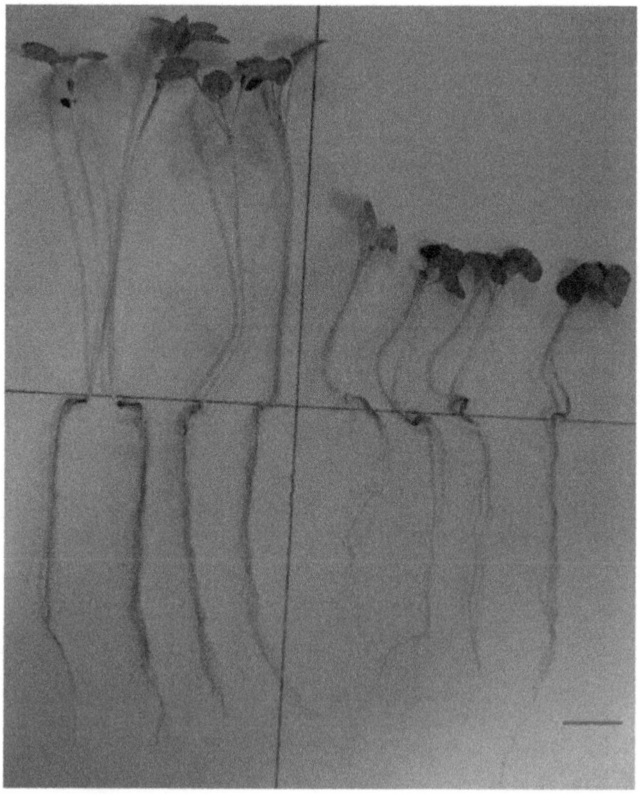

Figure 4. The effect of waterlogging on seedlings of B108 at the seedling stage.

the inhibition of shoots, the adventitious roots were induced by waterlogging stress (Figure 3). Despite of the decrease in weight of main roots under waterlogging, the WTC of root weight did not change linearly according to the tolerance ability of varieties. Besides, the WTC of shoot weight was also not significantly correlated with yield. Seedlings had several leaves at this stage (Figure 4), and the leaf growth was not decreased by waterlogging as the same as the length of shoots, leading to no significant correlation between WTC of shoot weight and yield.

The advantage and disadvantage of the three systems for screening waterlogging tolerance

The three systems tested had their both advantages and disadvantages. Although yield is the most important trait from an agronomical point of view, the examination of yield and yield related traits under waterlogging stress is the most time-consuming step in the selection and testing process. Besides, it is difficult to reproduce the field conditions in every experiment, including the soil, stress extent, and the weather, etc. The major problem in waterlogging-related studies is that the observed tolerance is specific to a given environment. Reproducible experiments are possible only if the genotypes are tested under controlled water and weather conditions. Some studies reporting the effect of waterlogging on yield traits were performed under controlled climatic conditions [39]. In such situations, all of the growth conditions can be controlled. However, these are cost intensive and the capacity of the experiments is limited. To guarantee the selection of suitable genetic sources for breeding and/or to clarify the role of stress-induced genes in the regulatory pathways, more accurate and convenient testing systems need to be elaborated.

Seedling test conducted in controlled conditions is the most used method to screen waterlogging tolerance and provide a lot of data about physiological or molecular aspects [17,40]. Our results suggest that some traits due to waterlogging stress were induced at early growth stages and could predict yield loss in the fields. Although the labor of growing seedlings in containers in greenhouse is much less compared to that in the field, it is still very costly and laborious. The biggest problem is that the variation of seedling traits under waterlogging compared to waterlogging and the difference among different varieties was not very large, despite the seedlings were waterlogged for 5 days.

It is fortunate to find WTC of some traits at seed germination stage significantly related with WTC of yield in our study. Germination tests are efficient for selecting lines from large numbers of accessions in different species. They require little investment and provide a large number of data within several days. This simple and rapid method can be used for mass testing.

If plants have a consistent pattern of waterlogging tolerance at all growth stages, the tolerant ones could be selected at any stage. Based on the ranking analysis (Table S4), although some varieties with median tolerance were found to be not consistently superior across the three growth stages, those varieties of extreme tolerance or sensitivity showed consistency in their tolerance during the ontogeny of the whole plant, suggesting that the screening at the early stage would provide a suitable basis for the initial selection of a large number of germplasm entries or breeding populations for waterlogging tolerance.

Conclusion

In summary, the variations in waterlogging tolerance with respect to yield were found among different varieties. These variations provide an important basis for the further development of breeding materials that are more tolerant to waterlogging stress. According to the fact that flash waterlogging restricts plant growth and growth is restored after removal of the stress.

The assessments of genotypes for physiological parameters at the three growth stages suggest that there were differences in waterlogging tolerance at all the development stages. Correlation analysis revealed that there was consistency in the waterlogging tolerance of the selected genotypes until maturity, and good waterlogging tolerance at the germination stage and seedling stage can guarantee tolerance in later stages. The waterlogging-tolerant plants could be selected at any stage, and the selections would be more effective at the seed germination stage. Our study provides a method for screening plants for waterlogging tolerance, which helps in the initial selection of a large number of germplasms or breeding populations for waterlogging tolerance and determine their potential utility in crop-improvement.

Supporting Information

Table S1 The effect of waterlogging on the 25 varieties in fields.

Table S2 The effect of waterlogging on the 25 varieties at seedling establishment stage.

Table S3 The effect of waterlogging on the 25 varieties at seedling stage.

Table S4 Performance-based ranking of the 25 varieties for yield significantly-related traits at seedling establishment stage and seedling stage.

Author Contributions

Conceived and designed the experiments: X. Zhang. Performed the experiments: X. Zou. Analyzed the data: CH. Contributed reagents/materials/analysis tools: LZ MX YC. Wrote the paper: X. Zou.

References

1. Voesenek L, Bailey-Serres J (2013) Flooding tolerance: O sensing and survival strategies. Current Opinion in Plant Biology.
2. Jackson MB, Colmer TD (2005) Response and adaptation by plants to flooding stress. Ann Bot 96: 501–505.
3. Kennedy RA, Rumpho ME, Fox TC (1992) Anaerobic metabolism in plants. Plant Physiol 100: 1–6.
4. Thomas AL, Guerreiro SM, Sodek L (2005) Aerenchyma formation and recovery from hypoxia of the flooded root system of nodulated soybean. Ann Bot 96: 1191–1198.
5. Nagai K, Hattori Y, Ashikari M (2010) Stunt or elongate? Two opposite strategies by which rice adapts to floods. Journal of Plant Research 123: 303–309.
6. Ahmed F, Rafii MY, Ismail MR, Juraimi AS, Rahim HA, et al. (2013) Waterlogging tolerance of crops: breeding, mechanism of tolerance, molecular approaches, and future prospects. Biomed Res Int 2013: 963525.
7. Amador ML, Sancho S, Bielsa B, Gomez-Aparisi J, Rubio-Cabetas MJ (2012) Physiological and biochemical parameters controlling waterlogging stress tolerance in Prunus before and after drainage. Physiol Plant 144: 357–368.
8. Yu M, Chen GY (2013) Conditional QTL mapping for waterlogging tolerance in two RILs populations of wheat. Springerplus 2: 245.
9. Setter TL, Waters I, Sharma SK, Singh KN, Kulshreshtha N, et al. (2009) Review of wheat improvement for waterlogging tolerance in Australia and India: the importance of anaerobiosis and element toxicities associated with different soils. Ann Bot 103: 221–235.
10. Vashisht D, Hesselink A, Pierik R, Ammerlaan JM, Bailey-Serres J, et al. (2011) Natural variation of submergence tolerance among Arabidopsis thaliana accessions. New Phytol 190: 299–310.
11. Pigliucci M, Kolodynska A (2002) Phenotypic plasticity and integration in response to flooded conditions in natural accessions of *Arabidopsis thaliana* (L.) Heynh (Brassicaceae). Ann Bot 90: 199–207.
12. Yu XQ, Luo N, Yan JP, Tang JC, Liu SW, et al. (2012) Differential growth response and carbohydrate metabolism of global collection of perennial ryegrass accessions to submergence and recovery following de-submergence. Journal of Plant Physiology 169: 1040–1049.
13. Cornelious B, Chen P, Chen Y, de Leon N, Shannon JG, et al. (2005) Identification of QTLs underlying water-logging tolerance in soybean. Molecular Breeding 16: 103–112.
14. Qiu F, Zheng Y, Zhang Z, Xu S (2007) Mapping of QTL associated with waterlogging tolerance during the seedling stage in maize. Ann Bot 99: 1067–1081.
15. Xu K, Xu X, Fukao T, Canlas P, Maghirang-Rodriguez R, et al. (2006) Sub1A is an ethylene-response-factor-like gene that confers submergence tolerance to rice. Nature 442: 705–708.
16. Hattori Y, Nagai K, Furukawa S, Song XJ, Kawano R, et al. (2009) The ethylene response factors SNORKEL1 and SNORKEL2 allow rice to adapt to deep water. Nature 460: 1026–1030.
17. Zou X, Tan X, Hu C, Zeng L, Lu G, et al. (2013) The Transcriptome of *Brassica napus* L. Roots under Waterlogging at the Seedling Stage. Int J Mol Sci 14: 2637–2651.
18. Kruse RL, Mehr DR (2008) Data management for prospective research studies using SAS software. BMC Med Res Methodol 8: 61.
19. Komatsu S, Makino T, Yasue H (2013) Proteomic and Biochemical Analyses of the Cotyledon and Root of Flooding-Stressed Soybean Plants. PLoS One 8.
20. Chen X, Pierik R, Peeters AJM, Poorter H, Visser EJW, et al. (2010) Endogenous Abscisic Acid as a Key Switch for Natural Variation in Flooding-Induced Shoot Elongation. Plant Physiology 154: 969–977.
21. Hattori Y, Nagai K, Furukawa S, Song XJ, Kawano R, et al. (2009) The ethylene response factors SNORKEL1 and SNORKEL2 allow rice to adapt to deep water. Nature 460: 1026–U1116.
22. Zheng CF, Jiang D, Dai TB, Jing Q, Cao WX (2009) [Effects of salt and waterlogging stress at post-anthesis stage on wheat grain yield and quality]. Ying Yong Sheng Tai Xue Bao 20: 2391–2398.
23. Christianson JA, Llewellyn DJ, Dennis ES, Wilson IW (2010) Global gene expression responses to waterlogging in roots and leaves of cotton (Gossypium hirsutum L.). Plant Cell Physiol 51: 21–37.
24. Magneschi L, Perata P (2009) Rice germination and seedling growth in the absence of oxygen. Ann Bot 103: 181–196.
25. Al-Ani A, Bruzau F, Raymond P, Saint-Ges V, Leblanc JM, et al. (1985) Germination, respiration, and adenylate energy charge of seeds at various oxygen partial pressures. Plant Physiol 79: 885–890.
26. Magneschi L, Perata P (2009) Rice germination and seedling growth in the absence of oxygen. Annals of Botany 103: 181–196.
27. Alpi A, Beevers H (1983) Effects of o(2) concentration on rice seedlings. Plant Physiol 71: 30–34.
28. Fukao T, Kennedy RA, Yamasue Y, Rumpho ME (2003) Genetic and biochemical analysis of anaerobically-induced enzymes during seed germination of Echinochloa crus-galli varieties tolerant and intolerant of anoxia. J Exp Bot 54: 1421–1429.
29. Ismail AM, Ella ES, Vergara GV, Mackill DJ (2009) Mechanisms associated with tolerance to flooding during germination and early seedling growth in rice (*Oryza sativa*). Ann Bot 103: 197–209.
30. Bailey-Serres J, Voesenek LACJ (2008) Flooding stress: Acclimations and genetic diversity. Annual Review of Plant Biology 59: 313–339.
31. Sauter M (2013) Root responses to flooding. Current Opinion in Plant Biology 16: 282–286.
32. Weitbrecht K, Muller K, Leubner-Metzger G (2011) First off the mark: early seed germination. Journal of Experimental Botany 62: 3289–3309.
33. Lee SC, Mustroph A, Sasidharan R, Vashisht D, Pedersen O, et al. (2011) Molecular characterization of the submergence response of the Arabidopsis thaliana ecotype Columbia. New Phytol 190: 457–471.
34. Zeng F, Shabala L, Zhou M, Zhang G, Shabala S (2013) Barley responses to combined waterlogging and salinity stress: separating effects of oxygen deprivation and elemental toxicity. Front Plant Sci 4: 313.
35. VanToai TT, St Martin SK, Chase K, Boru G, Schnipke V, et al. (2001) Identification of a QTL associated with tolerance of soybean to soil waterlogging. Crop Science 41: 1247–1252.
36. Shimono H, Konno T, Sakai H, Sameshima R (2012) Interactive Effects of Elevated Atmospheric CO2 and Waterlogging on Vegetative Growth of Soybean (Glycine max (L.) Merr.). Plant Production Science 15: 238–245.
37. Xu K, Xu X, Fukao T, Canlas P, Maghirang-Rodriguez R, et al. (2006) Sub1A is an ethylene-response-factor-like gene that confers submergence tolerance to rice. Nature 442: 705–708.
38. Qiu FZ, Zheng YL, Zhang ZL, Xu SZ (2007) Mapping of QTL associated with waterlogging tolerance during the seedling stage in maize. Annals of Botany 99: 1067–1081.
39. Zaidi PH, Rafique S, Singh NN (2003) Response of maize (*Zea mays* L.) genotypes to excess soil moisture stress: morpho-physiological effects and basis of tolerance. European Journal of Agronomy 19: 383–399.
40. Zou X, Jiang Y, Liu L, Zhang Z, Zheng Y (2010) Identification of transcriptome induced in roots of maize seedlings at the late stage of waterlogging. BMC Plant Biol 10: 189.

4

Molecular Genetic Diversity of Major Indian Rice Cultivars over Decadal Periods

Gangaprasad Choudhary[☯], Nagireddy Ranjitkumar[☯], Malathi Surapaneni[☯], Dondapati Annekitty Deborah, Abhilash Vipparla, Ghanta Anuradha, Ebrahimali Abubacker Siddiq, Lakshminarayana Reddy Vemireddy*

Institute of Biotechnology, Acharya NG Ranga Agricultural University, Rajendranagar, Hyderabad, Andhra Pradesh, India

Abstract

Genetic diversity in representative sets of high yielding varieties of rice released in India between 1970 and 2010 was studied at molecular level employing hypervariable microsatellite markers. Of 64 rice SSR primer pairs studied, 52 showed polymorphism, when screened in 100 rice genotypes. A total of 184 alleles was identified averaging 3.63 alleles per locus. Cluster analysis clearly grouped the 100 genotypes into their respective decadal periods i.e., 1970s, 1980s, 1990s and 2000s. The trend of diversity over the decadal periods estimated based on the number of alleles (*Na*), allelic richness (*Rs*), Nei's genetic diversity index (*He*), observed heterozygosity (*Ho*) and polymorphism information content (PIC) revealed increase of diversity over the periods in year of releasewise and longevitywise classification of rice varieties. Analysis of molecular variance (AMOVA) suggested more variation in within the decadal periods than among the decades. Pairwise comparison of population differentiation (*Fst*) among decadal periods showed significant difference between all the pairs except a few. Analysis of trends of appearing and disappearing alleles over decadal periods showed an increase in the appearance of alleles and decrease in disappearance in both the categories of varieties. It was obvious from the present findings, that genetic diversity was progressively on the rise in the varieties released during the decadal periods, between 1970s and 2000s.

Editor: Rajagopal Subramanyam, University of Hyderabad, India

Funding: The authors acknowledge ANGRAU (Acharya NG Ranga Agricultural University) for providing the financial and lab facilities under Rashtriya Krishi Vikas Yojana (RKVY). The funders had no role in study design, data collection and analysis, decision to publish, or preparation of the manuscript.

Competing Interests: The authors have declared that no competing interests exist.

* E-mail: vlnreddi@gmail.com

☯ These authors contributed equally to this work.

Introduction

Ever since the domestication of crop plants, man has been improving them giving selection emphasis to traits that suit his agro-ecological and socioeconomic needs. In rice, like many crops, selection preference has been for improvement of yield enhancing traits like compact panicle with more grains/panicle, large seed size, non-shattering habit etc. The selection process continued for centuries result in cultivars far different from the wild/weedy progenitor species in their habit and potential. Since the advent of the short statured high yielding varieties in the mid-sixties, the selection priority of breeders has been for higher stability or performance, need based crop duration, tolerance to various stresses and consumer preferred grain quality. Excessive breeding emphasis in this direction given during the last 50 years knowingly or unknowingly has led to some sort of genetic uniformity among the currently cultivated high yielding varieties. The improved germplasm is being excessively depended on for needed variability for progressive improvement of the crop with no precise knowledge of extent of exploitable variability beyond simply inherited traits. Also it is not clear to what extent breeding strategies *in vogue* have been facilitating to broaden or narrow the genetic diversity in the breeding nurseries of rice.

Precise knowledge of exploitable genetic diversity and genetic relatedness among the constituents of the improved gene pool is vital for meticulously planning and executing target directed breeding especially when genetic relatedness among the improved parental sources appears to become increasingly narrow. Unlike morpho-physiological traits used earlier to estimate genetic variability/relatedness, molecular markers have become quite handy in precisely understanding the extent of genetic divergence among varieties being chosen these days as parental sources in breeding programs. Many recent scientific papers tracing the trend of genetic diversity in crop varieties released over the years reveal the diversity levels to often fluctuate strongly from one time period to the successive periods [1–3] and sometimes show conflicting results [4,5]. By and large no clear pattern has emerged from such studies in the past, as both increasing and decreasing trends in diversity have been observed [6–8]. Diversity loss in rice during the last few centuries is obvious from the rapid decline of rice varieties from 400,000 before colonialism to 30,000 by mid-19th Century. This number has further come down during the era of the Green revolution as few high yielding varieties replaced thousands of native varieties [9]. Estimation of genetic diversity in varieties released in different crops during the last century reveals no clear trends [10]. In rice, both decreased [11] as well as increased trend of diversity [12] has been reported.

Since the introduction of semidwarf high yielding varieties in the mid 1960s, simultaneously launch of the All India Coordinated Rice Improvement Programmes (AICRIP) and active involvement of International Rice Research Institute (IRRI), Philippines facilitated rapid exchange of germplasm among breeders, institutes and between countries culminating in the release of around 1000 high yielding varieties in the country. Despite such large number of varieties developed using diverse germplasm, molecular marker based diversity analysis has shown the genetic base of Indian rice gene pool to be surprisingly narrow [13,14]. Moreover, with regard to trends of genetic diversity in major Indian rice cultivars, however, little work has been done Recently, hypervariable microsatellite markers evenly distributed in rice genome have been demonstrated to be quite effective in estimating genetic diversity [15] Keeping the foregoing, the present study was undertaken to understand the level and trend of genetic diversity in the high yielding varieties developed in India over decadal periods between 1970s and 2000s employing hypervariable microsatellite markers.

Materials and Methods

Plant Material

The experimental material comprised of 100 rice cultivars, which included 89 high yielding varieties released between 1970 and 2010 in different states of India and 11 traditional varieties (Table 1 and Table S1). They were grouped according to the decade of their release (herefrom regarded as "year of release-wise") for general cultivation across regions *viz.*, 1970s, 1980s, 1990s and 2000s. They were as well classified into "longevitywise" as differences in genetic diversity could be observed between them, especially when a variety is popular within and across decades.

DNA Extraction

Genomic DNA was isolated from 20 day-old seedlings germinated in sterile petri dishes lined with moist filter paper using the CTAB (Cetyl Try methyl Ammonim Bromide) method described [16] with some modifications. The purity and concentration of the isolated genomic DNA samples were estimated by UV-absorption spectrophotometer (Beckman DU 650 model) as per the procedure described by Sambrook [17]. Quantification of DNA was done by analyzing the purified DNA on 0.8% agarose gel with lambda (λ) *Hind* III DNA as standard. Based on the intensity and thickness of genomic DNA bands, as compared to lambda (λ) *Hind* III DNA, the concentration and quality of DNA in individual samples were determined.

SSR Marker Analysis

In all, 64 hypervariable microsatellite markers distributed on all the 12 chromosomes of rice were picked up from our previous study [15]. These markers covering 236.8 Mb of physical distance with average distance between them being 5.92 Mb. PCR reactions were carried out in 10 µl reaction volume containing of 10× PCR buffer (10 mM Tris-HCl pH 8.3, 50 mM KCl), 1.5 mM $MgCl_2$, 0.2 mM each dNTPs, 5 pmol of each forward and reverse primer, 0.5U of Taq DNA polymerase (NEB), and 5 ng of genomic DNA. Reactions were carried out in GenAmp PCR system 9700 (Applied Biosystems, USA) thermal cycler using the following temperature profile: an initial denaturation of 5 min at 94°C followed by 35 cycles of 45 s at 94°C, 45 s at 55°C and 1 min at 72°C, then a final extension of 5 min at 72°C. Amplification products were resolved on 3% metaphor agarose gels using a horizontal gel electrophoresis unit (CBS Scientific, USA). The DNA fragments were then visualized under UV-transilluminator and documented using ALPHA IMAGER gel documentation system (Alpha Innotech, USA) which was stored for further scoring and permanent records.

Data Analysis

Only clear and unambiguous bands of SSR markers were scored. The sizes of the amplified fragments were estimated with the help of Alpha image software by Gel documentation system using 100 bp DNA ladder (NEB) as size standard. Markers were scored for the presence (1) or absence (0) of the corresponding band among the genotypes. To measure the informativeness of the markers, polymorphism information content (PIC) for each of the SSR markers was computed according to the formula: $PIC = 1 - \Sigma Pi^2 - \Sigma\Sigma Pi^2 Pj^2$ where 'i' is the total number of alleles detected for SSR marker and 'Pi' is the frequency of the ith allele in the set of hundred genotypes investigated and $j = i+1$ [18].

Genetic diversity parameters *viz.*, number of alleles *(Na)*, observed heterozygosity (*Ho*), Shannon Index (*I*) and Nei's genetic diversity index (*He*) [19] were evaluated using POP-GENE v 1.31 (http://www.ualberta.ca/~fyeh). The allelic richness (*Rs*), is a measure of the number of alleles independent of the sample size, was measured using FSTAT [20]. Significant difference in genetic diversity parameters between each pair of the decadal periods was calculated using the Wilcoxon matched

Table 1. Rice varieties used in the present study.

Decadal Period	Released Rice Varieties	Total No.
Landraces	INRC10192, Lalnakanda, Hasansona, Solumpiket, Basmati370, Acharmati, Bate Aus, Aus Boro, Dular, Aus Bako, Azucena	11
1970s	Manoharsali, Taichung Native-1, Jaya, Tella Hamsa, Rajeswari, Annapoorna, Jyothi, PR-106, N-22 (Nagina-22), Surekha (WL-13400), Prabhat (MTU-3626), Swarnadhan, Masur, WGL3200	14
1980s	Annada, Rasi, SasyaSree, PLA1100, Swarna (MTU-7029), Himalaya-2, Kalinga III, Vikas, CO-43, Keshari, Parijat, VL Dhan-16, VLDhan-206, Pathara, Prasanna, Sonasali, Pothana, Sabita, HKR-120, Mandya Vijaya, Suraksha, Satya, ASD-17, CSR-10, Kasturi, Kanak, Pusa Basmati-1, Tikkana (NLR-27999), Pinakini (NLR-9672-96), Samba Mahsuri (BPT-5204), Vanaprabha	30
1990s	Krishnaveni (MTU-2077), Haryana Basmati, IR-64, Chandana, VLDhan-221, Kavya (WGL-48684), Sneha, Swarnamukhi (NLR-145), Badami, Pusa-44, Ghanteswari, Khandagiri, Nilagiri, Himalaya-2216, Mahi Sugandha, Taraori Basmati, Basmati-386, Krishna Hamsa, Nidhi, VLDhan-61, Triguna, CSR-13, Lalithagiri, Uydyagiri	24
2000s	Bharani (NLR-30491), Srikakulam Sannalu, Somasila (NLR-33358), Sravani (NLR-33359), Cottondora Sannalu (MTU1010), PR-115, Mugad Sugandha-1, Sumathi, Vandana, Yamini (CSR 30), Pusa sugandha-3, Super, Pusa-1121, WGL-32100, MTU-1061 (Indra), Taramati, Suganda samba, Vasumati, Sharbathi, Improved Samba mahsuri and Swarna sub1	21

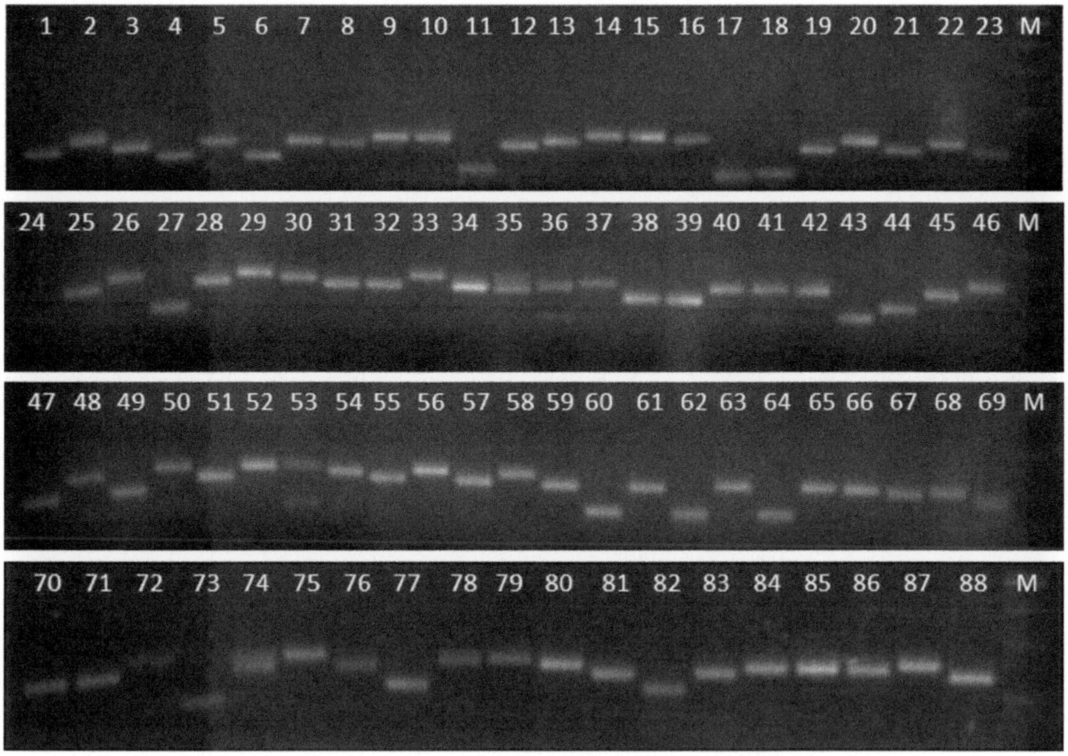

Figure 1. A representative gel picture of screening of rice varieties with RM562. For decoding of the numbers refer Table S1.

pairs test, a nonparametric alternative to the *t*-test (http://www.stattools.net/Wilcoxon_Pgm.php). In addition, number of rare alleles (number of alleles in less than 5% of the population) and high frequency or common alleles (number of alleles in more than 85% of the population) were estimated using Microsoft Excel 2007. The *F*-statistics (*Fst*) were used to analyze genetic differentiation of the varieties in all possible pairs of decades based on allelic discrepancy at each locus by the procedure of AMOVA (Analysis of Molecular Variance), besides the genetic variation within and among the populations using the software ARLEQUIN v 2.0 [21]. The appearance and disappearance of alleles and number of private alleles have been estimated using the CONVERT v 1.31software [22]. In order to overcome the problem of unequal sample size, resampling was done using an in-house script for R software [23]. The UNJ (Unweighted Neighbour Joining method) cluster analysis followed by bootstrap analysis with1000 permutations for total cultivars was carried out using DARwin 5.0.145 (http://darwin.cirad.fr/). The dendrogram based on unbiased genetic distances among decadal periods was constructed by UPGMA (Unweighted pair-group method with arithmetic average) employing POPGENE v 1.31.

Results and Discussion

Hypervariable Microsatellite Marker Analysis

Sixty four hvRM (hypervariable rice microsatellite) markers distributed evenly on all the 12 chromosomes were chosen to assess the genetic relatedness among the 100 genotypes. Fifty-two of them were found to show polymorphism (81.25%)(Table 2 and Figure 1). In all, 184 alleles were identified by amplification of the 52 polymorphic hvRMs with an average number of alleles of 3.6 per locus, with the number ranging from 3 (RM16416) to 7 (RM8207) (Table 2). This was significantly lower than the average number of alleles reported by Jain *et al* (7.8) [24], Spada *et al.* (7.2) [25] and Zhu *et al.* (4.37) [26] and higher than the Chuan-Guang and Gui-Quan [27]. Yu *et al.* [28] studied 193 rice accessions drawn from 26 countries using 101 SSR primer pairs and detected an average allele number of 6.3 per locus, which is also higher than the value reported here. Luce *et al.* [29] analyzed 419 rice accessions from the gene banks in five European countries using 16 SSR loci (different from the ones we selected) and reported an average of 9.1 alleles per locus. The higher value than the present study (3.6) could probably be as a result of larger number of accessions used by these authors. Of the 52 loci, 27 were with three alleles while 16 loci with four alleles (Figure 2). All the 52 markers revealed high PIC values the range being between 0.67 (RM16416) and 0.97 (RM14735) with more than 50% loci in the range of 0.8 to 0.9 (Figure 2). Some of the hvRM markers like RM11340, RM12548, RM13584, RM14270, RM14735, RM15580, and RM22273 showed high PIC values in our previous study as well [15]. The average PIC value estimated in the present study (0.87) is more than that of the previous studies by Giarrocco *et al.* [30] (0.69) and Jayamani *et al.* [31] (0.67) possibly because of hypervariable microsatellite markers used. Recently, it has been demonstrated that microsatellite markers of hypervariable nature would be more polymorphic than non-hypervariable markers [15] The *He* values ranged from 0.15 (RM12031) to 0.76 (RM14735) with an average of 0.59 and more than 40% of the loci were in the frequency of 0.7 to 0.8 (Figure 2). The average *He* value of the present study is slightly lower than the earlier studies [11,32] due to the inclusion of only major Indian rice cultivars, which largely belong to *indica* sub group.

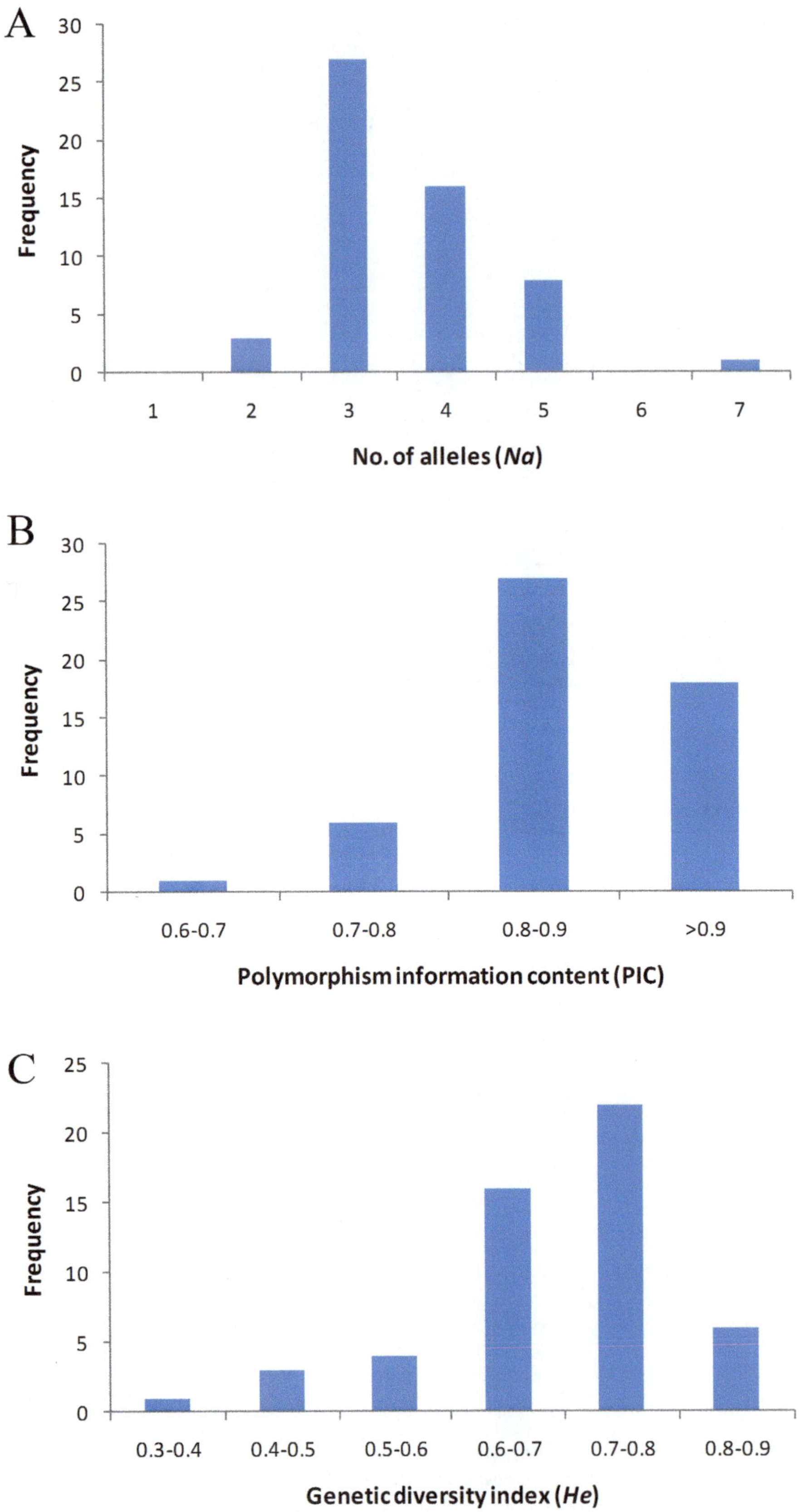

Figure 2. Distribution of number of alleles (*Na*) (A), polymorphism information content (PIC) (B) and genetic diversity (*He*) (C) estimated from 100 rice varieties using 52 hypervariable microsatellite loci.

Table 2. Molecular diversity parameters of the microsatellite markers used in the study.

Locus	Chr.	SSR motif	*Na*	*Rs*	PIC	*Ho*	*He*	*I*
RM562	1	(AAG)13	5	1.63	0.94	0.04	0.73	1.34
RM11313	1	(AAT)23	3	1.49	0.86	0.00	0.56	0.90
RM11340	1	(AT)42	4	1.47	0.89	0.07	0.50	0.89
RM11356	1	(AT)31	3	1.54	0.88	0.12	0.56	0.89
RM11597	1	(AT)42	3	1.48	0.9	0.00	0.57	0.92
RM12031	1	(AG)40	3	1.18	0.74	0.02	0.15	0.32
RM12253	1	(AAT)23	4	1.57	0.91	0.12	0.60	1.02
RM12292	1	(AT)38	3	1.62	0.92	0.14	0.66	1.09
RM12353	2	(AAT)32	4	1.40	0.86	0.16	0.56	0.91
RM12548	2	(AT)46	3	1.57	0.9	0.01	0.60	0.99
RM13131	2	(AT)32	3	1.44	0.91	0.01	0.55	0.90
RM13584	2	(AATC)5	3	1.57	0.88	0.05	0.63	1.05
RM14270	3	(AT)46	4	1.60	0.93	0.00	0.66	1.09
RM14735	3	(AT)42	5	1.60	0.98	0.06	0.76	1.52
RM14778	3	(AT)37	3	1.62	0.89	0.01	0.63	1.04
RM15004	3	(AAT)38	4	1.64	0.91	0.05	0.65	1.14
RM15580	3	(AT)50	3	1.56	0.85	0.02	0.59	0.95
RM16416	4	(AAT)12	3	1.32	0.69	0.02	0.34	0.53
RM16577	4	(AT)29	4	1.60	0.91	0.13	0.62	1.03
RM17405	4	(AAT)36	3	1.47	0.88	0.00	0.49	0.78
RM17669	4	(AT)30	3	1.58	0.9	0.03	0.58	0.95
RM5693	5	(AAT)18	4	1.59	0.78	0.59	0.61	1.08
RM5844	5	(AAT)20	5	1.68	0.94	0.01	0.68	1.24
RM5907	5	(AAT)19	3	1.50	0.86	0.01	0.48	0.84
RM18384	5	(AAG)22	3	1.42	0.9	0.26	0.40	0.80
RM18639	5	(AAT)17	5	1.66	0.78	0.81	0.68	1.26
RM19545	6	(AAT)21	4	1.58	0.91	0.00	0.58	0.94
RM20037	6	(AT)38	4	1.60	0.92	0.01	0.63	1.05
RM20710	6	(AT)46	5	1.70	0.97	0.49	0.76	1.52
RM21693	7	(AT)44	5	1.60	0.93	0.05	0.66	1.19
RM21941	7	(AAT)25	3	1.56	0.84	0.05	0.58	0.98
RM6965	7	(AAG)15	3	1.49	0.91	0.03	0.54	0.90
RM22250	8	(AAT)30	5	1.71	0.94	0.05	0.71	1.34
RM22544	8	(AAT)20	4	1.69	0.89	0.41	0.73	1.35
RM22565	8	(ACAT)15	4	1.67	0.9	0.08	0.68	1.26
RM22688	8	(AAT)28	3	1.34	0.74	0.01	0.38	0.56
RM22273	8	(AT)35	4	1.66	0.83	0.34	0.68	1.24
RM23017	8	(AAT)18	3	1.42	0.83	0.05	0.51	0.79
RM23036	8	(AGAT)15	4	1.62	0.9	0.18	0.64	1.11
RM23362	8	(AAG)19	5	1.60	0.94	0.46	0.66	1.22
RM23741	9	(AAT)28	4	1.59	0.78	0.85	0.59	0.99
RM24015	9	(AGAT)9	4	1.68	0.86	0.68	0.70	1.29
RM24044	9	(AAG)11	3	1.45	0.88	0.19	0.50	0.84
RM24260	9	(AAT)31	3	1.55	0.88	0.00	0.63	1.04
RM5708	10	(AAT)22	4	1.68	0.83	0.47	0.69	1.30
RM8207	10	(AAG)23	7	1.69	0.95	0.05	0.69	1.34
RM25262	10	(AAT)38	3	1.41	0.79	0.08	0.47	0.66
RM25969	11	(AAG)18	3	1.34	0.81	0.01	0.40	0.72
RM26190	11	(AGAT)13	3	1.47	0.92	0.01	0.59	0.96
RM26632	11	(AAAG)9	3	1.53	0.85	0.01	0.57	0.95

Table 2. Cont.

Locus	Chr.	SSR motif	*Na*	*Rs*	PIC	*Ho*	*He*	*I*
RM27840	12	(AAT)37	3	1.54	0.86	0.02	0.58	0.95
RM28279	12	(AATC)8	3	1.56	0.84	0.28	0.61	1.01
Mean			3.69	1.55	0.87	0.15	0.59	1.02
SD			0.90	0.11	0.06	0.22	0.11	0.24

Chr.- Chromosome; *Na*-Number of alleles; PIC-Polymorphism Information Content; *Rs*: Allelic richness; *Ho*- Observed heterozygosity; *He*- Nei's genetic diversity; *I*-Shannon Index; SD-Standard Deviation.

Overall Genetic Diversity in Popular Varieties of Rice

The unweighted neighbour-joining (UNJ) dendrogram constructed on the basis of genetic similarity matrix grouped the 100 genotypes into five clusters *viz.*, landraces, 1970s, 1980s, 1990s, and 2000s (Figure 3). The phylogeny tree reveals that appearing of some of the varieties released during one decade in another decade due to the presence of common parents in their pedigree. For instance, ASD17 and CSR10 belonging to the decadal period of the 1980s clustered with those of the 1970s as IR8 and Jaya of the 1970s being common in their parentage. Varieties of the decadal periods1990s and 2000s comprised largely of varieties released during their respective decades. However, varieties like Tikkana and Pinakini of the 1980s and Somasila and Srikakulam Sannalu of 2000s have been found to cluster with those of the 1990s. Interestingly, Basmati varieties irrespective of their year of release formed a separate sub-cluster. This result is in agreement with previous reports by Glaszman [33], Nagaraju *et al* [34] and Narshimulu *et al* [15]. Equally and interestingly the recently released varieties developed through marker-assisted breeding *viz.*, Swarna-sub1 and improved Samba mahsuri grouped with landrace cluster instead of the expected grouping with the 2000s decade where in their original parents *viz.*, Swarna and Samba mahsuri exist. This could be due to the presence of part of the donor genomes of FR13A, a flood resistant landrace the source for *sub-1* gene and SS1113, the source of bacterial blight resistance genes (*xa13, xa5* and *Xa21*) in improved Samba mahsuri, remaining in the improved versions of the varieties even after many backcrosses. Longevitywise clustering of the 100 genotypes too was in total agreement with the decadewise clustering. Even though some varieties were popular beyond the decade(s) of their release, they do not affect the clustering pattern of the genotypes as well as the genetic diversity of successive decadal periods. For instance, Tella Hamsa and Jyoti remain popular even as late as in the 2000s, albeit released decades back in the states of Andhra Pradesh and Kerala, respectively.

Genetic distance between each pair of decadal periods suggest that it was the highest between the landraces and the high yielding varieties of the 1990s (0.2715), while the lowest between the decades 1990s and 2000s (0.0905) in the year of releasewise group. In case of longevitywise group also genetic distance was highest between the landraces and the high yielding varieties of the 1990s (0.210) but lowest between those of the 1980s and 2000s (0.0237) (Table 3). A dendrogram constructed based on Nei's genetic distance also confirms the above findings as the genetic distance between landrace cultivars and the improved varieties released in 1990s being the highest (Figure 4).

Trends of Genetic Diversity within Decadal Period

Genetic diversity trend of varietal groups released during different decades since the release of semi-dwarf high yielding varieties (1970s to 2000s) along with sets of landraces and Basmati accessions was studied based on number of alleles (*Na*), Nei's genetic diversity index (*He*) and polymorphism information content (PIC) (Figure 5 and Table S2). As well, the diversity within decadal periods was estimated through separate analysis of year of releasewise as well as longevitywise varieties. The diversity trends as estimated through a set of parameters are as under:

Overall increasing trend of number of alleles (*Na*) was observed from landraces to varieties of 2000s in the categories of both year of releasewise and longevitywise classified rice varieties. *Na* values were increased by 13.38% and 13.87% from landraces to 2000s while it was 12.79% and 9.83% from 1970s to the 2000s, respectively for both the categories. On an average, landraces to total modern cultivars percentage of *Na* increase has been observed to be 9.02 for year of releasewise and 10.91 for longevitywise varieties. This is in contrast to the observation of Wei *et al.* [11], who reported decreasing trend of the *Na* score of 7.8% from the 1950s to the 1990s.

Genetic diversity index (*He*) also reveals that the genetic diversity has been increased approximately 10% from landraces to 2000s and landraces to modern cultivars in the year of releasewise category while it was 16.02% and 13.39% for longevitywise category, respectively. However, from 1970s to 2000s the trend of increase was 4.71% and 8.92% for year of releasewise and longevitywise varieties, respectively. The observed trends in the present study are in agreement with Mantegazza *et al.*, [12], who have reported increase of genetic diversity of Italian rice cultivars between 1850 and 2001. The same trend has been reported also by Morin *et al.*, [35]. However, these findings vary with the report of Wei *et al.*, [11] who observed 7% loss of genetic diversity between 1950s and 1990s.

Polymorphism information content (PIC) also shows increasing trend over decadal periods, though not to the level of *Na* and *He*. The percent increase of PIC values from landraces to all modern cultivars, from landraces to 2000s and from 1970s to 2000s as estimated by year of releasewise were 2.75, 4.25 and 5.77%, respectively, while the corresponding values estimated longevitywise are 4.17, 5.88 and 4.87%. Overall, the trends as measured by the parameters of *Na, He* and PIC appear to slightly slow down in the decadal period of 1990s but regains its increasing trend in that of 2000s.

The genetic diversity in Indian cultivars ($He = 0.54$) as estimated by the Nei's genetic diversity index in the present study is nearly equal to that of the representative world collection of *indica* varieties ($He = 0.55$). It is, however, overwhelmingly high as compared to the representative world collection of temperate *japonica* ($He = 0.39$) [36]. Thus, the findings of the present study confirms further, the *indica* varietal group to be much more genetically diverse than *japonica* group which is in agreement with many earlier reports [32,37].

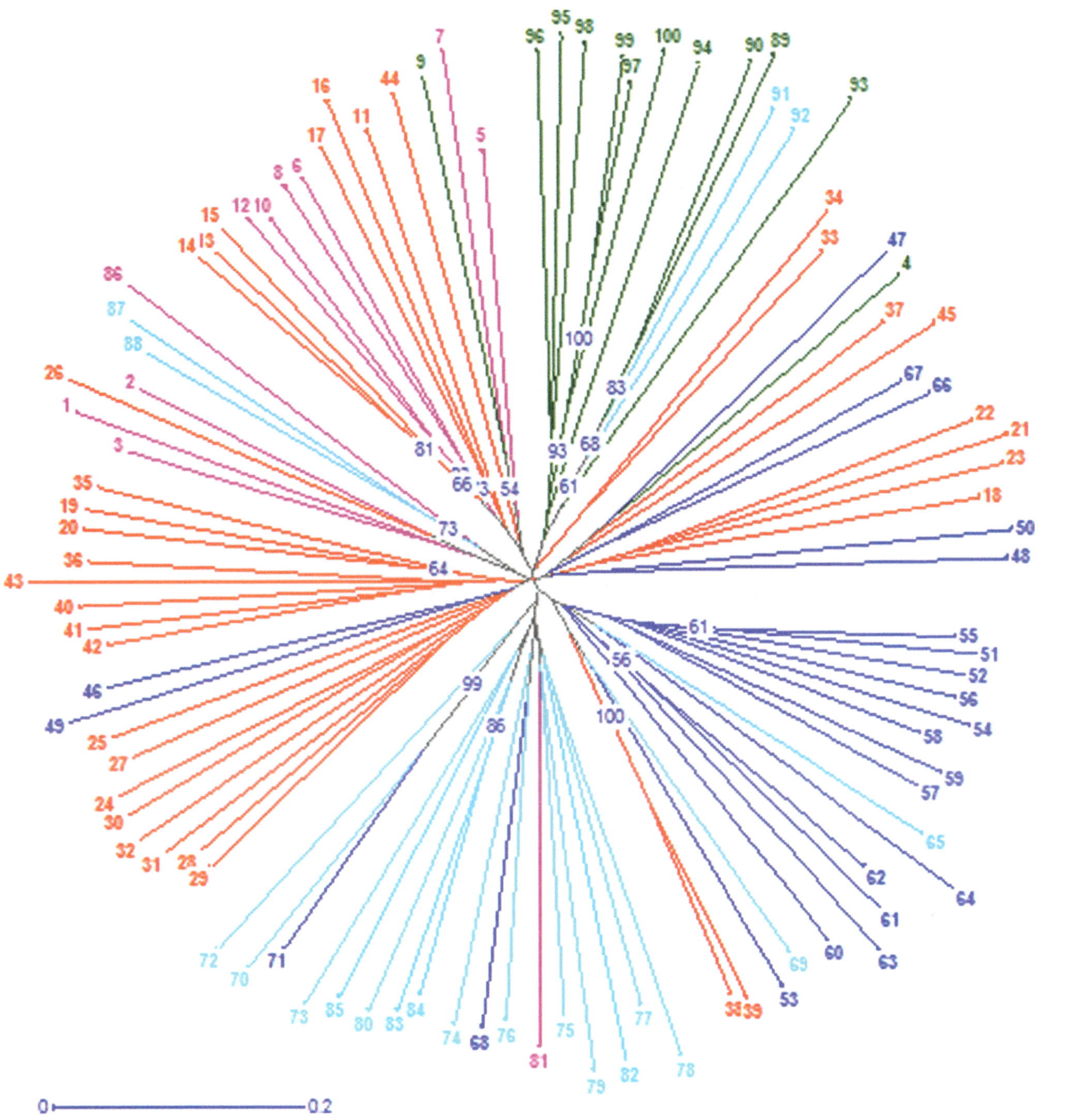

Figure 3. Genetic relationship among 100 rice varieties estimated using Unbiased Neighbour-Joining dendrogram of 52 hypervariable microsatellite loci. Landraces- green colored;1970s-pink colored;1980-red colored;1990s-dark blue colored;2000s-sky blue colored. Rice varieties were represented in numbers. For full details of the varieties refer Table S1.

Overall, the present investigation reveals a genetic diversity as estimated using different parameters such as allele number (*Na*), polymorphism information content (PIC) and genetic diversity index (*He*), to increase progressively in the advancing decadal periods from 1970s to 2000s and interestingly, it was higher as compared to even landraces. Use of genotypes, which cover longer periods always offer added advantage in temporal studies. In the present study only a few landraces which are being cultivated for their certain features by local farmers has been included to compare genetic diversity in varieties released before and after 1970s. The analysis of genetic diversity using various parameters clearly reveals the increasing trend from landraces to the decadal period 2000s. While this observation is in agreement with those of Mantegazza *et al* [12], who have reported steady increase in the levels of gene diversity in Italian rice germplasm, it was not so with the findings of Steele *et al* [38] who have reported no change in the level of diversity following the introduction of modern rice varieties. Wei *et al* [11] and Yuan *et al* [32] report, however, loss

Table 3. Nei's Unbiased measures of genetic distance among decadal periods of major Indian rice cultivars.

Decade	Landraces	1970s	1980s	1990s	2000s
Landraces	****	0.204	0.182	0.210	0.169
1970s	0.206	****	0.052	0.059	0.056
1980s	0.189	0.099	****	0.046	0.024
1990s	0.272	0.126	0.118	****	0.029
2000s	0.211	0.132	0.100	0.091	****

Lower diagonal: Year of releasewise; Upper diagonal: Longevitywise.

of more alleles in modern cultivars as compared to varieties released in the 1950s. Put together, the present findings and previous reports, it is reasonable to assume that global rice genetic diversity tend to increase over the decades by gaining as well as losing to different proportion of alleles. In the absence of phenotypic expression of alleles gained or lost in rice genome it would be difficult to conclude if they are of adaptive or agronomic value. Qualitative study of variation in gene diversity, over the decadal periods, as estimated by the total number of alleles that appeared and disappeared showed an increase in the appearance and decrease in the disappearance of alleles from landraces to 2000s in both the year of releasewise and longevitywise analysis (Figure 6). Though we presume that disappearance of certain alleles might be of deleterious nature and that appearance of new alleles to be of positive breeding value, it need not be so always. Thus, there is a need for still more precisely planned further study. As all the major cultivars of India are belonging to the *indica* subspecies of *Oryza sativa,* we could not compare with the *japonica* varieties in terms of allele appearance and disappearance. However, earlier [11] it was proved that more alleles loss has been observed in *indica* varieties than in *japonica.* This phenomenon can be attributed to the higher diversity nature of *indica* varieties compared to *japonica.*

Many reasons may be attributed to the progressively increased genetic diversity over the decadal periods between 1970s and 2000s. In the initial decades, the advent of semidwarf plant type based exotic high yielding varieties like Taichung (Native)-1 and IR8 and varieties derived from crosses involving them with traditional Indian varieties could have introduced large variability for yield, earliness, photo-insensitivity, seed non-dormancy etc. In the subsequent decades, the focus has been shifted towards the development of polygenic durable resistance to multiple pathogens, adaptation to diverse rice ecologies and consumer preference driven quality improvement. For instance, high yielding varieties of the initial decades were more susceptible to indigenous as well as introduced diseases and insect pests like bacterial blight, tungro virus, hoppers etc. warranting to exploit exotic cultivar and wild/weedy sources. Convergent breeding efforts have made most of the high yielding varieties of today multiple pest resistant. Given the challenges ahead which have all the potential to impede further progress in crop improvement, there is need to enrich the variability employing multiparent breeding and exploitation of hidden variability from progenitor species and primitive cultivars.

Trends of Genetic Diversity among Decadal Periods

Analysis of molecular variance (AMOVA) has revealed most of the variation was existed within decadal periods i.e., 92.12% and 96.66%, respectively for year of releasewise and longevitywise as against very low variation between the decadal periods i.e., 3.34% for longevitywise classified varieties and 7.88% for year of releasewise classification (Table 4). These results are in agreement with Yuan *et al.* [32]. Pairwise comparison of population differentiation (*Fst*) among decadal periods reveals significant genetic differentiation among all except between the landraces and the decadal period of 1970s as estimated by both year of releasewise and longevitywise analysis. Also there was no

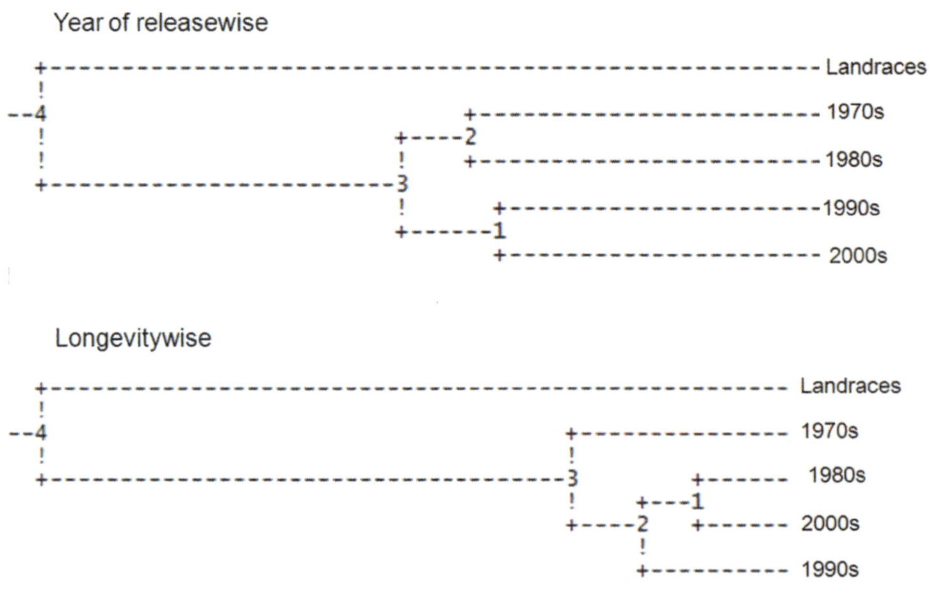

Figure 4. UPGMA dendrogram based on Nei's genetic distances using POPGENE v 1.31 showing the genetic relationship of rice genotypes among decadal periods. The genetic distances between different groups are as follows. Year of releasewise: 4 and Landraces - 10.96, 4 and 3–5.01, 3 and 2–0.97, 2 and 1970s - 4.97, 2 and 1980s- 4.97, 3 and 1–1.41, 1 and 1990s-4.52 and 1 and 2000s-4.52. Longevitywise: 4 and Landraces −9.55, 4 and 3–6.76, 3 and 1970s - 2.78, 3 and 2 -0.93, 2 and 1 - 0.66, 1 and 1980s - 1.18, 1 and 2000s - 1.18, 1 and 1990s - 1.85.

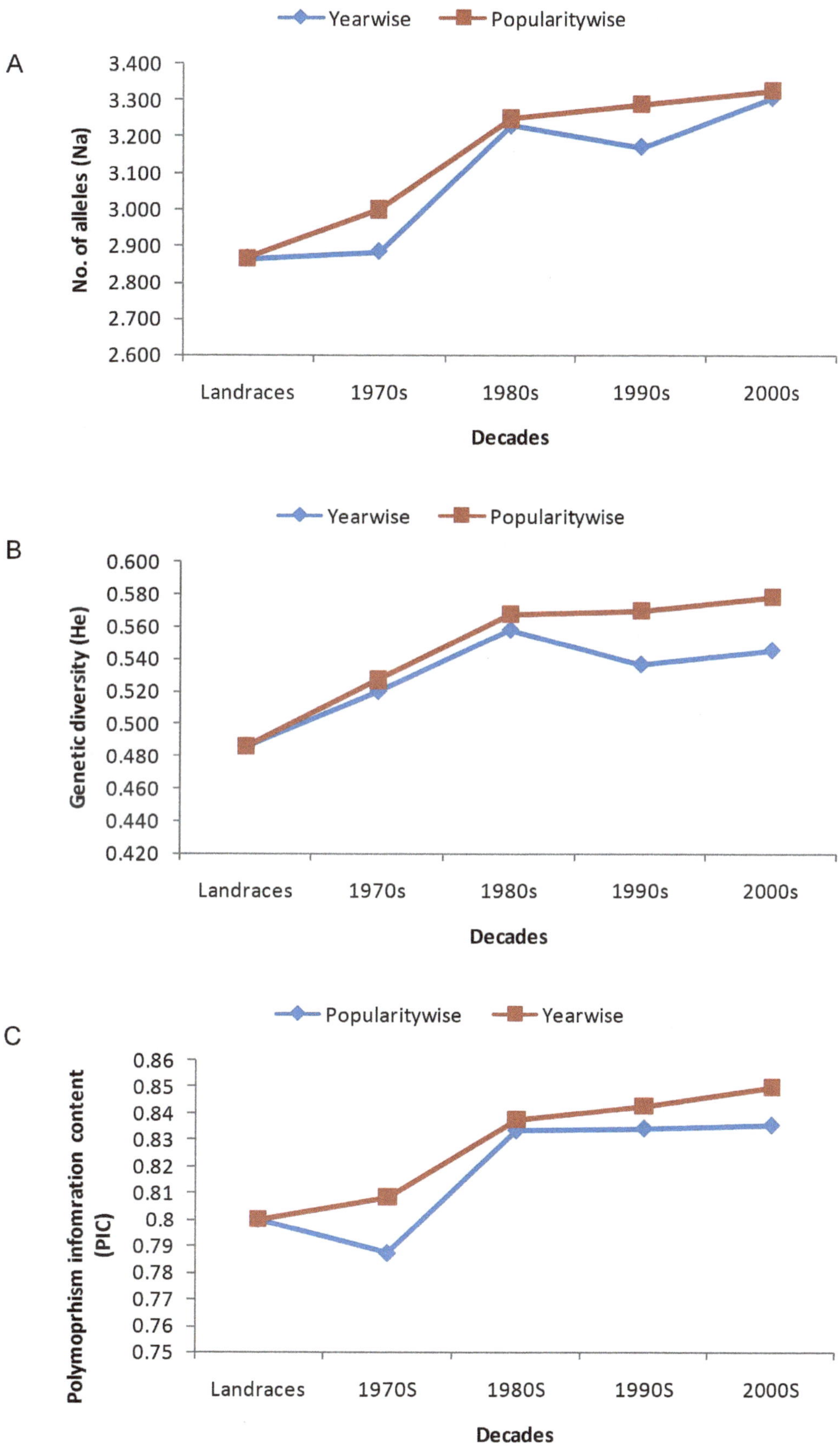

Figure 5. Changes in number of alleles (*Na*), Nei's genetic diversity (*He*) and PIC values over decadal periods.

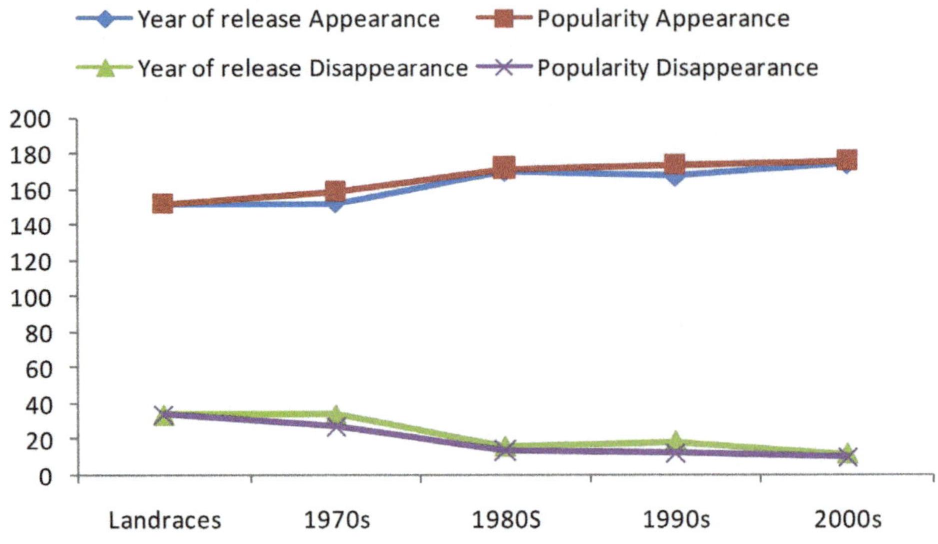

Figure 6. Appearance and disappearance of alleles over decadal periods.

significant differentiation between the 1970s and rest of the decades 1980s to 2000s in longevitywise analysis (Table 5).

Locus-by-locus AMOVA showed significant ($p<0.01$) differences in genetic variation among the decadal periods as exhibited by 14 loci *viz.*, RM11356, RM12031, RM13131, RM14735, RM15004, RM15580, RM5693, RM5907, RM18384, RM22554, RM22565, RM22688, RM23741, and RM8207. Interestingly, three loci viz., RM11313, RM12353, and RM6965 have been found to explain more than 20% of the variation among the decades (Table S3).

Analysis of Population Specific Alleles

Analysis for identification of unique population specific alleles has revealed all the decadal periods found to show specific alleles with the exception of the 1970s as estimated by year of releasewise as well as longevitywise classification. As expected, the landraces exhibited as many as six (RM12031, RM15004, RM5844, RM21693, RM22250 and RM23362) and seven (RM12031, RM15004, RM5844, RM21693, RM22250, RM23362 and RM8207) specific alleles compared to the decadal periods in both year of rleasewise and longevitywise categories, respectively (Table S4). While the decadal periods 1980s (RM18384) and 1990s (RM5708) having one specific allele each, the period 2000s is with two (RM5844 and RM23741) specific alleles. Of the 10 population specific alleles identified in all, five are located in genic regions (RM5844, RM18384, RM22250, RM23362 and RM23741) while rest in non-genic regions. Seven of the 10 alleles specific to populations are comprised of tri-nucleotide repeats, especially of AAT repeat motif.

Qualitative analysis of genetic diversity revealed that the landraces were comprising of more number of population specific alleles as compared to the modern high yielding varieties suggesting loss of some of the landrace specific alleles on account of intensive human selection over the decades. The present findings are in agreement with those of earlier workers [11,32] who have also reported loss of alleles in the progressively improved modern varieties. The reason can be attributed to conscious and simultaneous selection against alleles of adaptive value characteristic to landraces resulting in less number of certain alleles and selection for alleles of agronomic value in the modern cultivars.

Effect of Genic-SSRs, Non-genic-SSRs and Gene-specific Markers on Genetic Diversity Trends

Ever since the domestication of rice, man's breeding priority has been yield enhancement and "breeding out" the wild progenitor traits like lax panicle, shattering, awns, etc., and later the focus had been shifted towards "breeding in" the traits of plant architecture and resistance to biotic and abiotic stresses and acceptable cooking quality. In that long breeding process, except the genomic regions, which govern domestication related traits, a major part of the rice

Table 4. Analysis of molecular variance (AMOVA) of major Indian rice cultivars.

Sample	Source of variation	df	Variance components	Percentage of variation
Year of releasewise	Among populations	4	0.549**	7.88
	Within populations	195	6.426***	92.12
	Total	199	0	
Popularitywise	Among populations	4	0.231**	3.34
	Within populations	287	6.690***	96.66
	Total	291	0	

df: degrees of freedom; p>0.001-*** : p>0.01-**.

Table 5. Pairwise population differentiation (*Fst*) of major Indian rice cultivars among decadal periods.

Decade	Landraces	1970s	1980s	1990s	2000s
Landraces	0.00000	0.06822	0.05477***	0.08557***	0.05925***
1970s	0.05112	0.00000	0.0138	0.02299	0.02338
1980s	0.03997*	0.03957	0.00000	0.03998***	0.01241
1990s	0.13083***	0.07904***	0.10569***	0.00000	0.02389**
2000s	0.09004***	0.07927**	0.07507***	0.0665***	0.00000

Lower diagonal: Year of releasewise; Upper diagonal: Longevitywise.
$p>0.001$-***; $p>0.01$-**;$p>0.1$-*.

genome remains unchanged and hence large unfolded genetic diversity. To estimate genetic variability in crop improvement research, increasingly crop specific microsatellite markers are used. Since most of the microsatellites are non-coding, estimates made and inferences drawn based on them may not reflect true genomic diversity of a crop. To understand precisely the changing trends of genetic diversity in breeding material over periods, it is necessary to consider markers located both in genic and non-genic regions. In the present study, thus, microsatellite markers located in both genic and non-genic regions have been chosen to understand the level and trend of genetic diversity in varieties developed over the last four decades in comparison to the traditional varieties. The study using two marker types separately and together interestingly showed an increasing trend of genetic diversity (Figure S1, Table S5, Table S6). Analysis of gene specific marker data obtained from earlier studies also [39–41] clearly suggests an increasing trend of genetic diversity in Indian rice varieties developed over a period of four decades. In contrast, Qi *et al* [42] have reported initially declining trend of diversity in varieties developed between 1950 and 1980s and subsequently increasing trend as estimated based on both SSR markers and phenotypic traits in Chinese rice varieties.

Analysis of Genetic Diversity Based on Region, Ecology, Grain Size and Days to 50% Flowering Periodwise

When genetic diversity parameters viz., number of alleles (*Na*), number of effective alleles (*Ne*), Shannon index (*I*) and Nei's genetic diversity index (*He*) estimated according to their region of release in India (South, North and East), ecology (irrigated and rainfed), grain size (long and medium) and days to 50% flowering (early, medium early, medium and late) of a representative set of Indian rice varieties, observed variation was not correlating with their classification (Figure S2, Table S7). The reasons could be attributed to the repeated use of founder varieties such as T(N)1 and IR8 as well as their derivatives like Jaya, Rajeswari, Rasi, Prabhat, PR106 and Suraksha in the ancestry of many of the subsequently evolved varieties irrespective of their trait classification. Further, observation of pedigree records of some of the emerging new varieties released during the past two decades revealed that they were having some of the quality rice varieties such as Basmati370, Sona and BPT5204 (Samba mahsuri) as their donors but not the original founder varieties or their derivatives. The reason for this trend could be assumed as the changing consumer preferences towards quality rices. Keeping this trend in view breeders also being used quality rice varieties as the donors in their breeding programs.

Conclusions

Despite the perception that genetic variability has declined over the decades in the improved rice gene pool the present findings and previous reports suggest genetic diversity in the improved cultivars has been increasing over the decades, while losing some of the alleles which could be of unimportant and not relevant to the changing breeding objectives. However, it is important to understand the kind of alleles being gained and lost over the decades so as to plan our future breeding/selection strategies for directed improvement of the crop. The study has overall reveals that adequate diversity still exists in the advanced cultivar gene pool for steady varietal improvement. In the diversity conservation exercise being practiced at national and global levels, therefore, much emphasis is to be given to improved germplasm as that of the customarily conserved gene pool comprising wild/weedy species and landraces so that the genes of importance are conserved and profitably utilized to meet future challenges. It is, however, important to monitor periodically the level and trend of genetic diversity while pursuing aggressively rice breeding research.

Supporting Information

Figure S1 Changes in number of alleles *(Na)*, genetic diversity (*He*) over decadal periods using genic and non-genic SSRs along with gene-specific markers(previous studies).

Figure S2 Number of alleles *(Na)*, number of effective alleles (*Ne*), Shannon index (*I*) and Nei's genetic diversity (*He*) estimated in different groups of rice varieties i.e., region, ecology, grain size and days to 50% floweringwise.

Table S1 Details of the rice cultiavars used in the study.

Table S2 Summary statistics of genetic diversity parameters of major Indian rice cultivars.

Table S3 Locus by locus AMOVA among populations.

Table S4 Population specific alleles and their corresponding markers.

Table S5 List of genic and non-genic SSRs and their functions.

Table S6 Summary statistics of genetic diversity parameters of Indian rice cultivars using different microsatellite classes.

Table S7 Summary statistics for genetic diversity parameters of Indian rice varieties classified according to their region, ecology, grain size and days to 50% flowering.

Acknowledgments

We are very thankful to Dr.Mir for helping in resampling analysis using R-Software. Thanks are also due to the anonymous reviewers for their valuable comments for the improvement of the manuscript.

Author Contributions

Conceived and designed the experiments: LRV EAS. Performed the experiments: GC NR DAD AV. Analyzed the data: LRV MS. Contributed reagents/materials/analysis tools: GA. Wrote the paper: LRV EAS.

References

1. Christiansen MJ, Andersen SB, Ortiz R (2002) Diversity changes in an intensively bred wheat germplasm during the 20th century. Molecular Breeding 9: 1–11.
2. Tian QZ, Zhou RH, Jia JZ (2005) Genetic diversity trend of common wheat (*Triticum aestivum* L.) in China revealed with AFLP markers. Genetic Resources and Crop Evolution 52: 325–331.
3. White J, Law JR, MacKay I, Chalmers KJ, Smith JSC, et al. (2007) The genetic diversity of UK, US and Australian cultivars of Triticum aestivum measured by DArT markers and considered by genome. Theoretical and Applied Genetics 116: 439–453.
4. Huang X-Q, Wolf M, Ganal MW, Orford S, Koebner RMD, et al. (2007) Did Modern Plant Breeding Lead to Genetic Erosion in European Winter Wheat Varieties? Crop Science 47: 343.
5. Roussel V, Koenig J, Beckert M, Balfourier F (2004) Molecular diversity in French bread wheat accessions related to temporal trends and breeding programmes. TAG Theoretical and Applied Genetics 108: 920–930.
6. Fu Y-B, Peterson GW, Yu J-K, Gao L, Jia J, et al. (2006) Impact of plant breeding on genetic diversity of the Canadian hard red spring wheat germplasm as revealed by EST-derived SSR markers. Theoretical and Applied Genetics 112: 1239–1247.
7. Hazen SP, Zhu L, Kim H-S, Tang G, Ward RW (2002) Genetic diversity of winter wheat in Shaanxi province, China, and other common wheat germplasm pools. Genetic Resources and Crop Evolution 49: 439–448.
8. Hysing S-C, Säll T, Nybom H, Liljeroth E, Merker A, et al. (2008) Temporal diversity changes among 198 Nordic bread wheat landraces and cultivars detected by retrotransposon-based S-SAP analysis. Plant Genetic Resources 6.
9. Heal G, Walker B, Levin S, Arrow K, Dasgupta P, et al. (2004) Genetic diversity and interdependent crop choices in agriculture. Resource and Energy Economics 26: 175–184.
10. Wouw M, Hintum T, Kik C, Treuren R, Visser B (2010) Genetic diversity trends in twentieth century crop cultivars: a meta analysis. Theoretical and Applied Genetics 120: 1241–1252.
11. Wei X, Yuan X, Yu H, Wang Y, Xu Q, et al. (2009) Temporal changes in SSR allelic diversity of major rice cultivars in China. Journal of Genetics and Genomics 36: 363–370.
12. Mantegazza R, Biloni M, Grassi F, Basso B, Lu B-R, et al. (2008) Temporal Trends of Variation in Italian Rice Germplasm over the Past Two Centuries Revealed by AFLP and SSR Markers. Crop Science 48: 1832–1840.
13. Upadhyay P, Singh VK, Neeraja CN (2011) Identification of Genotype Specific Alleles and Molecular Diversity Assessment of Popular Rice (*Oryza sativa* L.) Varieties of India. International Journal of Plant Breeding and Genetics 5: 130–140.
14. Neeraja CN, Vemireddy LR, Malathi S, Siddiq EA (2009) Identification of alternate dwarfing gene sources to widely used Dee-Gee-Woo-Gen allele of sd1 gene by molecular and biochemical assays in rice (*Oryza sativa* L.). Electronic Journal of Biotechnology 12.
15. Narshimulu G, Jamaloddin M, Vemireddy LR, Anuradha G, Siddiq E (2011) Potentiality of evenly distributed hypervariable microsatellite markers in marker-assisted breeding of rice. Plant Breeding 130: 314–320.
16. Murray MG, Thompson WF (1980) Rapid isolation of high molecular weight plant DNA. Nucleic Acids Research 8: 4321–4326.
17. Sambrook J, Russell DW (2001) Molecular cloning: A laboratory manual. Cold Spring Harbour Laboratory Press, Cold Spring Harbour, New York.
18. Botstein D, White RL, Skolnick M, Davis RW (1980) Construction of a genetic linkage map in man using restriction fragment length polymorphisms. American Journal of Human Genetics 32: 314–331.
19. Nei M (1973) Analysis of gene diversity in subdivided populations. Proc Natl Acad Sci U S A 70: 3321–3323.
20. Goudet J (1995) FSTAT (Version 1.2): A Computer Program to Calculate F-Statistics. J Heredity 86: 485–486.
21. Excoffier L, Laval G, Schneider S (2005) Arlequin (version 3.0): an integrated software package for population genetics data analysis. Evol Bioinform Online 1: 47–50.
22. Glaubitz JC (2004) convert: A user-friendly program to reformat diploid genotypic data for commonly used population genetic software packages. Molecular Ecology Notes 4: 309–310.
23. Mir RR, Kumar J, Balyan HS, Gupta PK (2012) A study of genetic diversity among Indian bread wheat (*Triticum aestivum* L.) cultivars released during last 100 years. Genetic Resources and Crop Evolution 59: 717–726.
24. Jain S, Jain RK, McCouch SR (2004) Genetic analysis of Indian aromatic and quality rice (*Oryza sativa* L.) germplasm using panels of fluorescently-labeled microsatellite markers. Theoretical and Applied Genetics 109: 965–977.
25. Spada A, Mantegazza R, Biloni M, Caporali E, Sala F (2004) Italian rice varieties: historical data, molecular markers and pedigrees to reveal their genetic relationships. Plant Breeding 123: 105–111.
26. Zhu MY, Wang YY, Zhu YY, Lu BR (2004) Genetic diversity of rice landraces from Yunnan revealed by SSR analysis and its significance for conservation. J Huazhong Agri Univ 23: 187–191.
27. Chuan-Guang, Gui-Quan Z (2010) SSR analysis of genetic diversity and the temporal trends of major commercial inbred indica rice cultivars in south china in 1949–2005. Acta Agron Sin 36: 1843–1852.
28. Yu SB, Xu WJ, Vijayakumar CHM, Ali J, Fu BY, et al. (2003) Molecular diversity and multilocus organization of the parental lines used in the International Rice Molecular Breeding Program. Theoretical and Applied Genetics 108: 131–140.
29. Luce C, Noyer J, Tharreaeu D, Ahmadi N, Feyt H (2001) The use of microsatellite markers to examine the diversity of the genetic resources of rice (*Oryza sativa* L.) adapted to European conditions. Acta Horticulture 546: 221–235.
30. Giarrocco LE, Marassi MA, Salerno GL (2007) Assessment of the Genetic Diversity in Argentine Rice Cultivars with SSR Markers. Crop Science 47: 853–858.
31. Jayamani P, Negrão S, Martins M, Maçãs B, Oliveira MM (2007) Genetic Relatedness of Portuguese Rice Accessions from Diverse Origins as Assessed by Microsatellite Markers. Crop Science 47: 879–884.
32. Yuan X-P, Wei X-H, Hua L, Yu H-Y, Wang Y-P, et al. (2007) A Comparative Study of SSR Diversity in Chinese Major Rice Varieties Planted in 1950s and in the Recent Ten Years (1995–2004). Rice Science 14: 78–84.
33. Glaszmann JC (1987) Isozymes and classification of Asian rice varieties. Theoretical and Applied Genetics 74: 21–30.
34. Nagaraju J, Kathirvel M, Kumar RR, Siddiq EA, Hasnain SE (2002) Genetic analysis of traditional and evolved Basmati and non-Basmati rice varieties by using fluorescence-based ISSR-PCR and SSR markers. Proc Natl Acad Sci U S A 99: 5836–5841.
35. Morin SR, Calibo M, Garcia-Belen M, Pham J-L, Palis F (2002) Natural hazards and genetic diversity in rice. Agriculture and Human Values 19: 133–149.
36. Garris AJ, Tai TH, Coburn J, Kresovich S, McCouch S (2005) Genetic structure and diversity in *Oryza sativa* L. Genetics 169: 1631–1638.
37. Yonemaru J, Yamamoto T, Ebana K, Yamamoto E, Nagasaki H, et al. (2012) Genome-wide haplotype changes produced by artificial selection during modern rice breeding in Japan. PLoS One 7: e32982.
38. Steele KA, Gyawali S, Joshi KD, Shrestha P, Sthapit BR, et al. (2009) Has the introduction of modern rice varieties changed rice genetic diversity in a high-altitude region of Nepal? Field Crops Research 113: 24–30.
39. Ramkumar G, Sivaranjani AKP, Pandey MK, Sakthivel K, Shobha Rani N, et al. (2010) Development of a PCR-based SNP marker system for effective selection of kernel length and kernel elongation in rice. Molecular Breeding 26: 735–740.
40. Sundaram RM, Sakthivel K, Hariprasad AS, Ramesha MS, Viraktamath BC, et al. (2010) Development and validation of a PCR-based functional marker system for the major wide-compatible gene locus S5 in rice. Molecular Breeding 26: 719–727.
41. Singh A, Singh PK, Singh R, Pandit A, Mahato AK, et al. (2010) SNP haplotypes of the BADH1 gene and their association with aroma in rice (*Oryza sativa* L.). Molecular Breeding 26: 325–338.
42. Qi Y, Zhang D, Zhang H, Wang M, Sun J, et al. (2006) Genetic diversity of rice cultivars (*Oryza sativa* L.) in China and the temporal trends in recent fifty years. Chinese Science Bulletin 51: 681–688.

5

Association Analysis of Stem Rust Resistance in U.S. Winter Wheat

Dadong Zhang[1¤], Robert L. Bowden[2], Jianming Yu[3], Brett F. Carver[4], Guihua Bai[1,2*]

1 Department of Agronomy, Kansas State University, Manhattan, Kansas, United States of America, **2** USDA–ARS Hard Winter Wheat Genetics Research Unit, Manhattan, Kansas, United States of America, **3** Department of Agronomy, Iowa State University, Ames, Iowa, United States of America, **4** Plant Science Department, Oklahoma State University, Stillwater, Oklahoma, United States of America

Abstract

Stem rust has become a renewed threat to global wheat production after the emergence and spread of race TTKSK (also known as Ug99) and related races from Africa. To elucidate U.S. winter wheat resistance genes to stem rust, association mapping was conducted using a panel of 137 lines from cooperative U.S. winter wheat nurseries from 2008 and simple sequence repeat (SSR) and sequence tagged site (STS) markers across the wheat genome. Seedling infection types were evaluated in a greenhouse experiment using six U.S. stem rust races (QFCSC, QTHJC, RCRSC, RKQQC, TPMKC and TTTTF) and TTKSK, and adult plant responses to bulked U.S. races were evaluated in a field experiment. A linearization algorithm was used to convert the qualitative Stakman scale seedling infection types for quantitative analysis. Association mapping successfully detected six known stem rust seedling resistance genes in U.S. winter wheat lines with frequencies: *Sr6* (12%), *Sr24* (9%), *Sr31* (15%), *Sr36* (9%), *Sr38* (19%), and *Sr1RS*^{Amigo} (8%). Adult plant resistance gene *Sr2* was present in 4% of lines. *SrTmp* was postulated to be present in several hard winter wheat lines, but the frequency could not be accurately determined. *Sr38* was the most prevalent *Sr* gene in both hard and soft winter wheat and was the most effective *Sr* gene in the adult plant field test. Resistance to TTKSK was associated *with* nine markers on chromosome 2B that were in linkage disequilibrium and all of the resistance was attributed to the *Triticum timopheevii* chromosome segment carrying *Sr36.* Potential novel rust resistance alleles were associated with markers *Xwmc326-203* on 3BL, *Xgwm160-195* and *Xwmc313-225* on 4AL near *Sr7*, *Xgwm495-182* on 4BL, *Xwmc622-147* and *Xgwm624-146* on 4DL, and *Xgwm334-123* on 6AS near *Sr8.* *Xwmc326-203* was associated with adult plant resistance to bulked U.S. races and *Xgwm495-182* was associated with seedling resistance to TTKSK.

Editor: Zhengguang Zhang, Nanjing Agricultural University, China

Funding: The National Research Initiative Competitive Grants Coordinated Agricultural Project 2011- 68002-30029 from the USDA National Institute of Food and Agriculture. The funders had no role in study design, data collection and analysis, decision to publish, or preparation of the manuscript.

Competing Interests: The authors have declared that no competing interests exist.

* Email: Guihua.bai@ars.usda.gov

¤ Current address: Center for Public Health Genomics, University of Virginia, Charlottesville, Virginia, United States of America

Introduction

Stem rust (SR), caused by *Puccinia graminis* Pers.:Pers. f. sp. *tritici* Erikss. & E. Henn., historically was a destructive disease in wheat (*Triticum aestivum* L.) worldwide [1]. In the United States, SR occurred frequently from the 1920s to 1960s and caused yield losses up to 50% in severe epidemic years [2]. Since the late 1970s, major SR epidemics have not been reported due to the successful deployment of resistance genes in commercial wheat cultivars in conjunction with the eradication of common barberry (*Berberis vulgaris* L.) [2]. The emergence and spread of race TTKSK (also known as Ug99) and related races from Africa are of great concern because they have overcome many important resistance genes used in commercial production [1,3–6]. Race TTTTF, first detected in the U.S in 2000, is also virulent on a large number of important resistance genes [7]. Achieving more durable resistance will depend on deploying diverse combinations of race-specific qualitative resistance and/or race-nonspecific quantitative resistance genes [8].

Numerous SR resistance genes have been identified, but many have limited usefulness in agriculture [9,10]. Only six of the 30 genes listed from *T. aestivum* were effective against all tested races, and several of these conferred inadequate levels of resistance by themselves [9]. *Sr2* derived from *T. turgidum* is the only proven durable race-nonspecific SR resistance gene, although several newly identified adult plant resistance genes may eventually be shown to be durable [9,11]. Other important resistance genes from alien species include *Sr24*, *Sr31*, *Sr1RS*^{Amigo}, *Sr36* and *Sr38*. Although these genes have now been individually defeated by the new races [5,6,9], they are still useful in combinations. Other alien genes such as *Sr22, Sr25*, *Sr26*, *Sr35*, *Sr39*, *Sr40*, and *Sr44* remain effective against the new races but are not yet widely deployed due to concerns about linkage drag [9]. Alien chromosome segments are being shortened to reduce linkage drag for many sources of resistance [12,13].

An essential step in developing and deploying genetic resistance resources in U.S. winter wheat breeding programs is to understand the existing complement of SR resistance genes. Early gene postulation work based on characteristic low infection types (ITs) against a range of stem rust cultures identified resistance genes *Sr5*, *Sr6*, *Sr7b*, *Sr8*a, *Sr9a*, *Sr10*, *Sr11*, *Sr12*, *Sr17*, *Sr36*, *SrMcN*, and *SrTmp* in some U.S. wheat cultivars [14–16]. Jin and Singh [17] postulated the presence of *Sr6*, *Sr24*, *Sr31*, *Sr36*, $Sr1RS^{Amigo}$ or *SrTmp* in a set of 37 hard and 19 soft winter wheat cultivars. Yu et al. [18] detected marker alleles associated with *Sr2*, *Sr24*, *Sr36* and $Sr1RS^{Amigo}$ in a set of 31 U.S. wheat germplasm lines. Olson et al. [19] detected markers for *Sr24*, *Sr31*, *Sr36* and $Sr1RS^{Amigo}$ in a collection of 776 U.S. cultivars. Although these reports have overlapping results, they are each incomplete representations of the full complement of resistance genes.

Association studies have been used to discover and validate both major genes and quantitative trait loci in different plant species. In wheat, association mapping was used to identify SR resistance genes in CIMMYT spring wheat germplasm [18,20,21] and Ethiopian durum wheat [22,23] but using association mapping to study SR resistance in U.S. wheat has not been reported. This study analyzed a set of elite breeding lines from major U.S. winter wheat breeding programs using association mapping and a newly developed algorithm to convert complex Stakman IT scores [24] to a linear scale. Zhang et al. [25] previously described the genetic diversity, population structure, and linkage disequilibrium relationships in this population and it was successfully used for association mapping of resistance to *Soilborne wheat mosaic virus* [26]. Our objectives in this study were to: 1) evaluate the IT data conversion method for detecting stem rust resistance genes, 2) validate reliability of DNA markers linked to known SR resistance genes in U.S. winter wheat, 3) discover potential new genes and/or markers for wheat SR resistance, and 4) determine the composition and frequency of SR resistance genes in U.S. winter wheat elite breeding lines.

Materials and Methods

Plant materials

A set of 137 U.S. elite breeding lines and cultivars was selected from the 2008 USDA-ARS Southern (SRPN, n = 44) and Northern (NRPN, n = 28) Hard Winter Wheat Regional Performance Nurseries, and the USDA-ARS Uniform Eastern (UESRWWN, n = 34) and Southern Soft Red Winter Wheat Nurseries (USSRWWN, n = 31), after removal of sibling lines. These accessions included 72 hard winter wheat (HWW) and 65 soft winter wheat (SWW) lines (Table S1). Seed was provided by the breeding program at Oklahoma State University, Stillwater, OK.

Marker data

Leaf tissue was sampled from a single plant at the two-leaf stage, and DNA was extracted using the cetyltrimethyl ammonium bromide (CTAB) method [25]. PCR amplifications were performed in a DNA Engine Thermal Cycler (Bio-Rad Laboratories, Hercules, CA) with 12 µl PCR mixture containing 1.2 µl of 10× PCR buffer (Bioline, Taunton, MA), 2.5 mM of $MgCl_2$, 200 µM of each dNTP, 50 nM of forward tailed primer that was synthesized by adding 18 bp of M13 tail to 5′ end of each forward primer, 250 nM of reverse primer and 200 nM of M13 fluorescent-dye labeled primer, 0.6 U of *Taq* DNA polymerase, and about 80 ng template DNA. Specific PCR programs were used according to available published papers to amplify marker fragments for known genes; otherwise, a touchdown PCR program was used [25].

All accessions were genotyped for 289 markers (Table S2), including 272 SSRs distributed over all 21 chromosomes (http://wheat.pw.usda.gov/GG2/index.shtml) and 17 previously reported markers closely linked to SR resistance genes (Table 1). Wheat lines or cultivars with known genes were used as controls for identifying the correct PCR fragment sizes of each marker. PCR products were analyzed in an ABI DNA Analyzer (Applied Biosystems, Foster City, CA), and marker data were scored using GeneMarker version 1.6 (SoftGenetics LLC, State College, PA) and manually checked twice for accuracy. Alleles from each marker were recorded following Breseghello et al. [27]. Marker alleles were named by a combination of primer name and target fragment size (bp). For example, *Xscm9-241* is the marker allele for the 1B/1R translocation, where Xscm9 is the primer name and 241 is the fragment size in bp including the 18 bp M13 sequence. The number of alleles recorded for each primer is listed in Table S2. Fifty-eight percent of amplified SSR alleles were at less than 5% frequency and so were excluded from the association analysis. The number of remaining alleles for analysis was 1042.

Stem rust evaluation

All wheat accessions were evaluated for seedling resistance to races QFCSC, QTHJC, RCRSC, RKQQC, TPMKC, TTTTF and TTKSK in a greenhouse and for adult plant resistance to bulked U.S. races (QFCSC, QTHJC, RCRSC, RKQQC and TPMKC) in the field at the USDA Cereal Disease Laboratory in St. Paul, MN in 2008. Field disease ratings were based on the percentage infection of the stems using the modified Cobb scale [28] when susceptible controls reached 60–70% severity. Seedling IT was scored using the Stakman scale [24]. Details on plant culture, inoculation methods, and scoring methods for the greenhouse and field experiments were described [29].

To meet the data format required for association analysis, original seedling IT data were converted to a 0–9 linear disease scale as we described in a preliminary report [30]. Simple infection types were converted as follows: 0, 1−, 1, 1+, 2−, 2, 2+, 3−, 3 and 3+ were coded as 0, 1, 2, 3, 4, 5, 6, 7, 8 and 9, respectively. For lines with heterogeneous reactions, only the most prevalent IT was used. The semicolon symbol for hypersensitive fleck ";" was converted to 0. IT 4 was converted to 9. Special annotation code "S" for susceptible was converted to 9 and "S LIF" for low infection frequency was converted to 8. Special annotation codes "C" for extra chlorosis and "N" for extra necrosis were ignored. Double minus and double plus annotations were converted to single minus and single plus, respectively. Complex ranges such as ;12+ were first collapsed to ;2+. Then the first and last ITs of the range were converted and averaged with the first IT being double-weighted because the most prevalent IT is always listed first. Mesothetic reaction types X−, X, and X+ were converted to linearized scores of 4, 5, and 6, respectively. Y and Z mesothetic infection types were treated similarly to X. The conversion algorithm is implemented with examples as an editable Excel spreadsheet in Table S3. Each IT score was based on one replication comprising five to six seedlings per isolate, except for TTKSK, in which two replications were used for each accession and a mean value was used for association analysis.

Gene postulation

Named stem rust resistance genes present in each accession were postulated by the presence of diagnostic markers from published reports (Table 1). Expected ITs [10,14,31] and virulence/avirulence relationships were subsequently compared to observed ITs. For the purpose of gene postulation, the lower IT was assumed to be correct when ITs were heterogeneous. When

Table 1. List of markers associated with rust resistance genes, assigned chromosome, and number of alleles detected for each marker across 137 U.S. wheat accessions.

Entry	Marker	Chr.	No. of alleles	Gene	Positive Control	References
1	*XcsSr2*	3BS	3	*Sr2*	Scout 66	[57]
2	*Xsr2stm559*	3BS	8	*Sr2*	Scout 66	[58]
3	*XSr2X3B028F08*	3BS	2	*Sr2*	Scout 66	[59]
4	*Xcfa2019*	7AL	6	*Sr22*	Sr22Tb	[60]
5	*XSr24#12*	3DL	2	*Sr24*	Jagalene	[61]
6	*XSr24#50*	3DL	2	*Sr24*	Jagalene	[61]
7	*Xbarc71*	3DL	7	*Sr24*	Jagalene	[61]
8	*XSr26#43*	6AL	0	*Sr26*		[61]
9	*Xscm9*	1B/1A[a]	2	*Sr31, Sr1RS^{Amigo}*	Amigo	[45]
10	*Xgwm319*	2BS	3	*Sr36/Sr40*	Vista	[46]
11	*Xgwm374*	2BS	6	*Sr40*	RL6088	[62]
12	*Xwmc477*	2BS	6	*Sr36/Sr39/Sr40*	Vista	[46,62]
13	*Xwmc474*	2BS	14	*Sr40*	RL6088	[62]
14	*Xventriup.Ln2*	2AS	2	*Sr38/Yr17/Lr37*	Madsen	[47]
15	*Xcfd43*	2DS	6	*Sr6*		[48]
16	*Xwmc453*	2DS	12	*Sr6*		[48]
17	*Xgwm484*	2DS	24	*Sr6*		[48]

[a]*Xscm9* acts as a rye-specific SSR marker with two fragments amplified, 225 bp and 241 bp.
A fragment of 225 bp (forward primer tailed) indicates the T1RS•1BL chromosome, and resistance gene *Sr31* and 241 bp indicates the T1RS•1AL chromosome and gene *Sr1RS^{Amigo}*.

observed infection types were substantially higher than expected infection types, the postulated resistance gene was assumed to be absent. When the observed infection types were lower than expected infection types, presence of an additional unknown gene(s) was postulated and indicated by a "+".

Association analysis

Population structure (Q) was determined by STRUCTURE 2.2 [32] using 42 genome-specific markers across all arms of the 21 chromosomes. Six independent runs were conducted using the admixture model by assuming that individuals might have mixed ancestries. Then k, the number of subpopulations, ranging from 2 to 10 was evaluated using a burn-in length of 2×105 and run length of 1×105. The maximum likelihood of each k value, the variance among 10 runs and the pedigree information of each line were used to determine the optimal number of subpopulations. Individuals were assigned to a subpopulation if confidence estimated by the program was at least 0.5; if the confidence value was below 0.5, a combination of information on geographical origin, market class, and breeding history was considered to assign them to a reasonable subpopulation. Population structure information matrix Q (n×k), where n is the number of accessions assayed and k is the number of subpopulations defined, was used in association analyses.

The comparison among models was conducted following Yu et al. [33]. Pair-wise kinship (K) coefficients among the 137 accessions were estimated using two types of algorithms, 'KL' proposed by [34] and 'KR' suggested by [35,36] using the program SPAGeDi ver. 1.2 [37]. For these two algorithms, KR gives more weight to rare alleles and provides more power to detect genetic structure, whereas KL does not suffer bias in the presence of low-frequency alleles but provides low power to detect structure.

TASSEL version 2.1 [38] was used for model selection based on the 271 markers, excluding the 18 previously reported markers linked to stem rust resistance genes. The EMMA algorithm [39] and 'P3D' [40] were set during the process, then the *P*-values observed from each model were aligned against the expected *P*-values. The expected *P*-values were calculated as $r(x_m)/271$, where $r(x_m)$ is the rank of the *P*-value x_m observed for the *m*th marker locus. A mean of the squared differences (MSD) between observed and expected *P*-values of all marker loci was calculated as a measure for the deviation of the observed *P*-values from the expected distribution. A high MSD value indicated a high rate of empirical type I error [41]. The model with the smallest MSD was used for final association analysis.

Association analysis was conducted using PROC MIXED in SAS (SAS Institute, Cary, NC). Marker alleles with a frequency lower than 5% were excluded for calculation. The threshold for claiming significance of associations was set to $P<0.001$. A distance-based cluster analysis was conducted using PowerMarker v. 3.25 [42] and the unweighted pair group method with arithmetic mean (UPGMA) based on Nei distance [43].

If a chromosome region had more than three significant markers associated with a given trait, linkage disequilibrium (LD) was evaluated according to the frequency of target alleles using TASSEL 2.1 (http://www.maizegenetics.net/tasselx) with 1000 permutations. Marker order and genetic distance between markers on a wheat chromosome were adopted from previously established consensus maps [44].

Results

Stem rust resistance of U.S. winter wheat

Significant variation in the responses to different rust races was observed among the U.S. wheat accessions. At the seedling stage, a relatively high proportion of the accessions was resistant (converted scale values of 0 to 4) to QFCSC (58.4%) and QTHJC (40.1%), with at least one-third of accessions showing flecks (converted scale 0). For races RCRSC, RKQQC and TPMKC, a lower proportion of accessions (27–37%) showed resistance, with 12–17% of accessions having flecks. About 53% and 64% accessions were highly susceptible (IT>3) to races TTKSK and TTTTF, respectively, and less than 10% of accessions showed flecks. In the field experiment, about 25% of accessions showed negligible symptoms of rust infection, and 62% of accessions showed 40% or lower SEV to U.S. bulked races in adult plants. A total of 28 accessions showed resistance to all races tested in both seedling and adult stages.

Population structure and statistical model comparison

Structure analysis identified a high level of population structure in the association-mapping panel, and four was the optimal number of subpopulations, with three HWW and one SWW subpopulations. Further details about the population structure are listed by Zhang et al. [25]. The QK model had the smallest MSD values for all disease measurements in model tests, and thus provided the best control of the false positive rate among all models tested. Because QK^L had slightly smaller MSD value than QK^R for some measurements, it was selected for further association analysis.

Detection of known stem rust resistance genes

All races used in this study are avirulent to *Sr24*. Marker allele *XSr24#50-212* for *Sr24* was associated with resistance to all races except TPMKC (Table 2). All 13 accessions carrying the allele were uniformly resistant or moderately resistant to the six U.S. races, including TPMKC, and TTKSK, except that 'Wesley' and possibly 'TX03A0563' appeared to be phenotypically heterogeneous for *Sr24* (Table 2). The marker for *Sr24* was present in each of the four subpopulations (Table S1).

Marker alleles *Xscm9-241* and *Xscm9-225* are diagnostic for $Sr1RS^{Amigo}$ that resides on wheat-rye translocation T1AL·1RS and for *Sr31* on translocation T1BL·IRS, respectively [45]. *Xscm9-241* was associated with resistance to all races except QTHJC (Table 2). Twelve accessions with $Sr1RS^{Amigo}$ showed resistance in both seedling and adult stages, but 5 accessions appeared to be phenotypically heterogeneous. *Xscm9-225* for *Sr31* was significantly associated with resistance to all races except TTKSK (Table 2). Seventeen accessions carrying *Xscm9-225* were uniformly resistant to all tested U.S. races and four appeared to be heterogeneous (Table S1). *Xscm9-225* was the marker with the lowest $\log_{10}P$-value for seedling resistance to races RCRSC, TPMKC and TTTTF and for adult plant resistance.

Among the seven races tested, only QFCSC, QTHJC and TTKSK are avirulent to *Sr36*. Marker alleles *Xwmc477-176* and *Xgwm319-182* on chromosome 2B were reported to be diagnostic for *Sr36* [46]. In this study, both markers had identical results and were significantly associated with resistance to the three races and bulked U.S. races in the field (Table 2). Twelve of 15 accessions that carried the positive alleles showed near immunity to the three races at the seedling stage and high resistance to bulked U.S. races (Table S1). Accession 'G61505' was positive for both markers but susceptible to all races, indicating that *Sr36* was absent. 'GA991209-6E33' and 'India Exp.' carried the marker for *Sr31* in addition to the two markers for *Sr36* and were resistant to most races, but not to TTKSK, suggesting the absence of a functional *Sr36*. Thus, the two markers for *Sr36* produced false positives in the three accessions. All of the accessions with *Sr36* marker alleles for resistance were SWW.

Races QFCSC, QTHJC, RCRSC, RKQQC, and TPMKC are avirulent, and the others are virulent on *Sr38*. *Xventriup.ln2* is a marker linked to the *Sr38/Lr37/Yr17* gene cluster [47]. The marker was associated with resistance to QFCSC, QTHJC and RKQQC (Table 2). It was also associated with resistance to bulked isolates in the field. Twenty-four accessions exhibited uniform resistance, and two appeared to be heterogeneous for *Sr38* (Table S1).

Races QFCSC, RCRSC, and TPMKC are avirulent, and the others are virulent on *Sr6*. *Xcfd43-213* and *Xwmc453-130* linked to *Sr6* [48] showed significant association with resistance to QFCSC and TPMKC (Table 2). Twenty accessions had positive alleles for *Sr6*, and most showed high resistance to QFCSC and TPMKC; however, two accessions appeared to be heterogeneous, and three appeared to lack the phenotype for *Sr6*.

Marker *csSr2* for *Sr2* did not show significant association with resistance to any races in the association analysis, but it was detected in CO02W237, 'Snowmass', 'Thunder CL', 'Tiger' and 'Scout 66'. *Sr2* was present in two HWW subpopulations. Scout 66, an old HWW cultivar from Nebraska, is known to possess *Sr2* [10]. Tiger from Kansas State University and the other three from Colorado State University are newer HWW cultivars/lines. Two other markers for *Sr2* were less diagnostic than *csSr2*, and thus also not significant for any races tested.

Novel marker associations with resistance to TTKSK

Seven marker alleles (*Xgwm148-127*, *Xbarc91-null*, *Xwmc474-141*, *Xgwm374-193*, *Xgwm120-null*, *Xgwm47-163* and *Xwmc332-165*) on chromosome 2B were associated with seedling resistance to TTKSK, QFCSC and QTHJC. These seven closely linked markers showed significant linkage disequilibrium (LD) with diagnostic markers *Xwmc477-176* and *Xgwm319-182* for *Sr36* (Fig. 1). In most cases, the seven linked markers identified the same positive lines as the markers for *Sr36* (Table 3).

Marker allele *Xbarc181-194* on 1B was significantly associated with resistance to TTKSK and occurred only in the HWW accessions (Table S1). Among 13 accessions carrying this allele, three were susceptible to TTKSK; two had missing or contradictory phenotypic data; five lines also carry $Sr1RS^{Amigo}$ and the remaining three, CO03W043, CO03W139 and CO03064, had an IT of 2 to 2++ for TTKSK without carrying any known resistance gene.

Marker allele *Xgwm495-182* on 4BL was significantly associated with resistance to TTKSK (Table 2, Table S1). Five of the eight accessions carrying *Xgwm495-182* had no known gene for resistance to TTKSK. Two lines had *Sr36* and one had both *Sr24* and $Sr1RS^{Amigo}$.

Xbarc239-301 on 5DL was significantly associated with resistance to TTKSK, but nine of 15 positive accessions carried one or two other effective genes (Table S1). Three positive accessions were susceptible to all races, and two were susceptible to TTKSK. The lack of consistent association with resistance suggests that this association is spurious.

Novel marker associations with resistance to other races

Xgwm334-123 on 6AS was present in nine SRW lines and was associated with unexplained resistance to QFCSC and QTHJC in six accessions (Table S1). *Xgwm160-195* on 4AL and *Xwmc622-147* on 4DL were associated with resistance to RCRSC, RKQQC and TTTTF, and accessions with *Xwmc622-147* showed the lowest mean IT for TTTTF compared with all other marker alleles detected for this race. *Xgwm624-146* was associated with resistance to RKQQC. *Xwmc313-225* is tightly linked to *Xgwm160-195* and was also associated with resistance to RCRSC.

Table 2. Markers associated with known stem rust resistance genes and newly detected loci, and mean rust ratings of accessions carrying these marker alleles after inoculation with seven stem rust races at the seedling stage and bulked U.S. races at the adult stage.

	Allele	Location	Resistance Gene[b]	Seedling IT[d]							Adult plant
				QFCSC	QTHJC	TTKSK	RCRSC	RKQQC	TPMKC	TTTTF	Severity, %
Marker for known genes[a]	*SrXscm9-241*	T1RS·1AL	*$Sr1RS^{Amigo}$*	2.64	-	4.64	3.91	3.18	3.27	4.10	13.2
	Xscm9-225	T1RS·1BL	*Sr31*	2.14	1.75	-	2.81	3.85	2.39	3.10	13.0
	Xventriup.Ln2	2AS	*Sr38*	0.72	1.12	-	-	2.00	-	-	5.2
	Xwmc477-176	2BS	*Sr36*	0.64	1.20	1.73	-	-	-	-	15.5
	Xcfd43-213	2DS	*Sr6*	1.32	-	-	-	-	1.00	-	-
	Xsr24#50-212	3DL	*Sr24*	2.00	3.07	4.19	3.77	3.54	-	4.30	-
Newly detected marker associations	*Xbarc181-194*	1BL		-	-	6.14	-	-	-	-	-
	Xgwm95-133	2AS		-	0.25	-	2.38	-	-	3.13	-
	Xwmc702-203	2AS		-	2.00	-	-	-	3.28	-	12.6
	Xwmc326-203	3BL		-	-	-	-	-	-	-	16.4
	Xgwm383-213	3DL		-	-	-	2.38	-	-	-	-
	Xgwm160-195	4AL	*Sr7*	-	-	-	2.75	3.91	-	4.82	-
	Xwmc313-225	4AL	*Sr7*	-	-	-	3.40	-	-	-	-
	Xgwm495-182	4BL		-	-	4.25	-	-	-	-	-
	Xwmc622-147	4DL		-	-	-	2.86	3.57	-	2.43	-
	Xgwm 624-146	4DL		-	-	-	-	3.63	-	-	-
	Xgwm540-143	5BS		-	-	-	-	-	2.08	-	-
	Xbarc 239-301	5DL		-	-	4.47	-	-	-	-	-
	Xgwm334-123	6AS	*Sr8*	0.5	1.44	-	-	-	-	-	-
Negative alleles[c]	*Xgwm11-208*	1BS		-	-	-	8.18	8.24	7.19	8.76	50.3
	Xwmc116-385	7AL		-	-	-	7.43	7.79	7.21	-	-
	Xbarc91-144	2BS		6.13	6.40	8.67	-	-	-	-	50.0

[a]Means for seedling IT were calculated from the transformed 0 to 9 scale.
Only significant markers are shown ($P<0.001$).
[b]Postulated gene based on diagnostic marker or chromosome location.
[c]Not all negative alleles associated with susceptibility are listed.
[d]QFCSC (virulence/avirulence formula 5, 8a, 9a, 9d, 9g, 10, 17, 21, McN/6, 7b, 9b, 9e, 11, 24, 30, 31, 36, 38, Tmp, *$Sr1RS^{Amigo}$*), QTHJC (5, 6, 8a, 9b, 9d, 9g, 10, 11, 17, 21, McN/7b, 9a, 9e, 24, 30, 31, 36, 38, Tmp, *$Sr1RS^{Amigo}$*), RCRSC (5, 7b, 9a, 9b, 9d, 9g, 10, 17, 21, 36, McN/6, 8a, 9e, 11, 24, 30, 31, 38, Tmp, *$Sr1RS^{Amigo}$*), RKQQC (5, 6, 7b, 8a, 9a, 9b, 9d, 9g, 21, 36, McN/9e, 10, 11, 17, 24, 30, 31, 38, Tmp, *$Sr1RS^{Amigo}$*), TPMKC (5, 7b, 8a, 9a, 9d, 9e, 9g, 10, 11, 17, 21, 36, Tmp, McN/6, 9a, 9b, 24, 30, 31, 38, *$Sr1RS^{Amigo}$*), TTTTF (5, 6, 7b, 8a, 9a, 9b, 9d, 9e, 9g, 10, 11, 17, 21, 30, 36, 38, Tmp, McN/24, 31, *$Sr1RS^{Amigo}$*), and TTKSK (5, 6, 7b, 8a, 9a, 9b, 9d, 9e, 9g, 10, 11, 17, 21, 30, 31, 38, McN/24, 36, Tmp, *$Sr1RS^{Amigo}$*).

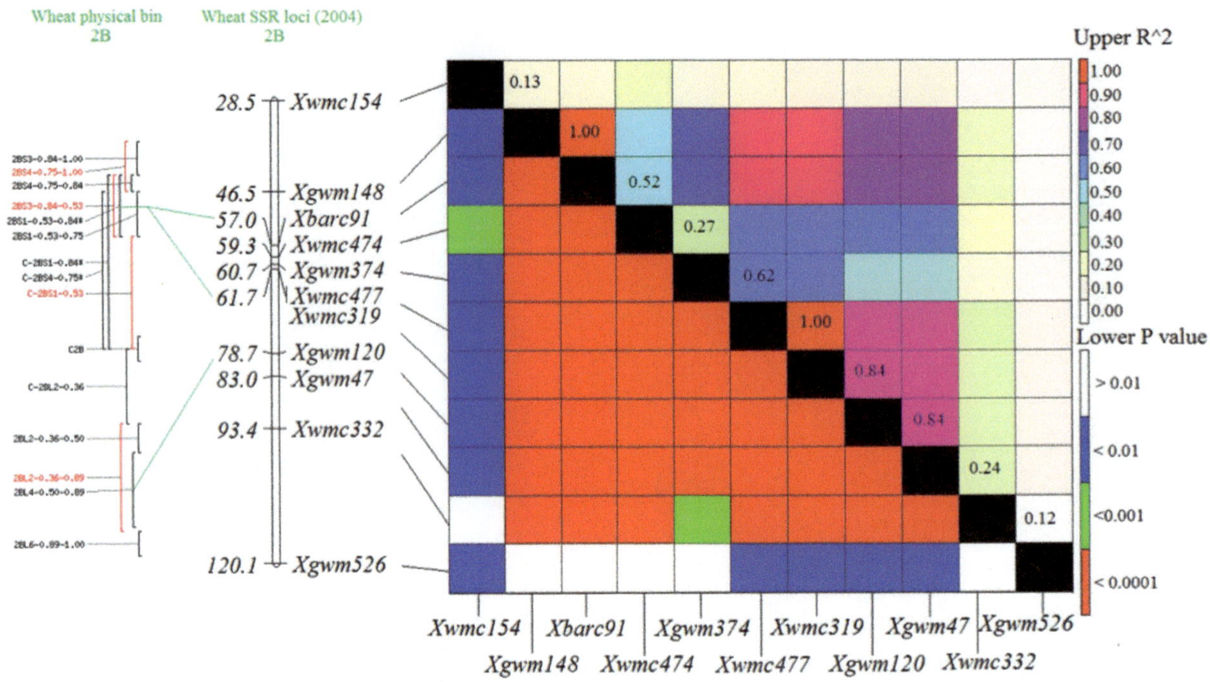

Figure 1. Physical bin map, linkage map, and linkage disequilibrium (LD) among 11 markers associated with resistance to race TTKSK on chromosome 2B. The upper diagonal indicates the LD level between markers reflected by R^2, and the bottom diagonal indicates the statistical significance of LD between markers as reflected by *P*-values.

Xwmc702-203 on 2AS was associated with resistance to QTHJC and TPMKC. Eight accessions carrying *Xgwm95-133* on chromosome 2AS were resistant to QTHJC, RCRSC and TTTTF, and this marker appeared to be linked to the allele with the highest level of resistance to QTHJC (IT '0;'). *Xwmc326-203* on 3BL was only associated with adult plant resistance in the field.

Several other markers showed significant association with resistance, but most of the accessions that carry these markers also carry other known resistance genes. For example, eight accessions with the *Xgwm383-213* marker allele on chromosome 3DL had the lowest average IT ('0;' to '2') to RCRSC, but six of them also carried either *Sr31* or *Sr24* (Table S1). Fourteen accessions carry *Xgwm540-143* on chromosome 5B, and seven of them had the markers for *Sr6*.

Several significant marker alleles were associated with high rust susceptibility (Table 2). For example, accessions with *Xwmc116-385* on chromosome 7A had an average IT higher than '3;' for RCRSC, RKQQC and TPMKC. Six accessions with *Xgwm11-208* allele on chromosome 1B showed high susceptibility to RCRSC, TTTTF, TPMKC and RKQQC in seedling and adult stages.

Discussion

Association mapping using the seedling IT linearization method described here (Table S3) successfully detected *Sr6, Sr24, Sr31, Sr36, Sr38*, and *Sr1RS*^{Amigo} in U.S. winter wheat lines (Table 2). Utilization of seven isolates with known race specificities and previously published markers for these resistance genes allowed estimation of error rates for 42 marker-phenotype associations. For 11 instances where positive marker associations were not expected to occur because races were virulent on the particular resistance gene, none were significantly associated with resistance. For 31 instances where a positive marker association was expected because races were avirulent on the resistance genes, 26 associations were significant. Although the number of tests was relatively small, the results demonstrate the utility of association mapping with linearized ITs. Letta et al. [22] also used our IT linearization algorithm and similar association analyses to successfully map stem rust seedling resistance in durum wheat. The Stakman IT scale [24] is very useful for precise qualitative descriptions of rust resistance phenotypes and is routinely used to score rust reactions of experimental lines and characterize specific resistance genes. However, the system allows nonlinear or compound ITs such as X+, ;1N, or 13- that are not amenable to quantitative analysis. The linearization algorithm allows qualitative IT data to be converted for quantitative analysis.

Expected ITs based on marker genotypes were compared with actual resistance phenotypes to assess the prediction reliability of the markers. After accounting for heterogeneity of some wheat lines, the genotypic and phenotypic data showed excellent agreement for *Sr24*, *Sr31*, *Sr38*, and *Sr1RS*^{Amigo} (Table S1). Each of these genes is on a non-recombining alien chromosome segment and markers were confirmed to be diagnostic in this U.S. winter wheat panel.

Marker *Xwmc477* was reported to be completely linked with *Sr36* on chromosome arm 2BS in two populations [46]. *Xgwm319* was also tightly linked at 0.9 cM distant in one population and completely linked in the other. In our study, marker alleles *Xwmc477-176* and *Xgwm319-182* for *Sr36* were positive for 15 lines. However, ITs indicated that *Sr36* was not present in 3 of the 15 lines (Table S1). Association mapping identified seven additional markers on 2B that were in linkage disequilibrium with markers for *Sr36* (Fig. 1). Based on similar ITs and race specificity (Table S1) and similar haplotypes among the lines (Table 3), the alien chromosome segment from *Triticum timopheevii* carrying *Sr36* was sufficient to explain the resistance associations on 2B. Susceptible lines 'G61505' and 'India Exp.' carried six or seven

Table 3. Phenotypes and haplotypes of accessions that carry positive alleles on chromosome 2B associated with resistance to TTKSK at the seedling stage.

Accession	Class	Gene postulation[1]	TTKSK-1	TTKSK-2	QFCSC	QTHJC	Xgwm148-127	Xbarc91-Null	Xwmc474-141	Xgwm374-193	Xwmc477-176	Xgwm319-182	Xgwm120-Null	Xgwm47-163	Xwmc332-165
G69202	SRW	Sr36, +	0	0	0;	0	+	+	+	+	+	+	+	+	+
P03112A1-7-14	SRW	Sr36, Sr38	0	0	0	0	+	+	+	−	+	+	+	+	+
INW0411	SRW	Sr31, Sr36$	0	0/;2	0	0	+	+	+	−	+	+	+	+	+
VA02W-555	SRW	Sr31, Sr36	0	0	0	0	+	+	+	−	+	+	+	+	+
VA04W-259	SRW	Sr36	0	0;	0	0	+	+	+	+	+	+	+	+	+
NC03-6228	SRW	Sr31$, Sr36	0	0	0	0	+	+	+	+	+	+	+	+	+
AR96077-7-2	SRW	Sr36	0	;1+	0	0	+	+	+	+	+	+	+	+	+
G61505	SRW	Sr36#	S	S	S	S	+	+	+	+	+	+	+	+	+
P04287A1-10	SRW	Sr36, +	2;	0;	0	0	+	+	−	−	+	+	+	+	+
G41732	SRW	Sr36, +	0	0	0	0/S	+	+	−	+	+	+	+	+	+
India Exp.	SRW	Sr31$, Sr36#	S	S	S/2	S/2-	+	+	−	+	+	+	+	+	+
NC04-15533	SRW	Sr36	0	0;	0;/;1	0	+	+	−	+	+	+	+	+	+
B030543	SRW	Sr36$	0	0/S	;	;	+	+	−	+	+	+	−	−	−
MD99W483-06-9	SRW	Sr31, Sr36	0	0	0	0	−	−	+	−	+	+	+	+	+
GA991209-6E33	SRW	Sr31, Sr36#, +	2+	2+ LIF	0	0	−	−	−	−	+	+	−	−	−

[1]"+" denotes extra resistance that could not be attributed to postulated genes in the lines for at least one of seven races; "#"denotes putative nonfunctional alleles where phenotypes did not confirm gene postulations based on diagnostic markers; and "$" denotes that the functional allele appeared to be heterogeneous.
"S" denotes susceptible infection type (IT) 3 or 4; "/" denotes heterogeneous, the predominant type given first; "LIF" denotes low infection frequency with fewer pustules.

positive alleles from this LD block, in addition to *Xwmc477-176* and *Xgwm319-182*. Therefore, the *Sr36* haplotype appears to be intact and loss of *Sr36* gene function is likely. In contrast, susceptible line 'GA991209-6E33' had the negatively associated alleles at seven markers, in addition to positive alleles at *Xwmc477-176* and *Xgwm319*-182. This suggests that the alien translocation segment was disrupted in this line. Tsilo et al. [46] found a susceptible line, 'CK9877', that showed a negative allele for *Xwmc477*, but positive for *Xgwm319*. Olson et al. [19] reported that the positive allele of *Xwmc477* was associated with resistance in 54 of 57 cases. They attributed the remaining three susceptible lines to recombination or heterogeneity of the seed sources. *Xwmc477* appears to be the best marker for *Sr36*, but it appears not to be completely diagnostic.

Markers *Xcfd43* and *Xwmc453* for *Sr6* were tightly linked and diagnostic for *Sr6* in a diverse set of 46 wheat lines [48]. In the present study, marker alleles *Xcfd43-213* and *Xwmc453-130* were both positive for 20 accessions, but ITs indicated that *Sr6* was not present in 3 of the 20 lines, including 'NE05430', 'NE05496', and 'Trego'. *Sr6* is on chromosome arm 2DS from common wheat and appears to have normal recombination rates [48]. The false positives are likely due to recombination between the resistance gene and the markers, which are 1.1 to 1.5 cM distant from *Sr6* [48].

Association mapping identified three potentially novel marker associations with resistance to race TTKSK for *Xbarc181-194* (1BL), *Xgwm495-182* (4BL), *Xbarc239-301* (5DL), and one with susceptibility to TTKSK for *Xbarc91-144* (Table 1). Njau et al. [49] mapped a stem rust resistance QTL in 'Pavon 76' to 1BL and inferred that it was the pleiotropic APR gene *Lr46/Yr29/Pm39*. However, that QTL was centered on *Xbarc80*, which is more than 50 cM distal from *Xbarc181*. Letta et al. [22] used association mapping to locate a stem rust resistance QTL near *Xbarc8* on 1BL in durum and suggested it was likely *Sr14*. However, *Xbarc8* is more than 10 cM proximal to *Xbarc181* and *Sr14* is not deployed in hexaploid wheat [10]. All 13 positive lines for *Xbarc181-194* were HWW and many were related to 'TAM 107' and/or Colorado experimental lines. Four Colorado lines showed good resistance to TTKSK, but contained no known resistance gene based on marker genotypes. They were previously postulated to carry *SrTmp* based on race specificity and infection phenotype (https://www.ars.usda.gov/SP2UserFiles/ad_hoc/54402000HardWinterWheatRegionalNurseryProgram/08SRPN.xls). Three of the four lines were positive for *Xbarc181-194*, which suggested that *Xbarc181-194* was associated with *SrTmp*. However, a gene thought to be *SrTmp* was recently mapped in the Colorado line 'Ripper' on 6DS [50]. It is therefore possible that the association with *Xbarc181-194* on 1BL is a spurious correlation between *SrTmp* and a marker that happens to be common in the same lines. In the present study, marker allele *Xgwm495-182* was associated with otherwise unexplained resistance to TTKSK in five lines. The magnitude of the effect of this locus appeared to be similar to *Sr24*, but it was not effective against other races (Table 2). Bhavani et al. [51] reported that a seedling stem rust resistance gene, temporarily designated *SrNing*, mapped near *Xgwm149* and *Xgwm495* on 4BL. *Xbarc239-301* on 5DL was also significantly associated with resistance to TTKSK, but most of the lines with *Xbarc239-301* carry either one or two known effective genes or are susceptible to TTKSK. Thus, the association with resistance is probably spurious. *Xbarc91-144* was associated with higher susceptibility to TTKSK. A null allele, *Xbarc91-null*, was part of the *Sr36* linkage block on 2BS (Fig. 1, Table 3). It is likely that *Xbarc91-144* detected the absence of *Sr36*. Therefore, the only novel association with seedling resistance to TTKSK that appears to be promising is *Xgwm495-182* on 4BL. Further work is needed to verify this marker association and possible relationship to *SrNing*.

Association mapping identified markers associated with resistance to races other than TTKSK on 2AS, 3BL, 3DL, 4AL, 4DL, 5BS, and 6AS, while markers were associated with susceptibility on 1BS and 7A (Table 2). *Xwmc702-203* and *Xgwm95-133* are tightly linked on 2AS and loosely linked to *Sr38*. Five of eight positive lines for *Xgwm95-133* and 10 of 19 positive lines for *Xwmc702-203* also carried *Sr38*, which may account for the association with resistance. However, *Xgwm95-133* was associated with resistance to TTTTF, which is virulent on *Sr38*. This suggests that *Xwmc702-203* and *Xgwm95-133* could be associated with a novel resistance gene on 2AS. Twenty-nine lines carried *Xwmc326-203* on 3BL, which was associated with resistance at the adult stage only. *Xwmc326-203* is distal on 3BL and is unlinked to the *Sr2* APR gene on 3BS. This marker was interesting because it was associated with stem rust APR in the winter wheat landrace variety, Kharkof. Markers *Xgwm160-195* and *Xwmc313-225* are tightly linked at the distal end of 4AL near *Sr7*. The race specificity is not consistent with allele *Sr7b* in the differential set, but the markers could be associated with a different allele of *Sr7*. *Xgwm383-213* on 3DL and *Xgwm540-143* on 5BS commonly occurred with other known genes and their effects are probably spurious. *Xwmc622-147*, which is 19 cM proximal to *Xgwm624-146* on 4DL, was associated with resistance to three different races. Both markers are distal to the pleiotropic locus *Lr67/Yr46/Sr55* on 4DL [52]. *Xwmc622-147* was interesting because it was associated with strong resistance to TTTTF. Marker allele *Xgwm334-123* was associated with resistance to two races. It is located at the tip of 6AS near *Sr8*. The race specificity was not consistent with *Sr8a*, so it could be associated with a different allele of *Sr8*. Marker *Xgwm11-208* on 1BS was associated with susceptibility and likely indicates the absence of *Sr31* and/or $Sr1RS^{Amigo}$. Marker *Xwmc116-385* on distal 7AL was associated with higher susceptibility to three races, but the explanation is unclear. The most promising novel marker associations for races other than TTKSK are a possible APR gene near *Xwmc326-203* on 3BL, *Xgwm160-195* and *Xwmc313-225* on 4AL near *Sr7*, *Xwmc622-147* and *Xgwm624-146* on 4DL, and *Xgwm334-123* on 6AS near *Sr8*.

The impetus for this study was to assess the complement of stem rust resistance genes in U.S. winter wheat accessions from regional cooperative nurseries. Nineteen accessions (7%) were postulated to have no resistance genes for stem rust (Table S1). After correcting for false positives, frequencies of stem rust resistance genes in the U.S. winter wheat panel were *Sr2* (4%), *Sr6* (12%), *Sr24* (9%), *Sr31* (15%), *Sr36* (9%), *Sr38* (19%), and $Sr1RS^{Amigo}$ (8%). *Sr2* was found only in HWW and *Sr36* was found only in SWW. Fifty-two accessions (38%) were postulated to have some degree of additional unexplained resistance to one or more races. *SrTmp* was previously postulated to be present in four lines in the panel, but the frequency of *SrTmp* could not be determined because our associated marker was questionable. Association mapping yielded markers on 3BL, 4AL, 4BL, 4DL, and 6AS that may be associated with additional resistance genes, but all of them need to be validated.

Our results were in general agreement with Jin and Singh [17], Olson et al. [19], and Yu et al. [18]. The biggest difference was that *Sr38* from *Aegilops ventricosa* was found to be the most prevalent stem rust resistance gene in both hard and soft U.S. winter wheat. *Sr38* may have been overlooked previously because the phenotype is often confusing. Jin et al. [31] listed the IT of *Sr38* as ;23 and McIntosh et al. [10] listed the IT as X with larger pustules toward the leaf base [10]. The commonly used *Sr38* differential, 'Trident', often shows ITs of 0; to ;1, which suggests that it carries an additional gene. In the present study, lines putatively carrying only *Sr38* typically had seedling ITs of 0;, ;, ;1, ;13-, ;13, ;3, or 3; (Table S1). To account for the mesothetic reaction and pattern of larger pustules at the base, the typical IT for *Sr38* would best be scored as Z according to the Stakman scale. Fortunately, the marker *Xventriup.Ln2* appears to be very diagnostic for *Sr38*. *Sr38* does not provide protection against TTTTF or the TTKSK group of races, but it is effective against other North American races, especially at the adult stage. *Sr38* was the most effective SR gene in the field test. The average severity for lines carrying only *Sr38* was 5.2%, which is less than half of the value for the next most effective gene, *Sr31* (Table 2). *Sr38* is completely linked with other valuable traits like resistance to leaf rust (*Lr37*) and stripe rust (*Yr17*) that help to explain the prevalence of the segment in U.S. winter wheat breeding lines [47].

There were several reasons why some SR genes might have remained undetected in this study. First, the seven races used were all virulent on *Sr5*, *Sr9d*, *Sr9g*, *Sr21*, *SrMcN*, and all but one were virulent on *Sr10* and *Sr17*. Second, resistance alleles with a low frequency might be overlooked in this study because the power to detect an association is a function of allele frequency [53]. That might have affected the ability to detect a significant marker association for *Sr2*, which was present in only 4% of lines. Third, only one set of phenotypic data was available for adult plant field severity. This may have reduced the power to detect APR genes *Sr2* and *Lr34/Yr18/Sr57*, which is known to be present in HWW [54]. Fourth, the number of markers was insufficient for thorough genomic coverage. Although multiple alleles were recorded for most markers, 58% of amplified SSR alleles were at less than 5% frequency and so were excluded from the association analysis. Fifth, the size of the association mapping panel was relatively small. Nevertheless, known SR resistance genes and some potentially new resistance alleles were significantly associated with markers, thus demonstrating that archived qualitative rust infection type data can be linearized and then mined by association mapping.

This study analyzed archived stem rust phenotypic data from cooperative regional winter wheat trials from 2008. More than one half of the wheat accessions in the study were highly susceptible to race TTKSK, with only about 10% of accessions showing a high level of resistance with an IT of '0' to ';'. Most of the effective resistance was attributable to *Sr24*, *Sr36*, and $Sr1RS^{Amigo}$. In the intervening period, all of three of these genes have been defeated by new virulent races from Africa [5,6,55], so the risk of exotic races to U.S. winter wheat remains high. Efforts are underway to combine existing resistance genes with new stem rust genes such as *Sr22*, *Sr26*, and *Sr35* that are effective against the new races [56]. Race TTTTF, which is indigenous to the U.S., is also virulent on all but a few resistance genes. Although it has not yet become prevalent, improved resistance to TTTTF should also be a priority for winter wheat breeders in the U.S. [7].

Supporting Information

Table S1 Linear alignment of neighbor-joining tree, rust rating data, significant positive markers, positive alleles ($P < 0.001$), and two most significant negative alleles.

Table S2 List of 271 markers, assigned chromosome, and number of alleles detected across 174 U.S. wheat accessions.

Table S3 Conversions of Stakman infection types to a linear scale.

Acknowledgments

The authors thank Yue Jin for providing stem rust phenotypic data and phenotype-based gene postulations as well as helpful comments. Mention of trade names or commercial products in this article is solely for the purpose of providing specific information and does not imply recommendation or endorsement by the U.S. Department of Agriculture. USDA is an equal opportunity provider and employer. Contribution number 14-217-J from the Kansas Agricultural Experiment Station.

Author Contributions

Conceived and designed the experiments: GB. Performed the experiments: DZ. Analyzed the data: DZ JY. Contributed reagents/materials/analysis tools: BFC GB JY RLB. Contributed to the writing of the manuscript: DZ RLB GB.

References

1. Singh R (2006) Current status, likely migration and strategies to mitigate the threat to wheat production from race Ug99 (TTKS) of stem rust pathogen. CAB Reviews: Perspectives in Agriculture, Veterinary Science, Nutr and Nat Resour 1: 1–13.
2. Leonard KJ, Szabo LJ (2005) Stem rust of small grains and grasses caused by *Puccinia graminis*. Mol Plant Pathol 6: 99–111.
3. Wanyera R, Kinyua MG, Jin Y, Singh RP (2006) The spread of stem rust caused by *Puccinia graminis* f. sp. *tritici*, with virulence on *Sr31* in wheat in eastern Africa. Plant Dis 90: 113–113.
4. Pretorius ZA, Singh RP, Wagoire WW, Payne TS (2000) Detection of virulence to wheat stem rust resistance gene *Sr31* in *Puccinia graminis*. f. sp. *tritici* in Uganda. Plant Dis 84.
5. Jin Y, Szabo LJ, Pretorius ZA, Singh RP, Ward R, et al. (2008) Detection of virulence to resistance gene *Sr24* within race TTKS of *Puccinia graminis* f. sp. *tritici*. Plant Dis 92: 923–926.
6. Jin Y, Szabo L, Rouse M, Fetch Jr T, Pretorius Z, et al. (2009) Detection of virulence to resistance gene *Sr36* within the TTKS race lineage of *Puccinia graminis* f. sp. *tritici*. Plant Disease 93: 367–370.
7. Jin Y (2005) Races of *Puccinia graminis* identified in the United States during 2003. Plant Dis 89: 1125–1127.
8. Singh RP, Hodson DP, Huerta-Espino J, Jin Y, Njau P, et al. (2008) Will stem rust destroy the world's wheat crop? In: Donald LS, editor. Advances in Agronomy: Academic Press. pp. 271–309.
9. Singh RP, Hodson DP, Huerta-Espino J, Jin Y, Bhavani S, et al. (2011) The emergence of Ug99 races of the stem rust fungus is a threat to world wheat production. Ann Rev Phytopath 49: 465–481.
10. McIntosh RA, Wellings CR, Park RF (1995) Wheat Rusts: an Atlas of Resistance Genes: CSIRO Publishing.
11. Herrera-Foessel SA, Singh RP, Lillemo M, Huerta-Espino J, Bhavani S, et al. (2014) *Lr67/Yr46* confers adult plant resistance to stem rust and powdery mildew in wheat. Theor Appl Genet 127: 781–789.
12. Mago R, Zhang P, Bariana H, Verlin D, Bansal U, et al. (2009) Development of wheat lines carrying stem rust resistance gene *Sr39* with reduced *Aegilops speltoides* chromatin and simple PCR markers for marker-assisted selection. Theor Appl Genet 119: 1441–1450.
13. Dundas IS, Anugrahwati DR, Verlin D, Park R, Bariana H, et al. (2007) New sources of rust resistance from alien species: meliorating linked defects and discovery. Crop Past Sci 58: 545–549.
14. Roelfs A, McVey D (1979) Low infection types produced by *Puccinia graminis* f. sp. *tritici* and wheat lines with designated genes for resistance. Phytopathology 69: 722–730.
15. Roelfs AP, Groth JV (1980) A comparison of virulence phenotypes in wheat stem rust populations reproducing sexually and asexually. Phytopathology 70: 855–862.
16. McVey DV (1992) Genes for rust resistance in International Winter Wheat Nurseries XII through XVII. Crop Sci 32: 891–895.
17. Jin Y, Singh RP (2006) Resistance in U.S. wheat to recent Eastern African isolates of *Puccinia graminis* f. sp. *tritici* with virulence to resistance gene *Sr31*. Plant Dis 90: 476–480.
18. Yu L-X, Liu S, Anderson JA, Singh RP, Jin Y, et al. (2010) Haplotype diversity of stem rust resistance loci in uncharacterized wheat lines. Mol Breed 26: 667–680.

19. Olson EL, Brown-Guedira G, Marshall DS, Jin Y, Mergoum M, et al. (2010) Genotyping of U.S. wheat germplasm for presence of stem rust resistance genes *Sr24*, *Sr36* and *Sr1RS*^{Amigo}. Crop Science 50: 668–675.
20. Yu LX, Lorenz A, Rutkoski J, Singh RP, Bhavani S, et al. (2011) Association mapping and gene-gene interaction for stem rust resistance in CIMMYT spring wheat germplasm. Theor Appl Genet 123: 1257–1268.
21. Crossa J, Burgueno J, Dreisigacker S, Vargas M, Herrera-Foessel SA, et al. (2007) Association analysis of historical bread wheat germplasm using additive genetic covariance of relatives and population structure. Genetics 177: 1889–1913.
22. Letta T, Olivera P, Maccaferri M, Jin Y, Ammar K, et al. (2014) Association mapping reveals novel stem rust resistance loci in durum wheat at the seedling stage. Plant Gen 7: 1 doi: 10.3835/plantgenome2013.08.0026
23. Letta T, Maccaferri M, Badebo A, Ammar K, Ricci A, et al. (2013) Searching for novel sources of field resistance to Ug99 and Ethiopian stem rust races in durum wheat via association mapping. Theor Appl Genet 126: 1237–1256.
24. Stakman EC, Stewart PM, Loegering W (1962) Identification of physiologic races of *Puccinia graminis var. tritici*. US Department of Agriculture, Agricultural Research Service. pp. Publ. E617.
25. Zhang D, Bai G, Zhu C, Yu J, Carver BF (2010) Genetic diversity, population structure, and linkage disequilibrium in U.S. elite winter wheat. Plant Gen 3: 117–127.
26. Zhang D, Bai G, Hunger RM, Bockus WW, Yu J, et al. (2011) Association study of resistance to soilborne wheat mosaic virus in U.S. winter wheat. Phytopathology 101: 1322–1329.
27. Breseghello F, Sorrells ME (2006) Association mapping of kernel size and milling quality in wheat (*Triticum aestivum* L.) cultivars. Genetics 172: 1165–1177.
28. Roelfs A, Singh R, Saari E (1992) Rust Diseases of Wheat: Concepts and Methods of Disease Management.
29. Rouse MN, Olson EL, Gill BS, Pumphrey MO, Jin Y (2011) Stem rust resistance in *Aegilops tauschii* germplasm. Crop Science 51: 2074.
30. Zhang D, Bowden R, Bai G (2011) A method to linearize Stakman infection type ratings for statistical analysis. In: McIntosh R, editor. Proceedings Borlaug Global Rust Initiative 2011 Technical Workshop. Saint Paul, Minnesota, U.S.A: Borlaug Global Rust Initiative. pp. 170.
31. Jin Y, Singh RP, Ward RW, Wanyera R, Kinyua M, et al. (2007) Characterization of seedling infection types and adult plant infection responses of monogenic *Sr* gene lines to race TTKS of *Puccinia graminis* f. sp. *tritici*. Plant Dis 91: 1096–1099.
32. Pritchard JK, Stephens M, Donnelly P (2000) Inference of population structure using multilocus genotype data. Genetics 155: 945–959.
33. Yu J, Pressoir G, Briggs WH, Vroh Bi I, Yamasaki M, et al. (2006) A unified mixed-model method for association mapping that accounts for multiple levels of relatedness. Nat Genet 38: 203–208.
34. Loiselle BA, Sork VL, Nason J, Graham C (1995) Spatial genetic structure of a tropical understory shrub, *Psychotria officinalis* (Rubiaceae). Am J Bot 82: 1420–1425.
35. Ritland K (1996) Estimators for pairwise relatedness and individual inbreeding coefficients. Genet Res 67.
36. Ritland K (2000) Marker-inferred relatedness as a tool for detecting heritability in nature. Mol Ecol 9: 1195–1204.
37. Hardy OJ, Vekemans X (2002) SPAGeDi : a versatile computer program to analyse spatial genetic structure at the individual or population levels. Mol Ecol Notes 2: 618–620.
38. Bradbury PJ, Zhang Z, Kroon DE, Casstevens TM, Ramdoss Y, et al. (2007) TASSEL: software for association mapping of complex traits in diverse samples. Bioinformatics 23: 2633–2635.
39. Kang HM, Zaitlen NA, Wade CM, Kirby A, Heckerman D, et al. (2008) Efficient control of population structure in model organism association mapping. Genetics 178: 1709–1723.
40. Zhang Z, Ersoz E, Lai CQ, Todhunter RJ, Tiwari HK, et al. (2010) Mixed linear model approach adapted for genome-wide association studies. Nat Genet 42: 355–360.
41. Stich B, Mohring J, Piepho HP, Heckenberger M, Buckler ES, et al. (2008) Comparison of mixed-model approaches for association mapping. Genetics 178: 1745–1754.
42. Liu K, Muse SV (2005) PowerMarker: an integrated analysis environment for genetic marker analysis. Bioinformatics 21: 2128–2129.
43. Nei M (1973) Analysis of gene diversity in subdivided populations. Proc Natl Acad Sci U S A 70: 3321–3323.
44. Somers DJ, Isaac P, Edwards K (2004) A high-density microsatellite consensus map for bread wheat (*Triticum aestivum* L.). Theor Appl Genet 109: 1105–1114.
45. Saal B, Wricke G (1999) Development of simple sequence repeat markers in rye (*Secale cereale* L.). Genome 42: 964–972.
46. Tsilo TJ, Jin Y, James AA (2008) Diagnostic microsatellite markers for the detection of stem rust resistance gene *Sr36* in diverse genetic backgrounds of wheat. Crop Sci 48: 253–261.
47. Helguera M, Khan IA, Kolmer J, Lijavetzky D, Zhong-qi L, et al. (2003) PCR assays for the *Lr37-Yr17-Sr38* cluster of rust resistance genes and their use to develop isogenic hard red spring wheat lines. Crop Sci 43: 1839–1847.
48. Tsilo TJ, Chao S, Jin Y, Anderson JA (2009) Identification and validation of SSR markers linked to the stem rust resistance gene *Sr6* on the short arm of chromosome 2D in wheat. Theor Appl Genet 118: 515–524.
49. Njau PN, Bhavani S, Huerta-Espino J, Keller B, Singh RP (2013) Identification of QTL associated with durable adult plant resistance to stem rust race Ug99 in wheat cultivar 'Pavon 76'. Euphytica 190: 33–44.
50. Lopez-Vera EE, Nelson S, Singh RP, Basnet BR, Haley SD, et al. (2014) Resistance to stem rust Ug99 in six bread wheat cultivars maps to chromosome 6DS. Theoret Appl Genet 127: 231–239.
51. Bhavani S, Singh RP, Huerta-Espino J (2010) Mapping of Ug99 effective stem rust resistance genes *Sr45* and *SrNing* on chromosomes 1DS and 4BL in CIMMYT wheat germplasm. In: Singh D, Jin Y, editors. BGRI 2010 Technical Workshop. St. Petersburg, Russia. pp. 231.
52. Herrera-Foessel S, Singh R, Huerta-Espino J, Rosewarne G, Periyannan S, et al. (2012) *Lr68* : a new gene conferring slow rusting resistance to leaf rust in wheat. Theor Appl Genet 124: 1475–1486.
53. Brachi B, Faure N, Horton M, Flahauw E, Vazquez A, et al. (2010) Linkage and association mapping of *Arabidopsis thaliana* flowering time in nature. PLoS Genet 6: e1000940.
54. Kolmer JA, Singh RP, Garvin DF, Viccars L, William HM, et al. (2008) Analysis of the *Lr34/Yr18* rust resistance region in wheat germplasm. Crop Sci 48: 1841.
55. Olivera PD, Jin Y, Rouse M, Badebo A, Fetch T, et al. (2012) Races of *Puccinia graminis* f. sp. *tritici* with combined virulence to *Sr13* and *Sr9e* in a field stem rust screening nursery in Ethiopia. Plant Dis 96: 623–628.
56. Bernardo AN, Bowden RL, Rouse MN, Newcomb MS, Marshall DS, et al. (2013) Validation of molecular markers for new stem rust resistance genes in U.S. hard winter wheat. Crop Sci 53: 755–764.
57. Mago R, Brown-Guedira G, Dreisigacker S, Breen J, Jin Y, et al. (2011) An accurate DNA marker assay for stem rust resistance gene *Sr2* in wheat. Theor Appl Genet 122: 735–744.
58. Hayden MJ, Kuchel H, Chalmers KJ (2004) Sequence tagged microsatellites for the *Xgwm533* locus provide new diagnostic markers to select for the presence of stem rust resistance gene *Sr2* in bread wheat (*Triticum aestivum* L.). Theor Appl Genet 109: 1641–1647.
59. McNeil MD, Kota R, Paux E, Dunn D, McLean R, et al. (2008) BAC-derived markers for assaying the stem rust resistance gene, *Sr2*, in wheat breeding programs. Mol Breed 22: 15–24.
60. Khan RR, Bariana HS, Dholakia BB, Naik SV, Lagu MD, et al. (2005) Molecular mapping of stem and leaf rust resistance in wheat. TAG Theor Appl Genet 111: 846–850.
61. Mago R, Bariana HS, Dundas IS, Spielmeyer W, Lawrence GJ, et al. (2005) Development of PCR markers for the selection of wheat stem rust resistance genes *Sr24* and *Sr26* in diverse wheat germplasm. Theor Appl Genet 111: 496–504.
62. Wu S, Pumphrey M, Bai G (2009) Molecular mapping of stem-rust-resistance gene *Sr40* in wheat. Crop Sci 49: 1681–1686.

6

A Unique Mutation in a MYB Gene Cosegregates with the Nectarine Phenotype in Peach

Elisa Vendramin[1]ⓘ, Giorgio Pea[2]ⓘ, Luca Dondini[3], Igor Pacheco[4], Maria Teresa Dettori[1], Laura Gazza[1], Simone Scalabrin[5], Francesco Strozzi[2], Stefano Tartarini[3], Daniele Bassi[4], Ignazio Verde[1]*, Laura Rossini[2,4]*

1 Consiglio per la Ricerca e la Sperimentazione in Agricoltura - Centro di Ricerca per la Frutticoltura (CRA-FRU), Rome, Italy, **2** Parco Tecnologico Padano, Via Einstein, Loc. C.na Codazza, Lodi, Italy, **3** Università di Bologna, DipSA, Bologna, Italy, **4** Università degli Studi di Milano, DiSAA, Milan, Italy, **5** IGA Technology Services, Udine, Italy

Abstract

Nectarines play a key role in peach industry; the fuzzless skin has implications for consumer acceptance. The peach/nectarine (*G/g*) trait was described as monogenic and previously mapped on chromosome 5. Here, the position of the *G* locus was delimited within a 1.1 cM interval (635 kb) based on linkage analysis of an F_2 progeny from the cross 'Contender' (C, peach) x 'Ambra' (A, nectarine). Careful inspection of the genes annotated in the corresponding genomic sequence (Peach v1.0), coupled with variant discovery, led to the identification of MYB gene *PpeMYB25* as a candidate for trichome formation on fruit skin. Analysis of genomic re-sequencing data from five peach/nectarine accessions pointed to the insertion of a LTR retroelement in exon 3 of the *PpeMYB25* gene as the cause of the recessive glabrous phenotype. A functional marker (indelG) developed on the LTR insertion cosegregated with the trait in the CxA F_2 progeny and was validated on a broad panel of genotypes, including all known putative donors of the nectarine trait. This marker was shown to efficiently discriminate between peach and nectarine plants, indicating that a unique mutational event gave rise to the nectarine trait and providing a useful diagnostic tool for early seedling selection in peach breeding programs.

Editor: Cameron Peace, Washington State University, United States of America

Funding: This work was supported by the Ministero delle Politiche Agricole Alimentari e Forestali-Italy (MiPAAF www.politicheagricole.it) through the project 'DRUPOMICS' (grant DM14999/7303/08) and by an Italian grant to DB funded by private and public agencies "MAS.PES: apricot and peach breeding by molecular-assisted selection". The funders had no role in study design, data collection and analysis, decision to publish, or preparation of the manuscript.

Competing Interests: Dr. Simone Scalabrin, one of the authors of the manuscript, is currently affiliated with IGA Technology Services.

* E-mail: ignazio.verde@entecra.it (IV); laura.rossini@unimi.it (LR)

ⓘ These authors contributed equally to this work.

Introduction

Peach (*Prunus persica* L. Batsch) is one of the most important fruit crops in temperate regions with about 21 million tons produced worldwide and Italy, with over 1.6 million tons, is the second producer after China (FAOSTAT 2011, http://faostat.fao.org/). The presence or absence of skin pubescence (fuzziness) is one of the commercial characteristics used to classify peach fruits along with flesh features (color, adhesion and texture) and fruit shape. Nectarines, characterized by the absence of fruit trichomes, are widely cultivated and play an important role in world peach production (30% in Italy, http://agri.istat.it/, 2013; 20% in USA, http://www.nass.usda.gov/, 2013) and may be associated with decreased allergenic properties. In *P. persica* two major allergens, Pru p 1 and Pru p 3, are known as responsible for the oral allergy syndrome (OAS) [1]. Indeed, the Pru p 3 protein was undetectable in the nectarine 'Rita Star' suggesting that this may be considered as a hypoallergenic variety [2]. Interestingly, in *Humulus lupulus* and in *Nicotiana tabacum*, genes encoding proteins highly similar to Pru p 1 and Pru p 3 are mainly expressed in trichomes [3,4].

The peach/nectarine character is monogenic (*G/g*) with nectarine recessive to fuzzy fruit [5]. The *G* locus was mapped in the distal part of linkage group (LG) 5 [6,7] spanning a region from 15,126,681 to 16,315,341 (1.189 Mb) of pseudomolecule 5 of the peach reference genome (Peach v1.0) [8]. Peach originated in North-West China and was domesticated there about 4,000–5,000 years ago [8]. From China it spread westwards reaching Persia following the Silk Road, was introduced to Rome in the first century BC and then disseminated to all the Roman Empire [9]. Nectarines have been known in China for over 2,000 years [10] and have been reported in most of the oases of the Tarim Basin (China) and along the Silk Road trade routes in Central Asia and the Caucasus [9,11]. The means and timing of their introduction in Europe are not clear. Likely, Romans did not know this type of peaches [9], but nectarines have been described by several botanists in Europe since the Renaissance period [9]. Old European nectarine varieties include 'Lord Napier', 'Precoce di Croncels' and 'Galopin'. In Southern Italy, traditional local white nectarines, called 'Sbergie' (Sicily) or 'Merendelle' (Calabria), have been cultivated since the 16th century [12]. Cluster analysis suggested that these local accessions are distinct from the western nectarine germplasm, pointing to a putative different origin of this group of cultivars [13]. Historically, nectarines have had little impact in China's peach industry [11], and nowadays there are no reports of traditional nectarine cultivars available in China [14]. The timing of introduction of nectarines to the United States (US) is controversial: their cultivation is reported in the early 20th century, although a newspaper article (New York Gazette March 28, 1768, p. 3) described nectarines being grown in the US prior to the War of Independence. Modern nectarine breeding started in the US in the middle of the 20th century. In 1942, Anderson introduced the nectarine 'Le Grand' using the accession 'Quetta', discovered near the homonymous city in India (now part of Pakistan) in 1906 [15], as the source of the nectarine trait. Other known sources of the nectarine trait used in modern western breeding programs were 'Goldmine' and 'Lippiatt' discovered in New Zealand in 1900 and 1916, respectively [15]. These latter three genotypes are acknowledged as donors of most of the current nectarine cultivars widespread in US and Europe. Modern Japanese breeding programs have extensively used two old European nectarines, 'Precoce di Croncels' and 'Lord Napier', and modern US cultivars [16]. In the last decades, the trait was introduced to Chinese breeding programs directly from western accessions or indirectly using Japanese material [17].

Trichomes are hair-like appendages that derive from the differentiation of epidermal cells and are classified based on their morphology (unicellular or multicellular), and secretory abilities (glandular or non-glandular) [18,19]. Trichomes may develop on several plant organs (leaf, fruit, seed, etc.). They play an important role in protecting plants against biotic and abiotic stresses [20–22] and can also hold a direct economic relevance. Aromatic substances are often synthesized by glandular trichomes, for example in aromatic plants, such as peppermint (*Mentha piperita*) [23] and basil (*Ocimum basilicum*) [24]. Cotton (*Gossipium hirsutum*) seed fibers are classified as non-glandular trichomes and represent one of the most highly expanded plant cell types [25]. In peach fruit, trichomes are non-glandular and unicellular and first appear on the ovary as early as four weeks before anthesis as observed in the peach 'Contender' [26]. By the time of physiological ripening most fruit skin trichomes are dead cells [27]. In *Arabidopsis* a number of genes involved in trichome formation and development have been identified by mutant analyses [28] and transcriptome profiling [29], revealing a complex regulatory network. Several transcription factors interact during trichome initiation and formation: in particular members of the R2R3-MYB class are known to act as positive regulators [30–33], while single-repeat MYB proteins function in negative control [34–36]. Mutations in the R2R3-MYB gene *GLABRA1* (*GL1*) result in glabrous plants in *A. thaliana* [37], *A. lyrata* [37,38] and other Brassicaceae species [39]. In *Gossipium hirsutum*, *GhMYB25*, which encodes an R2R3-MYB factor, is involved in the differentiation of ovule epidermal cells into cotton fibers, as well as in the formation of leaf thricomes [40].

The aims of the present study were to precisely map the *G* locus, identify a candidate gene and develop a reliable marker for the nectarine phenotype (glabrous fruit). To these ends, we used an F_2 population from a cross between the peach 'Contender' (C) and the nectarine 'Ambra' (A) [41,42] to develop a Single Nucleotide Polymorphism (SNP) map around the *G* locus. Analysis of the corresponding region in the peach genome sequence (Peach v1.0) [8] led to the identification of an R2R3-MYB gene as a candidate for trichome formation in peach fruit. A functional marker (indelG) developed on this gene provides a useful tool for early seedling selection for the peach/nectarine trait in breeding programs.

Materials and Methods

Plant materials & DNA extraction

An F_2 population of 305 seedlings derived from the cross between the peach 'Contender' (C) and the nectarine 'Ambra' (A), segregating for the peach/nectarine trait, [41,42] (CxA F_2), was used to develop a SNP map around the *G* locus. The trees were located in a farm belonging to the Municipality of Castel San Pietro (Bologna, Emilia Romagna, Italy) leased to ASTRA (latitude: from 44°24′44.18″N to: 44°24′30.08″N; longitude: from 11°35′47.21″E, to: 11°36′2.00″E). No specific permission was required because Daniele Bassi is the curator of the peach germplasm collection grown there and no endangered or protected species were involved.

Trees were planted on their own roots with a spacing of 1 m within and 4 m between rows and trained as slender spindle (one stem with short lateral scaffolds). Pruning was performed yearly and standard cultural practices were applied. Scoring of the peach/nectarine trait was carried out in two seasons to confirm correct scoring of the phenotype.

Ninety-five *P. persica* genotypes, 46 peaches and 49 nectarines grown at the CRA-FRU experimental farm (Rome, Italy) (except for 'Galopin' and 'Lord Napier' grown at Ivalsa-CNR, Follonica, Italy), were analyzed to validate the functional marker. For each accession, phenotype, pedigree, geographical origin and putative donor of the nectarine trait are reported in Table 1. DNA was extracted from leaf tissue using the DNeasy Plant Mini Kit (Qiagen GmbH, Hilden, Germany) as per manufacturer's protocol and quantified with NanoDrop (Thermo Scientific, Waltham, MA, USA).

Linkage map

In total 305 individuals from the CxA F_2 progeny were analyzed to map the *G* locus using an upgraded version of the CxA map [41,42] covering LG 5 from base 10,192,138 to base 17,544,073 of the peach genome sequence pseudomolecule 5 and spanning the already known *G* interval [7]. To refine the position of the *G* locus, SNPs located in this region were selected from those identified through analysis of re-sequencing data from the CxA F_1 individual [42] (biosample SRS335631, run SRR502997). About 300 bp of the SNP-flanking sequence were downloaded from the IGA peach Gbrowse (http://www.appliedgenomics.org/) and the Mass ARRAY Assay Design 3.1 software was used to design multiplex assays for SNP analysis [43]. SNP genotyping was performed using the iPLEX Gold technology available for Sequenom platforms (Sequenom, Inc., San Diego, CA, USA). SNP markers and the *G* locus (scored as a dominant phenotypic marker) were mapped using Joinmap 3.0 [44] with a minimum LOD score of 10 for grouping; the Kosambi mapping function [45] was used to convert recombination frequencies into map distances. Based on this map, three new SNP markers (S5_15988499, S5_15865556 and S5_15866258) were developed in the *G* locus interval (scaffold_5, from 15,853,006 to 16,488,104; see Results and Discussion) and genotyped on eight informative recombinants by sequencing 200 bp encompassing the SNP (primer sequences shown in Table 2). Standard PCRs were performed using GoTaq Green Master Mix (Promega, Madison, WI, USA). Each PCR reaction contained 1 X GoTaq Green Master Mix, 0.4 µM of each primer, 20 ng template DNA and sterile Milli-Q water to a final volume of 25 µl. The PCR protocol consisted in an initial step at 95°C (5 min), followed by 40 cycles at 95°C (30 s), 60°C (30 s) and 72°C (1 min), and a final elongation at 72°C (5 min). PCR products were purified with ExoSapIT (Amersham PharmaciaBiotech, Uppsala, Sweden) and sequenced with the Big Dye Terminator v3.1 Cycle Sequencing Kit (Applied Biosystems, Foster City, CA, USA). After ethanol precipitation, sequencing products were mixed with 15 µl of HiDi formamide and subjected to capillary electrophoresis in an ABI Prism 3730 DNA Analyzer (Applied Biosystems, Foster City, CA,USA). Genotyping was performed by

Table 1. List of 95 peach/nectarine varieties representing the diversity of cultivated *P. persica*.

Accession name	Genotype	Phenotype	Pedigree	Geographical origin[a]	Putative donor of nectarine allele[b]
Ambra	197-197	nectarine	Stark Red Gold x Mayfire	Italy (B)	Lippiatt/Goldmine
Aniversario	197-197	nectarine	–	Argentina (B)	-/-
Big Top	197-197	nectarine	–	USA (B)	-/-
Branca	197-197	nectarine	Goldmine x Pala	Brazil (B)	Goldmine/Goldmine
California	197-197	nectarine	P60-30 x Fantasia	Italy, 1994 (B)	Quetta/-
Centenaria	197-197	nectarine	Docura 2 op	Brazil (B)	Goldmine/Goldmine
Chiyodared	197-197	nectarine	Hirakutsared x Nectared 5	Japan (B)	Lord Napier or Precoce di Croncels/-
Crasiommolo Rosso	197-197	nectarine	–	Italy (L)	-/-
Crimson Gold	197-197	nectarine	Nectarine selection x July Elberta	USA, 1967 (B)	-/-
Sabrina	197-197	nectarine	–	Spain (B)	-/-
Fairlane	197-197	nectarine	(Le Grand x Sun Grand) x Fantasia	USA, 1973 (B)	Lippiatt/Quetta
Fantasia	197-197	nectarine	Gold king x p101-24 (=Red King op)	USA, 1969 (B)	Quetta/Quetta
Firebrite	197-197	nectarine	Flavortop x (Red King F_2)	USA, 1974 (B)	Quetta/-
Fuzador	197-197	nectarine	((Greensboro op x Elberta) x JHHale) F_2	France, 1973 (B)	-/-
Galopin	197-197	nectarine	–	England, 1869 (L)	-/-
Gold Mine	197-197	nectarine	–	New Zeland, 1900 (B)	
Golden Grand	197-197	nectarine	Le Grand x (Le Grand x Kim)	USA, 1954 (B)	Quetta/Lippiatt
Jacquotte	197-197	nectarine	–	France, 1979 (B)	-/-
Laura	197-197	nectarine	–	USA, 1995 (B)	
JinXia	197-197	nectarine	(Okubo x Okitsu) x Okitsu F_2	China (B)	Lord Napier/Precoce di Croncels
Lord Napier	197-197	nectarine	–	Belgium, 1859 (L)	-/-
Madonna Di Agosto	197-197	nectarine	–	Italy (L)	-/-
Madonna Di Giugno	197-197	nectarine	–	Italy (L)	-/-
Magali	197-197	nectarine	–	France, 1988 (B)	-/-
Max	197-197	nectarine	–	Italy, 1995 (B)	-/-
Mayfire	197-197	nectarine	Armking op	USA, 1984 (B)	Goldmine/Goldmine
Nectaross	197-197	nectarine	Stark Red Gold x Le Grand	Italy, 1983 (B)	Lippiatt/Quetta
Nettarina Pendula	197-197	nectarine	–	Italy (L)	-/-
Nico	197-197	nectarine	–	Italy (B)	-/-
Phn 91-12	197-197	nectarine	–	USA (B)	-/-
Phn 91-14	197-197	nectarine	–	USA (B)	-/-
Phn 91-17	197-197	nectarine	–	USA (B)	-/-
Quetta	197-197	nectarine	–	Pakistan, 1906 (L)	
Ricci 2	197-197	nectarine	Stark Red Gold x Tastyfree	Italy, 1993 (B)	Lippiatt/Quetta
Romamer 1	197-197	nectarine	–	Romania, 1983 (B)	-/-
Romamer 2	197-197	nectarine	City 29-245 x RR48-153	Romania, 1983 (B)	-/-
Russian Nectarine 592-81	197-197	nectarine	–	Ukrayne, 1980 (L)	-/-
Russian Nectarine 598-81	197-197	nectarine	–	Ukrayne, 1980 (L)	-/-
Shizukured	197-197	nectarine	Okitsu x NJN17	Japan (B)	Lord Napier or Precoce di Croncels/-
Silver Lode	197-197	nectarine	(Goldmine x Rio Oso Gem) x (Goldmine x July Elberta)	USA, 1951 (B)	Goldmine/Goldmine
Sirio	197-197	nectarine	Flamekist x Fantasia	Italy, 1987 (B)	Quetta/Quetta
Snow Queen	197-197	nectarine	–	USA, 1975	-/-
Souvenir Nikitski	197-197	nectarine	Lola op	Ukrayne, 1988 (B)	-/-
StarkRedgold	197-197	nectarine	Sun Grand op	USA (B)	Lippiatt/Lippiatt
Summer Beauty	197-197	nectarine	Red Diamond x Sun Grand	USA, 1979 (B)	Quetta/Lippiatt
Vania	197-197	nectarine	–	Italy, 1990 (B)	-/-
ZeeGlo	197-197	nectarine	(Red Grand op) x (Sun Grand x Merril Gem)	USA, 1988 (B)	Quetta/Lippiatt
Zephyr	197-197	nectarine	–	France, 1992 (B)	-/-

Table 1. Cont.

Accession name	Genotype	Phenotype	Pedigree	Geographical origin[a]	Putative donor of nectarine allele[b]
Zincal 5	197-197	nectarine	–	Spain (B)	-/-
Acireale	941-941	peach	–	Italy (L)	
Amber Gold	941-941	peach	Red Grand x Royal May	USA, 1973 (B)	
Amsden	941-941	peach		USA, 1868 (B)	
Autumnglo	941-197	peach	(Candoka x Tennessee Natural) x Merril Fiesta	USA, 1977 (B)	-
BaekmiJosaeng	941-941	peach	Mishima x Nunome Wase	Republic of Korea, 1983 (B)	
Baldagenais	941-197	peach	–	France	-
Changbang Mutant	941-941	peach	–	Republic of Korea, 1986 (B)	
Chiyomaru	941-941	peach	–	Japan, 1988 (B)	
Chui Huang Tao	941-941	peach	–	China (B)	
Ciccio Petrino	941-941	peach	–	Italy	
City 32-82	941-197	peach	–	USA (B)	-
Contender	941-941	peach	Wiblo x [Norman x (Candor x (Summercrest x Redhaven))]	USA, 1987 (B)	
Cp 88/2	941-941	peach	–	Mexico (B)	
Early Gold	941-941	peach	–	Japan, 1980 (B)	
Elberta	941-941	peach	Chinese Cling op (perhaps x Early Crawford)	USA, 1889 (B)	
CxA F_1	941-197	peach	Contender x Ambra	Italy (B)	Lippiatt or Goldmine
Fairtime	941-197	peach	(Rodeo x Kirkman Gem) op	USA, 1968 (B)	
Fayette	941-941	peach	Fay Elberta x (Fireglow x (Fireglow x Hiley))	USA, 1966 (B)	
FeiChing Bai Li 17	941-941	peach	–	China (B)	
FeiChing Tao	941-941	peach	–	China, 1909 (B)	
Fidelia	941-197	peach	(O'Henry x Giant Babcock) x (May Grand x Sam Houston)	USA, 1986 (B)	-
Frau Maria Rudolf	941-941	peach	Top Red Delicious op	Germany (B)	
Grosse Mignonne	941-941	peach	–	France, 1667 (L)	
Higama	941-941	peach	Japanese seed op	France, 1970 (B)	
Hwando 1	941-941	peach	–	Republic of Korea, 1977 (B)	
J H Hale	941-941	peach	putative selfpollination of Elberta	USA (B)	
JingYu	941-197	peach	Okubo x Okitsu	China (B)	Lord Napier or Precoce di Croncels
KurakataWase	941-941	peach	–	Japan (B)	
PLove2- 2N	941-941	peach	di-haploids of Lovell	USA, 1882 (B)	
O'Henry	941-197	peach	Merrill Bonanza op	USA, 1970 (B)	-
Pantao 20-58	941-941	peach	–	China, 1869 (B)	
Pillar	941-941	peach	–	Japan (B)	
Redhaven	941-941	peach	Halehaven x Kalhaven	USA, 1940 (B)	
Reginella II	941-941	peach	–	Italy (B)	
Rou Tao	941-941	peach	–	China (L)	
Russotto	941-941	peach	–	Italy (B)	
Sahua Hong Pantao	941-941	peach	–	China (L)	
Sanguinella	941-941	peach	–	Italy (L)	
Shen Zhou Mi Tao	941-941	peach	–	China (B)	
Siberian C	941-941	peach	selection of a China seedling	Canada, 1967 (B)	
Summer Pearl	941-197	peach	[(Candoka x Tennessee natural op) x ((Candoka x Flaming Gold) x NJN5 op)] x (PI119844 x White Hale)	USA, 1979 (B)	-
TaturaDawn	941-941	peach	Levis selfpollination	Australia, 1960 (B)	
Yoshihime	941-197	peach	(Nakatsu Hakuto x Nunome Wase) x Akatsuki	Japan, 1990 (B)	-
Yu Bai	941-941	peach	–	China (L)	
Yumyeong	941-941	peach	Yamato-Wase x Nunome Wase	Republic of Korea, 1977 (B)	

Table 1. Cont.

Accession name	Genotype	Phenotype	Pedigree	Geographical origin[a]	Putative donor of nectarine allele[b]
ZansetsuShidare	941-941	peach	-	Japan (B)	

[a](B) = breeding materials (L) = landrace.
[b]- = unknown donor of the nectarine allele.
For each accession the following information is reported: name, genotype at the indelG marker (941 bp reference peach allele, 197 bp nectarine allele carrying the retrotransposon insertion), phenotype, pedigree, country of origin and year of release/discovery when known, putative donor of the nectarine allele in *P.persica* if pedrigree information was available.

visual inspection of the resulting electropherograms using 4PEAKS freeware (Nucleobytes Inc.).

Variant discovery from NGS data

In order to identify genetic variants putatively involved in the control of the nectarine trait publicly available paired-end (PE) whole-genome re-sequencing data of *P. persica* accessions from study SRP013437 [8] were downloaded from the NCBI Sequence Read Archive (SRA) [46]. Five accessions were considered for this study: 'Bolero' (biosample SRS335629, run SRR501836), 'OroA' (biosample SRS335635, run SRR502986), 'Lovell' Clone PLov2-2N (biosample SRS335634, run SRR502985), 'Quetta' (biosample SRS335636, runs SRR502989 and SRR502987) and F_1 'Contender' × 'Ambra' (biosample SRS335631, run SRR502997). 'Quetta' was included as the reference nectarine accession. The CxA F_1 individual originated the CxA F_2 population used to map the *G* locus. The peach 'Lovell' Clone PLov2-2N is the doubled haploid used to generate the reference peach genome sequence (Peach v1.0) [8], providing an internal control for false variant calling. Finally, the peaches 'Bolero' and 'OroA' were chosen as controls for nectarine segregation, as it has been demonstrated that the nectarine trait does not segregate in the 'Bolero' × 'OroA' F_1 population [42].

SRA data of each run were dumped in fastq format using the *fastq-dump* tool of NCBI sratoolkit v2.1.16 software (http://www.ncbi.nlm.nih.gov/Traces/sra/sra.cgi?view = software), splitting forward and reverse paired reads for each sample into two separate files. Reads were quality filtered on a single sample basis using Trimmomatic v0.22 [47], first trimming leading and trailing bases below a quality threshold of 20, and then removing trimmed reads shorter than 24 bp or having an average quality below 20 (calculated on 8 bp long sliding windows). For each sample, only reads passing the quality filtering as matching pairs were retained and aligned to the whole *P. persica* reference genome Peach v1.0 using the Burrows-Wheeler Alignment Tool (BWA v0.6.2) [48]. The *aln* (IS linear-time algorithm) and *sampe* (all default options except *-n* 25 *-N* 25) commands were applied, respectively, for finding suffix array (SA) coordinates of each individual read and to convert them to chromosomal coordinates and to pair the reads. The resulting alignment SAM files were converted by Picard Tools version 1.77 (http://picard.sourceforge.net/) to sorted BAM files compliant to the Genome Analysis Toolkit (GATK) format, using the tools *CleanSam*, *SamFormatConverter* and *AddOrReplaceReadGroups*. GATK-compliant BAM files were submitted to GATK version 2.3–3 [49] for pre-processing procedures, i.e. indel realignment, duplicate removal and base quality score recalibration (BQSR). The data table needed for the recalibration step in BQSR was manually generated upon validated SNP data from the Peach 9K chip array [50]. Variant discovery procedures were then applied using whole-genome recalibrated alignments of all five samples simultaneously. Genotypes for SNP and small INDEL variants were called through the GATK *HaplotypeCaller* tool applying hard filtering parameters [51]. Structural variants were also independently called on the same recalibrated alignment data by Pindel software v0.2.4t [52] following standard procedures.

Reads from the resequencing of 'Quetta' (biosample SRS335636, runs SRR502989 and SRR502987) were also analyzed using the CLC genomic workbench (6.0.1). The 75 bp pair-end fragments were trimmed for quality retaining only nucleotides with Phred values higher than 30, and trimmed reads were aligned against the *P. persica* reference genome (Peach v1.0) [8] using the read mapping tool. Only reads with over 90% identity over at least 92% of their length were mapped on the reference. All variant discovery searches were limited to the locus *G* mapping interval defined by informative recombinants in the C×A F_2 mapping population (scaffold_5, from 15,853,006 to 16,488,104, see Results and Discussion).

Validation of candidate variant

To validate the putative variant individuated among the resequenced genotypes long-range PCRs were performed on five nectarines genotypes ('Quetta', 'Goldmine', 'Madonna di Agosto', 'Stark Red Gold' and 'Ambra') and on the peach 'Contender', with primers Seq16F and Seq4R (Table 3) designed flanking the putative insertion, using Herculase DNA Polymerase (Agilent Technologies, Santa Clara, CA, USA). Each reaction contained 1x Herculase reaction buffer, 0.3 mM dNTP mix, 0.5 µM each Seq16F and Seq4R primer, 3% DMSO, 2.5 U Herculase polymerase, 300 ng template DNA, and sterile Milli-Q water to a final volume of 50 µl. The following PCR protocol was performed on a Esco Swift Maxi thermocycler (Esco GB Ltd, Downton, UK) or an Applied Biosystems 2720 Thermal Cycler (Applied Biosystems, Foster City, CA, USA): 95°C for 5 min; 28 cycles of 95°C (1 min), 59°C (1 min), 72°C (12 min) followed by a step at 72°C for additional 12 min. All PCR amplicons were checked on 1% agarose gel in an overnight run in SB buffer. A standard ethidium bromide staining was used for band visualization.

For restriction analysis, 5 µl of long-range PCR products from 'Quetta', 'Goldmine', 'Madonna di Agosto', 'Stark Red Gold' and 'Ambra' were digested with *EcoR*I and *Hind*III (Fermentas, Vilnius, Lithuania) in a single overnight reaction at 37°C. Master mix was calculated by double digest tool (http://www.thermoscientificbio.com/webtools/doubledigest/) with 0.5 U/sample of each restriction enzyme and buffer R. Restriction products were separated on 1% agarose gels and stained with ethidium bromide for band visualization.

The 'Quetta' long-range PCR product was first purified with the Macherey-Nagel PCR clean-up kit (Carlo Erba reagents, Italy) and quantified by Picogreen (Quant-iT PicoGreen dsDNA kit,

Table 2. SNP markers.

Markers[a]	cM[b]	Peach v1.0 position (nt)	SNP allele[c]	Forward primer	Reverse primer	SNP Extension primer [d]
S5_10192138	0	10,192,138	G/A	GATGAATGGGTGAAGGTAAG	TTCCGCAAAAAAAAACATATC	ttaagGGGTGAAGGTAAGTTTGCACA
S5_11640083	4.2	11,640,083	T/G	CCTACTACACAATTGCCTTA	GTCGGTCGTCAGTTTTTTTG	cgacCTTGTAGATTCTAATGGAAGTA
S5_12847567	11.4	12,847,567	T/A	TGCGGATTTTTCTTAGCTAC	CTCTTTCCCAATCTCAATCG	AATCGCATTGTTGAGAC
S5_13449464	14.3	13,449,464	G/C	CGGTGATTGATATGATGATG	CCACTCAAATTGCCTTTCCC	GTGATTGATATGATGATGATT-TATAT
S5_13852617	16.7	13,852,617	A/C	ATACCTATGTTCACTCCCCG	TTGGCTGGTAAGGTTGTTGG	TGTACTGATGTGTGAATCTAATG
S5_14584095	24.3	14,584,095	G/C	AAGTTGTTCCAGTGGCAACC	TATAGTGGGGCTGGAATCTG	cccGGCTGGAATCTGTTCTCTCAGAC
S5_14894563	25.5	14,894,563	C/A	GCAGAGGAATTTTTCCCTAC	TTAGGGAGGGAGCTATGTTG	gGTTGGAGGTATTTGGGC
S5_14949332	26.0	14,949,332	T/C	AGAAGATGTGGTTCCAGAGG	TTCTCGATCCGGAAGGAGAT	TCCCAGATCCAAGACCC
S5_15731107	28.4	15,731,107	T/G	CTGTTGTAAGACAGGTTTGG	AACATGCTTGCGGCTTCGTC	GGCTTCGTCCATACGCC
S5_15853006	28.8	15,853,006	T/A	TAGTTTGTCTGTCAAACCGC	CCGAGAAGACTGAAGAGTTG	GCTATTAAAGACATTAGAGATGA
S5_15865556	*29.1*	15,865,556	A/T	GGTTGGGGCCTCTGTATTCT	TAAAGGCAACCACATTGCAG	
S5_15866258	*29.1*	15,866,258	A/G	TCAGCTTGTCCATGGCATTA	GCCGTAAAGGCTTTCCTCTC	
S5_15988499	*29.4*	15,988,499	C/T	GCCGTGAAGTGGAGTTCTCT	GATTCTCACTCTGCTCCTGTCT	
S5_16488104	29.9	16,488,104	A/G	TTCGCATTCATTAGTTCAC	CAGGTTTGTGAGTTTGCTTG	TACTAAACGGAAGCTATGT
S5_17544073	36.2	17,544,073	G/A	GCCATCTCTCTGTTTCTCTG	GTAGTATCAGCCGACTGTAG	cTAACATACATGACATGACATACA-CCC

[a]SNP detection was performed by Sequenom MassArray technology, except for the SNPs in italics that were genotyped by Sanger sequence.
[b]the genetic distances estimated by the analysis of informative recombinant plants are reported in italics.
[c]the first nucleotide correspond to the reference allele in the peach genome (Peach v1.0).
[d]lower case bases correspond to the tails added to the SNP extension primer for Sequenom MassArray analysis.
For each marker the following information is reported: marker name, position in cM with respect to the map in Figure 1, position in bp with respect to the peach genome sequence (Peach v1.0), SNP allele and the primer sequences used.

Life Technology, US) in preparation for sequencing using the Illumina MiSeq platform with a 150 bp paired end sequencing strategy. Preparation of Nextera XT library was performed with 1 ng of genomic DNA according to the Nextera XT protocol (Ver. Oct 2012, rev C). Briefly, the DNA was fragmented in 5 μl of Amplicon Tagment Mix and 10 μl of Tagment DNA buffer (Illumina, San Diego, CA, USA). Tagmentation reactions were performed by incubation at 55°C for 5 min followed by neutralization with 5 μl of Neutralize Tagment Buffer for 5 min. Tagmented DNA (25 μl) was used as the template in a 50 μl limited-cycle PCR (12 cycles) and processed as outlined in the Nextera XT protocol. Amplified DNA was purified using 90 μl of AMPure XP beads then normalized with 45 μl of combined Library Normalization beads/additives. In preparation for cluster generation and sequencing, the normalized library was diluted in hybridization buffer and heat denatured. Due to the low diversity of the library a phiX spike-in (30%) was added to the final denatured 10 pM library. The sample was sequenced using the MiSeq Personal Sequencer (Illumina Inc., San Diego, CA, USA) running MiSeq Control Software Version 2.0.

A total of 20 M reads were obtained and analyzed using the CLC genomic workbench (6.0.1). Trimming and De Novo assembly tools were used with default parameters. The assembly obtained was filtered retaining only contigs with a length greater

Table 3. List of Primers.

Name	Sequence 5′->3′	Start Position on Scaffold 5 (nt)	Len	Notes
Seq16F	ATTCCGCTCCTCATGTAGTACA	15,898,340	22	upstream the LTR insertion
Seq4R	CCAAATAAACCACCACCTACTCTGTTA	15,899,297	27	downstream the LTR insertion
Seq15F	GCTGTAGGCTAAGGTGGAGAA	15,898,078	21	RT-PCR primer designed on exon 2 of *PpeMYB25*
Seq15R	ACTGAGCAATGTGGCTGAGA	15,898,586	20	RT-PCR primer designed on exon 3 of *PpeMYB25*
ppaRPII_RT-F	TGAAGCATACACCTATGATGATGAAG		26	RT-PCR housekeeping, [53](Tong et al, 2009)
ppaRPII_RT-R	CTTTGACAGCACCAGTAGATTCC		23	RT-PCR housekeeping, [53](Tong et al, 2009)
indelG-F	CTTGCACCTGAGTTCGATTCCG	15,898,324	22	upstream the indel in *PpeMYB25*
indelG-1R	GGCTTCAATGGCAGAACAAGG		21	within left border of the indel in *PpeMYB25*
indelG-2R	GCAGGTGGTGGAGATTCATTCAT	15,899,264	23	downstream the indel in *PpeMYB25*

Name, sequence, position on Peach v1.0 and length of primers used to perform long-range PCR, RT-PCR and indelG assay.

than 200 bp and formed by more than 500 reads. The filtered contigs were mapped using BWA v0.7.5a [53] with the MEM algorithm against the Peach v1.0 genome. Identification of conserved domains on contigs was performed using HMMER v3.1 [54] against the PFAM [55] database v27.0. Blast search of contigs was done using NCBI-BLAST+v2.2.27 [56] against the NT database downloaded from NCBI FTP site on Oct 23 2013.

Expression analysis

Total RNA was extracted from floral buds collected from 'Contender' and 'Ambra' at different developmental stages (seven, five, four and one week before anthesis) using the RNeasy Plant Mini Kit (Qiagen GmbH, Hilden, Germany) and treated with DNAse I (Sigma-Aldrich, St. Louis, MI, USA) following manufacturers' instructions; 1 µg of RNA was reverse-transcribed using the GoScript Reverse Transcription System (Promega, Fitchburg, WI, USA) with oligo $(dT)_{15}$ according to the manufacturer's protocol. For reverse transcription analysis, primers were designed on exon 2 (Seq15F) and exon 3 (Seq15R) of the *PpeMYB25* gene and the RNA polymerase II sequence was used as reference gene [57] (Table 3). For RT-PCR, 1 µL of cDNA was used with 1x GoTaq Green Master Mix, 0.4 µM of each primer and sterile Milli-Q water to a final volume of 10 µl. The PCR protocol consisted of an initial denaturation at 95°C for 2 min, followed by 35 cycles at 95°C (20 s), 62°C (20 s) and 72°C (30 s), followed by a final elongation at 72°C (5 min). PCR products were checked on 1% agarose gel at 5 V/cm in TBE buffer. A standard ethidium bromide staining was used for band visualization.

Functional marker design and genotyping

To perform association studies for the nectarine trait and provide a tool for marker assisted breeding (MAB) a codominant marker (indelG), consisting of a three-primer PCR assay (primers indelG-F, indelG-1R and indelG-2R; Table 3), was developed for genotyping of the candidate insertion: two outer primers (one forward and one reverse), designed on opposite sides of the insertion were combined in a single reaction with an inner reverse primer (designed on the reconstructed left sequence of the insertion).

PCRs were carried out in 10 µl containing 10 ng of template DNA, 1x PCR buffer, 1.5 mm $MgCl_2$, 200 µM each dNTP, 0.2 µM each primers and 0.5 U of Platinum Taq DNA polymerase (Life Technologies, Carlsbad, CA, USA). Amplifications were performed on a Veriti thermal cycle (Life Technologies, Carlsbad, CA, USA) with the following temperature profile: 95°C (5 min) followed by 35 cycles at 94°C (30 s), 61°C (30 s), 72°C (30 s) and a final extension at 72°C (10 min). PCR products were separated on an ethidium bromide stained 1% agarose gel. The 305 seedlings of the CxA F_2 population, the parents and the hybrid CxA F_1, as well as 46 peach and 49 nectarine accessions, were analyzed (Table 1).

Results and Discussion

Mapping of the G locus

Scoring of the CxA F_2 progeny revealed the presence of 246 peach and 59 nectarine plants, indicating a slight distortion from the expected 3:1 segregation ($\chi^2 = 5.41$, $p<0.05$ with 1 d.f.). In agreement with a previous report [7], the peach/nectarine phenotype was regarded as a dominant trait (*G* locus). A total of 12 SNPs, covering about 7 Mb of LG 5 around the *G* locus, were mapped on the whole progeny (Figure 1). Consistent with the observed distortion of the phenotypic trait, a skewed segregation was also found for all the markers around the *G* locus from S5_15731107 to S5_16488104 (Figure 1). The *G* locus was placed between markers S5_15853006 and S5_16488104 within an interval 1.1 cM (635 kb, Figure 1). Three additional SNPs inside the *G* region (S5_15865556, S5_15866258 and S5_15988499) were also successfully mapped by genotyping of informative recombinant plants. However none of them were useful to refine the position of *G* locus (data not shown).

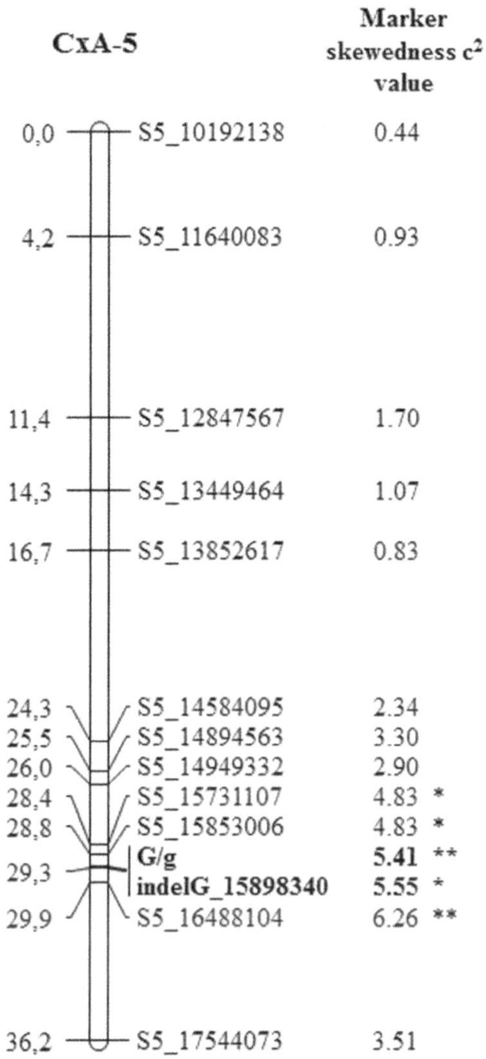

Figure 1. LG 5 CxA map around the *G* locus. Linkage map obtained from analysis of the CxA F_2 progeny. On the left side distances are indicated in cM; on the right the marker name, the physical position on Peach v1.0 and marker skewedness are reported. The peach/nectarine locus and the indelG marker are shown in bold.

Variant identification in the locus G genomic region

In order to identify variants putatively underlying the nectarine phenotype, genome-wide recalibrated alignment data from five *P. persica* accessions were examined in detail around the *G* locus (scaffold_5, from 15,853,006 to 16,488,104). In this region, GATK *HaplotypeCaller* detected 291 SNP and indel variants above the chosen minimum phred-scaled score quality threshold of 200, out of which 67 mapped within predicted genes in the peach genome (Peach v1.0) [8]. Of the latter, 20 variants, distributed in seven genes, were heterozygous in CxA F_1 with non-segregating joint genotype combinations in 'Bolero' and 'OroA'. In the nectarine 'Quetta' the same analysis identified two variants, both indels, homozygous for the non-reference allele and located in two distinct genes (*ppa023682m* and *ppa023143m*). Both of them were heterozygous in CxA F_1, where no other variants were found in these two genes, and were thus considered as putative candidates for the *G* locus. The first candidate variant is a deletion of 3 motives in a known $(AAAC)_6$ microsatellite at position 16,463,040 which maps within the only intron of the predicted gene *ppa023628m* (best *Arabidopsis thaliana* blastx match AT3G29575.1, ABI five binding protein *AFP3*, e-value 5×10^{-18}). ABI Five Binding Proteins (AFPs) are members of a small plant-specific protein family, characterized by three conserved domains of unknown function. AFPs act as negative regulators of ABA signaling [58] and have no known involvement in trichome formation. Next we focused on the second variant, reported by *Haplotype Caller* as two distinct insertions of 76 and 49 bp on the left and right side, respectively, of an $(AC)_3$ motif at position 15,898,458. This INDEL variant maps to the last exon (exon 3) of the predicted gene *ppa023143m* (best *Arabidopsis thaliana* blastx match AT5G15310.1, MYB domain protein 16, AtMYB16, e-value 1×10^{-73}). Similarity with R2R3-MYB transcription factors known to control epidermal cell differentiation [40,59] (see below) pointed to this gene as a likely candidate for the peach/nectarine trait [54]. This second variant was also detected by Pindel software in C×A F_1 and 'Quetta' samples only, as a large insertion compared to the reference sequence. In particular, this large insertion at position 15,898,458 was supported in Pindel by a total of 31 reads, 17 overhanging on the left side of the insertion (10 in C×A F_1 and 7 in 'Quetta') and 14 on its right side (5 in C×A F_1 and 9 in 'Quetta'). The presence of this insertion in the nectarine allele is also supported by analysis of 'Quetta' resequencing data using CLC Genomic Workbench. Within the considered mapping interval (from 15,853,006 to 16,488,104) a total of 94,789 reads (9.15 million nucleotides) were aligned against the reference genome sequence, 61% of which in pairs and the remaining as single reads due to unexpected mapping distances, mate inversion, unmapping or mapping in other contigs. In agreement with Pindel results, the third exon of the *ppa023143m* gene showed a dramatic reduction of paired-end distances and an increase of single reads at position 15,898,458 (Figure 2), compatible with a large insertion in 'Quetta' compared to the 'Lovell' reference sequence. Due to this insertion only single reads could align in the region; the software reports the lack of paired reads assigning the value zero to the paired-end distance and increasing to 100% the percentage of single reads (Figure 2).

Validation and reconstruction of a long insertion in exon 3 of gene ppa023143m

The physical presence of the long insertion within exon 3 of gene *ppa023143m* was confirmed by long-range PCR using a primer pair designed on intron 2 (Seq16F) and exon 3 (Seq4R) flanking the insertion site. An amplification product of about 7 kb was obtained in five nectarines, 'Madonna di Agosto', 'Quetta', 'Stark Red Gold', 'Goldmine' and 'Ambra' (Figure 3). In contrast, in a peach genotype ('Contender') the same primer pair gave an amplification product of 960 bp (data not shown). 'Quetta' and 'Goldmine' are two donors of the trait in modern breeding and 'Stark Red Gold' is known to carry the nectarine allele of 'Lippiat', the third donor of the trait. 'Madonna di Agosto' belongs to a group of landraces not directly related to modern breeding germplasm [12,13]. The double digestion of the five amplicons, with *Eco*RI/*Hind*III, shows the same restriction pattern for all the accessions (Figure 3) suggesting that a unique mutational event gave origin to the nectarine trait present in the modern nectarine germplasm as well as in the local southern Italian ecotypes.

The amplicon obtained in 'Quetta' was sequenced by Next Generation Sequencing (NGS). Following filtering and assembly 90% of the reads were collected in three major contigs. Contig_1 (GenBank accession number KJ150676), formed by a consensus sequence of 5,836 bp, was mapped on scaffold 3 of peach genome v1.0 and showed a perfect match for 5,713 bp (scaffold_3:13,409,926.13,415,638) corresponding to the predicted LTR_1684 region (http://services.appliedgenomics.org/gbrowse/prunus_public/). The missing 123 bp that do not align on scaffold 3 showed a perfect match on scaffold 5 (scaffold_5:15,898,458.15,898,581), confirming that this contig represents an insertion in the third exon of predicted gene *ppa023143m*. When the LTR_1684 region, found in correspondence of Contig_1 alignment on scaffold 3, was submitted to CENSOR [60] (Release

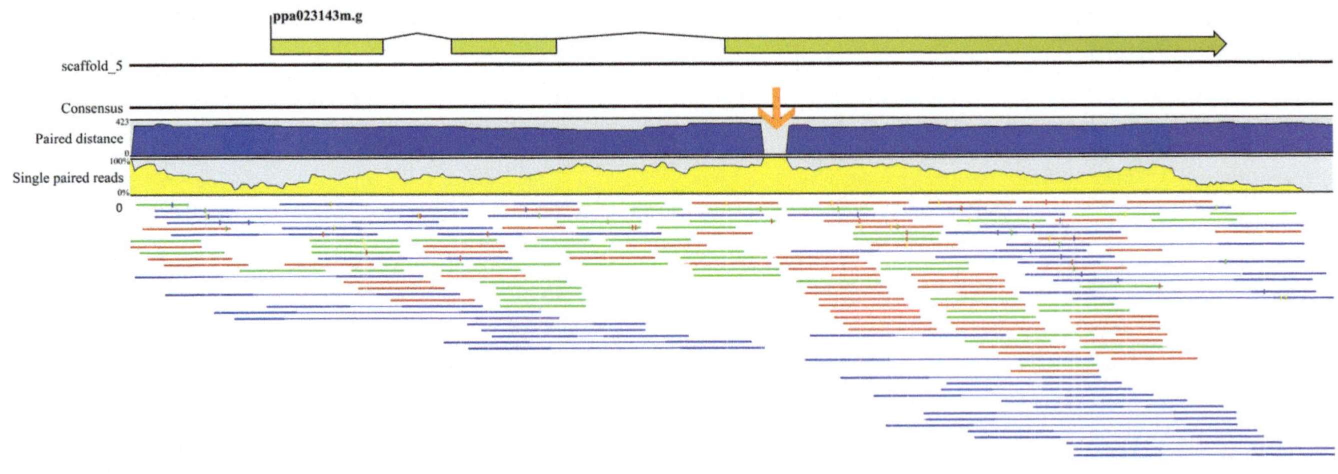

Figure 2. Alignment of Quetta reads against a 635 kb interval of Peach v1.0 pseudomolecule 5. Alignment results of reads, obtained by the resequencing of 'Quetta', against the peach genome region identified by the mapping interval in LG5 (from 15,853,006 bp to 16,488,104 bp). Top panel: intron-exon structure of *ppa023143m*. Central panel: plot of 'Quetta' paired-end distance (blue) and frequencies of single reads (yellow) at the *ppa023143m* locus. Bottom panel: blue lines are paired reads, green and red lines correspond to single reads with missing mate on the right and left side, respectively. The orange arrow points to the putative insertion inside exon 3 of *ppa023143m*.

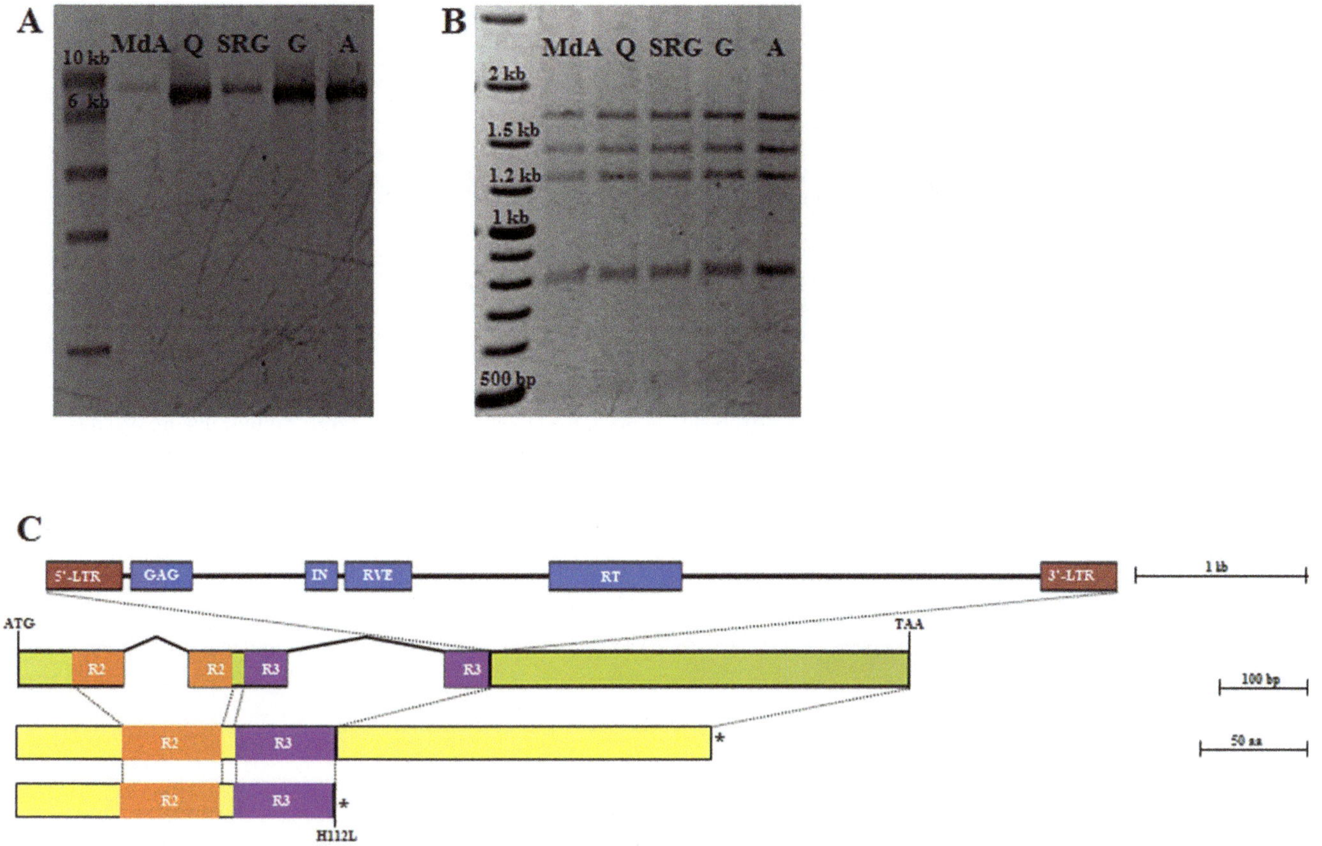

Figure 3. Variant discovery in *PpeMYB25* (annotation refinement of *ppa023143m*). Five nectarine genotypes ('Madonna di Agosto', MdA; 'Quetta', Q; 'Stark Red Gold', SRG; 'Goldmine', G; 'Ambra', A) were analyzed to confirm the presence of the insertion within exon 3 of *PpeMYB25*. (A) Long-range amplification products reveal for all the accessions a fragment of about 7 kb (compared to 960 bp expected from the reference genome). (B) Double digestion results of the long-range PCR products show the same pattern for all the genotypes. (C) Position and structure of the Ty-*copia* retrotransposon deduced by the by the NGS analysis of 'Quetta' long-range amplicon. The insertion results in a truncated version of the R2R3-MYB protein.

18.01, 16 Jan 2013) high similarity was found to a 6,033 bp annotated Ty1-*copia* retrotransposon (Copia-24_FV-I) from strawberry (*Fragaria vesca*). A conserved protein domain analysis on the complete sequence of Contig_1 also revealed the presence of different domains: UBN2 (gag-polypeptide of LTR copia-type) from position 1,354 to 1,689 of Contig_1, GAG-pre-integrase domain from position 2,329 to 2,511, RVE (Integrase core domain) from position 2,551 to 2,928 and RVT2 (Reverse transcriptase domain) from position 3,694 to 4,431. The four domains identified showed a high prediction confidence, with E-values of 3.8e-16, 4.3e-13, 6.8e-28 and 1.6e-98 respectively. All these predicted domains are typical of retrotransposable elements. The remaining two contigs, Contig_2 and Contig_3, 300 and 876 bp respectively, were also mapped on the peach genome. Both were split on scaffold 5 (scaffold_5:15,898,340.15,898,361; scaffold_5: 15,898,458. 15,899,262) in exon 3 of predicted gene *ppa023143m* and on scaffold 3 in the same LTR_1684 region of Contig_1. Thus, these contigs span the point of insertion of the Ty-*copia* retroelement within exon 3 of the gene. Flanking this insertion point, we found a characteristic Target Site Duplication (TSD) $(AC)_3$ [61] produced by the retroelement upon insertion into a new site. Together these results confirm the insertion of a Ty1-*copia* retrotransposon in the 'Quetta' allele of gene *ppa023143m*.

A BLASTN search of the retrotransposon sequence of Contig_1 against the peach reference genome returned 5 highly similar hits (from 87.1% to 100% sequence identity), two on chromosome 3, and one each on chromosomes 4, 7 and 8. All five hits were about 6 kb long and were precisely delimited by highly similar (from 97.7% to 100.0%) LTR sequences, each flanked by characteristic Target Site Duplications, thus confirming the existence of other copies of this LTR-retroelement in the reference peach genome. In particular LTRs found in the 'Quetta' allele of *ppa023143m* are identical to those present in scaffold 3 (scaffold_3:14, 488,093.14,488,522) and scaffold 4 (scaffold_4:27,503,652.27, 504,081). These data are consistent with previous analyses showing that 12.6% of LTR-retrotransposons have identical LTRs, indicating recent and ongoing retrotransposition activity in the peach genome [8].

Transposable elements (TEs) are known to cause many kinds of genetic variations in plants and played an important role in plant evolution and domestication [62]. In a survey of allelic variants at 60 genes involved in crop domestication and diversification, 15% were caused by TE insertions [63]. For example, an LTR insertion in a regulatory region of the *teosinte branched1 (tb1)* gene resulted in overexpression of the gene, causing the conversion from highly branched wild teosinte to the single culm architecture of domesticated maize [64,65]. However, the most common effect of TE insertion is the loss of gene function [57] and recessive TE-

```
PpeMYB25   15 RGPWTPEEDRKLLDFIQLHGHGSWSSLPQKAGLKRCGKSCRLRWRNYLRPDIK
GhMYB25    14 KGPWTPEEDQKLLAYIEQHGHGSWRALPLKAGLQRCGKSCRLRWINYLRPDIK
AmMYBML1   14 RGPWTPEEDQKLLSYIQEHGHGSWRALPSKAGLQRCGKSCRLRWSNYLRPDIK

PpeMYB25   68 RGNFSLHEDQTIIQLHALLGNRWSAIAAHLPKRTDNEIKNYWNTHL
GhMYB25    67 RGKFSLQEEQTIIQLHALLGNRWSAIATHLPKRTDNEIKNYWNTHL
AmMYBML1   67 RGKFSLQEEQAIIQLHAFLGNRWSAIATHLPKRTDNEIKNYWNTHL
```

Figure 4. Aminoacid alignment of the R2 and R3 MYB repeat sequences. MYB domains (pfam00249) of peach PpeMYB25, cotton GhMYB25 (ACJ07153.1, [39]) and *Antirrhinum* AmMYBML1 (CAB433991.1, [54]) were aligned using the Muscle on line tool at EBI (http://www.ebi.ac.uk/Tools/msa/muscle/). Graphic display of the alignment was obtained using BoxShade (http://www.ch.embnet.org/software/BOX_form.html). Black shaded residues are identical, grey shaded residues are similar. Coordinates in the protein sequences are indicated.

induced mutations have played an important role in plant domestication [62]; examples include "sticky" foxtail millet [66], Mendel's wrinkled peas [67], seedless apple [68], fruit color in grape [69–71] and peach flesh [72,73]. In some cases, multiple independent mutational events were selected as demonstrated by insertion of TEs in the *waxy* locus resulting in the "sticky" phenotype in foxtail millet [66]. Another example is the yellow peach phenotype, which is associated with three different mutational events (an LTR insertion, a SNP and a frameshift mutation at a microsatellite locus) that occurred independently in a carotenoid cleavage dioxygenase gene directly involved in pigment degradation [72,73]. In contrast with the situation for "sticky" foxtail millet [66] and peach color [72,73] our results suggest that a unique mutational event has originated the nectarine phenotype, i.e. the loss of trichomes in peach fruit. The occurrence of unique mutations affecting single genes selected by humans during domestication and diversification of crop species is not rare. In a recent review of 60 such genes, 26 display a unique mutational event selected and spread by humans [63].

Gene ppa023143m encodes an R2R3-MYB transcription factor putatively involved in trichome formation

Careful inspection of predicted gene *ppa023143m* led us to reannotate the coding sequence (CDS) extending exon 3 compared to the Peach v1.0 annotation [8]. The reannotated CDS is predicted to encode a peptide of 330 aminoacids showing similarity to the R2R3-MYB transcription factors GhMYB25 from allotetraploid cotton *Gossypium hirsutum* (58.4% similarity) [40] and MIXTA-like1 from *Antirrhinum* (AmMYBML1, 55.3% similarity) [59] (Figure 4). In eukaryotes, MYB factors represent one of the largest and most functionally diverse gene families, which dramatically expanded in plants [74–77] playing a central role in a variety of processes from plant development to responses to biotic and abiotic stresses [74,78,79]. MYB family members share a highly conserved DNA binding domain (the MYB domain) usually composed of up to three amino acid repeats (R1, R2, R3) [80]. *GhMYB25* and *AmMYBML1* belong to R2R3-MYB sub-group 9 [40,77] along with other genes involved in regulating petal epidermal cell shape [81]. *AmMYBML1* plays a role in trichome differentiation in the corolla tube of the *Antirrhinum* flower [59]. *GhMYB25*, normally expressed at the time of fiber initiation in the outer integument of ovules, is differentially expressed between fibreless mutants and normal lined cotton [40,82,83] and its altered expression affects seed, fiber and trichome development in transgenic cotton [40]. Considering the known role of these homologues in trichome development, *ppa023143m* is a strong candidate for the peach/nectarine phenotype and was named *PpeMYB25*. The insertion of the Ty1-*copia* retrotransposon in exon 3 of the *PpeMYB25* gene introduces an H112L substitution and a premature stop codon (TAA), resulting in a peptide of 112 aminoacids precisely truncated at the C-terminal end of the R3 MYB domain. GhMYB25 and AmMYBML1 share the distinctive C-terminal motif of R2R3-MYB sub-group 9 [40,59,76] along with *PpeMYB25* and other genes involved in regulating epidermal cell shape [81]. Analysis of *Antirrhinum* R2R3-MYB genes from subgroup 9 indicated that the C-terminal domain folds as an amphipathic α-helix with putative transactivation ability [84]. An insertional mutant in the MIXTA gene, resulting in loss of this C-

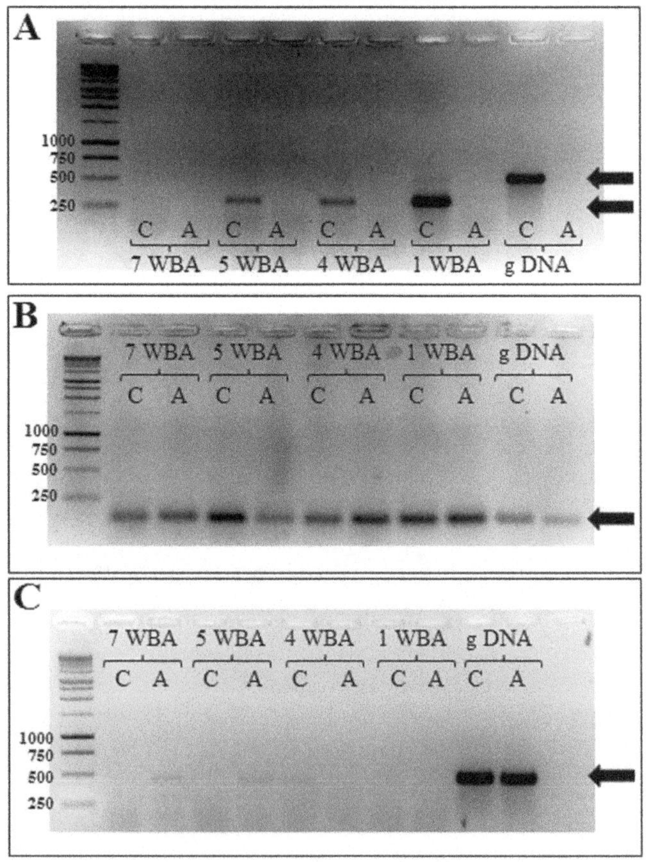

Figure 5. Expression analysis of *PpeMYB25* in peach and nectarine flower buds. (A) The expression patterns of the R2R3-MYB gene were evaluated in 'Contender' [C] and 'Ambra' [A] buds at seven, five, four and one week before anthesis (WBA). Genomic DNA of the two cultivars was also tested as a control. The same samples were analyzed for expression of *RPII* as standard (B) and checked for DNA contamination (C).

terminal region has been shown to cause recessive phenotypic alteration in epidermal cell differentiation [85]. The recessiveness of the nectarine trait indicates that it corresponds to a loss-of-function mutation. According to this reasoning and by analogy with observations in cotton, we propose that the observed insertion in the *PpeMYB25* gene results in a non-functional form of the MYB transcription factor that normally promotes trichome formation in fuzzy peaches. If this were correct, all nectarines should be non-functional homozygous mutants at this gene.

In order to evaluate the timing of *PpeMYB25* transcript expression with respect to trichome development, RT-PCR analyses were performed on 'Contender' and 'Ambra' floral buds sampled at seven, five, four and one week before anthesis. A forward primer designed on exon 2 was used in combination with a reverse primer on exon 3 (downstream of the LTR insertion) in order to evaluate the expression profiles of the gene. Expression was evident in 'Contender' from five weeks before anthesis, just before trichomes begin to appear [26] and continued through to one week before anthesis accompanying trichome differentiation (Figure 5). In contrast, the expression of *PpeMYB25* was never visible in 'Ambra' floral buds, consistent with the presence and the position of the large insertion (Figure 5). Together sequence and expression analyses support the proposed involvement of *PpeMYB25* gene in promoting trichome differentiation. Mutational events in regulatory genes have played a major role during crop domestication and breeding; in a recent overeview of domestication and diversification genes, 37 out of 60 (~62%) encoded transcription factors [63].

Association study and origin of the nectarine trait in peach germplasm

The genotype at the LTR retrotransposon insertion in gene *PpeMYB25* was assessed in the CxA F_2 population as well as in a panel of 95 peach and nectarine accessions by means of a functional marker based on a three primers PCR assay (indelG) (Figure 6). As expected, the marker co-segregated with the *G* locus in the CxA F_2 progeny (Figure 1), with all the nectarines displaying a unique fragment of 197 bp. Similarly, all nectarines in the germplasm panel were characterized by a fragment of the same length (Figure 6). Peach accessions fell into two categories: those homozygous for the reference allele (941 bp) and those heterozygous (197 bp, 941 bp). Taking into account pedigree information [15,86], we confirmed all the known heterozygotes ('Autumnglo', 'Fairtime', 'Fidelia', 'Jing Yu', 'O'Henry', 'Summer Pearl' and CxA F_1, Table 1) and we also detected three previously unknown heterozygous genotypes ('Baldagenais', 'City 32–82' and 'Yoshihime') carrying the nectarine allele originated from the LTR insertion. The complete association between the indelG marker and the trait confirms the presence of the Ty1-*copia* retrotransposon within exon 3 of the *PpeMYB25* gene in all the nectarines analyzed.

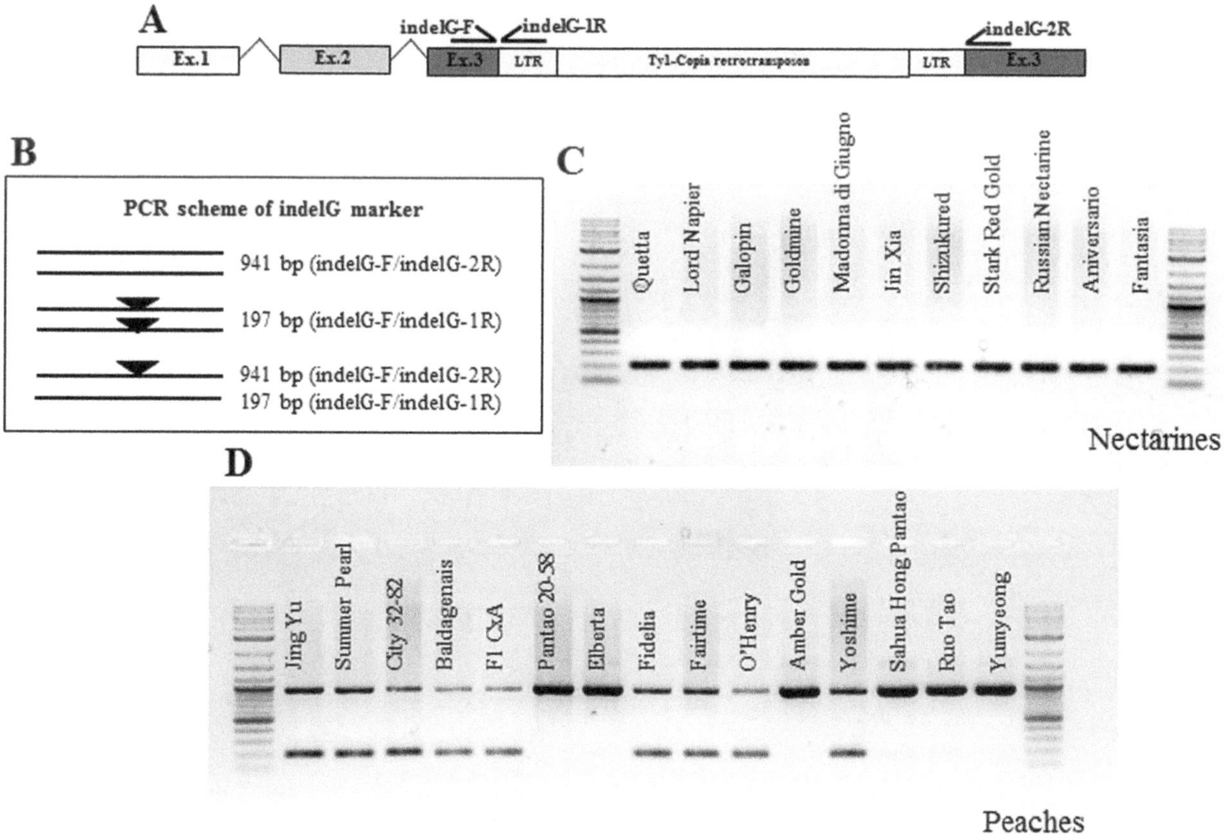

Figure 6. Functional Marker indelG. A marker assay was developed based on sequence information on the *PpeMYB25* gene and the Ty1-*copia* insertion. Three primers were designed to discriminate peach and nectarine genotypes (A, B). A panel of nectarines including the putative donors of the trait, show a unique fragment of about 200 bp (C). A set of peaches, of diverse pedigree and origins (Table 1) (D), shows homozygous or heterozygous patterns.

Modern western nectarines trace back to three founders discovered at the beginning of the last century ('Quetta' in Pakistan, and 'Lippiatt' and 'Goldmine' in New Zealand). The founders and their modern descendants show the presence of the retrotransposon insertion. The insertion is also present in Southern Italian traditional landraces ('Madonna di Giugno' and 'Madonna di Agosto') cultivated since the 16th century [12]. In addition to these genotypes, we confirmed the presence of the retrotransposon in several non-related accessions including old European landraces ('Lord Napier' and 'Galopin') [15], modern Asian cultivars (Chinese and Japanese) and different nectarine cultivars of unknown pedigree (from Italy, East Europe, USA, Mexico and South America). The modern Asian nectarines analyzed in this study ('Chiyodared', 'Shizukured' and 'Jin Xia') have two old European landraces, 'Lord Napier' and 'Precoce di Croncels', as donors of the nectarine trait. No traditional nectarine landraces have been reported in China [14] and all modern Chinese nectarines inherited the trait from western germplasm [17].

Together these results indicate that all known nectarine germplasm derives from a unique mutational event in *PpeMYB25* selected and spread by humans during peach dissemination and breeding.

Conclusions

Nectarines play an important role in the peach industry. In the present study, using a candidate gene approach coupled with fine mapping and NGS-based variant discovery, we provide strong evidence that the transcription factor gene *PpeMYB25* acts as a positive regulator of trichome formation in peach fruit. The insertion of a Ty1-*copia* retrotransposon within the third exon of *PpeMYB25* was identified as the putative cause of a loss-of-function mutation underlying the nectarine phenotype, further supporting the importance of transposition in plant genome evolution and phenotypic variability in domesticated crops. Finally, the development of a functional marker, indelG, provides an efficient diagnostic tool for the early selection of the peach/nectarine trait in marker assisted breeding (MAB).

Acknowledgments

We wish to thank S. Foschi (CRPV, Cesena, Italy) and M. Lama (ASTRA Innovazione, Faenza, Italy) for their valuable contribution in field tree management, as well as C. Cantini (Ivalsa-CNR Follonica, Italy) for providing leaf tissue for DNA analyses of the accessions 'Galopin' and 'Lord Napier'. We are grateful to D.S. Horner (DBS, Università di Milano, Italy) for critical comments and English revision of the manuscript. The authors thank the Centre for Applied Biomedical Research (CRBA) of Bologna for valuable contributions to the Sequenom analyses. The authors also thank Fondazione Cassa di Risparmio in Bologna for supporting CRBA. We acknowledge the Department of Rare Books and Special Collections, Princeton University Library (USA) for providing photo duplication of the New York Gazette March 28, 1768.

Author Contributions

Conceived and designed the experiments: GP ST IV LR. Performed the experiments: EV GP LD IP LG MTD. Analyzed the data: EV GP LD IP SS FS ST IV LR. Contributed reagents/materials/analysis tools: DB ST IV LR. Wrote the paper: EV GP ST IV LR. Critically revised the manuscript: LD IP MTD SS FS DB LG.

References

1. Pastorello E, Farioli L, Pravettoni V, Scibilia J, Mascheri A, et al. (2011) Pru p 3-sensitised Italian peach-allergic patients are less likely to develop severe symptoms when also presenting IgE antibodies to Pru p 1 and Pru p 4. Int Arch Allergy Immunol 156: 362–372.
2. Botton A, Vegro M, De Franceschi F, Ramina A, Gemignani C, et al. (2006) Different expression of Pp-LTP1 and accumulation of Pru p 3 in fruits of two *Prunus persica* L. Batsch genotypes. Plant Sci 171: 106–113.
3. Wang G, Tian L, Aziz N, Broun P, Dai X, et al. (2008) Terpene biosynthesis in glandular trichomes of hop. Plant Physiol 148: 1254–1266.
4. Harada E, Kim JA, Meyer AJ, Hell R, Clemens S, et al. (2010) Expression profiling of tobacco leaf trichomes identifies genes for biotic and abiotic stresses. Plant Cell Physiol 51: 1627–1637.
5. Blake MA (1932) The JH Hale peach as a parent in peach crosses. Proc Natl Acad Sci USA 29: 131–136.
6. Dirlewanger E, Cosson P, Boudehri K, Renaud C, Capdeville G, et al. (2006) Development of a second-generation genetic linkage map for peach [*Prunus persica* (L.) Batsch] and characterization of morphological traits affecting flower and fruit. Tree Genet Genomes 3: 1–13.
7. Le Dantec L, Cardinet G, Bonet J, Fouché M, Boudehri K, et al. (2010) Development and mapping of peach candidate genes involved in fruit quality and their transferability and potential use in other Rosaceae species. Tree Genet Genomes 6: 995–1012.
8. Verde I, Abbott AG, Scalabrin S, Jung S, Shu S, et al. (2013) The high-quality draft genome of peach (*Prunus persica*) identifies unique patterns of genetic diversity, domestication and genome evolution. Nat Genet 45: 487–494.
9. Faust M, Timon B (1995) Origin and dissemination of peach. John Wiley. Janik J, editor Hortticultural Reviews, Volume 17.
10. Yoon J, Liu D, Song W, Liu W, Zhang A, et al. (2006) Genetic diversity and ecogeographical phylogenetic relationships among peach and nectarine cultivars based on Simple Sequence Repeat (SSR) markers. J Amer Soc Hort Sci 131: 513–521.
11. Hesse CO (1975) Peach In: Advances in fruit breeding. Purdue Uni. Janick, J, Moore J, editor West Lafayette, Ind.
12. Venuto A (1516) De agricultura opusculum. Edizioni dell'Orso ISBN 978-88-6274-029-6.
13. Marchese A, Tobutt KR, Caruso T (2005) Molecular characterisation of Sicilian *Prunus persica* cultivars using microsatellites. J Hortic Sci Biotechnol 80: 121–129.
14. Okie WR, Bacon T, Bassi D (2008) Fresh Market Cultivar Development in: Peach, Botany, Production and Uses. CAB Intern. Lane DR and Bassi D, editor.
15. Okie WR (1998) Handbook of peach and nectarine varieties. U.S. Dept. of Agriculture, Agricultural Research Service; National Technical Information Service, distributor.
16. Yoshida M (1994) Horticulture in Japan. In: Kitagawa, H, Iwahori, S, Yakuwa, T, Konishi, K, editor. Peach. Tokyo: Asakura Publishing Co., Ltd. pp. 32–37.
17. Tian JB, Cheng F, Cheng H, Han F, Dong B (2012) Review and perspect of nectarine breeding in China. Acta Hortricolture 962: 97–103.
18. Esau K (1977)Anatomy of seed plants, 2nd edn. Wiley. New York.
19. Uphof JCT (1962) Plant Air, Encyclopedia of plant anatomy. Gerbruder. Zimmerman, W; Ozenda, PG, editor Berlin.
20. Jeffree CE (2006) The fine structure of the plant cuticle. In 'Biology of the plant cuticle'. Blackwell. Riederer M, Müller C, editor Oxford.
21. Stavrianakou S, Liakopoulos G, Miltiadou D, Markoglou A, Ziogas B, et al. (2010) Antifungal and antibacterial capacity of extracted material from non-glandular and glandular leaf hairs applied at physiological concentrations. Plant Stress 4: 25–30.
22. Xia Y, Yu K, Navarre D, Seebold K, Kachroo A, et al. (2010) The glabra1 mutation affects cuticle formation and plant responses to microbes. Plant Physiol 154: 833–846.
23. Lange BM, Wildung MR, Stauber EJ, Sanchez C, Pouchnik D, et al. (2000) Probing essential oil biosynthesis and secretion by functional evaluation of expressed sequence tags from mint glandular trichomes. Proc Natl Acad Sci 97: 2934–2939.
24. Iijima Y, Davidovich-Rikanati R, Fridman E, Gang DR, Bar E, et al. (2004) The biochemical and molecular basis for the divergent patterns in the biosynthesis of terpenes and phenylpropenes in the peltate glands of three cultivars of basil. Plant Physiol 136: 3724–3736.
25. Wilkins TA, Rajasekaran K, Anderson DM (2000) Cotton Biotechnology. CRC Crit Rev Plant Sci 19: 511–550.
26. Creller MA, Werner DJ (1996) Characterizing the novel fruit surface morphology of 'Marina' Peach Using Scanning Electron Microscopy. 121: 198–203.
27. Fernández V, Khayet M, Montero-Prado P, Heredia-Guerrero JA, Liakopoulos G, et al. (2011) New insights into the properties of pubescent surfaces: peach fruit as a model. Plant Physiol 156: 2098–2108.
28. Hülskamp M (2004) Plant trichomes: a model for cell differentiation. Nat Rev Mol Cell Biol 5: 471–480.
29. Jakoby MJ, Falkenhan D, Mader MT, Brininstool G, Wischnitzki E, et al. (2008) Transcriptional profiling of mature *Arabidopsis* trichomes reveals that NOECK encodes the MIXTA-like transcriptional regulator MYB106. Plant Physiol 148: 1583–1602.
30. Oppenheimer DG, Herman PL, Sivakumaran S, Esch J, Marks MD (1991) A myb gene required for leaf trichome differentiation in *Arabidopsis* is expressed in stipules. Cell 67: 483–493.

31. Payne CT, Zhang F, Lloyd AM (2000) GL3 encodes a bHLH protein that regulates trichome development in *Arabidopsis* through interaction with GL1 and TTG1. Genetics 156: 1349–1362.
32. Zhang F, Gonzalez A, Zhao M, Payne CT, Lloyd A (2003) A network of redundant bHLH proteins functions in all TTG1-dependent pathways of *Arabidopsis*. Development 130: 4859–4869.
33. Wang S, Chen JG (2008) *Arabidopsis* transient expression analysis reveals that activation of GLABRA2 may require concurrent binding of GLABRA1 and GLABRA3 to the promoter of GLABRA2. Plant Cell Physiol 49: 1792–1804.
34. Wada T, Tachibana T, Shimura Y, Okada K (1997) Epidermal cell differentiation in arabidopsis determined by a Myb homolog, CPC. Science (80) 277: 1113–1116.
35. Schellmann S, Schnittger A, Kirik V, Wada T, Okada K, et al. (2002) TRIPTYCHON and CAPRICE mediate lateral inhibition during trichome and root hair patterning in *Arabidopsis*. EMBO J 21: 5036–5046.
36. Kirik V, Simon M, Wester K, Schiefelbein J, Hulskamp M (2004) ENHANCER of TRY and CPC 2 (ETC2) reveals redundancy in the region-specific control of trichome development of *Arabidopsis*. Plant Mol Biol 55: 389–398.
37. Hauser MT, Harr B, Schlotterer C (2001) Trichome distribution in *Arabidopsis thaliana* and its close relative *Arabidopsis lyrata*: molecular analysis of the candidate gene GLABROUS1. Mol Biol Evol 18: 1754–1763.
38. Kivimäki M, Kärkkäinen K, Gaudeul M, Løe G, Agren J (2007) Gene, phenotype and function: GLABROUS1 and resistance to herbivory in natural populations of *Arabidopsis lyrata*. Mol Ecol 16: 453–462.
39. Li F, Zou Z, Yong HY, Kitashiba H, Nishio T (2013) Nucleotide sequence variation of GLABRA1 contributing to phenotypic variation of leaf hairiness in *Brassicaceae* vegetables. Theor Appl Genet 126: 1227–1236.
40. Machado A, Wu Y, Yang Y, Llewellyn DJ, Dennis ES (2009) The MYB transcription factor GhMYB25 regulates early fibre and trichome development. Plant J 59: 52–62.
41. Pirona R, Eduardo I, Pacheco I, Da Silva Linge C, Miculan M, et al. (2013) Fine mapping and identification of a candidate gene for a major locus controlling maturity date in peach. BMC Plant Biol 13: 166.
42. Eduardo I, Pacheco I, Chietera G, Bassi D, Pozzi C, et al. (2011) QTL analysis of fruit quality traits in two peach intraspecific populations and importance of maturity date pleiotropic effect. Tree Genet Genomes 7: 323–335.
43. Jurinke C,van den Boom D, Cantor CR, Köster H (2002) The use of MassARRAY technology for high throughput genotyping in: chip technology. Hoheisel J, Brazma A, Büssow K, Cantor CR, Christians FC, et al, editors Berlin, Heidelberg: Springer Berlin Heidelberg.
44. Van Ooijen JW, Voorrips RW (2002) Joinmap 3.0, Software for the calculation of genetic linkage maps. Wageningen, The Netherlands: Plant Research International.
45. Kosambi DD (1943) The estimation of map distances from recombination values. Ann Eugen 12: 172–175.
46. Wheeler DL, Barrett T, Benson DA, Bryant SH, Canese K, et al. (2008) Database resources of the National Center for Biotechnology Information. Nucleic Acids Res 36: D13–21.
47. Lohse M, Bolger AM, Nagel A, Fernie AR, Lunn JE, et al. (2012) RobiNA: a user-friendly, integrated software solution for RNA-Seq-based transcriptomics. Nucleic Acids Res 40: W622–W627.
48. Li H, Durbin R (2009) Fast and accurate short read alignment with Burrows-Wheeler transform. Bioinformatics 25: 1754–1760.
49. McKenna A, Hanna M, Banks E, Sivachenko A, Cibulskis K, et al. (2010) The Genome Analysis Toolkit: a MapReduce framework for analyzing next-generation DNA sequencing data. Genome Res 20: 1297–1303.
50. Verde I, Bassil N, Scalabrin S, Gilmore B, Lawley CT, et al. (2012) Development and evaluation of a 9K SNP array for peach by internationally coordinated SNP detection and validation in breeding germplasm. PLoS One 7: e35668.
51. DePristo MA, Banks E, Poplin R, Garimella K V, Maguire JR, et al. (2011) A framework for variation discovery and genotyping using next-generation DNA sequencing data. Nat Genet 43: 491–498.
52. Ye K, Schulz MH, Long Q, Apweiler R, Ning Z (2009) Pindel: a pattern growth approach to detect break points of large deletions and medium sized insertions from paired-end short reads. Bioinformatics 25: 2865–2871.
53. Li H, Durbin R (2010) Fast and accurate long-read alignment with Burrows-Wheeler transform. Bioinformatics 26: 589–595.
54. Finn RD, Clements J, Eddy SR (2011) HMMER web server: interactive sequence similarity searching. Nucleic Acids Res 39: W29–37.
55. Punta M, Coggill PC, Eberhardt RY, Mistry J, Tate J, et al. (2012) The Pfam protein families database. Nucleic Acids Res 40: D290–301.
56. Morgulis A, Coulouris G, Raytselis Y, Madden TL, Agarwala R, et al. (2008) Database indexing for production MegaBLAST searches. Bioinformatics 24: 1757–1764.
57. Tong Z, Gao Z, Wang F, Zhou J, Zhang Z (2009) Selection of reliable reference genes for gene expression studies in peach using real-time PCR. BMC Mol Biol 10: 71.
58. Lopez-Molina L, Mongrand S, Kinoshita N, Chua NH (2003) AFP is a novel negative regulator of ABA signaling that promotes ABI5 protein degradation. Genes Dev 17: 410–418.
59. Perez-Rodriguez M, Jaffe FW, Butelli E, Glover BJ, Martin C (2005) Development of three different cell types is associated with the activity of a specific MYB transcription factor in the ventral petal of *Antirrhinum majus* flowers. Development 132: 359–370.
60. Kohany O, Gentles AJ, Hankus L, Jurka J (2006) Annotation, submission and screening of repetitive elements in Repbase: RepbaseSubmitter and Censor. BMC Bioinformatics 7: 474.
61. Kumar A (1996) The adventures of the Ty1-*copia* group of retrotransposons in plants. Trends Genet 12: 41–43.
62. Lisch D (2013) How important are transposons for plant evolution? Nat Rev Genet 14: 49–61.
63. Meyer RS, Purugganan MD (2013) Evolution of crop species: genetics of domestication and diversification. Nat Rev Genet 14: 840–852.
64. Studer A, Zhao Q, Ross-Ibarra J, Doebley J (2011) Identification of a functional transposon insertion in the maize domestication gene tb1. Nat Genet 43: 1160–1163.
65. Doebley JF, Gaut BS, Smith BD (2006) The molecular genetics of crop domestication. Cell 127: 1309–1321.
66. Kawase M, Fukunaga K, Kato K (2005) Diverse origins of waxy foxtail millet crops in East and Southeast Asia mediated by multiple transposable element insertions. Mol Genet Genomics 274: 131–140.
67. Bhattacharyya MK, Smith AM, Ellis THN, Hedley C, Martin C (1990) The wrinkled-seed character of pea described by Mendel is caused by a transposon-like insertion in a gene encoding starch-branching enzyme. Cell 60: 115–122.
68. Yao JL, Dong YH, Morris BAM (2001) Parthenocarpic apple fruit production conferred by transposon insertion mutations in a MADS-box transcription factor. Proc Natl Acad Sci USA 98: 1306–1311.
69. Kobayashi H, Goto-Yamamoto N, Hirochika H (2004) Retrotransposon-induced mutations in grape skin Color. Science (80) 304: 982.
70. Cadle-Davidson MM, Owens CL (2008) Genomic amplification of the Gret1 retroelement in white-fruited accessions of wild vitis and interspecific hybrids. Theor Appl Genet 116: 1079–1094.
71. Shimazaki M, Fujita K, Kobayashi H, Suzuki S (2011) Pink-colored grape berry is the result of short insertion in intron of color regulatory gene. PLoS One 6: e21308.
72. Falchi R, Vendramin E, Zanon L, Scalabrin S, Cipriani G, et al. (2013) Three distinct mutational mechanisms acting on a single gene underpin the origin of yellow flesh in peach. Plant J.: 76 (2):175-187.
73. Adami M, Franceschi P, Brandi F, Liverani A, Giovannini D, et al. (2013) identifying a carotenoid cleavage dioxygenase (ccd4) gene controlling yellow/white fruit flesh color of peach. Plant Mol Biol Report 31: 1166–1175.
74. Dubos C, Stracke R, Grotewold E, Weisshaar B, Martin C, et al. (2010) MYB transcription factors in *Arabidopsis*. Trends Plant Sci 15: 573–581.
75. Wilkins O, Nahal H, Foong J, Provart NJ, Campbell MM (2009) Expansion and diversification of the *Populus* R2R3-MYB family of transcription factors. Plant Physiol 149: 981–993.
76. Stracke R, Werber M, Weisshaar B (2001) The R2R3-MYB gene family in *Arabidopsis thaliana* R. Curr Opin Plant Biol 4: 447–456.
77. Jia L, Clegg MT, Jiang T (2004) Evolutionary dynamics of the DNA-binding domains in putative R2R3-MYB genes identified from rice subspecies indica and japonica genomes. Plant Physiol 134: 575–585.
78. Zhang L, Zhao G, Jia J, Liu X, Kong X (2012) Molecular characterization of 60 isolated wheat MYB genes and analysis of their expression during abiotic stress. J Exp Bot 63: 203–214.
79. Glover B, Perez-Rodriguez M, Martin C (1998) Development of several epidermal cell types can be specified by the same MYB-related plant transcription factor. Development 125: 3497–3508.
80. Rosinski JA, Atchley WR (1998) Molecular evolution of the Myb family of transcription factors: evidence for polyphyletic origin. J Mol Evol 46: 74–83.
81. Martin C, Bhatt K, Baumann K, Jin H, Zachgo S, et al. (2002) The mechanics of cell fate determination in petals. Philos Trans R Soc Lond B Biol Sci 357: 809–813.
82. Wu Y, Machado AC, White RG, Llewellyn DJ, Dennis ES (2006) Expression profiling identifies genes expressed early during lint fibre initiation in cotton. Plant Cell Physiol 47: 107–127.
83. Lee JJ, Hassan OSS, Gao W, Wei NE, Kohel RJ, et al. (2006) Developmental and gene expression analyses of a cotton naked seed mutant. Planta 223: 418–432.
84. Jaffé FW, Tattersall A, Glover BJ (2007) A truncated MYB transcription factor from *Antirrhinum majus* regulates epidermal cell outgrowth. J Exp Bot 58: 1515–1524.
85. Noda K, Glover BJ, Linstead P, Martin C (1994) Flower colour intensity depends on specialized cell shape controlled by a Myb-related transcription factor. Nature 369: 661–664.
86. Aranzana MJ, Abbassi EK, Howad W, Arús P (2010) Genetic variation, population structure and linkage disequilibrium in peach commercial varieties. BMC Genet 11: 69.

7

Geographical Gradient of the *eIF4E* Alleles Conferring Resistance to Potyviruses in Pea (*Pisum*) Germplasm

Eva Konečná[1,3], Dana Šafářová[2], Milan Navrátil[2], Pavel Hanáček[1,3], Clarice Coyne[4], Andrew Flavell[5], Margarita Vishnyakova[6], Mike Ambrose[7], Robert Redden[8], Petr Smýkal[9]*

1 Department of Plant Biology, Mendel University in Brno, Brno, Czech Republic, **2** Department of Cell Biology and Genetics, Palacky University in Olomouc, Olomouc, Czech Republic, **3** CEITEC MENDELU, Mendel University in Brno, Brno, Czech Republic, **4** Western Regional Plant Introduction Station - USDA, Pullman, Washington, United States of America, **5** Division of Plant Sciences, University of Dundee at James Hutton Institute, Invergowrie, United Kingdom, **6** Vavilov Institute of Plant Industries, Saint Petersburg, Russian Federation, **7** John Innes Centre, Norwich, United Kingdom, **8** Australian Grains Genebank, Horsham, Victoria, Australia, **9** Department of Botany, Palacky University in Olomouc, Olomouc, Czech Republic

Abstract

Background: The eukaryotic translation initiation factor 4E was shown to be involved in resistance against several potyviruses in plants, including pea. We combined our knowledge of pea germplasm diversity with that of the *eIF4E* gene to identify novel genetic diversity.

Methodology/Principal findings: Germplasm of 2803 pea accessions was screened for *eIF4E* intron 3 length polymorphism, resulting in the detection of four $eIF4E^{A\text{-}B\text{-}C\text{-}S}$ variants, whose distribution was geographically structured. The $eIF4E^{A}$ variant conferring resistance to the P1 PSbMV pathotype was found in 53 accessions (1.9%), of which 15 were landraces from India, Afghanistan, Nepal, and 7 were from Ethiopia. A newly discovered variant, $eIF4E^{B}$, was present in 328 accessions (11.7%) from Ethiopia (29%), Afghanistan (23%), India (20%), Israel (25%) and China (39%). The $eIF4E^{C}$ variant was detected in 91 accessions (3.2% of total) from India (20%), Afghanistan (33%), the Iberian Peninsula (22%) and the Balkans (9.3%). The $eIF4E^{S}$ variant for susceptibility predominated as the wild type. Sequencing of 73 samples, identified 34 alleles at the whole gene, 26 at cDNA and 19 protein variants, respectively. Fifteen alleles were virologically tested and 9 alleles ($eIF4E^{A\text{-}1\text{-}2\text{-}3\text{-}4\text{-}5\text{-}6\text{-}7}$, $eIF4E^{B\text{-}1}$, $eIF4E^{C\text{-}2}$) conferred resistance to the P1 PSbMV pathotype.

Conclusions/Significance: This work identified novel *eIF4E* alleles within geographically structured pea germplasm and indicated their independent evolution from the susceptible $eIF4E^{S1}$ allele. Despite high variation present in wild *Pisum* accessions, none of them possessed resistance alleles, supporting a hypothesis of distinct mode of evolution of resistance in wild as opposed to crop species. The Highlands of Central Asia, the northern regions of the Indian subcontinent, Eastern Africa and China were identified as important centers of pea diversity that correspond with the diversity of the pathogen. The series of alleles identified in this study provides the basis to study the co-evolution of potyviruses and the pea host.

Editor: João Pinto, Instituto de Higiene e Medicina Tropical, Portugal

Funding: This work was supported by the Ministry of Agriculture IQ91A229 project. CC was supported by USDA (United States Department of Agriculture) project #5348-2100 = 017-00D. The funders had no role in study design, data collection and analysis, decision to publish, or preparation of the manuscript.

Competing Interests: The authors have declared that no competing interests exist.

* E-mail: petr.smykal@upol.cz

Introduction

Crop genetic diversity is an important pre-requisite for improving crop traits through breeding, particularly since the presence of closely related and genetically uniform varieties provides an ideal genetic environment for disease epidemics to occur, as evidenced by several historical and also recent events (the 1846 potato blight in Ireland, the 1970 corn blight in the USA, or the 1999 wheat rust in Africa). Available crop genetic resources function as reservoirs of often yet undiscovered allelic variants that provide an opportunity for genetic improvement of a given cultivated species [1,2]. However, the identification of specific, often rare traits requires specific screening and testing of entire, often very large collections. This is a time- and resource-intensive process. This situation improves, however, once a respective underlying gene is identified, especially in the case of monogenic traits. Various available genomic technologies [3–7] can be applied to uncover such variation. Such screening of wild and cultivated germplasm for allelic variation of identified resistance genes has been receiving increased attention [2,8], as it can be efficiently substituted for phenotypic characterization.

Pea is an ancient legume crop, originating and domesticated in the Middle East and Mediterranean regions, from where its cultivation spread to today's Russia, and westwards through the Danube Valley and/or through ancient Greece and Rome into Europe. Pea likewise spread eastward into Persia, India and China [9]. Archaeological evidence indicates that the first instance of pea dates back to 8000 B.C. [10] in the Near East. Later, during the Stone and Bronze Ages, the existence of pea was documented in Europe and India. From there, it entered China by the first century B.C. [11]. Wild *Pisum sativum* subsp. *elatius* and subsp.

sativum are found naturally in Europe, Northwestern Asia and in temperate Africa, while *P. fulvum* range is restricted to the Middle East. *Pisum abyssinicum* is found in Ethiopia and Yemen [12,13]. There are 25 larger sized germplasm collections preserving pea diversity, which collectively hold around 72,000 accessions, while an additional 27,000 accessions are maintained in 146 smaller collections around the world [14,15]. The diversity of these collections has been studied using both by morphological descriptors and agronomical traits as well as molecular markers (reviewed in Smýkal et al. [14]), and diversity core collections were formed thereafter [16–19]. These studies showed that although *Pisum* is a comparably small genus containing two or three species, it has a wide and structured diversity, showing a range of degrees of relatedness that reflect taxonomic identifiers, eco-geography and breeding gene pools [13,17].

Along with abiotic stresses, plant pathogens are a major detriment to agriculture and threaten global food security. The use of genetic resistance is considered to be the most effective and sustainable strategy for controlling plant pathogens in agricultural practice, as it is environmentally friendly, targets specific pathogens, and provides reliable protection without additional labor or material costs [20]. Long before plants were domesticated and grown as monocultures, plant pathogens were co-evolving with wild plants growing in mixed-species communities. Evolution has continued to occur within domesticated plants growing as selected genotypes in denser populations than in the wild. Furthermore, the domestication of wild plants has distributed crops far from their places of origin [21,22], and the introduction of pathogens eventually accompanied this distribution. This co-evolutionary process shaped both plants and their pathogens, including viruses [23–25].

Pea seed borne mosaic virus (PSbMV), member of the genus *Potyvirus*, has been since identification in Czechoslovakia [26] reported worldwide and causes serious yield losses in a broad spectra of legumes including the most economically important like pea, lentil, faba bean, and chickpea [27]. The virus causes various symptoms, depending on the host and virus isolate/pathotype, such as the downward rolling of leaflets, the transient clearing and swelling of leaf veins, chlorotic mosaics, stunting, and delayed flowering. PSbMV is transmitted between plants in a non-persistent manner by aphids and then infects seeds [28]. Due to its seed-borne transmission, PSbMV presents a serious phytosanitary risk both for germplasm maintenance [29] and seed production.

The pea genome contains two virologicaly defined clusters of recessive resistance genes that are responsive to various potyviruses. One cluster (on pea linkage group II) includes *bcm, cyv-1, mo, sbm-2* and *sbm-3* loci, conferring resistance to *Bean common mosaic virus* (BCMV) *Clover yellow vein virus* (ClYVV), *Bean yellow mosaic virus* (BYMV-S s pathotype) and *Pea seed-borne mosaic virus* (PSbMV, pathotype P2) respectively. The second cluster (on pea linkage group VI) includes *cyv-2, wlv* and *sbm-1* loci, conferring resistance to *Clover yellow vein virus* (ClYVV), *Bean yellow mosaic virus* (BYMV-W pathotype) and the *Pea seed-borne mosaic virus* (PSbMV, pathotype P1), respectively [30,31]. With exception of *cyv-2* and *sbm-1* loci, shown to be identical [32,33], it is not clear if these clusters are closely linked separate genes or the same gene with alleles of different specificity.

The pea *eIF4E* gene was identified as a susceptibility factor corresponding to the recessive resistance gene, *sbm-1* locus, and a homologue *eIF(iso)4E* is presumed to be *sbm-2* locus [31–35]. PSbMV has been well-studied genomically [36–38] and viral P3-6K1 and VPg proteins have been identified as PSbMV determinants [30] responsible for physical interaction with host eIF4E or eIF(iso)4E proteins and critical for viral infection. Studies on pepper and *Arabidopsis* suggest that potyviruses may selectively use either eIF4E and/or eIF(iso)4E proteins to achieve infection [39,40].

Recessive resistance of pea to PSbMV corresponds with the matching-allele model, explaining the interaction between potyviruses and plant hosts [41]. The mutant resistant allele of pea *eIF4E* (named *eIF4E^{A}* in this study) differs from its wild-type (sensitive) counterpart by five non-conservative amino acids [32,31,42]. Natural variation and functional analysis have revealed evidence of co-evolution between *eIF4E* and potyviral VPg [3,43,44]. The *eIF4E* allelic diversity has been systematically screened in various crop collections, such as pepper [5,45,46], melon [4], tomato [47] and barley [3]. A possible link between the spread of potyviruses and the origin of agriculture was demonstrated by Gibbs et al. [48], who clearly showed that the human-mediated spread of crop hosts was followed by the further diversification of viruses.

In this study, we have combined our knowledge of pea germplasm diversity with that of the *eIF4E* resistance gene and systematically screened 2803 accessions with known geographical origins including *Pisum* species with the aim of identifying allelic diversity present within the broader pea germplasm held in *ex situ* collections.

Results

Geographical distribution of eIF4E variants

A total of 2803 pea accessions were screened for *eIF4E* intron 3 length polymorphism using two sets of primer combinations, resulting in the detection of four respective *eIF4E$^{A\text{-}B\text{-}C\text{-}S}$* variants (Table 1, Fig. 1). The first corresponded to the already known *eIF4E^{A}* resistance [31] variant (amplified 243 bp fragment with primer combination A and 536 bp with primer combination B), the second to the susceptible *eIF4E^{S}* [31] (293 and 586 bp fragments), the third to the novel *eIF4E^{B}* (293 and 536 bp fragments) and the fourth to the novel *eIF4E^{C}* (293 and 592 bp fragments). The resistance *eIF4E^{A}* variant was found in seven accessions from Ethiopia-Sudan, fourteen accessions from India, five accessions from Pakistan-Nepal and one accession each from China, Russia and Afghanistan. In addition, this variant was detected in fifteen USA and Canadian, two South American and eleven European modern varieties or breeding lines, which have the Ethiopian line PI193835 in their pedigree. In total, the *eIF4E^{A}* variant was detected in 53 accessions, which represent 1.9% of those tested. The *eIF4E^{B}* variant was abundant in accessions of Ethiopian (29%), Afghan (23%), Nepal-Indian (20%), Caucasus (6.4%) and particularly Chinese (36.5%) origin (Table 2, Fig. 2). Moreover, it was found in five (e.g. 20%) accessions of South American origin. In contrast, this variant was under-represented in landraces of European origin, except in the Balkans (4 acc.), and absent from modern pea varieties. The *eIF4E^{C}* variant was the most frequent in Iberian peninsula (25%) and India-Nepal (16.6%), followed by Balkan (9%), Ethiopia (6%), and Afghan (6.2%) locations, and occurring in total of 91 (3.2%) accessions (Table 2). Due to bulking of 10 plants per sample in the case of Chinese origin accessions (ATFCC), we detected high proportion (15%) of sample heterogeneity. This heterogeneity was not possible to test in USDA, IPK, CGN and JIC samples, as they originated from single plants, while in CzNPC and VIR samples (originated again from bulks of 10 plants per samples) prevailed susceptible alleles. Finally, the susceptible *eIF4E^{S}* variant was found in 53.8% to 100% proportion within studied regions, in total 2331 accessions (83%). The lowest occurrence was in 145 studied accessions from India (53.8%), followed by China (63%) and Ethiopia (63%). In contrast, within the 1145 analyzed European origin accessions, including modern pea varieties, it was predominant (97%) (Fig. 2, Table 2). It is clear that there is bias towards European (1145 acc.) compare to other geographical regions, affecting likely allele distribution resulting in S allele over-representation.

Isolation of eIF4E sequences from the selected accessions

In order to assess the genetic diversity within the four classes (A, B, C and S) and to test whether intron 3 lengths correspond to DNA and amino acid haplotypes; we sequenced 32, 25, 9 and 7 accessions with respective *eIF4E*$^{A-B-C-S}$ variants from selected geographically diverse regions: Turkey, Nepal, Pakistan, Afghanistan, Ethiopia and China (Fig. 2, Table 2, 3, Table S2). The total length of genomic DNA clones of *eIF4E*$^{A-B-C-S}$ variants from ATG to stop codons was 2102, 2080, 2208, 1973–2892 bp long, respectively (Table S2), and length polymorphism was largely conferred by 50 and 56 bp insertions/deletions of minisatellite-like repeat sequences located at 102 and 421 bp respectively from the beginning of intron 3 (Fig. 1, Table 1). The extensive variation of *eIF4E* ^{S} was due to wild *Pisum* samples, especially of more distant subspecies (*P. abyssinicum*) and species (*P. fulvum*). Beside intron 3 polymorphism, extensive nucleotide variation was found; altogether, there were 156 SNPs identified in 73 sequences (Fig. 1, Table S2) resulting in 34 alleles at whole gene (exones-introns), 26 at cDNA and 19 alleles at protein levels, respectively. Of these 156 SNPs, 62 were unique, e.g. occurred only in once within the sequenced samples.

There were four different sequences of susceptible *eIF4E*^{S} found within 25 sequenced samples of the cultivated *Pisum sativum* gene pool and three additional sequences within 7 samples of wild pea species (Table 3, Table S2). Translation of computationally spliced cDNA has resulted in four (*eIF4E*$^{S-1-2-3 \text{ and } 7}$) protein variants within the cultivated genepool and an additional three (*eIF4E*$^{S-4-5-6}$) in *P. fulvum* and *P. abyssinicum* (Fig. 3, Table S2). The typically resistant *eIF4E*$^{A-1}$ allele [31,39], as represented by the PI193835 accession from Ethiopia (genebank sequence number GU289735), differed from the *eIF4E*$^{S-1}$ allele (represented by JI194, genebank sequence number KF053441) by five nucleotide (G185T, C218A, C221A in exon 1, G410A in exon 2 and T687G in exon 3) exchanges, while the *eIF4E*$^{A-6}$ allele (PI269818 accession, genebank sequence number KF053455) differed by an additional two nucleotide (G217C in exon 1, C993T in intron 3) exchanges, one AGC triplet (AGC229-231) deletion, and one single base (C768T) deletion, in addition to 1200 bp intron 3 length similar to susceptible 1201 bp (Table 2, Table S2). Within the 28 sequenced accessions with *eIF4E*^{B} intron length variant, we detected seven alleles, both at nucleotide (Fig. 1) and amino acid (Fig. 3) levels. The three *eIF4E*$^{B-1-2-3}$ alleles found in nine sequenced samples differed from the *eIF4E*^{S} alleles in 40 SNPs and one nucleotide insertion, resulting in four DNA and three protein variants. The deduced amino acid sequence of the *eIF4E*$^{B-1}$ allele differed from the susceptible *eIF4E*^{S} allele at three amino acid (M207I and LD218-219QE) exchanges in exons 4 and 5. Two accessions from Turkey (JI1370, JI1090) differed by an additional three SNPs in exon 1, leading to two amino acid exchanges, V23D and V49A, resulting in two *eIF4E*$^{B-2-3}$ alleles (Table 3). We have detected two *eIF4E*^{C} alleles among seven sequenced samples. The *eIF4E*$^{C-2}$ allele of the VIR1859 accession, besides having a 1257bp long intron 3 due to its extra copy of a second 56 bp repeat, differed by two (A56T in exon 1, G1872A in intron 3) nucleotides from the *eIF4E*^{S} allele, leading to single amino acid (N19I) exchange at exon 1 (Fig. 3). The remaining six sequences of *eIF4E*^{C} were identical to the *eIF4E*^{S} allele, except for a 56 bp insertion in intron 3 and a single nucleotide exchange at intron 3 in PI505122 and PI639981, accessions from Albania and Bulgaria (Table 2, S2). We were interested in establishing which, if any, of the *eIF4E A, B* or *C* alleles can be found in wild pea. It turned out that all analyzed wild pea accesions displayed the *eIF4E*^{S} variant as assessed by intron 3 length. Similarly, sequencing analysis of the seven selected accessions had shown that all *P. sativum* subsp. *elatius*, except for *P. fulvum* JI1007, JI1010 and *P. abyssinicum* JI1632 accessions with one or two amino acid exchanges, had the typically susceptible *eIF4E*$^{S-1}$ allele (Table 2, 3, S2). All had substantial sequence polymorphism compare to cultivated pea except for *P. sativum* subsp. *elatius* JI2630 from the Crimea.

Haplotype network analysis

Haplotype network analysis was used to reveal relationships between sequences. This showed the clear separation of *Pisum sativum* subsp. *elatius* (JI3157, JI1091, JI2630), *Pisum fulvum* (JI1007, JI1010) and *Pisum abyssinicum* (JI1632) accessions from cultivated germplasm (Fig. 4). The only exception was *Pisum sativum* subsp.

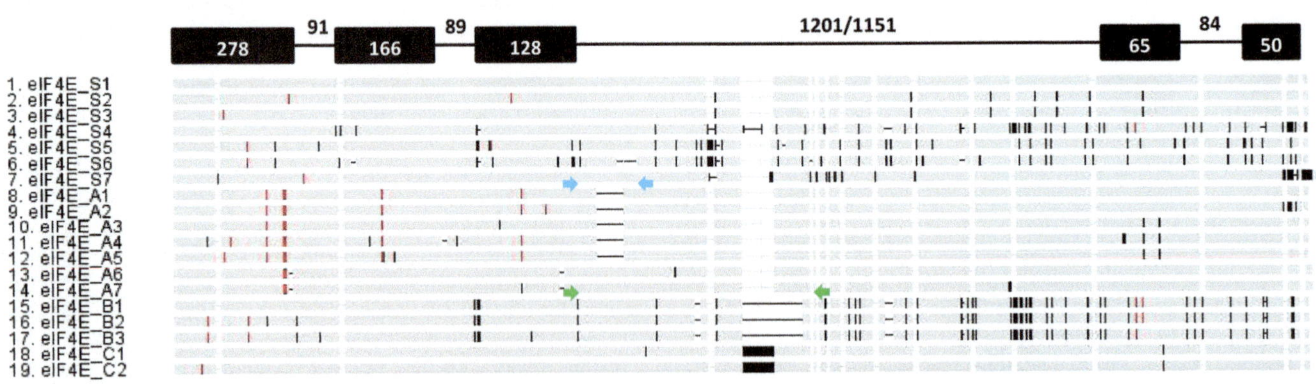

Figure 1. Schematic representation of sequence alignment of the all 19 identified protein *eIF4E* alleles. Four principal *eIF4E* variants identified by intron 3 polymorphism are designated A,B,C and S, while numbers indicate the respective allelic variant. Black bars indicate polymorphism nucleotides within both exons and introns, while red bars indicate polymorphism leading to amino acid exchanges. Horizontal lines indicate insertions/deletions. The heading line indicates nucleotide numbers and exon (solid black boxes)-intron (lines) positions and sizes. Blue arrows indicate primer A combination (Ps-eIF4E-750F and Ps-eIF4E-586gR) and green arrows indicate primer B combination (Ps-eIF4E-750F and Ps-eIF4E-1270R).

Figure 2. Geographical distribution of four *eIF4E* alleles expressed as percentage of total. Comparison of allele frequencies in 13 geographical regions, as detailed in Table 2.

elatius JI261 (genebank sequence number KF053440) from Turkey, which had a sensitive allele at the amino acid level, while exhibiting a transitory stage towards the $eIF4E^B$ allele at the nucleotide level (Table S2). Moreover, intron 3 was 36 bp longer, with a length of 1187 bp. Interestingly, the JI261 sequence was identical to $eIF4E^S$ until position 956 of intron 3, from which point forward it was similar to the $eIF4E^B$ allele. An important difference was the absence of the exon 4 and 5 exchanges, where it was again identical to the susceptible $eIF4E^S$ allele. The JI261 sequence occurred between the $eIF4E^S$ and $eIF4E^B$ alleles, with 23 mutational steps. A further 24 mutations separated the $eIF4E^B$ alleles of JI1194 (genebank sequence number KF053439) of Afghanistan, ATC7134 (KF053433) of China, and JI1090 (KF053434) and JI1370 (KF053432) of Turkey (Fig. 4). The $eIF4E^A$ alleles identified in Indian accessions (PI347464, PI347494, PI378158) proved to be derived from Ethiopian (PI193835, PIS479), Afghan (PIS468) and Indian (PI356991, ATC1044) accessions, and were identical. A further cluster contained Chinese (ATC6931, ATC6927) and Indian (PI347484, PI347422 and PI347328) origin S alleles, both at nucleotide and amino acid levels (Fig. 4).

Testing for response to PSbMV infection

The functional status of the 15 alleles of all four detected variants was verified through biological testing of their response to P1 PSbMV infection. All 19 of the pea accessions bearing $eIF4E^{A-1}$ tested showed resistance to viral infection i.e. the same phenotype as the non-infected or resistant controls (Table S2). DAS-ELISA testing proved negative, indicating the absence of viral coat protein. Six samples had $eIF4E^{A-2-3-4-5-6-7}$ alleles with single (PI116056, PI193584, JI1788) or multiple (PI269818, PI269774) amino acid exchanges, but none of these changed the expression of resistance in these accessions. The eight tested accessions bearing the $eIF4E^{B-1}$ allele split into two phenotypic groups. Three accessions (ATC7134 from China and CGN3311, PIS477 from Pakistan, all $eIF4E^{B-1}$) were resistant to the P1 pathotype of PSbMV, as the virus was not able to replicate and spread. The second group of accessions (ATC7140, ATC6928 from China, CGN3319 from Pakistan, all $eIF4E^{B-1}$; and JI1090, $eIF4E^{B-2}$, from Turkey), showed the typical susceptibility symptoms of infection, vein clearing and chlorotic leaf mosaics, leaf size decrease, internode shortening, and sporadically mild leaf-rolling, 14 days after inoculation by PSbMV. The symptoms were comparable

Table 1. Allele designation based on intron 3 length polymorphism, indicating size of obtained fragments with primer combinations A and B, number of repeat motives, intron 3 difference and total length.

eIF4E variant	Primer combination A product (bp)	Primer combination B product (bp)	copies of 50 bp repeat	copies of 56 bp repeat	intron 3 difference	intron 3 length (bp)
S	293	586	2	2	0	1201
A	243	586	1	2	−50 bp	1151
B	293	536	2	1	−70 bp	1131
C	293	592	2	3	+56 bp	1257

Table 2. Summary of germplasm screening of 2803 pea accessions, indicating four principal *eIF4E* intron 3 length variants (A-B-C and S), distribution in total numbers and percentage over 13 geographical regions.

Regions						Percentage				
	Number of accesions	A variant	B variant	C variant	S variant		A variant	B variant	C variant	S variant
Ethiopia - East Africa	349	7	101	21	220		2.0	28.9	6.0	63.0
Afghanistan-Pakistan	178	1	41	11	125		0.6	23.0	6.2	70.2
India-Nepal	145	14	29	24	78		9.7	20.0	16.6	53.8
China-Mongolia	271	1	99	1	170		0.4	36.5	0.4	62.7
Turkey-Syria	64	0	3	1	60		0.0	4.7	1.6	93.8
Israel-Lebanon-Jordan	28	0	7	1	20		0.0	25.0	3.6	71.4
Balkan	75	1	4	7	63		1.3	5.3	9.3	84.0
Russia-Caucassus	295	1	19	2	273		0.3	6.4	0.7	92.5
Iberian Peninsula	36	0	2	9	25		0.0	5.6	25.0	69.4
Australia-New Zealand	11	0	0	0	11		0.0	0.0	0.0	100.0
South America	25	2	5	2	16		8.0	20.0	8.0	64.0
Europe	1145	15	15	7	1108		1.3	1.3	0.6	96.8
USA-Canada	181	11	3	5	162		6.1	1.7	2.8	89.5
Total	**2803**	**53**	**328**	**91**	**2331**	**Average**	**1.9**	**11.7**	**3.2**	**83.2**

Table 3. List of sequenced 73 accessions assigned to identified 19 protein alleles of four *eIF4E* $^{A\text{-}B\text{-}C\text{-}S}$ intron 3 length variants with indicated country of origin.

eIF4E allele	Number of accessions with given allele	Accession(s)	Origin
eIF4E^{S-1}	24	JI 182, JI 1785, JI 1030, JI 190, PI 357290, JI 193, JI 194, JI 205, JI 1107, JI 1085, JI 2065, JI 1845, JI 1532, JI 2607, JI 1121, JI 1756, JI 2571, JI 3001, JI 2630, JI 1091, JI 3157, JI 261, ATC_7173, ATC_6927	Nepal, Nepal, Iran, Sudan, Macedonia, Sudan, Sudan, Russia, Nepal, Turkey, Tanzania, Greece, Hindukush, Lybia, Nepal, Nepal, Gruzie, Iran, Crimea, Greece, Turkey, Turkey, China, China
eIF4E^{S-2}	1	PI 347328	India
eIF4E^{S-3}	3	PI 347422, PI 347484, ATC_6931	India, India, China
eIF4E^{S-4}	1	JI 1632	Ethiopia
eIF4E^{S-5}	1	JI 1007	Israel
eIF4E^{S-6}	1	JI 1010	Israel
eIF4E^{S-7}	1	JI 2646	Malawi
eIF4E^{A-1}	19	JI 967, JI 467, JI 1790, ATC_1044, PIS_468, PIS_479,CGN_3302, PI 116056, PI 193584,PI 249645, PI 356991, PI 356992, PI 347494, PI 347492, JI 1260, PI 193586, JI 2643, JI 1787, JI 1788	Ethiopia, Europe, Afghanistan, Afghanistan, Afghanistan, Ethiopia, Ethiopia, India, Ethiopia, India, India, India, India, India, India, Ethiopia, Malawi, Nepal, Nepal
eIF4E^{A-2}	1	JI 1546	Ethiopia
eIF4E^{A-3}	1	PI 193835	Ethiopia
eIF4E^{A-4}	1	PI 378158	India
eIF4E^{A-5}	1	PI 347464	India
eIF4E^{A-6}	1	PI 269818	England
eIF4E^{A-7}	1	PI 269774	England
eIF4E^{B-1}	7	ATC_7134, CGN_3311, CGN_3319, ATC_7140, PIS_477, JI 1194, ATC_3275	China, Pakistan, Pakistan, China, Hindukush-Pakistan, Afghanistan, China
eIF4E^{B-2}	1	JI 1370	Turkey
eIF4E^{B-3}	1	JI 1090	Turkey
eIF4E^{C-1}	6	JI 267, PI 505122, PI 639981, JI1109, JI1104, JI1108	Greece, Albania, Bulgaria, Nepal, Nepal, Nepal
eIF4E^{C-2}	1	VIR 1589	Russia, Sverdlovsk

	19	23	34	49	62*	73-74*	77	86	107*	108	115	148	163	169*	186	207	218-219
S allele																	
eIF4E^{S-1}	N	V	E	V	W	AA	S	S	G	A	H	L	I	N	N	M	LD
eIF4E^{S-2}	N	V	E	V	W	AA	R	S	G	A	H	L	T	N	N	M	LD
eIF4E^{S-3}	N	V	A	V	W	AA	S	S	G	A	H	L	I	N	N	M	LD
eIF4E^{S-4}	N	V	E	V	W	AA	S	S	G	A	H	L	I	N	N	I	LD
eIF4E^{S-5}	N	V	E	A	W	AA	S	S	G	A	H	W	I	N	N	M	LD
eIF4E^{S-6}	N	V	E	A	W	AA	S	S	G	A	H	L	I	N	N	M	LD
eIF4E^{S-7}	N	V	E	V	W	AA	S	C	G	A	H	L	I	N	N	M	LD
A allele																	
eIF4E^{A-1}	N	V	E	V	L	DD	S	S	R	A	H	L	I	K	N	M	LD
eIF4E^{A-2}	N	V	E	V	L	DD	S	S	R	A	H	L	I	K	H	M	LD
eIF4E^{A-3}	N	V	E	V	L	DD	S	S	R	A	H	L	I	N	N	M	LD
eIF4E^{A-4}	N	A	E	V	L	DD	S	S	R	A	H	L	I	K	N	M	LD
eIF4E^{A-5}	N	V	A	V	L	DD	S	S	R	P	R	L	I	K	N	M	LD
eIF4E^{A-6}	N	V	E	V	W	PD	_	S	G	A	H	L	I	N	N	M	LD
eIF4E^{A-7}	N	V	E	V	W	PD	_	S	G	A	H	L	I	K	N	M	LD
B allele																	
eIF4E^{B-1}	N	V	E	V	W	AA	S	S	G	A	H	L	I	N	N	I	QE
eIF4E^{B-2}	N	D	E	A	W	AA	S	S	G	A	H	L	I	N	N	I	QE
eIF4E^{B-3}	N	D	E	A	W	AA	S	S	G	A	H	L	I	N	N	I	LD
C allele																	
eIF4E^{C-1}	N	V	E	V	W	AA	S	S	G	A	H	L	I	N	N	M	LD
eIF4E^{C-2}	I	V	E	V	W	AA	S	S	G	A	H	L	I	N	N	M	LD

Figure 3. Summarized amino acid exchanges of identified 19 *eIF4E* alleles. Asterix indicates the position of exchanges previously identified as crucial for resistance.

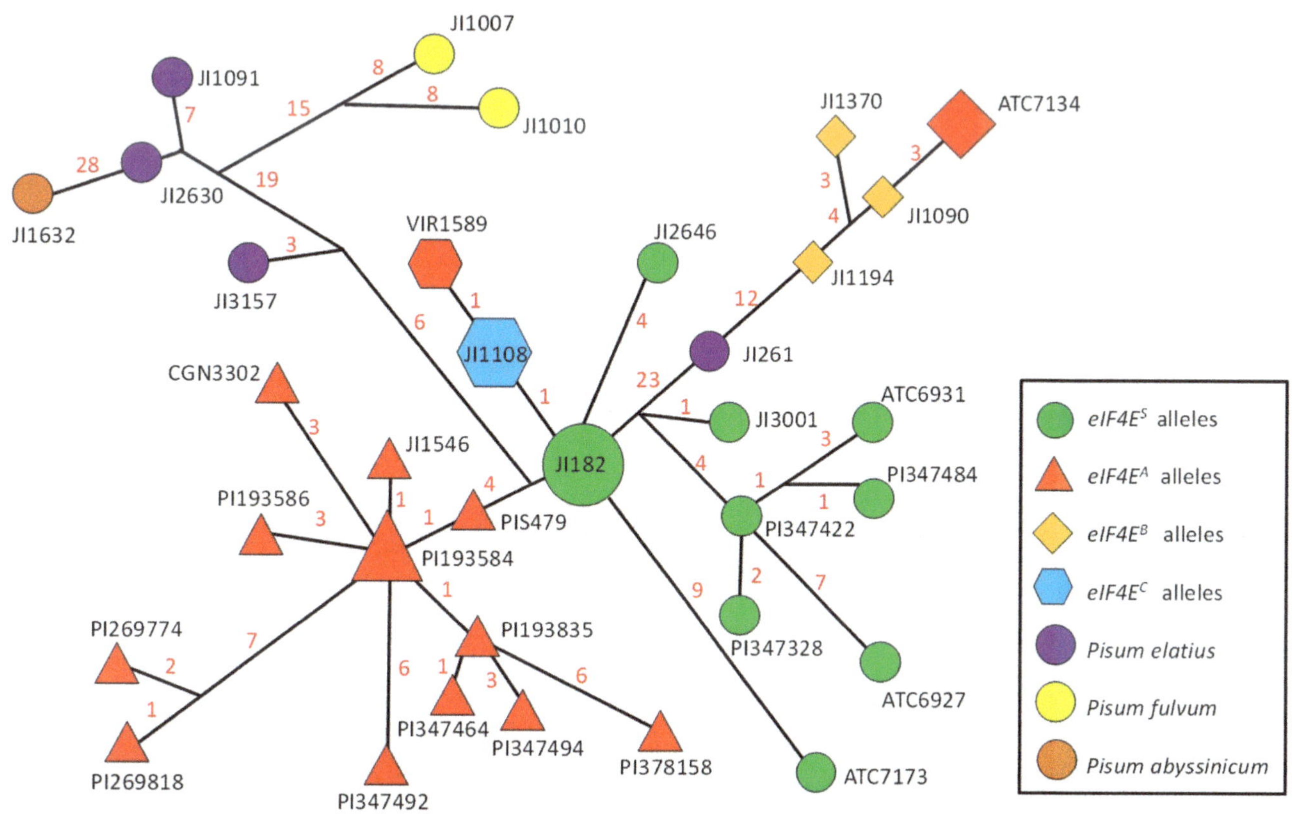

Figure 4. Haplotype network of 34 alleles identified at *eIF4E* whole gene level, using median-joining network algorithm, implemented in NETWORK. It is based on total of 156 SNP characters, and excludes 50, 56 bp indels in intron 3, as these would compromise analysis. Red numbers above lines indicate mutational steps. The shortest tree has 235 mutations. Triangles indicate A, squares B, hexagons C and circles S variants as detailed in Table 1. Red colour of symbols indicates accessions tested resistant to P1 PSbMV. Size of symbols is proportional to number of accessions with given haplotype.

with the response of the sensitive genotypes with the *eIF4E*$^{S-1}$ allele. All tested *eIF4E*$^{S-1}$ (18 accesions), *eIF4E*$^{S-2}$ (1 acc.) and *eIF4E*$^{S-3}$ (3 acc.) displayed sensitive response to P1 PSbMV pathotype (Table S2). We have not been able to test remaining *eIF4E*$^{S-4,5,6 \text{ and } 7}$ alleles, due to the shortage of homogenous plant material. Finally, of the three tested accessions with the *eIF4E*$^{C-1-2}$ alleles, only VIR1589 from the Sverdlovsk region of Russia was resistant, while JI1104 and JI1108 from Nepal proved susceptible (Table S2). This corresponded to an amino acid exchange at position N19I in exon 1 between the *eIF4E*$^{C-1}$ and *eIF4E*$^{C-2}$ alleles (Table S2).

Discussion

In this study, we combined our knowledge and access to a wide range of pea germplasm [13,17] with an interest in PSbMV resistance [31] and focused our analysis on the geographical distribution of selected, easily detectable alleles of the *eIF4E* gene using the polymorphism of intron 3.

eIF4E gene structure and diversity

The pea *eIF4E* gene is comprised of five exons, interrupted by four intron sequences, with a total length of 2.1 kb [31], which is spliced into a 687 nucleotide long (open reading frame) mRNA [32]. The size polymorphism of intron 3 is caused by 50 and 56 bp insertions of minisatellite-like repeat sequences [31], separated by 259 bp of common intron sequence. This polymorphism led to the initial mapping of the *sbm-1* locus [35] and the development of a co-dominant marker for breeding [31]. Previous sequencing of 43 accessions of pea varieties showed four haplotypes of intron 1, eight haplotypes of the intron 2 and three haplotypes of intron 4, while 17 haplotypes were detected in intron 3 sequences ([31] and unpublished). Since all, except PI269774 (Sankia) and PI269818 (Aa134), previously tested susceptible and resistant accessions differed by 50 or 56 bp indels in intron 3 [31], we exploited this length polymorphism to screen a wider pea germplasm population. This led to the identification of novel polymorphisms in four principal variants (*eIF4E*^{S}, *eIF4E*$^{A-B-C}$). Another functional *eIF4E* resistance allele found in the PI269818 and PI269774 accessions was not directly tested in this study, as its polymorphism is more difficult to analyze. It is clear that this type of analysis does not provide a comprehensive dataset, as more subtle, single-nucleotide mutations are missed, as documented by sequenced samples, whereas in 73 sequences, 19 *eIF4E* alleles were found (Fig. 1, Table 3, S1).

The mutant *eIF4E*$^{A-1}$ allele of pea gene *eIF4E* differs from its *eIF4E*$^{S-1}$ counterpart in five non-conservative amino acid exchanges in and around the cap-binding pocket of pea *eIF4E*, which impacts infection by PSbMV [32]. Previously reported mutations in pea *eIF4E* were localized on the β1 (W62L), β1-β2 (AA73-74DD), β3-β4 (G107R) and β5 (N169K) loops [42]. Positions of the newly identified amino acid (V23D, V49A and LD218-219QE) changes in *eIF4E*$^{B-1-2-3}$ alleles do not correspond with the known mutations leading to resistance and are not localized on the variable β loops regions, with the exception of the M207I change located on the α3-β7 loop. The only resistance determinant mapped to the β7 loop is for BaYMV infection in barley [54]. In the *Capsicum* species, several *eIF4E* haplotypes with amino acid changes were reported, some of which included amino acid changes in positions 218 and 219 similar to those in the pea *eIF4E*^{B} allele. Systematic testing of individual mutations by *in planta* assay showed that only a specific combination of mutations leads to resistance [42]. This hypothesis confirms previously reported amino acid changes in pea eIF4E in PI269818 and PI269774 accessions [31] with *eIF4E*$^{A6-7}$ alleles, which have amino acid changes in the cap-binding pocket [31] and lack one (S78) amino acid (Fig. 3, Table 3 and Table S2, Supporting information). Amino acid changes in the loop near the cap-recognition pocket were shown to be directly associated with resistance to potyviruses in pepper, lettuce, and pea [35,55]. Moreover, not all *eIF4E* proteins encoded by the resistance alleles are defective in their ability to bind the m7-GTP cap, suggesting that disrupted cap binding is not always required for potyvirus resistance [56]. Thus mutations conferring *sbm-1* resistance to PSbMV act combinatorially [42]. The interactions between particular *eIF4E* and particular viral VPgs are highly specific for host-virus interactions. Higher plants are unique in that they encode two distinct isoforms of eIF4F that have both overlapping and isoform-specific roles, eIF4F protein complex, which contains eIF4E and eIF4G, and eIF(iso)4F protein complex, which contains eIF(iso)4E and eIF(iso)4G [25]. Although these two complexes seem to be equivalent for the *in vitro* translation, they differ in their *in vivo* expression patterns and specificity for cellular mRNAs and likely also viral RNA. Potyviruses could selectively use either *eIF(iso)4E* or *eIF4E* to infect plants, while some are able to use both [39]. In pepper, the combination of *eIF4E* (*pvr2* locus on chromosome 4) and *eIF(iso)4E* (*pvr6* locus on chromosome 3) resistance alleles showed a complementary effect [45]. Also in pea, the *eIF(iso)4E* homologous gene exists, mapped to LG II ([35] and close to the *sbm-2* locus (not shown). However, its direct involvement in potyvirus resistance still needs to be demonstrated. Thus we cannot exclude the possibility that interaction between eIF4E and eIF(iso)4E proteins may also play a role in pea. This might partly explain the discrepancies in virological testing of *eIF4E*^{B} accessions. Finally, the result of the interaction, the resistance or tolerance, depends not only on the genetic background but also on the virus pathotype [43,57]. It would be interesting to test region-specific potyvirus isolates [58] on the alleles identified in this study. The current phylogeny of potyviruses indicates that the progenitor probably infected plants growing in southwestern Eurasia and northern Africa, which evolved from a rymovirus by acquiring the ability to be transmitted by aphids [48]. It is notable that we did not find the resistant *eIF4E*^{A} variant in any of the 146 tested wild pea samples, nor have we been able to detect any intermediate alleles between the susceptible *eIF4E*$^{S-1}$ and resistant *eIF4E*$^{A-1}$, except in JI261 accessions already separated by 23 mutations between *eIF4E*$^{S-1}$ and *eIF4E*$^{B-1}$. Such an intermediate allele was found for *Pm3* powdery mildew resistance in wheat from the Himalayan Range, supporting the hypothesis that a recent evolution of these alleles took place in the hexaploid wheat gene pool [8]. This is a departure from the situation with other diseases for which resistance can often be found within the wild genepool, in contrast to the domestic gene pool. i.e. pathogens overcame resistance in the domestic gene pool. Haplotype analysis indicated the separate evolution of *eIF4E*^{A} alleles, with Ethiopian and Indian accessions being the most distantly related, while Chinese and some Afghan and Nepalese accessions were closer to *eIF4E*$^{S-1}$ (Fig. 4). Thus the evolution of the *eIF4E* gene is demonstrated in the available pea genepool.

Geographical distribution of eIF4E variants versus PSbMV diversity

Pea genotypes resistant to PSbMV were identified by Hagedorn and Gritton [59] in two Ethiopian lines (PI193586 and PI193835). Later, Hampton [60] found several resistant accessions from India. Since they are all identical in sequence [31], it raises the question of their origin. Hampton [60] proposed that India may have been the center of origin for the virus, and consequently also the center of PSbMV-immune germplasm. Despite this resistant germplasm, pea has not been systematically bred to include PSbMV resistance until recently; as a consequence, the vast majority of pea varieties are susceptible to this virus [31], especially when contrasted with landraces where the possibility of both natural and human selection for resistance exists. A relationship between pea host and PSbMV pathogen diversity in northern Pakistan and Afghanistan has been proposed [60,61]. More recently, phylogenetic analysis of PSbMV isolates from Australia and China grouped them on separate clades [55]. These results suggest the operation of co-evolutionary forces, which could be tested using the *eIF4E* alleles identified in this study. Interestingly, similar results were found in a study of barley *eIF4E* haplotype diversity, which was also found to be considerably higher in Central and East Asia, both regions with a long history of the bymovirus disease [3]. It is intriguing to see that in different crops, the resistance diversity center is outside the original place of domestication, especially in Central and Eastern Asia. It might be hypothesized that only some environments may be conducive for pathogen development in crops, hence co-evolution may be geographically limited. This agrees with results showing that the *eIF4E* gene is under positive selection pressure [3,44]. There is strong selection for the $eIF4E^B$ allele in China and less elsewhere (Fig. 2, Table 2), but no A or C alleles in China, possibly consistent with a genetic bottleneck in the later migration of pea to China, compared with central/western Asia [18]. Moreover, assignment of Chinese accessions to provinces showed preferential occurrence of B alleles in the autumn/winter-sown provinces of Henan, Anhui, Hubei and Sichuan (Table S1) that experience high rainfall, long growing seasons and cold, frost-prone vegetative phases. Conversely, S alleles predominated in spring-sown types grown in northern provinces with low rainfall and high, variable temperatures that result in severe drought [62]. Analysis of pea germplasm diversity, using retrotransposon-based insertion polymorphism [13,14,63] showed that although there is substantial genetic diversity present, it is only partially geographically structured. Especially in case of the $eIF4E^B$ alleles, there is substantial clustering of Indian and Ethiopian accessions, indicating a close relationship between materials from these areas (Fig. 5). The number of mutations makes independent origin unlikely, unless stringent positive selection is operating, as detected by Cavatorta [44]. We have also found evidence of positive selection in a set of 27 tested pea cDNA haplotypes (not shown). As proposed for barley *eIF4E*, the explanation for the unusually high overall degree of *eIF4E* sequence variation and haplotype diversity may be that these are important for adaptation to different local habitats [3]. The second, more plausible scenario is introgression and maintenance in populations by selective advantage. Based on the number of accesions with detected $eIF4E^A$ and $eIF4E^B$ alleles in the northern Indian subcontinent versus Ethiopia/eastern Africa, we speculate that these more likely originated in the Africa and were then brought by oversea trade to Afghanistan-Pakistan-Nepal-India region. The Silk Route, Amber Road and trans-Saharan trade routes were all instrumental in establishing links between Africa, India and beyond. This scenario, while not proven, is supported by a haplotype analysis network, where Indian $eIF4E^A$ alleles were shown to be derived from Ethiopian accessions (Fig. 4). There is bias towards European origin accesions, comprising 40.8% of samples, followed by Ethiopia and East Africa (12.4%) and Russia (10.5%) which consequently affects proportion (57%) of S alleles (Table 2). However, these samples were included to provide more complete view on $eIF4E^A$ diversity, including modern pea varieties. However this affects only percentage in total dataset and not within studied regions (Fig. 2, Table 2).

Various approaches have been applied to identify variants of the *eIF4E* gene in several crops. These showed different levels of diversity [3,43,46,47]. In the most comprehensive study, 1090 barley samples from 84 countries screened using the high-resolution melting PCR method of cDNA, led to the identification of 47 *eIF4E* haplotypes [3]. In our study, we have identified 34 alleles on exon-intron level with 156 single-nucleotide polymorphic sites in total of 73 sequences. This suggests that substantial diversity within the *eIF4E* gene can be expected within available pea germplasm, which might be fully revealed only by sequencing. It is notable that none of the 146 tested wild pea samples had $eIF4E^A$ or $eIF4E^B$ alleles, but at the amino acid level, all corresponded to the susceptible $eIF4E^{S\text{-}1}$ allele, despite substantial polymorphism both in introns and exons, indicating selection for functionality. The discovery of the intermediate allele in wild *P. sativum* subsp. *elatius* JI261 from Turkey is similar to data obtained in *Arabidopsis thaliana* [25], which suggests a distinct mode of evolution of resistance in wild species as opposed to crop species and supports the scenario in which potyviruses spread at the advent of agriculture, which brought plants into dense monocultures [48]. Very little is known about the selective forces that drive viral evolution in natural ecosystems, which contrasts with the more detailed population genetics studies in crop plants that have revealed the importance of mutation rates, recombination, genetic drift and migration in virus evolution [64]. Although we can not fully excluded the possibility, that JI261 allele could have resulted from domestic to wild introgression and not independent mutations, the analysis of 45 RBIP loci (Fig. 5) clustered it with typical *P. s.* subsp. *elatius* accessions [63]. The existence of resistance alleles only in the domesticated pea genepool leads us to speculate that the mutation originated during early cultivation. The series of alleles identified in this study provide an excellent basis for testing various potyviruses and pathotypes in which to study the co-evolution of potyviruses and their pea host.

Materials and Methods

Plant material

The 2803 pea accessions used in this study were drawn from the following collections: the Czech National Pea Collection (CzNPC, 1252 accessions); the USDA core collection (384); 119 accessions of Chinese origin from the Australian Temperate Field Crops Collection [(ATFCC) now reconstituted as Australian Grains Genebank, (AGG)], the Vavilov Institute (VIR) Russian Federation (69); the Centre for Genetic Resources (CGN), The Netherlands (14); Leibniz Institute of Plant Genetics and Crop Plant Research (IPK), Germany (35); and John Innes Centre (JIC), UK (836). The permission from the relevant institutions was obtained to access the collections, and the respective pea germplasm accessions were donated for this study under Standard Material Transfer Agreement of germplasm resources. Although we specifically targeted local landraces rather than globally grown, modern varieties, the analyses of 1252 accessions from the Czech National Pea collection represent the diversity present in the genepool of cultivated pea in the 19th to 21st centuries [16,49]. This collection is composed of 972 commercial varieties, 226 breeding lines and 54 landraces, which originate largely from Europe (925), the former Soviet Union (177), and the USA and Canada (77). In contrast, the John Innes Centre pea collection has, in addition to varieties (1071) and breeding lines (61), a large proportion of landraces (600), mutant stocks (585) and wild peas (445). Geographically, this collection is dominated by accessions of

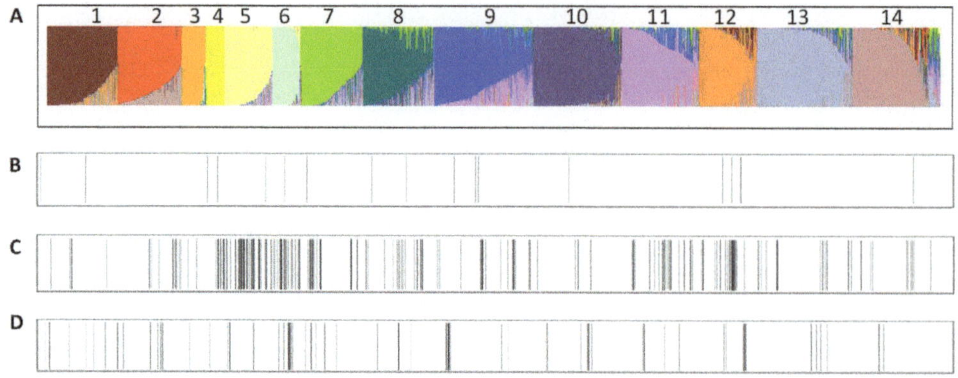

Figure 5. Visualization of genetic relationship of four identified *eIF4E* variants on the background of 14 BAPS identified clusters of 3,029 JIC accessions genotyped by 45 retrotransposon-insertion loci [13,63]. Notably, clusters 3 and 4 contain 145 acc. of wild *P. fulvum, P. s.* subsp. *elatius* and *P. abyssinicum*; cluster 7 contains 95 acc. from Afghanistan; cluster 8 has 225 acc. from Ethiopia; and cluster 9 contains 247 acc. from India and Ethiopia, while remaining clusters are more diverse (A). 18 accessions with the $eIF4E^{A}$ (B), 241 accessions with the $eIF4E^{B}$ (C), 81 accessions with the $eIF4E^{C}$ variant (D).

Ethiopian (388), Mediterranean (199), Indian (53) and Chinese (37) origin. The Australian ATFCC and USDA pea collections are partially complementary. The former includes the diversity core set of Chinese origin [18], of which 119 were studied here, while the latter has over 6000 accessions, of which a core set of 384 [19] was analyzed in this study. Finally, 69 accessions selected from the VIR collection originated in the Caucasus and Central Asia (Turkmenistan, Kazakhstan, Georgia, and Armenia) as well as in North Africa (Morocco and Algeria). It is worth mentioning that although Ethiopia is known as the origin and occurrence for *P. abyssinicum*, the accessions used in this study were mainly *P. sativum* subsp. *sativum*. This set was complemented with wild *P. fulvum* (9), *P. sativum subsp. elatius* (86), *P. sativum subsp. sativum* (19) (formerly *P. humile/P.syriacum*) (see Smýkal et al. [13] for taxonomical classification) and *P. abyssinicum* (32) (Table S1). These represent both primary and secondary centers of pea diversity as well as primary and secondary gene pools.

DNA isolation and PCR analysis

Young leaves were harvested from ten (in the case of the Czech, ATFCC, CGN, IPK and VIR collections) or single (in the case of the JIC and USDA collections) randomly chosen plants per accession and stored at −80°C. Genomic DNA was isolated and PCR performed using standard protocols [31]. Products were resolved on 1.5% TBE agarose gel and visualized using ethidium bromide staining under UV-light.

eIF4E variants amplification

Intron 3 length polymorphism was used as screening criteria, as developed by Smýkal et al. [31]. In short, differences in intron 3 length were assayed using PCR with the following pairs of primers: *Ps*-eIF4E-750F (5′-GGACTAAGAATGCTTCAAATGAAGCTGC-3′) and *Ps*-eIF4E-586gR (5′-GAATCATTTAAGAAGCTCGTGAAGTG-3′) primers (combination A, nested within the combination of B) that amplified 243 bp in the resistant and 293 bp in the susceptible accessions [31]; and (combination B) *Ps*-eIF4E-750F and *Ps*-eIF4E-1270R (5′-ATTCTCGATCACACTAGCCCCCTCC-3′) [35] primers that amplified 536 bp versus 586 bp fragment. The combination of these two assays resulted in the detection of respective $eIF4E^{A\text{-}B\text{-}C\text{-}S}$ variants (Fig. 1, Table 1). PCR amplification of 1000 bp product from start codon to intron 3 was carried out using primers *Ps*-eIF4E-1F start codon (5′-ATGGTTGTAGAAGACACCCCCAAATC-3′) and *Ps*-eIF4E-586gR. Primers *Ps*-eIF4E-794F (5′-GCTAGATGGTTGTTATGATGTTTATCAG-3′) and *Ps*-eIF4E-2188R stop codon (5′-TTGCTAGTTTGCTACCATGTAAGAACG-3′) were used to amplify a 1500 bp PCR product spanning the rest of *eIF4E* gene from intron 3 to exon 5. The PCR products were purified according to Werle et al. [50] protocol and sequenced using a BigDye Terminator kit (Applied Biosystems, UK) by Macrogene (Amsterdam, The Netherlands).

Bioinformatics

Primer design and restriction analyses were performed using FastPCR software version 5.1.83 (PrimerDigital Ltd., Finland). The DNA sequences were viewed and edited using Sequence Scanner version 2.0 (Applied Biosystems, UK). CLUSTALW alignment was performed using BioEdit version 7.09.0 [51]. The haplotype alignments were performed using a median-joining network algorithm, implemented in NETWORK 4.5.1.6 [52]. To reveal the genetic relationship of samples, we used previously made Bayesian clustering analysis of the genetic diversity of 3,029 JIC accessions genotyped by 45 retrotransposon-insertion loci [13]. Identified clusters were used as framework to visualize individual *eIF4E* alleles within the set of 836 JIC accessions analyzed in this study. Single nucleotide polymorphism mapping to reference sensitive allele was done using Geneious 6.1.6 analysis software (Biomatters, USA).

Virological testing

Plants were grown in a substrate Klassman no. 4 (Klasmann-Deilmann GmbH, Germany) in a growth chamber (Microclima 1000, Snijders Scientific, Holland) under a 16/8-h and 22/18°C day/night cycle. Evaluation of resistance/susceptibility to the PSbMV pathotype P-1 was conducted through mechanical inoculation using isolate PSB117CZ [31,53]. All together, 50 accessions were tested. Ten plants per accession were tested in same block, under the same conditions. Sensitive pea cultivars, Merkur and Raman, were used as a sensitive and B99 as a resistant controls [31]. Symptoms were observed at one-week intervals, and systemic infection was confirmed three weeks after infection with DAS-ELISA (Loewe Biochemica, Germany).

Supporting Information

Table S1 List of studied accessions, divided by 7 studied germplasm collection with indicated accession number, name, origin and *eIF4E* variant.

Table S2 Table of sequenced 73 accessions indicating single-nucleotide polymorphic (SNP) sites in exons and introns (SNP sheet) as well as resulting amino acid exchanges (protein sheet). Species name, germplasm, country of origin and *eIF4E* allele asignment are shown. Length of all introns and exons is shown in base pairs, with variable intron 3 being highlighted. Results of virological testing with P1 PSbMV are indicated as positive, susceptible (S) and resistant (R) reactions, respectively.

Acknowledgments

We thank Eva Fialová and Jana Veselá for technical assistance and numerous collegues for discussion over earlier versions of the manuscript.

Author Contributions

Conceived and designed the experiments: PS. Performed the experiments: EK PH CC DŠ PS. Analyzed the data: EK PS DŠ MN. Contributed reagents/materials/analysis tools: AF MA MV RR. Wrote the paper: EK MA RR CC PS.

References

1. Tanksley SD, McCouch SR (1997) Seed banks and molecular maps: Unlocking genetic potential from the wild. Science 277: 1063–1066.
2. Bhullar NK, Street K, Mackay M, Yahiaoui N, Keller B (2009) Unlocking wheat genetic resources for the molecular identification of previously undescribed functional alleles at the *Pm3* resistance locus. Proc Nat Academy of Sci USA 106: 9519–9524.
3. Hofinger BJ, Russell JR, Bass CHG, Baldwin T, dos Reis M, et al. (2011) An exceptionally high nucleotide and haplotype diversity and a signature of positive selection for the *eIF4E* resistance gene in barley are revealed by allele mining and phylogenetic analyses of natural populations. Mol Ecology 20: 3653–3668.
4. Nieto C, Piron F, Dalmais M, Marco CF, Moriones E, et al. (2007) EcoTILLING for the identification of allelic variants of melon eIF4E, a factor that controls virus susceptibility. BMC Plant Biology 7: 34.
5. Ibiza VP, Cañizares J, Nuez F (2010) EcoTILLING in *Capsicum* species: searching for new virus resistance. BMC Genomics 11: 631.
6. Reeves PA, Panella LW, Richards CHM (2012) Retention of agronomically important variation in germplasm core collections: implications for allele mining. Theor Appl Genet 124: 1155–1171.
7. Robaglia C, Caranta C (2006) Translation initiation factors: a weak link in plant RNA virus infection. Trends in Plant Sci 11: 40–45.
8. Bhullar NK, Zhang Z, Wicker T, Keller B (2010) Wheat gene bank accessions as a source of new alleles of the powdery mildew resistance gene *Pm3*: a large scale allele mining project. BMC Plant Biol 10: 88.
9. Chimwamurombe PM, Khulbe RK (2011) Domestication. In: Pratap A, Kumar J, editors. Biology and Breeding of Food Legumes. MA, USA: CABI, Cambridge. pp. 19–34.
10. Zohary D, Hopf M (1993) Domestication of Plants in the Old World—The Origin and Spread of Cultivated Plants in West Asia, Europe, and the Nile Valley. Oxford: Clarendon Press.
11. De Candolle A (1882) Origin of cultivated plants. Whitefish MT: Kesinger Publishing LCC.
12. Maxted N, Ambrose M (2001) *Peas* (*Pisum* L.). In: Maxted N, Bennett SJ, editors. Plant Genetic Resources of Legumes in the Mediterranean. Dordrecht: Kluwer Academic Publishers. pp. 181–190.
13. Smýkal P, Kenicer G, Flavell AJ, Corander J, Kosterin O, et al. (2011) Phylogeny, phylogeography and genetic diversity of the *Pisum* genus. Plant Gen Res 9: 4–18.
14. Smýkal P, Aubert G, Burstin J, Coyne C, Ellis N, et al. (2012) Pea (*Pisum sativum* L.) in the genomic era. MDPI Agronomy 2: 74–115.
15. Smýkal P, Coyne C, Redden R, Maxted N (2013) Peas. In: Singh M, Bisht IS, editors. Genetic and Genomic Resources for Grain Legume Improvement. London: Elsevier Insights. pp. 41–80.
16. Smýkal P, Hybl M, Corander J, JarkovskyJ, Flavell AJ, et al. (2008) Genetic diversity and population structure of pea (*Pisum sativum* L.) varieties derived from combined retrotransposon, microsatellite and morphological marker analysis. Theor Appl Genet 117: 413–424.
17. Jing R, Ambrose MA, Knox MR, Smykal P, Hybl M, et al. (2012) Genetic diversity in European *Pisum* germplasm collections. Theor Appl Genet 125: 367–380.
18. Zong X, Redden R, Liu Q, Wang S, Guan J, et al. (2009) Analysis of a diverse global Pisum sp. collection and comparison to a Chinese local *P. sativum* collection with microsatellite markers. Theor Appl Gen 118: 193–204.
19. Kwon SJ, Brown AF, Hu J, McGee RJ, Watt CA, et al. (2012) Genetic diversity, population structure and genome-wide marker-trait association analysis emphasizing seed nutrients of the USDA pea (*Pisum sativum* L.) core collection. Genes & Genomics 34: 305–320.
20. Wang A, Krishnaswamy S (2012) Eukaryotic translation initiation factor 4E-mediated recessive resistance to plant viruses and its utility in crop improvement. Mol Plant Pat 13: 795–803.
21. Vavilov NI (1949-1950) The phytogeographic basis of plant breeding. In Chester KS trans. The origin, variation, immunity, and breeding of cultivated plants. Waltham MA: Chronica Botanica. pp. 13–54.
22. Abbo S, Lev-Yadun S, Gopher A (2012) Plant domestication and crop evolution in the Near East: on events and processes. Crit Rev in Plant Sci 31: 241–257.
23. Lovisolo O, Hull R, Rosler O (2003) Coevolution of viruses with hosts and vectors and possible paleontology. Adv Virus Res 62: 325–379.
24. Jones EI, Ferriere R, Bronstein JL (2009) Eco-evolutionary dynamics of mutualists and exploiters. American Naturalist 174: 780–794.
25. Le Gall O, Aranda MA, Caranta C (2011) Plant resistance to viruses mediated by translation initiation factors. In: Caranta C, Aranda MA, Tepfer M & López-Moya J, editors. Recent Advances in Plant Virology. Norfolk: Caister Academic Press. pp. 177–194.
26. Musil M (1966) Über das Vorkommen des Virus des Blattrollens der Erbse in der Slowakei (Vorlaufige Mitteilung). Biologia 21: 133–138.
27. Latham LJ, Jones RAC (2001) Alfalfa mosaic and pea seed-borne mosaic viruses in cool season crop, annual pasture, and forage legumes: susceptibility, sensitivity, and seed transmission. Austr J Agri Res 52: 771–790.
28. Hampton RO, Mink GI (1975) Pea seed-borne mosaic virus. CMI/AAB Descriptions of Plant Viruses. 146 n.
29. Alconero R, Weeden NF, Gonsalves D, Fox DT (1985) Loss of genetic diversity in pea germplasm by the elimination of individuals infected by pea seedborne mosaic virus. Ann of App Bio 106: 357–364.
30. Hjulsager CK, Olsen BS, Jensen DM, Cordea MI, Krath BN, et al. (2006) Multiple determinants in the coding region of pea seed-borne mosaic virus P3 are involved in virulence against *sbm-2* resistance. Virology 355: 52–61.
31. Smýkal P, Šafářová D, Navrátil M, Dostálová R (2010) Marker assisted pea breeding: *eIF4E* allele specific markers to pea seed-borne mosaic virus (PSbMV) resistance Mol Breeding 26: 425–438.
32. Bruun-Rasmussen M, Moller IS, Tulinius G, Hansen KR, Lund OS, et al. (2007) The same allele of translation initiation factor 4E mediates resistance against two Potyvirus spp. in *Pisum sativum*. Mol Plant Microbe Interact 20: 1075–1082.
33. Andrade M, Abe Y, Nakahara KS, Uyeda I (2009) The *cyv-2* resistance to Clover yellow vein virus in pea is controlled by the eukaryotic initiation factor 4E. J Gen Plant Pathol 75: 241–249.
34. Johansen EI, Lund OS, Hjulsager CK, Laursen J (2001) Recessive resistance in *Pisum sativum* and Potyvirus Pathotype resolved in a gene-for-cistron correspondence between host and virus. J Virology 75: 6609–6614.
35. Gao Z, Eyers S, Thomas C, Ellis N, Maule A (2004) Identification of markers tightly linked to sbm recessive genes for resistance to pea seed-borne mosaic virus. Theor Appl Genet 109: 488–494.
36. Olsen BS, Johansen IE (2001) Nucleotide sequence and infectious cDNA clone of the L1 isolate of Pea seed-borne mosaic potyvirus. Archives in Virology 146: 15–25.
37. Johansen IE, Rasmussen OF, Heide M, Borkhardt B (1991) The complete nucleotide sequence of pea seed-borne mosaic virus RNA. J Gen Virology 72: 2625–2632.
38. Johansen IE, Keller KE, Dougherty WG, Hampton RO (1996) Biological and molecular properties of a pathotype P-1 and a pathotype P-4 isolate of pea seed-borne mosaic virus. J Gen Virology 77: 1329–1333.
39. Ruffel S, Gallois J-L, Moury B, Robaglia C, Palloix A, et al. (2006) Simultaneous mutations in translation initiation factors eIF4E and eIF(iso)4E are required to prevent Pepper veinal mottle virus infection of pepper. J GenVirol 87: 2089–2098.
40. Sato M, Nakaharaa K, Yoshii M, Ishikawa M, Uyeda I (2005) Selective involvement of members of the eukaryotic initiation factor 4E family in the infection of *Arabidopsis thaliana* by potyviruses. FEBS Letters 579: 1167–1171.
41. Fraile A, Garcia-Arenal F (2010) The coevolution of plants and viruses: resistance and pathogenicity. Adv Virus Res 76: 1–32.
42. Ashby AJ, Stevenson CEM, Jarvis GE, Lawson DM, Maule AJ (2011) Structure-Based Mutational Analysis of eIF4E in Relation to sbm1 Resistance to Pea Seed-Borne Mosaic virus in pea. PLoS ONE 6: e15873.
43. Charron C, Nicolai M, Gallois JL, Robaglia C, Moury B, et al. (2008) Natural variation and functional analyses provide evidence for co-evolution between plant eIF4E and potyviral VPg. Plant J 54: 56–68.

44. Cavatorta JR, Savage AE, Yeam I, Gray SM, Jahn MM (2008) Positive Darwinian selection at single amino acid sites conferring plant virus resistance. J Mol Evol 67: 551–559.
45. Rubio M, Nicolaï M, Caranta C, Palloix A (2009) Allele mining in the pepper gene pool provided new complementation effects between pvr2-eIF4E and pvr6-eIF(iso)4E alleles for resistance to pepper veinal mottle virus. J GenVirol 90: 2808–2814.
46. Jeong HJ, Kwon JK, Pandeya D, Hwang J, Hoang NH, et al. (2012) A survey of natural and ethyl methane sulfonate-induced variations of eIF4E using high-resolution melting analysis in Capsicum. Mol Breeding 29: 349–360.
47. Rigola DJ, van Oeveren J, Janssen A, Bonne A, Schneiders H, et al. (2009) High throughput detection of induced mutations and natural variation using KeyPoint technology. PLoS ONE 4: e4761.
48. Gibbs AJ, Ohshima K, Phillips MJ, Gibbs MJ (2008) The Prehistory of Potyviruses: Their Initial Radiation Was during the Dawn of Agriculture. PLoS ONE 3: e2523.
49. Cieslarová J, Hýbl M, Griga M, Smýkal P (2012) Molecular Analysis of Temporal Genetic Structuring in Pea (*Pisum sativum* L.) Cultivars Bred in the Czech Republic and in Former Czechoslovakia Since the Mid-20th Century. Czech J Gen and Plant Breeding 48: 61–73.
50. Werle E, Schneider C, Renner M, Völker M, Fiehn W (1994) Convenient single-step, one tube purification of PCR products for direct sequencing. Nucleic Acids Research 22: 4354–4355.
51. Hall TA (1999) BioEdit: a user-friendly biologicalsequence alignment editor and analysis program for Windows 95/98/NT. Nucl Acids Symp Ser 41: 95–98.
52. Bandelt HJ, Forster P, Röhl A (1999) Median-joining networks for inferring intraspecific phylogenies. Mol Biol Evol 16: 37–48.
53. Šafářová D, Navrátil M, Petrusová J, Pokorný R, Piaková Z (2008) Genetic and biological diversity of the pea seed-borne mosaic virus isolates occurring in Czech Republic. Acta Virologica 52: 53–57.
54. Stein N, Perovic D, Kumlehn J, Pellio B, Stracke S, et al. (2005) The eukaryotic translation initiation factor 4E confers multiallelic recessive bymovirus resistance in *Hordeum vulgare* (L.). Plant J 42: 912–22.
55. Nicaise V, German-Retana S, Sanjuan R, Dubrana MP, Mazier M, et al. (2003) The eukaryotic translation initiation factor 4E controls lettuce susceptibility to the potyvirus Lettuce mosaic virus. Plant Physiol 132: 1272–1282.
56. Kang BC, Yeam I, Frantz DJ, Murphy JF, Jahn MM (2005) The *pvr1* locus in pepper encodes a translation initiation factor eIF4E that interacts with tobacco etch virus VPg. Plant J 42: 392–405.
57. Hjulsager CK, Lund OS, Johansen E (2002) A new pathotype of pea seedborne mosaic virus explained by the properties of the p3-6k1 and viral genome-linked protein (VPg) coding regions. Mol Plant Microbe Interact 15: 169–171.
58. Wylie SJ, Coutts BA, Jones RAC (2011) Genetic variability of the coat protein sequence of pea seed-borne mosaic virus isolates and the current relationship between phylogenetic placement and resistance groups. Arch Virol 156: 1287–1290.
59. Hagedorn DJ, Gritton ET (1973) Inheritance of resistance to the pea seed-borne mosaic virus. Phytopathology 62: 1130–1133.
60. Hampton RO (1986) Geographic origin of pea seed-borne mosaic virus: a hypothesis. Pisum Newsletter 18: 22–26.
61. Ali A, Randles JW (2001) Genomic heterogeneity in Pea seed-borne mosaic virus isolates from Pakistan, the centre of diversity of the host species, *Pisum sativum*. Arch Virology 146: 1855–1870.
62. Ling Li, Redden RJ, Zong X, Berger JD, Bennett SJ (2013) Ecogeographic analysis of pea collection sites from China to determine potential sites with abiotic stresses. Genet Res Crop Evol 60: 1801–1815.
63. Jing R, Vershinin A, Grzebyta J, Shaw P, Smýkal P, et al. (2010) The genetic diversity and evolution of field pea (*Pisum*) studied by high throughput retrotransposon based insertion polymorphism (RBIP) marker analysis. BMC Evol Biol 10: 44.
64. Roossinck MJ (2011) The big unknown: plant virus biodiversity. Current Opinions in Virology 1: 63–67.

8

TASSEL-GBS: A High Capacity Genotyping by Sequencing Analysis Pipeline

Jeffrey C. Glaubitz[1]*, Terry M. Casstevens[1], Fei Lu[1], James Harriman[1], Robert J. Elshire[1¤], Qi Sun[2], Edward S. Buckler[1,3]

1 Institute for Genomic Diversity, Cornell University, Ithaca, New York, United States of America, **2** Biotechnology Resource Center Bioinformatics Facility, Cornell University, Ithaca, New York, United States of America, **3** USDA Agricultural Research Service, Ithaca, New York, United States of America

Abstract

Genotyping by sequencing (GBS) is a next generation sequencing based method that takes advantage of reduced representation to enable high throughput genotyping of large numbers of individuals at a large number of SNP markers. The relatively straightforward, robust, and cost-effective GBS protocol is currently being applied in numerous species by a large number of researchers. Herein we describe a bioinformatics pipeline, TASSEL-GBS, designed for the efficient processing of raw GBS sequence data into SNP genotypes. The TASSEL-GBS pipeline successfully fulfills the following key design criteria: (1) Ability to run on the modest computing resources that are typically available to small breeding or ecological research programs, including desktop or laptop machines with only 8–16 GB of RAM, (2) Scalability from small to extremely large studies, where hundreds of thousands or even millions of SNPs can be scored in up to 100,000 individuals (e.g., for large breeding programs or genetic surveys), and (3) Applicability in an accelerated breeding context, requiring rapid turnover from tissue collection to genotypes. Although a reference genome is required, the pipeline can also be run with an unfinished "pseudo-reference" consisting of numerous contigs. We describe the TASSEL-GBS pipeline in detail and benchmark it based upon a large scale, species wide analysis in maize (*Zea mays*), where the average error rate was reduced to 0.0042 through application of population genetic-based SNP filters. Overall, the GBS assay and the TASSEL-GBS pipeline provide robust tools for studying genomic diversity.

Editor: Nicholas A. Tinker, Agriculture and Agri-Food Canada, Canada

Funding: This work was supported by the National Science Foundation (www.nsf.gov) under the Plant Genome Research Program (PGRP) (grant numbers DBI-0820619 and IOS-1238014) and the Basic Research to Enable Agricultural Development (BREAD) project (ID:IOS-0965342), as well as by the USDA-ARS (www.usda.gov). The funders had no role in study design, data collection and analysis, decision to publish, or preparation of the manuscript.

Competing Interests: The authors have declared that no competing interests exist.

* E-mail: jcg233@cornell.edu

¤ Current address: AgResearch Limited, Grasslands Research Centre, Palmerston North, New Zealand

Introduction

The advent of next generation sequencing has elicited a revolution in biology buoyed by an advancing tidal wave of raw sequence data[1–4]. By combining the power of next generation sequencing with reduced representation[5], which focuses sequencing resources on the ends of restriction fragments, it is now possible to quickly genotype unprecedented numbers of samples even in large genome species[6–8]. Several low cost, high throughput methods that combine next generation sequencing with reduced-representation have been developed (e.g., [9–19]). Because of its relative simplicity and robustness, the genotyping by sequencing (GBS) method of Elshire et al. [12] or close derivatives thereof have already been applied in numerous species by many researchers (e.g., [7,17,20–31]). For the first time, generation of copious quantities of genotypic data for genetic experiments is no longer a bottleneck. Instead, the new bottleneck is the efficient bioinformatics analysis of the vast and ever-expanding sea of data. Opportunities to apply markers to breeding or conservation biology are now often limited only by the availability of appropriate bioinformatics tools.

To address this bioinformatics bottleneck, we implemented a GBS analysis pipeline in the Java program TASSEL[32] (version 4) that is specifically tailored to the GBS protocols of Elshire et al. [12] or Poland et al. [20]. However, the TASSEL-GBS pipeline is not limited to the specific restriction enzymes utilized in those protocols: it currently accepts 15 single restriction enzymes and 15 restriction enzyme pairs, and new enzymes are easily added. Furthermore, the TASSEL-GBS pipeline should work on nearly any restriction enzyme and barcoding approach (e.g., [33]), provided that sequence reads commence with the barcode immediately followed by the remnant of the restriction enzyme cut site (Figure 1A). Compared to other available pipelines for similar purposes [16,25,34–38] the TASSEL-GBS pipeline is specifically designed to efficiently handle large quantities of data from large numbers of samples: to date, we have analyzed more than 45,000 maize samples. The TASSEL-GBS pipeline was designed for species with a reference genome; however, it is possible to use incomplete genome assemblies consisting of numerous contigs as a pseudo-reference. For species without a reference genome, an alternative approach, appropriate for small to medium scale studies, has already been implemented in TASSEL [22].

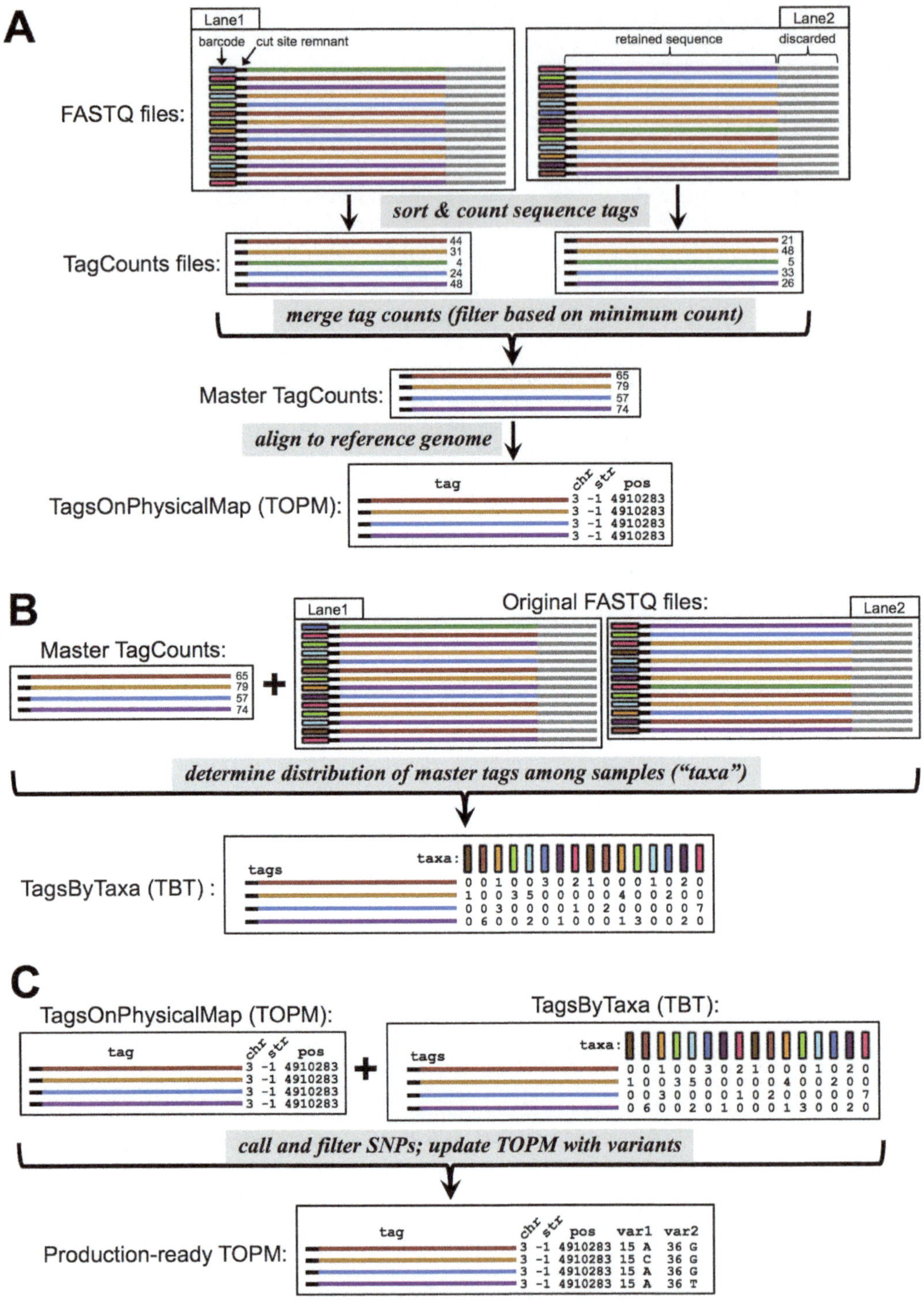

Figure 1. Schematic representation of the TASSEL-GBS Discovery Pipeline. (A) Barcoded sequence reads are processed and collapsed into a set of unique sequence tags, with one TagCounts file produced per input FASTQ file. The separate TagCounts files are then merged to form a "master" TagCounts file, which retains only those tags present at or above an experiment-wide minimum count. This master tag list is then aligned to the reference genome and a TagsOnPhysicalMap (TOPM) file is generated, containing the genomic position of each tag with a unique, best alignment. (B) The barcode information in the original FASTQ files is then used to tally the number of times each tag in the master tag list is observed in each sample ("taxon") and these counts are stored in a TagsByTaxa (TBT) file. (C) The information recorded in the TOPM and TBT is then used to discover SNPs at each "TagLocus" (set of tags with the same genomic position) and filter the SNPs based upon the proportion of taxa covered by the TagLocus, minor allele frequency, and inbreeding coefficient (F_{IT}). For each retained SNP, the allele represented by each tag in the corresponding TagLocus is recorded in the TOPM file, along with its relative position in the locus. The end product of the Discovery Pipeline is a "production-ready" TOPM that can then be used by the Production Pipeline to call SNPs.

Herein, we describe the TASSEL-GBS pipeline in detail, as implemented in TASSEL 4, and we benchmark the pipeline based upon a large-scale, species-wide analysis of maize (*Zea mays*). There were three primary motivations behind the development of this software: (1) Support the use of GBS in high-throughput, accelerated plant breeding, (2) Accommodate the high genomic diversity that is frequently encountered in species critical to agriculture and conservation, and (3) Provide an analysis platform that can be run in many contexts and with modest computational resources, such as those typically available in the developing world.

Terminology

A read is a single sequence in the FASTQ output file generated by the GBS assay

A ***good, barcoded read*** is a sequence read with a perfect match to one of the barcodes provided in a barcode key file and with no N's in the sequence following the barcode up to the trim length. Under the current implementation, reads are trimmed to 64 bp (not including the barcode).

A ***tag*** refers to a unique sequence (excluding the barcode) up to a specified length (currently 64 bp) from one or more "good, barcoded reads". A given tag is typically observed in numerous good, barcoded reads of identical sequence (up to the trim length).

For our purposes, a ***taxon*** refers to a nameable entity from which one or more DNA samples can be taken.

Design Considerations

Separation of SNP discovery and production SNP calling

Genomics-assisted, accelerated plant breeding usually consists of two separate phases of analysis: a survey of genetic diversity within a species or large breeding program to discover useful markers, followed by usage of these markers to rapidly advance generations. During the advancement cycle, time is of the essence, as thousands of samples need to be processed as quickly as possible so that decisions can be made for the next breeding cycle. The TASSEL-GBS software, by its division into Discovery (Figure 1) and Production pipelines (Figure 2), mirrors the two phases of accelerated plant breeding.

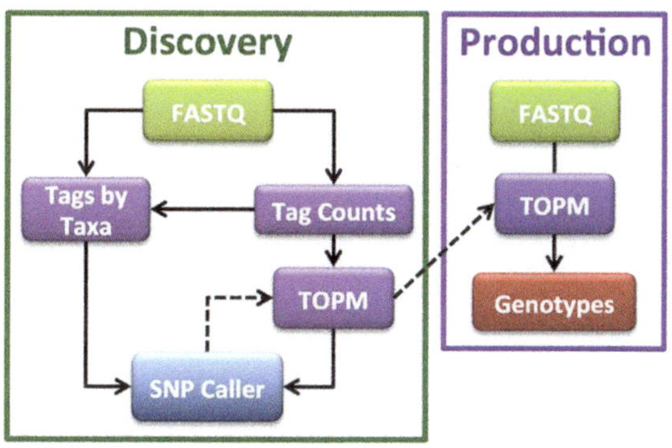

Figure 2. Relationship between the TASSEL-GBS Discovery and Production pipelines. The Discovery Pipeline is run periodically on all FASTQ files generated to date in a species, and the ascertained and filtered SNPs are stored in a "production-ready" TOPM. The Production pipeline utilizes this production-ready TOPM to quickly call SNPs either for the original samples in the Discovery Build, or for subsequent, post-Discovery samples.

The aim of the Discovery Pipeline (Figure 1) is to use the cumulative sequence data from all available samples run to date in a species (or breeding population) to discover SNPs. These SNPs are stored in a "TagsOnPhysicalMap" (TOPM) data structure, containing all of the potentially useful, unique, sequence tags, the genomic positions of the subset of the tags with unique best alignment positions, and the alleles that each useful tag represents for each discovered SNP. The "production-ready" TOPM (populated with variants for each useful tag) can then be in used in the single-step Production Pipeline to quickly produce genotypes, by determining which useful tags are present in each sample. A production-ready TOPM will be applicable to a breeding population as long as the genetic diversity present in the founders of the breeding population is well represented in the individuals that comprised the corresponding Discovery Build.

Our general approach is to periodically perform a comprehensive "Discovery Build" including all samples run to date in our study species. Performing a Discovery Build on a large number of samples is a multistep process that, depending on the number of samples, can require considerable computing resources (or time). For example, the most recent Discovery Build that we performed in maize (AllZeaGBSv2.6), comprising 31,978 samples (plus 758 blank negative controls, where TE buffer was substituted in place of a DNA sample), took 495 CPU-hours on 64 core Linux machine with 512GB of RAM (where each core was a 1.4 GHz AMD Opteron Processor 6272), plus additional time for staging of all of the input FASTQ files, etc. The Production Pipeline, in contrast, provides an avenue by which genotypes for the set of SNPs discovered in the most recent Discovery Build can be quickly generated for new samples. Running the Production Pipeline on a single FASTQ file containing sequence reads from 95 or 383 samples (plus a blank negative control) requires approximately 1 CPU-hour on a MacBook Pro with a 2.6 GHz Intel Core i7 processor and 16GB of RAM running OS X.

Favoring calling a large number of SNPs versus depth per SNP

In a GBS assay, the tradeoff, for a given genome size, between number of SNPs genotyped and the depth of coverage at each SNP is controlled by the level of multiplexing and the choice of restriction enzyme(s) [12,20]. One of our primary motivations for performing GBS (in maize and other organisms) is to enable GWAS, which requires a high density of markers, so that each causative polymorphism stands a reasonable chance of being in LD with one or more markers [39]. Hence, we favor increasing the number of markers at the expense of depth and thus designed the TASSEL-GBS pipeline with low coverage data in mind. The resultant missing data and under-calling of heterozygotes can be compensated for by redundant coverage of haplotypes at high marker density, facilitating imputation.

The likelihood of success of this imputation-based strategy depends on the number and length of homozygous, identical by descent (IBD) segments present in the study population. This, in turn, depends on the demographic history of the population. Imputation is least challenging in biparental populations consisting of RILs [40]. Imputation can also be relatively straightforward in a set of homozygous inbred individuals descending from a limited number of founders (e.g., modern maize lines; [24]). Even in outcrossing species, extensive homozygous IBD stretches can be present if a population bottleneck of sufficient severity occurred at some point in the demographic history of the study population

[41]. Such bottlenecks are common in outcrossing crop species, associated with domestication (e.g., [42]), modern improvement (e.g., [43]), or the development of a breeding population [44]. For unrelated (or distantly related) individuals (either inbred or outcrossed), large sample sizes improve the prospect of successful imputation: the more individuals genotyped, the more likely a given haplotype will be present in multiple individuals in homozygous form. Marker density and depth of coverage per marker need only be high enough to permit recognition of these homozygous IBD segments.

The TASSEL-GBS pipeline is thus optimized for low sequencing depth (0.5 to 3×) over a large number of markers in a large sample of individuals. However, it is flexible enough for the analysis of higher coverage data either from a small genome species, or from a large genome species where a lower level of multiplexing, replicate runs of the same library preps, and/or a less frequently cutting restriction enzyme (or enzyme combination) are used.

Capacity for large numbers of markers and samples

Our analysis strategy favoring large numbers of markers scored at low depth in a large sample of individuals requires the TASSEL-GBS pipeline to be able to handle very large data structures (Table 1). The largest data structure is the "TagsByTaxa" (TBT) object, which records the observed depth in each individual for each potentially useful sequence tag (where "taxa" in "TagsByTaxa" refers to an individual sample from a particular GBS library prep). Our most recent Discovery Build in maize (AllZeaGBSv2.6) comprised 97.5 million potentially useful tags from 31,978 samples (plus 758 blank negative controls), with the depth for each tag in each sample recorded as a single byte (prior to compression). In order to efficiently store and retrieve data from a matrix of this size (3.2 TB of uncompressed, raw data), we used the HDF5 storage format (http://www.hdfgroup.org). We also implemented a rapid and efficient run length compression algorithm to further decrease the storage size. At low depth and with high genetic diversity (numerous sequence tags per locus), the TBT is a sparse data matrix consisting mostly of zeros; our run length compression algorithm takes advantage of this. The sequence tags themselves are stored in a binary format requiring only 16 bytes to hold 64 bases (2 bits per base), prior to compression. Thus, 64 base tags can be conveniently held in two "longs" in Java. The 64 base upper limit on tag lengths in the current implementation of TASSEL-GBS will be lifted in the near future, which will be helpful for study organisms with limited diversity.

Capability to run on modest computing infrastructure

GBS provides an unprecedented opportunity for genomic markers to be used by researchers working in numerous species, including researchers in the developing world. Many of these potential users do not have access to big memory computers or large clusters. The TASSEL-GBS Discovery and Production pipelines can be run on a Linux, Mac or Windows computer with 8–16 GB of RAM. The main demand with respect to the amount of RAM required is the number good, barcoded reads in a typical FASTQ file produced from GBS, which currently ranges from 200–300 million. The small memory footprint required by the TASSEL-GBS pipeline, in relation to its high capacity in terms of number of markers and individual samples, renders it useful to a broad array of users without access to sophisticated computing infrastructure, a target user group which may include breeders carrying out genomic selection experiments.

Avoiding redundant alignment of identical reads

Genotyping approaches based on whole genome, rather than reduced-representation, sequencing typically align all or most of the reads produced to the reference genome prior to calling SNPs (e.g., [36]). In contrast, the TASSEL-GBS pipeline first collapses all of the reads into a master tag list containing all of the sequence tags present at or above a user-specified minimum count, tallied across all of the samples in the Discovery Build (Figure 1A). Each tag in this master tag list is then aligned to the reference genome. This strategy dramatically reduces the computation time devoted to alignment, and permits the use of more computationally expensive alignment algorithms. As the master tag list for our most recent maize Discovery Build (AllZeaGBSv2.6) consisted of 97.5 million tags, distilled from more than 46.8 billion sequence reads, a 392-fold reduction in computational time devoted to alignment was achieved in our case.

Favoring allelic redundancy over quality scores

Quality scores produced by the Illumina base caller are strongly negatively correlated with position in the read. In most whole genome sequencing approaches the position of a particular SNP in different reads is essentially random. In contrast, a GBS SNP has a consistent position in each read, as all of the GBS tags from a particular genomic location (a "TagLocus") that are used to discover and call SNPs originate from the same restriction enzyme cut site and have the same strand orientation. Consequently, the more distal SNPs in a GBS tag tend to have lower quality scores. If quality scores were used to filter GBS reads, these more distal SNPs might end up with lower depth of coverage. Furthermore, the quality scores frequently are not indicative of true quality [45,46]. Hence, rather than using quality scores to filter out bad reads, the TASSEL-GBS pipeline instead relies on the number of times a given tag has been observed as an indicator of sequence quality. GBS sequence tags that occur in a minimum, user-specified, number of reads across all of the samples in a Discovery Build are deemed as potentially useful and are kept for further processing (alignment to the genome and SNP calling). Illumina

Table 1. Size of the key data structures used by the TASSEL-GBS pipeline for a recent maize "Discovery Build" (AllZeaGBSv2.6).

Data Structure	Data Points	Compressed Size	Uncompressed size
Sequencing Files	4,679 Gnt[1]	3.9 TB	11.6 TB
Tags by Taxa (TBT)	3.2 trillion[2]	82.0 GB	3,198 GB
Tags on Physical Map (TOPM)	10.2 billion[3]	6.44 GB	14.0 GB

[1]Giganucleotides
[2]Read depths for 97,502,532 tags across 32,736 taxa (including 758 blank negative controls.
[3]105 data points per tag (with each base counted as one data point) times 97,502,532 tags.

quality scores are ignored, and therefore do not need to be tracked throughout the pipeline.

Population genetic-based filtering of putative SNPs

Putative SNPs from GBS may be of low quality for multiple reasons. The sequencing error rate for a SNP may be high because of its distance from the read start and/or its immediate sequence context [47,48]. Alternatively, paralogous sequence tags from different loci may be mistakenly aligned to a single TagLocus, resulting in spurious SNPs. To detect and filter out error-prone SNPs, the TASSEL-GBS pipeline relies on population-genetic parameters such as the minor allele frequency (MAF) and, in particular, the inbreeding coefficient (or "index of panmixia"), F_{IT}. Filtering based upon minimum MAF can remove spurious SNPs arising solely from sequencing error. Artifactual SNPs originating from paralogous tags will tend to be excessively heterozygous and can thus be distinguished on the basis of low F_{IT}.

These population-genetic filters are most powerful if a substantial proportion of the samples consist of inbred lines. Among inbred samples, both error-prone SNPs and spurious SNPs originating from paralogous tags will appear to be excessively heterozygous. The Discovery SNP caller in the TASSEL-GBS pipeline allows the user to specify which samples are highly inbred, and uses this subset of inbreds to calculate F_{IT} and apply the minimum F_{IT} filter. Additional, related, filters can also be applied enforcing a minimum "inbred coverage" (proportion of the inbred samples to be non-missing at the SNP) and maximum "inbred heterozygosity score" (= *nInbredHets*/[*nInbredsGT1ReadHomoMin* + *nInbredHets* + 0.5], where *nInbredHets* is the number of inbred taxa that are scored as heterozygous for the SNP, and *nInbredsGT1ReadHomoMin* is the number of inbred taxa with a read depth >1 for the SNP that are scored as minor allele homozygotes).

Implementation

Discovery Pipeline

The Discovery Pipeline consists of multiple steps, with each step (with the exception of alignment to the reference genome) being carried out by a TASSEL "plugin" that can be run from the TASSEL 4 Standalone command line interface. Detailed documentation on the function and usage of each individual plugin is available at www.maizegenetics.net/tassel/docs/TasselPipelineGBS.pdf. Rather than describe each individual plugin, here we describe the main functions of the pipeline (Figure 1) and their key features.

In order to have maximal power to discover and filter SNPs, we advocate running the Discovery Pipeline (i.e., performing a "Discovery Build") at the species-wide level, with all samples sequenced to date, across multiple FASTQ files. Each FASTQ file contains GBS data from multiple samples distinguished by DNA barcodes at the beginning of each read [12]. We currently perform GBS at 384 plex (3072 samples per flowcell), and run each GBS library prep on one or more flowcell lanes depending on the desired sequencing depth.

Collapsing reads into a master tag list. The TASSEL-GBS Discovery Pipeline first reads through each of the available FASTQ files and generates one output "TagCount" file per input FASTQ file (Figure 1A). Each output, binary TagCount file contains a sorted list of all the unique sequence tags observed in the corresponding FASTQ file, the length of each tag in bases, and the number of times each tag was observed. This list can be used as a key-value map where the tag sequence and length together serve as the key and the corresponding tag count serves as the value. Each nucleotide is encoded in two bits allowing tag sequences to be stored in Java longs (64 bits, or 32 bases per long). Degenerate bases or N's are not permitted and quality scores are not retained. Potentially chimeric sequences are eliminated by trimming the sequence at the corresponding restriction enzyme site, if present.

After this initial pass through all of the available FASTQ files, the individual TagCounts files are merged into a single, master TagCount file containing a list of all tags of interest for a species. Only tags occurring at or above a (user-specified) minimum number of reads across all of the FASTQ files in the experiment are retained in the output master tag list. The more times a particular tag has been observed, the less likely it contains a sequencing error. The minimum tag count controls the tradeoff between the amount of sequencing errors admitted into the analysis versus the minimum allele frequency of interest. We do not expect to eliminate all sequencing errors at this step. We are usually able to filter out or correct most of them in subsequent steps of the pipeline, or in further, downstream processing customized to the biology of the study population(s). Furthermore, tags containing one or more sequencing errors can still be useful to score SNPs at other, non-error positions.

Alignment of tags to the reference genome. Alignment of each tag of interest in the master tag list to the reference genome is carried out with third party software. To facilitate this, the master tag list file is converted from TagCount format into FASTQ format (with fake, uniformly high quality scores). Currently, SAM format [36] output alignment files produced by the free software programs Bowtie2 [49] or BWA [36] can be read by the TASSEL-GBS pipeline and converted into a "TagsOnPhysicalMap" (TOPM) file that can be used for SNP calling.

The TOPM contains all of the tags present in the master TagCount file and genomic positions for the subset of tags that align to a unique best position in the genome. Retention of all of the master tags in the TOPM, even those without unique best genomic positions, affords future incorporation of additional information regarding the positions of unplaced tags, either from alternative aligners or from genetic mapping evidence. The TOPM is sorted by tag sequence and functions as a key-value map where a tag sequence can be used as a key to retrieve its corresponding physical position. It is also possible to programmatically traverse the tags within a TOPM in their physical position order through the use of an in-memory, primitive treemap. At the SNP calling step of the Discovery Pipeline (see below), the alleles represented by each useful tag at each useful SNP ("variants") are added to the TOPM. Three file formats of the TOPM are supported: text, binary, and HDF5.

Determining the distribution of tags across individual samples. With a TOPM file available, the next ingredient needed to discover and call SNPs is a matrix that records the number of times each tag in the master tag list was observed in each DNA sample, which we refer to as a TagsByTaxa (TBT) file (Figure 1B). In order to maximize the capacity of our pipeline, we currently store the TBT in HDF5 format (http://www.hdfgroup.org). The HDF5 format facilitates extremely fast read and write access to large data sets. To construct a TBT, the DNA sample of origin of each good, barcoded read in the set of input FASTQ files is determined based upon its barcode. If a good, barcoded read matches one of the tags in the master tag list, its depth in the appropriate taxon is incremented in the output TBT, up to a maximum depth of 127. Since the matrix is often extremely sparse, a custom run length encoding compression algorithm was implemented that provides a high level of compression, with minimal reduction in access speed.

SNP discovery and initial filtering. SNP discovery (Figure 1C) is performed for each set of tags that align to the exact same starting genomic position and strand, where the starting genomic position of a tag is defined by the cut site remnant at the beginning of the tag. Such tags, originating from the same restriction enzyme cut site and with the same orientation (but not necessarily of the same length), collectively comprise a "TagLocus". To call SNPs and ensure that indels are handled consistently, a *de novo* multiple sequence alignment of all the tags in each TagLocus is performed using the BioJava 3.0 API [50], which implements the CLUSTAL W algorithm [51]. For each SNP in the resulting "TagLocusAlignment", the allele represented by each tag is determined and the TBT file is consulted to tally the observed depths of each allele in each taxon. The genotype of the SNP in each taxon is then determined either by a binomial likelihood ratio method of quantitative SNP calling (the default; for details see Supplementary Text S1) or, optionally, following the method of Hohenlohe et al. [52].

After genotypes are obtained for a potential SNP, initial filtering is then performed based upon user settings for minimum minor allele frequency, and for a minimum coefficient of panmixia, or inbreeding relative to the entire population, F_{IT}, (where $F_{IT} = 1 - Ho/He$, Ho = observed heterozygosity, He = expected heterozygosity = $2q(1-q)$, and q = minor allele frequency). Error-prone SNPs and spurious SNPs from paralogy often appear excessively heterozygous, with lower F_{IT} than expected. If the user supplies a "pedigree file" that indicates the expected inbreeding coefficient (F) of each taxon, then only inbred taxa, with an expected inbreeding coefficient greater than or equal to the user-specified minimum coefficient of panmixia (*minF* parameter), are used in the calculation of F_{IT}. Inbred samples, available in many crop species and model organisms, can greatly add to the power of this filter. If enough inbred samples are available, then, additional filtering of SNPs can then be optionally performed enforcing a minimum "inbred coverage" (proportion of the inbred samples to be non-missing at the SNP) and a maximum "inbred heterozygosity score" (defined above).

To illustrate the effectiveness of these population genetic-based SNP filters, as applied in our most recent maize Discovery Build (AllZeaGBSv2.6, with 31,978 samples), we focused on the subset of 5,254 samples from the maize Nested Association Mapping (NAM) population [53]. The NAM population is a series of 25 biparental, F_2-derived RIL families all with a common female parent, the inbred line B73. Error-prone SNPs can be identified in a biparental family through their tendency, when they are in fact not segregating, to *appear* to be weakly polymorphic, but with segregation ratios significantly deviating from the 1:1 expectation. In contrast, non-segregating SNPs in a given family that display no spurious polymorphism are free of genotypic error in that family. The availability of the 25 biparental NAM families provides tremendous power to detect error-prone SNPs, and thus to examine the effectiveness of our filters.

We used the 5,254 NAM RIL samples to estimate error rates for three alternative sets of chromosome 10 SNPs discovered in the full set of 31,978 maize samples comprising our AllZeaGBSv2.6 Discovery Build. The three sets of chromosome 10 SNPs were obtained after application of three different filtering regimes: (1) no filters other than MAF >= 0.001, (2) a minimal filter only for MAF >= 0.01, and (3) our "standard" maize Discovery Build filters of MAF >= 0.001, minimum F_{IT} in inbred samples of 0.8, inbred coverage >0.15, and inbred heterozygosity score <0.21. The SNPs were discovered and filtered based upon all 31,978 maize samples (including the NAM RILs) and their allele frequencies were then separately calculated in each of the 25 NAM families. To minimize sampling error, allele frequency was only estimated for a particular SNP-NAM family combination if at least 19 RILs in that family had non-missing genotypes for the SNP (n >= 19). Thus, by "SNP-NAM family combination" we mean a particular SNP (e.g., "S10_2918") in a particular NAM family (e.g., "B73 × B97") that has at least 19 non-missing genotypes, regardless of whether it is polymorphic or not within that family. Hence, for each SNP, allele frequencies were calculated in 25 or fewer families, depending on the amount of missing data in each family. NAM family-specific minor allele calls for a SNP were classified as errors if the family-specific MAF was greater than zero but less than 0.25, and the SNP significantly deviated from 1:1 segregation in that family at $p<0.001$ (binomial test). The overall error rate for a SNP was then estimated as the total number of these error calls divided by the total number of calls for that SNP in NAM families with n >= 19 where the SNP significantly deviated from 1:1 segregation at $p<0.001$ (including the monomorphic SNPs). A pictorial explanation of this method of estimating error rates, using a single NAM family for illustrative purposes, is provided in Supplementary Text S2.

Compared to non-filtered SNPs (Figures 3A and 3B), application of our "standard" maize Discovery Build filters (MAF >= 0.001, minimum F_{IT} in inbred samples of 0.8, inbred coverage >0.15, inbred heterozygosity score <0.21) (Figures 3E and 3F) greatly increased the proportion of SNP-NAM family combinations displaying either appropriate 1:1 segregation (Figure 3F), or no polymorphism at all (Figure 3E). In contrast, a minimal filter based only on MAF (MAF >= 0.01) (Figures 3C and 3D) was far less effective at removing error-prone SNPs than our standard filters. Furthermore, our standard filters clearly shifted the distribution of error rates (estimated from the NAM samples) toward zero (Figure 4) and reduced the mean error rate (Table 2) relative to either no filtering (other than MAF >= 0.001) or minimal filtering (only for MAF >0.01). Extremely low estimates of mean and median error rates after application of our standard filters (0.0042 and zero respectively; Table 2) indicate that, for the most part, highly reliable SNP genotypes are produced by the GBS assay and the TASSEL-GBS pipeline, at least for inbred samples (where under-calling of heterozygotes due to low coverage is not an issue).

Error prone SNPs that are not removed by our standard filters (e.g., polymorphic SNPs with family specific MAFs <0.25 in Figure 3F) can be easily removed from biparental RIL families by filtering for appropriate allele frequencies and/or based on their relatively low levels of linkage disequilibrium (LD) with neighboring SNPs. In addition, error-prone SNPs identified in biparental families can be excluded from analyses of the remaining, non-biparental samples in a Discovery Build. If biparental populations are not available in your study species, it should be possible to use half-sib (e.g., open pollinated) families, or any population having a prior expected allele frequency range for polymorphic markers, to filter out error-prone SNPs. Alternatively, recently bottlenecked populations with high levels of extended LD can be used to filter out error-prone SNPs, which should display relatively low levels of LD with their neighboring SNPs.

After SNP calling and filtering, the input TOPM is updated with variants, and this "production-ready" TOPM is then saved to a new file (Figure 1C). For each SNP that passed the filtering step, the allele that is represented by each tag in the corresponding TagLocus is recorded in the "production-ready" TOPM, as well as the relative position of the SNP with respect to the genomic position of the TagLocus. The production TOPM produced by our AllZeaGBSv2.6 Discovery Build contained 955,690 useful SNPs.

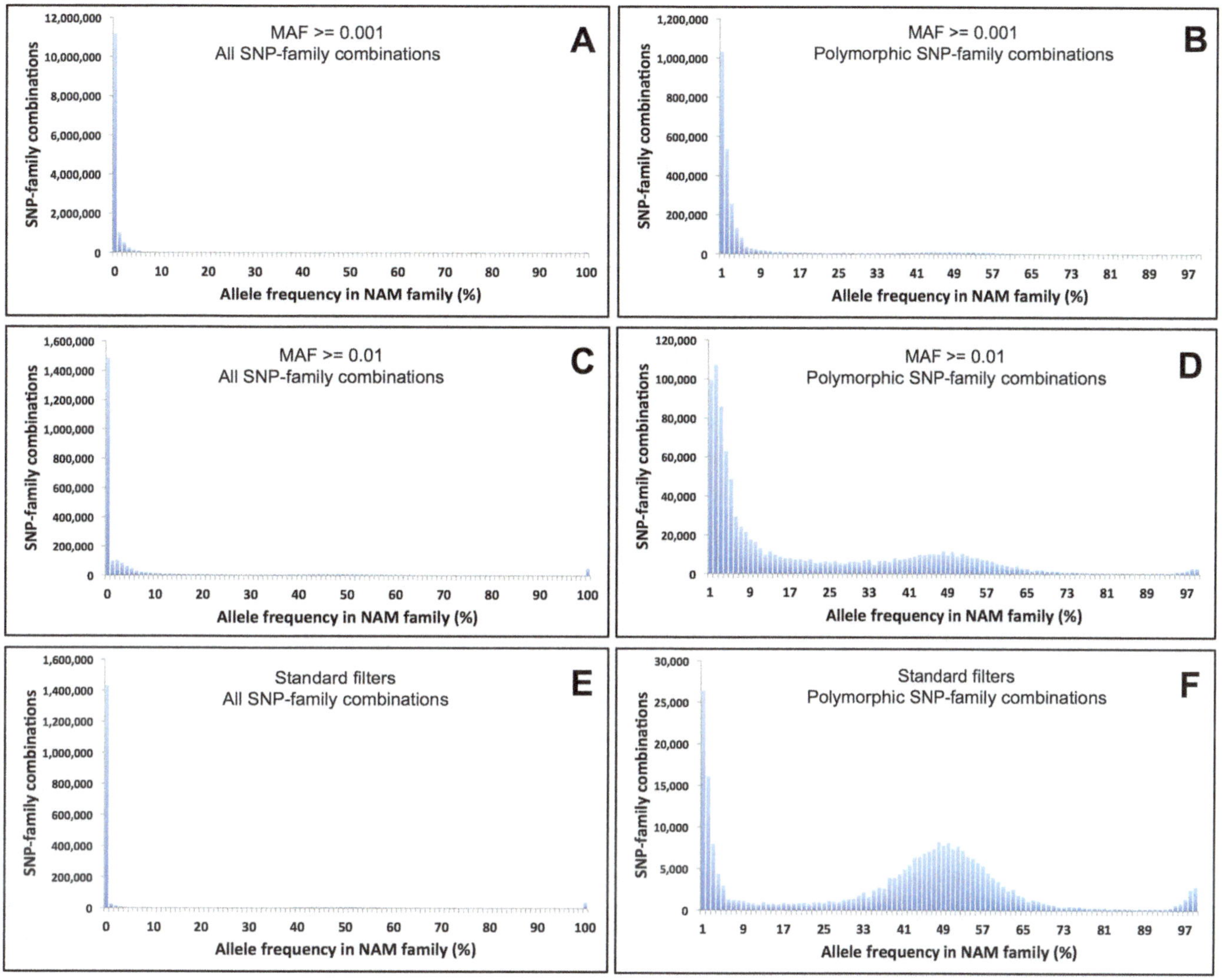

Figure 3. Within NAM family allele frequency distributions of chromosome 10 SNPs after different levels of filtering. Allele frequencies were calculated in each of the 25 Nested Association Mapping (NAM) families (collectively comprising 5,254 RILs) after application of the filters to the entire set of 31,978 maize samples in the AllZeaGBSv2.6 build. Allele frequencies were only estimated in a NAM family if at least 19 RILs had non-missing genotypes. Each histogram shows the allele frequency distribution for all the SNP-NAM family combinations with n >= 19. (A, B) No filter other than minimum MAF of 0.001. (C, D) A minimal filter only for MAF >= 0.01. (E, F) "Standard" maize build filters of MAF >= 0.001, minimum F_{IT} in inbred samples of 0.8, inbred coverage >0.15, and inbred heterozygosity score <0.21. (A, C, E) All SNP-family combinations: the error-free, monomorphic SNP-family combinations dwarf the segregating SNPs in all three cases. (B, D, F) Polymorphic SNP-family combinations only: omitting the monomorphic SNP-family combinations permits visualization of the remaining allele frequency distribution.

Production Pipeline

In contrast to the multiple-step Discovery Pipeline, the Production Pipeline consists of a single step, utilizing the production-ready TOPM generated by the Discovery Pipeline to produce genotypes directly from input FASTQ files (Figure 2). The Production Pipeline determines the taxon of origin of each good, barcoded sequence read in each input FASTQ file and then checks if the read matches one of the useful tags in the production-ready TOPM. In this manner, allelic depths for each useful SNP in the TOPM are recorded for each taxon, allowing quantitative SNP calling to be performed, again either by our own binomial likelihood ratio method (for details see Supplementary Text S1) or, optionally, according to the method of Hohenlohe et al. [52]. If the GBS library preps for some samples have been run in replicate on multiple flow cell lanes (to increase depth), the corresponding allelic depth information is tallied across the replicates prior to SNP calling. Genotype files are produced in HapMap format as well as in or our own custom HDF5 format (which also records allelic depth). The ability to convert from this custom HDF5 format into VCF format [54], which also retains allelic depth, will be added to the TASSEL GUI in the near future.

Downstream Processing

Further processing of the genotype files, such as sub-setting out specific taxa or genomic regions of interest, filtering SNPs or taxa based upon coverage, or filtering of SNPs based on minor allele frequency, can be performed either with the TASSEL 4 GUI or the TASSEL 4 Standalone command line interface (the ability to filter for SNPs that are in LD with their neighbors will be added soon). Depending on the genome size, the exact molecular protocol and restriction enzyme used, and the sequence depth obtained, missing data can be common and actual heterozygotes

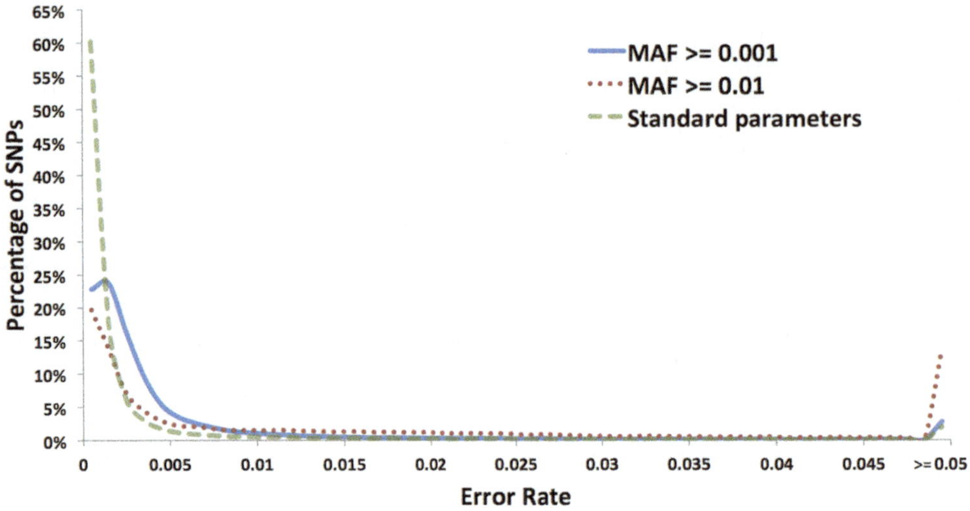

Figure 4. Error rate distribution of chromosome 10 SNPs for different levels of filtering. Error rates in the AllZeaGBSv2.6 Discovery build SNP calls were estimated using the NAM biparental families. NAM family-specific minor allele calls were defined as errors if the family-specific MAF was greater than zero but less than 0.25, and the SNP significantly deviated from 1:1 segregation in that family at $p<0.001$.

can be substantially under-called as homozygotes. Hence, imputation of missing data (and, possibly, phasing of the final genotypes) is usually desirable. Since the optimal imputation approach depends greatly upon the biology of the species and the experimental design, this topic is beyond the scope of this paper. Numerous, general purpose imputation tools are already available [55]. Custom imputation approaches that we are developing for our maize experimental populations will be the subject of future publications.

Hardware Needs and Installation

The TASSEL-GBS pipeline is part of the TASSEL package, which is written in Java, so it can be run on Linux, Mac, or Windows operating systems. A minimum of 8 GB of RAM is required (at least 16 GB is recommended). Detailed installation instructions are provided at www.maizegenetics.net/tassel as well as in the TASSEL-GBS pipeline documentation (www.maizegenetics.net/tassel/docs/TasselPipelineGBS.pdf). The source code is available from sourceforge.net/p/tassel/code/ci/master/tree, JUnit tests at sourceforge.net/p/tassel/maizegenetics4-test/ci/master/tree and a relatively small test data set at www.maizegenetics.net/tassel/GBSTestData.tar. The current version of TASSEL-GBS as described herein is implemented in TASSEL V4.3.5. Although we generally recommend using the latest version, users can revert to V4.3.5 or other versions by following the instructions posted in the following document www.maizegenetics.net/tassel/docs/UpdatingTasselStandaloneUsingGit.pdf, under the heading "To Update Packages to Older Releases…". The TASSEL-GBS pipeline will soon be available in TASSEL 5; major improvements to the pipeline (e.g., full VCF format support, allowing tag lengths greater than 64 bp, storage of tag depths per individual up to 10,000 rather than the current maximum of 127) will be implemented there.

Table 2. Comparison of error rates for chromosome 10 SNPs from the AllZeaGBSv2.6 build for different levels of filtering by the Discovery SNP caller.

Filter[1]	nSNPs	nSNPsTested[2]	avgErrorRate[3]	avg nSegregating[4]
MAF >= 0.001	694,517	680,623	0.00681	12,899
MAF >= 0.01	149,480	136,296	0.02218	12,818
Standard[5]	78,627	78,506	0.00420	7,192

[1]Filters applied to the entire build (31,978 non-blank samples)
[2]Minimum sample size of 19 in at least one maize Nested Association Mapping (NAM) family
[3]Average error rates estimated from 5,254 NAM RILs. Median error rates were zero for all three filters.
[4]Average number of chromosome 10 SNPs with n >= 19 and MAF between 0.25 and 0.75 across the 25 NAM families.
[5]MAF >= 0.001, minimum F_{IT} in inbred samples of 0.8, inbred coverage >0.15, inbred heterozygosity score <0.21.

Strengths and Weaknesses

Strengths

The strengths of GBS and the TASSEL-GBS pipeline are the large number of markers potentially produced (depending on the biology of the species and the choice of restriction enzymes), low cost and minimal startup cost, and integration of SNP discovery with SNP calling.

The potentially large number of markers available from GBS makes GWAS feasible in study populations where linkage disequilibrium (LD) extends far enough so that causative polymorphisms stand a reasonable chance of being in LD with one or more markers. Alternatively, the large number of markers facilitates accurate projection of haplotypes from a set of more densely genotyped reference haplotypes [56] derived from whole genome sequencing (WGS) (e.g., [57,58]). This projection strategy is especially effective if the WGS reference haplotypes are representative of the founders of the study population [59].

Compared to alternative high-density marker technologies such as SNP arrays, GBS is relatively inexpensive, particularly if low coverage data suffices for the purpose of your study. Startup costs for GBS are minimal, as startup involves only (1) testing that your one of your candidate restriction enzymes (or enzyme pairs) produces a suitable GBS library, and (2) optimization of the ratio of sample DNA to the PCR adapters [12]. In contrast, startup for a SNP array involves ascertainment of SNPs in a small discovery panel and assay design for each individual SNP.

The common practice of using of a small panel of individuals to discover SNPs for inclusion in a SNP array introduces an ascertainment bias that can severely distort estimates of key population genetic parameters gauging genetic diversity, subpopulation differentiation and relatedness among individuals [60–62]. In contrast, the TASSEL-GBS Discovery pipeline integrates SNP discovery with SNP calling, using all available samples to date, and thus avoids the ascertainment bias that would arise from a small discovery panel. This type of ascertainment bias will also be minimal for new samples run through the Production Pipeline, provided that their genetic diversity is well-represented among the samples included in the Discovery Build. However, there might still be some subtle biases in either pipeline, caused by factors such as null alleles [63,64], alignment to the reference, and the use of inbreds only (rather than the full set of samples) to filter SNPs for F_{IT}.

Weaknesses

The main weakness of the GBS assay, when conducted at low coverage, is the amount of missing data. However, numerous imputation approaches are currently available [55] and yet more are currently in development, for a wide range of biological scenarios. As discussed above, the most appropriate imputation method and the probability of imputation success depends upon the biology of the study population. For some purposes, such as estimation of population allele frequencies [65], kinship, relatedness, and genetic distance, phylogenetic reconstruction [22], or germplasm quality control [24], imputation of missing data is usually not necessary.

Conclusions

The TASSEL-GBS pipeline for identifying and calling SNPs from next-generation, genotyping by sequencing data fulfills our design criteria better than any existing pipeline. It has a capacity for very large analyses involving tens of thousands of samples, yet can also be run at smaller scales. The pipeline permits rapid processing of the data, yet has a relatively modest memory footprint, allowing it to be run on desktop or laptop computers. This increases its usability by researchers in developing countries who may lack access to sophisticated computing resources. The separation of SNP discovery and genotyping into two phases reduces potential ascertainment biases and, more importantly, makes the TASSEL-GBS pipeline highly suitable for use in a genomics-assisted, accelerated breeding context, where rapid turnaround times from tissue collection to genotypes are essential. Furthermore, the high density of markers potentially available from the GBS assay should enable accurate genomic prediction over multiple generations.

Acknowledgments

We thank Yang (Jon) Zhang for contributions to the code for VCF output of genotypes and Sara J. Miller for assistance with manuscript formatting. We also thank the numerous users of the TASSEL-GBS pipeline (Katie Hyma in particular) who have made useful suggestions or have reported bugs.

Author Contributions

Conceived and designed the experiments: ESB JCG QS. Performed the experiments: ESB JCG TMC FL JH QS RJE. Analyzed the data: JCG ESB TMC QS RJE. Wrote the paper: JCG ESB.

References

1. Metzker ML (2010) Sequencing technologies - the next generation. Nat Rev Genet 11: 31–46. doi:10.1038/nrg2626.
2. Shendure J, Lieberman Aiden E (2012) The expanding scope of DNA sequencing. Nat Biotechnol 30: 1084–1094. Available: http://www.ncbi.nlm.nih.gov/pubmed/23138308.
3. Edwards D, Batley J, Snowdon RJ (2013) Accessing complex crop genomes with next-generation sequencing. Theor Appl Genet 126: 1–11. Available: http://www.ncbi.nlm.nih.gov/pubmed/22948437. Accessed 11 November 2013.
4. Kilpinen H, Barrett JC (2013) How next-generation sequencing is transforming complex disease genetics. Trends Genet 29: 23–30. Available: http://www.ncbi.nlm.nih.gov/pubmed/23103023. Accessed 11 November 2013.
5. Altshuler D, Pollara VJ, Cowles CR, Van Etten WJ, Baldwin J, et al. (2000) An SNP map of the human genome generated by reduced representation shotgun sequencing. Nature 407: 513–516. Available: http://www.ncbi.nlm.nih.gov/pubmed/11029002.
6. Davey JW, Hohenlohe PA, Etter PD, Boone JQ, Catchen JM, et al. (2011) Genome-wide genetic marker discovery and genotyping using next-generation sequencing. Nat Rev Genet 12: 499–510. Available: http://www.ncbi.nlm.nih.gov/pubmed/21681211. Accessed 6 November 2013.
7. Poland JA, Rife TW (2012) Genotyping-by-Sequencing for Plant Breeding and Genetics. Plant Genome J 5: 92. Available: https://www.crops.org/publications/tpg/abstracts/5/3/92. Accessed 11 November 2013.
8. Narum SR, Buerkle CA, Davey JW, Miller MR, Hohenlohe PA (2013) Genotyping-by-sequencing in ecological and conservation genomics. Mol Ecol 22: 2841–2847. Available: http://www.ncbi.nlm.nih.gov/pubmed/23711105. Accessed 7 November 2013.
9. Van Orsouw NJ, Hogers RCJ, Janssen A, Yalcin F, Snoeijers S, et al. (2007) Complexity reduction of polymorphic sequences (CRoPS): a novel approach for large-scale polymorphism discovery in complex genomes. PLoS One 2: e1172. Available: http://www.pubmedcentral.nih.gov/articlerender.fcgi?artid=2048665&tool=pmcentrez&rendertype=abstract. Accessed 11 November 2013.
10. Van Tassell C, Smith T, Matukumalli L (2008) SNP discovery and allele frequency estimation by deep sequencing of reduced representation libraries. Nat Methods 5: 247–252. Available: http://www.nature.com/nmeth/journal/v5/n3/abs/nmeth.1185.html. Accessed 2013 Nov 11.
11. Baird NA, Etter PD, Atwood TS, Currey MC, Shiver AL, et al. (2008) Rapid SNP discovery and genetic mapping using sequenced RAD markers. PLoS One 3: e3376. Available: http://www.pubmedcentral.nih.gov/articlerender.fcgi?artid=2557064&tool=pmcentrez&rendertype=abstract. Accessed 2013 Nov 7.
12. Elshire RJ, Glaubitz JC, Sun Q, Poland JA, Kawamoto K, et al. (2011) A Robust, Simple Genotyping-by-Sequencing (GBS) Approach for High Diversity Species. PLoS One 6: e19379. Available: http://www.pubmedcentral.nih.gov/articlerender.fcgi?artid=3087801&tool=pmcentrez&rendertype=abstract. Accessed 2011 Jul 18.
13. Andolfatto P, Davison D, Erezyilmaz D, Hu TT, Mast J, et al. (2011) Multiplexed shotgun genotyping for rapid and efficient genetic mapping. Genome Res 21: 610–617. Available: http://www.pubmedcentral.nih.gov/articlerender.fcgi?artid=3065708&tool=pmcentrez&rendertype=abstract. Accessed 2013 Nov 8.
14. Wang S, Meyer E, McKay JK, Matz M V (2012) 2b-RAD: a simple and flexible method for genome-wide genotyping. Nat Methods 9: 808–810. Available: http://www.ncbi.nlm.nih.gov/pubmed/22609625. Accessed 2013 Nov 8.
15. Truong HT, Ramos AM, Yalcin F, de Ruiter M, van der Poel HJA, et al. (2012) Sequence-based genotyping for marker discovery and co-dominant scoring in germplasm and populations. PLoS One 7: e37565, . Available: http://www.pubmedcentral.nih.gov/articlerender.fcgi?artid=3360789&tool=pmcentrez&rendertype=abstract. Accessed 2013 Nov 8.
16. Monson-Miller J, Sanchez-Mendez DC, Fass J, Henry IM, Tai TH, et al. (2012) Reference genome-independent assessment of mutation density using restriction enzyme-phased sequencing. BMC Genomics 13: 72, . Available: http://www.pubmedcentral.nih.gov/articlerender.fcgi?artid=3305632&tool=pmcentrez&rendertype=abstract. Accessed 2013 Nov 11.
17. Chen Q, Ma Y, Yang Y, Chen Z, Liao R, et al. (2013) Genotyping by genome reducing and sequencing for outbred animals. PLoS One 8: e67500. Available: http://www.pubmedcentral.nih.gov/articlerender.fcgi?artid=3715491&tool=pmcentrez&rendertype=abstract. Accessed 2013 Nov 11.

18. Morishige DT, Klein PE, Hilley JL, Sahraeian SM, Sharma A, et al. (2013) Digital genotyping of sorghum – a diverse plant species with a large repeat-rich genome. BMC Genomics 14: 448. Available: http://www.pubmedcentral.nih.gov/articlerender.fcgi?artid=3716661&tool=pmcentrez&rendertype=abstract. Accessed 2013 Nov 10.
19. Stolle E, Moritz RFA (2013) RESTseq–efficient benchtop population genomics with RESTriction Fragment SEQuencing. PLoS One 8: e63960. Available: http://www.pubmedcentral.nih.gov/articlerender.fcgi?artid=3656931&tool=pmcentrez&rendertype=abstract. Accessed 2013 Nov 11.
20. Poland JA, Brown PJ, Sorrells ME, Jannink J-L (2012) Development of high-density genetic maps for barley and wheat using a novel two-enzyme genotyping-by-sequencing approach. PLoS One 7: e32253. Available: http://www.pubmedcentral.nih.gov/articlerender.fcgi?artid=3289635&tool=pmcentrez&rendertype=abstract. Accessed 2013 Nov 11.
21. Morris GP, Ramu P, Deshpande SP, Hash CT, Shah T, et al. (2013) Population genomic and genome-wide association studies of agroclimatic traits in sorghum. Proc Natl Acad Sci U S A 110: 453–458. Available: http://www.pubmedcentral.nih.gov/articlerender.fcgi?artid=3545811&tool=pmcentrez&rendertype=abstract. Accessed 2013 Nov 8.
22. Lu F, Lipka AE, Elshire RJ, Glaubitz JC, Cherney J, et al. (2013) Switchgrass genomic diversity, ploidy and evolution: novel insights from a network-based SNP discovery protocol. PLoS Genet 9: e1003215.
23. Maron LG, Guimarães CT, Kirst M, Albert PS, Birchler JA, et al. (2013) Aluminum tolerance in maize is associated with higher MATE1 gene copy number. Proc Natl Acad Sci U S A 110: 5241–5246. Available: http://www.pubmedcentral.nih.gov/articlerender.fcgi?artid=3612656&tool=pmcentrez&rendertype=abstract. Accessed 2013 Nov 11.
24. Romay MC, Millard MJ, Glaubitz JC, Peiffer JA, Swarts KL, et al. (2013) Comprehensive genotyping of the USA national maize inbred seed bank. Genome Biol 14: R55. Available: http://www.pubmedcentral.nih.gov/articlerender.fcgi?artid=3707059&tool=pmcentrez&rendertype=abstract. Accessed 2013 Nov 7.
25. Sonah H, Bastien M, Iquira E, Tardivel A, Légaré G, et al. (2013) An improved genotyping by sequencing (GBS) approach offering increased versatility and efficiency of SNP discovery and genotyping. PLoS One 8: e54603. Available: http://www.pubmedcentral.nih.gov/articlerender.fcgi?artid=3553054&tool=pmcentrez&rendertype=abstract. Accessed 2013 Nov 8.
26. De Donato M, Peters SO, Mitchell SE, Hussain T, Imumorin IG (2013) Genotyping-by-sequencing (GBS): a novel, efficient and cost-effective genotyping method for cattle using next-generation sequencing. PLoS One 8: e62137. Available: http://www.pubmedcentral.nih.gov/articlerender.fcgi?artid=3656875&tool=pmcentrez&rendertype=abstract. Accessed 2013 Nov 11.
27. Rutkoski JE, Poland J, Jannink J-L, Sorrells ME (2013) Imputation of unordered markers and the impact on genomic selection accuracy. G3 (Bethesda) 3: 427–439. Available: http://www.pubmedcentral.nih.gov/articlerender.fcgi?artid=3583451&tool=pmcentrez&rendertype=abstract. Accessed 2013 Nov 11.
28. Kim S-I, Tai TH (2013) Identification of SNPs in closely related Temperate Japonica rice cultivars using restriction enzyme-phased sequencing. PLoS One 8: e60176, . Available: http://www.pubmedcentral.nih.gov/articlerender.fcgi?artid=3608622&tool=pmcentrez&rendertype=abstract. Accessed 2013 Nov 11.
29. Ly D, Hamblin M, Rabbi I, Melaku G, Bakare M, et al. (2013) Relatedness and Genotype × Environment Interaction Affect Prediction Accuracies in Genomic Selection: A Study in Cassava. Crop Sci 53: 1312. Available: https://www.crops.org/publications/cs/abstracts/53/4/1312. Accessed 2013 Nov 11.
30. Saintenac C, Jiang D, Wang S, Akhunov E (2013) Sequence-based mapping of the polyploid wheat genome. G3 (Bethesda) 3: 1105–1114. Available: http://www.ncbi.nlm.nih.gov/pubmed/23665877. Accessed 2013 Nov 7.
31. White TA, Perkins SE, Heckel G, Searle JB (2013) Adaptive evolution during an ongoing range expansion: the invasive bank vole (Myodes glareolus) in Ireland. Mol Ecol 22: 2971–2985. Available: http://www.ncbi.nlm.nih.gov/pubmed/23701376. Accessed 2013 Nov 7.
32. Bradbury PJ, Zhang Z, Kroon DE, Casstevens TM, Ramdoss Y, et al. (2007) TASSEL: software for association mapping of complex traits in diverse samples. Bioinformatics 23: 2633–2635. Available: http://www.ncbi.nlm.nih.gov/pubmed/17586829. Accessed 2011 Aug 9.
33. Mascher M, Wu S, Amand PS, Stein N, Poland J (2013) Application of Genotyping-by-Sequencing on Semiconductor Sequencing Platforms: A Comparison of Genetic and Reference-Based Marker Ordering in Barley. PLoS One 8: e76925. Available: http://www.pubmedcentral.nih.gov/articlerender.fcgi?artid=3789676&tool=pmcentrez&rendertype=abstract. Accessed 2013 Nov 7.
34. Catchen JM, Amores A, Hohenlohe P, Cresko W, Postlethwait JH (2011) Stacks: building and genotyping Loci de novo from short-read sequences. G3 (Bethesda) 1: 171–182. Available: http://www.pubmedcentral.nih.gov/articlerender.fcgi?artid=3276136&tool=pmcentrez&rendertype=abstract. Accessed 2013 Nov 7.
35. Catchen J, Bassham S, Wilson T, Currey M, O'Brien C, et al. (2013) The population structure and recent colonization history of Oregon threespine stickleback determined using restriction-site associated DNA-sequencing. Mol Ecol 22: 2864–2883. Available: http://www.ncbi.nlm.nih.gov/pubmed/23718143. Accessed 2013 Nov 11.
36. Li H, Handsaker B, Wysoker A, Fennell T, Ruan J, et al. (2009) The Sequence Alignment/Map format and SAMtools. Bioinformatics 25: 2078–2079. Available: http://www.pubmedcentral.nih.gov/articlerender.fcgi?artid=2723002&tool=pmcentrez&rendertype=abstract. Accessed 2013 Nov 6.
37. Li R, Li Y, Fang X, Yang H, Wang J, et al. (2009) SNP detection for massively parallel whole-genome resequencing. Genome Res 19: 1124–1132. Available: http://www.pubmedcentral.nih.gov/articlerender.fcgi?artid=2694485&tool=pmcentrez&rendertype=abstract. Accessed 2013 Nov 6.
38. DePristo MA, Banks E, Poplin R, Garimella KV, Maguire JR, et al. (2011) A framework for variation discovery and genotyping using next-generation DNA sequencing data. Nat Genet 43: 491–498. Available: http://www.pubmedcentral.nih.gov/articlerender.fcgi?artid=3083463&tool=pmcentrez&rendertype=abstract. Accessed 2013 Nov 6.
39. Rafalski JA (2010) Association genetics in crop improvement. Curr Opin Plant Biol 13: 174–180. Available: http://www.ncbi.nlm.nih.gov/pubmed/20089441. Accessed 2013 Nov 8.
40. Xie W, Feng Q, Yu H, Huang X, Zhao Q, et al. (2010) Parent-independent genotyping for constructing an ultrahigh-density linkage map based on population sequencing. Proc Natl Acad Sci U S A 107: 10578–10583. Available: http://www.pubmedcentral.nih.gov/articlerender.fcgi?artid=2890813&tool=pmcentrez&rendertype=abstract. Accessed 2013 Nov 11.
41. Li H, Durbin R (2011) Inference of human population history from individual whole-genome sequences. Nature 475: 493–496. Available: http://www.pubmedcentral.nih.gov/articlerender.fcgi?artid=3154645&tool=pmcentrez&rendertype=abstract. Accessed 2013 Nov 6.
42. Harter AV, Gardner KA, Falush D, Lentz DL, Bye RA, et al. (2004) Origin of extant domesticated sunflowers in eastern North America. Nature 430: 201–205. Available: http://www.ncbi.nlm.nih.gov/pubmed/15241413.
43. Ofori A, Becker HC, Kopisch-Obuch FJ (2008) Effect of crop improvement on genetic diversity in oilseed Brassica rapa (turnip-rape) cultivars, detected by SSR markers. J Appl Genet 49: 207–212. Available: http://www.ncbi.nlm.nih.gov/pubmed/18670055.
44. Cowling WA (2013) Sustainable plant breeding. Plant Breed 132: 1–9. Available: http://doi.wiley.com/10.1111/pbr.12026. Accessed 2013 Nov 11.
45. Dohm JC, Lottaz C, Borodina T, Himmelbauer H (2008) Substantial biases in ultra-short read data sets from high-throughput DNA sequencing. Nucleic Acids Res 36: e105. Available: http://www.pubmedcentral.nih.gov/articlerender.fcgi?artid=2532726&tool=pmcentrez&rendertype=abstract. Accessed 2013 Nov 6.
46. Eren AM, Vineis JH, Morrison HG, Sogin ML (2013) A filtering method to generate high quality short reads using illumina paired-end technology. PLoS One 8: e66643. Available: http://www.pubmedcentral.nih.gov/articlerender.fcgi?artid=3684618&tool=pmcentrez&rendertype=abstract. Accessed 2013 Nov 6.
47. McElroy KE, Luciani F, Thomas T (2012) GemSIM: general, error-model based simulator of next-generation sequencing data. BMC Genomics 13: 74. Available: http://www.pubmedcentral.nih.gov/articlerender.fcgi?artid=3305602&tool=pmcentrez&rendertype=abstract. Accessed 2013 Nov 11.
48. Allhoff M, Schönhuth A, Martin M, Costa IG, Rahmann S, et al. (2013) Discovering motifs that induce sequencing errors. BMC Bioinformatics 14 Suppl 5: S1. Available: http://www.pubmedcentral.nih.gov/articlerender.fcgi?artid=3622629&tool=pmcentrez&rendertype=abstract. Accessed 2013 Nov 11.
49. Langmead B, Salzberg SL (2012) Fast gapped-read alignment with Bowtie 2. Nat Methods 9: 357–359. Available: http://www.pubmedcentral.nih.gov/articlerender.fcgi?artid=3322381&tool=pmcentrez&rendertype=abstract. Accessed 2013 Nov 7.
50. Prlić A, Yates A, Bliven SE, Rose PW, Jacobsen J, et al. (2012) BioJava: an open-source framework for bioinformatics in 2012. Bioinformatics 28: 2693–2695. Available: http://www.pubmedcentral.nih.gov/articlerender.fcgi?artid=3467744&tool=pmcentrez&rendertype=abstract. Accessed 2013 Nov 10.
51. Thompson JD, Higgins DG, Gibson TJ (1994) CLUSTAL W: improving the sensitivity of progressive multiple sequence alignment through sequence weighting, position-specific gap penalties and weight matrix choice. Nucleic Acids Res 22: 4673–4680. Available: http://www.pubmedcentral.nih.gov/articlerender.fcgi?artid=308517&tool=pmcentrez&rendertype=abstract.
52. Hohenlohe PA, Bassham S, Etter PD, Stiffler N, Johnson EA, et al. (2010) Population genomics of parallel adaptation in threespine stickleback using sequenced RAD tags. PLoS Genet 6: e1000862. Available: http://www.pubmedcentral.nih.gov/articlerender.fcgi?artid=2829049&tool=pmcentrez&rendertype=abstract. Accessed 2013 Nov 7.
53. McMullen MD, Kresovich S, Villeda HS, Bradbury P, Li H, et al. (2009) Genetic properties of the maize nested association mapping population. Science 325: 737–740. Available: http://www.ncbi.nlm.nih.gov/pubmed/19661427. Accessed 2011 Jun 14.
54. Danecek P, Auton A, Abecasis G, Albers CA, Banks E, et al. (2011) The variant call format and VCFtools. Bioinformatics 27: 2156–2158. Available: http://www.pubmedcentral.nih.gov/articlerender.fcgi?artid=3137218&tool=pmcentrez&rendertype=abstract. Accessed 2013 Nov 6.
55. Marchini J, Howie B (2010) Genotype imputation for genome-wide association studies. Nat Rev Genet 11: 499–511. Available: http://www.ncbi.nlm.nih.gov/pubmed/20517342. Accessed 2013 Nov 6.

56. Howie B, Marchini J, Stephens M (2011) Genotype imputation with thousands of genomes. G3 (Bethesda) 1: 457–470. Available: http://www.pubmedcentral.nih.gov/articlerender.fcgi?artid = 3276165&tool = pmcentrez&rendertype = abstract. Accessed 2013 Nov 11.
57. Abecasis GR, Auton A, Brooks LD, DePristo MA, Durbin RM, et al. (2012) An integrated map of genetic variation from 1,092 human genomes. Nature 491: 56–65. Available: http://www.pubmedcentral.nih.gov/articlerender.fcgi?artid = 3498066&tool = pmcentrez&rendertype = abstract. Accessed 2013 Nov 6.
58. Chia J-M, Song C, Bradbury PJ, Costich D, de Leon N, et al. (2012) Maize HapMap2 identifies extant variation from a genome in flux. Nat Genet 44: 803–807. Available: http://dx.doi.org/10.1038/ng.2313. Accessed 2012 Nov 6.
59. Tian F, Bradbury PJ, Brown PJ, Hung H, Sun Q, et al. (2011) Genome-wide association study of leaf architecture in the maize nested association mapping population. Nat Genet 43: 159–162. Available: http://www.ncbi.nlm.nih.gov/pubmed/21217756. Accessed 2011 Jun 13.
60. Eller E (2001) Effects of ascertainment bias on recovering human demographic history. Hum Biol 73: 411–427. Available: http://muse.jhu.edu/journals/hub/summary/v081/81.5-6.eller.html. Accessed 2013 Nov 11.
61. Clark AG, Hubisz MJ, Bustamante CD, Williamson SH, Nielsen R (2005) Ascertainment bias in studies of human genome-wide polymorphism. Genome Res 15: 1496–1502. Available: http://www.pubmedcentral.nih.gov/articlerender.fcgi?artid = 1310637&tool = pmcentrez&rendertype = abstract. Accessed 2013 Nov 6.
62. Albrechtsen A, Nielsen FC, Nielsen R (2010) Ascertainment biases in SNP chips affect measures of population divergence. Mol Biol Evol 27: 2534–2547. Available: http://www.pubmedcentral.nih.gov/articlerender.fcgi?artid = 3107607&tool = pmcentrez&rendertype = abstract. Accessed 2013 Nov 11.
63. Gautier M, Gharbi K, Cezard T, Foucaud J, Kerdelhué C, et al. (2013) The effect of RAD allele dropout on the estimation of genetic variation within and between populations. Mol Ecol 22: 3165–3178. Available: http://www.ncbi.nlm.nih.gov/pubmed/23110526. Accessed 2013 Nov 11.
64. Arnold B, Corbett-Detig RB, Hartl D, Bomblies K (2013) RADseq underestimates diversity and introduces genealogical biases due to nonrandom haplotype sampling. Mol Ecol 22: 3179–3190. Available: http://www.ncbi.nlm.nih.gov/pubmed/23551379. Accessed 2013 Nov 7.
65. Alex Buerkle C, Gompert Z (2013) Population genomics based on low coverage sequencing: how low should we go? Mol Ecol 22: 3028–3035. Available: http://www.ncbi.nlm.nih.gov/pubmed/23174005. Accessed 2013 Nov 6.

Selection of Reference Genes for Gene Expression Studies in Siberian Apricot (*Prunus sibirica* L.) Germplasm Using Quantitative Real-Time PCR

Jun Niu[1], Baoqing Zhu[1], Jian Cai[1], Peixue Li[2], Libing Wang[3], Huitang Dai[2], Lin Qiu[2], Haiyan Yu[3], Denglong Ha[2], Haiyan Zhao[2], Zhixiang Zhang[1], Shanzhi Lin[1]*

1 College of Biological Sciences and Biotechnology, College of Nature Conservation, National Engineering Laboratory for Tree Breeding, Key Laboratory of Genetics and Breeding in Forest Trees and Ornamental Plants, Ministry of Education, Beijing Forestry University, Beijing, China, **2** Jigongshan National Nature Reserve, Xingyang, China, **3** Research Institute of Forestry, Chinese Academy of Forestry, Beijing, China

Abstract

Quantitative real time reverse transcription polymerase chain reaction has been applied in a vast range of studies of gene expression analysis. However, real-time PCR data must be normalized with one or more reference genes. In this study, eleven putative consistently expressed genes (*ACT, TUA, TUB, CYP, DNAj, ELFA, F-box27, RPL12, GAPDH, UBC* and *UBQ*) in nine Siberian Apricot Germplasms (including much variability) were evaluated for their potential as references for the normalization of gene expression by NormFinder and geNorm programs. From our studies, *ACT, UBC, CYP, UBQ* and *RPL12* as suitable for normalization were identified by geNorm, while *UBC* and *CYP* as the best pair by NormFinder. Moreover, *UBC* was selected as the most stably expressed gene by both algorithms in different Siberian Apricot seed samples. We also detected that a set of three genes (*ACT, CYP* and *UBC*) by geNorm as control for normalization could lead to accurate results. Furthermore, the expression levels of oleosin gene were analyzed to validate the suitability of the selected reference genes. These obtained experimental results could make an important contribution to normalize real-time PCR data for gene expression analysis in Siberian Apricot Germplasm.

Editor: Sara Amancio, ISA, Portugal

Funding: This research was supported by the National Natural Sciences Foundation of China (No. J1103516), the Central Public-Interest Scientific Institution Basal Research Fund (RIF2013-02) (No. 201004001) and the Chinese Key Technology Research and Development Program of Twelfth Five-Year Plan (No. 2011BAD22B08). The funders had no role in study design, data collection and analysis, decision to publish, or preparation of the manuscript.

Competing Interests: The authors have declared that no competing interests exist.

* Email: szlin@bjfu.edu.cn

These authors contributed equally to this work.

Introduction

Quantitative real time reverse transcription polymerase chain reaction (RT-qPCR) has replaced the classical reverse transcription-polymerase chain reaction with quantitative accuracy, high sensitivity and high-throughput characteristics, thus becoming the most common method for detection and quantification of mRNA transcription levels of target genes [1]. However, obtaining accurate and reliable quantitative gene expression results is difficult. This is due to experimental variation, such as differences in amount and quality of starting material, quantity and quality of RNA, and enzymatic efficiencies during reverse transcription [2]. Because reference genes undergo the same preparation steps as the target gene, by selecting one or more reference genes used for normalization in real-time PCR analysis this issue can be avoided. These reference genes under selection could potentially stabilize the experimental variability that typically occurs as a result of the various steps of the experimental procedure [3]. Certainly, the expression levels of reference genes should remain constant between the cells of different varieties, tissues, lifecycle phases and experimental conditions, otherwise, it can lead to erroneous results in quantification of the interesting gene [4].

The housekeeping genes, such as 18S rRNA, actins (*ACT*), cyclophilin (*CYP*), tubulin (*TUA* and *TUB*) and glyceral-dehyde-3-phosphate dehydrogenase (*GAPDH*), were typically used as reference genes for normalization of mRNA expression owing to their stable expression [5,6]. However, studies have shown that the expressions of those commonly used reference genes will not display a stable expression under some experimental conditions, for example, a reference gene may be stably expressed in one organism but unsuitable for normalization of gene expression in another [7]. Therefore choosing suitable reference genes for normalization relative to certain experimental materials and conditions is crucial. In recent years, many statistical algorithms, such as NormFinder [8], BestKeeper [9] and geNORM [10], have been extensively applied to evaluate ideal reference genes for normalizing real-time PCR data according to the specific tissue type and experimental conditions [11,12].

Siberian Apricot (*Prunus sibirica* L.), belonging to the Rosaceae, is a deciduous shrub native to temperate, continental, mountainous climates including regions of northern and north-eastern China, eastern and southeastern portions of Mongolia, Eastern Siberia and the Maritime Territory of Russia [13]. It grows in temperate climates and thrives with abundant sunlight, low temperatures, strong winds, low rainfall and poor soil. In China, Siberian Apricot is one of the most economically and ecologically important tree species owing to its plentiful resource, fast growth, annual seed production more than 192,500 tons, high oil content in seed kernel, and a wide range of uses [14]. Although the 26S ribosomal RNA gene was used as the sole reference gene for examining carotenoid biosynthesis in apricot during a recent expression survey [15], the results have not been entirely validated.

The aim of this study was to select and evaluate the stability of 11 candidate reference genes using different Siberian Apricot Germplasm seeds in China by statistical software (NormFinder and geNorm). The optimal reference genes might be applied in further studies on the Siberian Apricot gene expression for selecting high-quality seeds at the molecular and genetic level. Furthermore, the expression levels of oleosin gene were assessed using different reference genes, to validate the selection of candidate reference genes.

Materials and Methods

Ethics Statement

All materials are in a natural state, and specific permission was not required for these locations/activities. No protected plant species was sampled for this research.

Tissue collection

Based on an overall investigation of Siberian Apricot in China, 9 superior trees (Donglin, Lanxian, Fuxing, Zhidan, Keqi, Tianjing, Yanji, Zhenyuan and Minghe) have been selected from each tree population (Figure S1). Selections were based on phenotypic assessment of agronomic morphological characteristics of economic interest, including seed yields per plant, kernel rate of seed, number of fruit clusters per plant, number of fruits per cluster, disease resistance, and so forth. A global positioning system (Garmin GPSMAP76) was used to mark the position of the collected accessions. Although the experimenters did not study the effect of each single factor on candidate reference genes stability, an overall assessment indicated a fair degree of variability among the trees analyzed, at least on climatic and environmental factors (Table 1). Table 1 shows the specific location, meteorological data and edaphic condition of the Siberian Apricot Germplasms. Fully ripened fruits of Siberian Apricot were harvested at maturity stage in the 2011 harvest season. The sarcocarp of all Siberian Apricot seeds was removed, and they were immediately immersed in liquid nitrogen and stored at −80°C until RNA extraction.

RNA extraction and cDNA preparation

Total RNA was isolated from tissue samples using the RNeasy Fibrous Tissue Mini Kit (QIAGEN) according to the manufacturer's recommendations. The RNase-free DNase kit (Promega) was subsequently used for eradication of DNA contamination in total RNA preparations. Extracted RNA was qualified and quantified using a Nanodrop ND-1000 Spectrophotometer (Nanodrop Technologies, Wilmington, DE, USA) and all the samples showed a 260/280 nm ratio from 1.9 to 2.1. Equal amounts of total RNA (1.5 μg) in all samples were reverse transcribed respectively using the Reverse transcription System (Promega) in a 20 μl reaction using oligo-dT primers, according to the manufacturer's instructions, and then the cDNA was diluted 1:5 with nuclease-free water before being used as templates in the qPCR process.

Table 1. The data of material location, meteorological data and edaphic condition.

Germplasm*	longitude/°	latitude/°	Altitude/m	Annual averaged temperature/°C	mean temperature (Jan)/°C	mean temperature (July)/°C	Annual rainfall/ mm	Non-frost/d	Sunshine duration/h	Soil
Donglin	131°08.086′	44°08.200′	382	4.9	−14.6	21.5	530	151	2322	Dark Brow Soil
Keqi	117°32.361′	43°24.030′	1303	2.5	−25.0	20.0	380	95	2825	Cinnamon Soil
Yanji	129°23.102′	42°50.209′	326	5.7	−14.4	21.3	533	140	2317	Dark Brow Soil
Fuxing	121°22.121′	42°09.387′	289	7.0	−11.6	24.2	494	150	2623	Cinnamon Soil
Tianjing	117°23.952′	40°09.231′	194	11.3	−5.5	26.0	518	195	2778	Cinnamon Soil
Lanxian	111°29.357′	38°08.920′	1655	6.8	−9.7	21.0	500	126	2861	Raw Soil
Zhidan	108°45.741′	36°49.025′	1366	7.8	−7.9	21.5	525	142	2313	Loessal Soil
Minghe	102°53.545′	36°11.090′	2043	9.0	−6.4	19.8	292	198	2458	Chestnut Soil
Zhenyuan	107°13.601′	35°35.159′	1280	9.9	−5.0	20.8	515	155	2350	Red Clay

*All of the trees are full bearing period.

Selection of Siberian Apricot sequences and PCR primer design

To identify Siberian Apricot homologs of internal genes, such as *ACT*, *TUA*, *TUB*, *CYP*, chaperone protein danJ (*DNAj*), elongation factor 1-alpha (*ELFA*) and polyubiquitin (*UBQ*), we queried expressed sequence tag (EST) databases with *Prunus amygdalus*, *P. maritima* and *P. avium* nucleotide sequence using nucleotide blast, then homologous comparison gene sequences were performed to confirm the function of selected apricot ESTs. The rest of apricot reference genes could be found in NCBI nucleic acid database. The amplification primers for real-time PCR were designed using the Primer Premier 5 software with melting temperatures between 58 to 62°C, and the absence of secondary structures was verified by the UNAFold program (http://eu.idtdna.com/UNAFold) according to D'haene et al [16]. Most of the primer pairs were targeted to different exons or spanned 2 exons in order to avoid false positives during amplification due to potential residual genomic DNA in the templates. Specificity of each sequence primer pair was verified by conventional RT-PCR under the same conditions as described below for real-time RT-PCR, after that, products were confirmed by gel electrophoresis on 1.5% agarose gel and visualized after staining with ethidium bromide.

Real-time RT-PCR and data analyses

Real-time amplification reactions using SYBR Green were performed with the ABI 7500 Fast Real-Time PCR System (Applied Biosystems). Reactions were prepared in a 20 µl volume containing: 4 µl of template, 0.8 µl of each amplification primer (0.4 µM), 0.4 µl of ROX Reference DyeII, 10 µl of 2×SYBR premix Ex Taq^{TM}II (TaKaRa Biotechnology) and 6 µl of dH_2O. Each PCR reaction was performed in triplicate and template-free negative controls were performed simultaneously.. Amplifications were performed starting with an initial step of 95°C for 30 s, followed by 40 cycles of denaturation at 95°C for 3 s and primer annealing at 60°C for 30 s. Melting curve analysis ranging from 60°C to 95°C with temperatures increasing in increments of 0.2°C every 10 s was performed for all PCR products using ABI Prism Dissociation Curve Analysis Software to confirm the occurrence of specific amplification peaks. The PCR efficiency (E) was estimated from the data obtained from the exponential phase of each individual amplification plot and the equation ($E = 10^{slope}$) [17]. Negative controls consisting of nuclease-free water instead of template, and reverse transcriptase controls prepared by substituting reverse transcriptase for nuclease-free water in the cDNA synthesis step were included in all analyses for each primer pair. A RT-qPCR checklist listing additional technical information as proposed by Bustin et al. [18,19] is provided in Table S1.

Two publicly available software programs, geNorm and NormFinder were used to analyze Cq values which were collected from the ABI PRISM 7500 Sequence Detection System. Cq values were converted into relative quantities for geNorm and NormFinder taking into account the PCR efficiency of the primer pairs. These values were then used as software inputs and were analyzed as specified by Vandesompele et al [10] and Andersen et al [8].

Normalization of the target gene

The oleosin gene (GenBank: AY962832), which encodes an oil body associated protein used as a stabilizer of oil bodies, was used for validating the impact of the use of inappropriate reference genes on the gene expression analysis. The primers were designed with Primer 5 software design (Forward: GCCACCAGGGGCTGACCAA and Reverse: CCTTCCCAGCCAACTTATGACG). The qPCR procedure was carried out following the same parameters used for the analysis of reference genes. The relative expression level of the target gene was calculated with different normalization factors in nine Siberian Apricot Germplasms employing the stable and unstable single gene and the reference genes recommended by geNorm and NormFinder program.

Results

Primer selection and amplification specificity and efficiency

Eleven Siberian Apricot reference genes commonly used as internal genes for plant gene expression studies were found via screening the Siberian Apricot EST database. Those genes were *ACT*, *TUA*, *TUB*, *CYP*, *DNAj*, *ELFA*, *UBQ*, *GAPDH*, ubiquitin-conjugating enzyme (*UBC*), ribosomal protein L12 (*RPL12*) and f-box protein 27 (*F-box27*) most of which represent different functional classes and gene families, with the possible exception of *TUA and TUB*; *UBC* and *UBQ*. Based on information about *Fragaria vesca subsp* genes, these potential internal genes intron/exon structures were then fabricated, drawing from other plant genome sequences. Then PCR primers were designed on different exons or spanning an exon-exon junction (Table 2).

Agarose gel electrophoresis and melting curve analyses were performed following the RT-qPCR experiment, to determine specificity of primers designed in the current study. The primer pairs all amplified the expected size of the single PCR product (Figure 1). In addition, the specificity of amplicon was confirmed by the presence of a single peak during the melt curve and sequencing analysis (Figure S2). Average amplification efficiency varied from 99.2% and 105.9% (Table 3).

It was verified by PCR in both cDNA and gDNA whether these amplification primers were available. The primers for *TUA*, *DNAj*, *ELFA* and *F-box27* were unable to get amplified bands when using gDNA as template, and the amplifications of *ACT*, *TUB*, *GAPDH* and *UBC* genes with a gDNA template were longer than those obtained with a cDNA template. Thus those PCR primers were identified on different exons or spanning an exon-exon junction. In addition, the specific amplified bands of *CYP*, *RPL12* and *UBQ*, had the same size in cDNA and gDNA (Figure 1) as expected (Table 2). Although the experimental result of *ACT* gene deviated from our forecast in which the amplification of *ACT* should have had the same size, the remaining amplification primers on the list were exactly as we expected them to be.

Expression levels of the reference genes

To exclude any artificial errors in real-time PCR analysis of all the 11 candidate reference genes, three technical repetitions were performed for real-time PCR using gene-specific primers in the same cDNA pool. Meanwhile non-template controls were performed in parallel with each template and primer combination. The results showed that the single PCR product was amplified by each primer combination of the 11 candidate reference genes form various cDNA templates (Figure S2). Quantification cycle (Cq) values (the number of cycles needed for the fluorescence to reach a specific threshold level of detection) were determined in order to make comparison among each PCR run. The 11 candidate reference genes showed a relatively wide range of expression level from the lowest mean Cq value in *TUA* (23.36) to the highest in *DNAj* (32.56) with the most lying between 24 and 28 across all tested samples (Figure 2).

Table 2. Description of Siberian Apricot candidate reference genes.

Gene[a]	Homologous species	Homologous Locus	% homology	E vuale	Primers location[b]
ACT	*Fragaria vesca subsp*	XM_004294460.1	90%	0	N/F
TUA	*Fragaria vesca subsp*	XM_004289661.1	91%	0	3/4
TUB	*Fragaria vesca subsp*	XM_004294187.1	88%	0	2/3
CYP	*Fragaria vesca subsp*	XM_004289796.1	87%	2e-174	N/F
DNAj	*Populus trichocarpa*	XM_002316443.1	86%	1e-121	2–3/4
ELFA	*Fragaria vesca subsp*	XM_004309835.1	92%	0	1–2/2
GAPDH	*Vitis vinifera*	XM_002263109.2	84%	0	9/10
RPL12	*Fragaria vesca subsp.*	XM_004291528.1	87%	6e-156	N/F
UBC	*Fragaria vesca subsp.*	XM_004287962.1	84%	5e-162	4/5
UBQ	*Fragaria vesca subsp*	XM_004294293.1	87%	0	N/F

[a]*ACT, TUA, TUB, CYP, DNAj, ELFA* and *UBQ* were ESTs based on the other species reference gene sequence determined via BLASTN.
[b]The prediction of exon, the previous digits indicated the site of primer forward and the back correspond to primer reverse. N/F meaning no intron or limited ESTs could not conjecture the intron.

Expression stability of reference genes

In order to identify and rank the most suitable reference genes based on their expression stability, the entire Cq dataset was analyzed using two different statistical approaches (geNorm and NormFinder).

The expression stability value of the 11 candidate reference genes was calculated by the geNorm program and then the reference genes *TUB* and *RPL12* were identified as the two most stably-expressed genes (Figure 3A), according to their M values defined as the average pairwise variation of that gene relative to all other potential reference genes in a set of cDNA samples. Since two pairs of reference genes (*TUB* and *TUA*; *UBQ* and *UBC*) were suspected to be coregulated, we removed both *TUB* and *UBQ* from analysis. The occurrence of a shift in ranking after the removal of those suspected coregulated genes (*TUB* and *UBQ*) showed that *ACT* and *UBC* were the best pair (Figure 3B). To determine whether the two pairs of reference genes were coregulated genes, *TUB* and *UBQ* were independently removed (Figure 3C and Figure 3D). The resulting ranking indicated that *TUB* and *TUA* were coregulated. In general, selecting more than two reference genes as an internal control can help significantly with the correction experimental error, providing get more reliable results. This is critical for accurate quantification of the genes, especially small expression differences in gene expression studies [20]. In order to select the ideal number of reference genes required for accurate normalization, the pairwise variation (Vn/Vn+1) was calculated by the geNorm algorithm. Vn values were calculated by stepwise inclusion of more reference genes until the (n+1) gene made no significant contribution to the newly calculated normalization factor [21]. According to the analysis, a value of 0.15 is usually regarded as the selection threshold value, for example $V2/3<0.15$ means that 2 reference genes could be used for normalization [10]. In our present study, V5/6 $(0.148)<0.15<$ V4/5 (0.151) suggested that five reference genes were the best option for accurate normalization (Figure 4). However, we found the value of V3/4 (0.151) to be very close to 0.15, and three reference genes were used for attempted normalization. Finally, *UBC*, *ACT*, *CYP*, *UBQ* and *RPL12* were identified as the most suitable reference genes, and *UBC*, *CYP* and *ACT* were tentatively selected for normalization by geNorm.

For a more robust analysis, we used the NormFinder approach, which is less affected by correlated expression compared with geNorm. A low stability value is indicative of stable expression; the NormFinder analysis provided roughly the same ranking of the candidate genes as geNorm without *TUB* (Figure 5). Based on our data, NormFinder identified the pair of *UBC* and CYP as the best

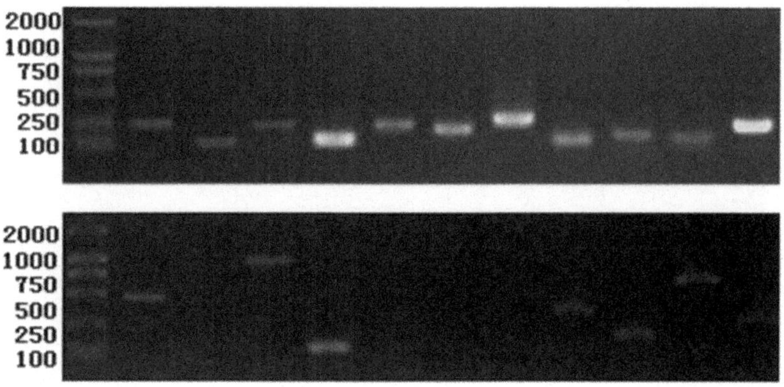

Figure 1. Performance of the amplification primers. Amplicons obtained by real-time PCR using cDNA (up) and gDNA (down) as template and electrophoresis using agarose gel (1.5%). The amplification primers from left to right are *ACT, TUA, TUB, CYP, DNAj, ELFA, F-box27, RPL12, GAPDH, UBC* and *UBQ*.

Table 3. Reference genes and their primer sequences used for real-time PCR.

Abbreviation	Gene name	Accession No.	Primer forward(F)/reverse(R)	Primer reverse	Amplicon size (bp)	E^b (%)	R^{2b}
ACT	Actin	CV049956.1	ACATTGTTCTTAGTGGTGGGTC	AGATTCGTCATACTCTGCCTTT	209	100.5±4.8	0.994
TUA	Alpha-tubulin	CV046479.1	TTGACATTGAGCGACCCACC	TCACATCCACATTCAGAGCACC	108	100.5±2.8	0.997
TUB	Beta-tubulin	CV045221.1	CTTGACAATGAAGCCCTCTATGA	AGTAAGAGGAGCAAAGCCCAC	219	104.6±5.7	0.995
CYP	Cyclophilin	CV046015.1	CAACGGATCTCAGTTCTTCGTCTGC	GACCCAACCTTCTCGATGTTCTTCA	120	101.8±2.1	0.998
DNAj	Chaperone protein dnaJ	CV052151.1	GGTGGACACGACCCATTTGA	ACCTGACTTTGACCCTTTACCC	219	105.1±4.7	0.998
ELFA	Elongation factor 1-alpha	CV046439.1	ACTGGAACCTCACAGGCTGAC	GGAGTAGTGGCATCCATCTTGTTA	170	105.9±3.6	0.993
F-box 27	F-box protein 27	EU836687.1	CGTGGAGTGATTTGATTGGC	AAGTTTGGGTGGTGGAGGC	106	99.8±5.5	0.997
GAPDH	Glyceraldehyde-3-phosphate	JN786944.1	ATGTCTTTCCGTGTTCCTACTGT	TTTCCCTCAGACTCCTCCTTG	116	100.4±3.5	0.993
RPL12	Ribosomal protein L12	U93168.1	CGATCCCTCACAGGTCGTCG	TCCAGTCGTTGGCGGTCTCC	149	99.2±3.3	0.990
UBC	Ubiquitin-conjugating enzyme	AF008910.1	GAGACCAGCAATAACCGTGAA	TCTTGTACTCCGTGGCATCCT	128	99.2±4.2	0.997
UBQ	Polyubiquitin	CB821710.1	CTCTGACTGGCAAGACCATAACA	CCACGGAGACGAAGGACAA	205	102.0±1.9	0.998

[b]PCR efficiency (E) and correlation coefficients (R^2) were calculated by LinRegPCR method.

one combination of reference genes in different Siberian Apricot Germplasms, and *F-box27, ELFA, DNAj* ranked consistently poorly.

Expression profiling of oleosin

The expression level of the oleosin gene was performed as an example to show the effect of the choice of reference genes on the expression profiling of other genes. The oleosin expression was analyzed in different Siberian Apricot Germplasms by using the various combinations of top reference genes indicated by each of the three methods: *UBC*, *CYP* and *ACT*, and *CYP*, *UBQ*, *ACT*, *RPL12* and *UBC* as indicated by geNorm; *CYP* and *UBC* as indicated by NormFinder. In addition, we also used the optimal (*UBC*) and the inadequate (*F-box27*) genes as reference genes. As shown in Figure 6, normalization using the unstable reference *F-box27* gene led to erroneous estimation of the target gene. When the *UBC* was used as a unique reference, the expressions of oleosin in Tianjing and Yanji were significantly overestimated. The results from geNorm (three and five genes) and NormFinder had a better parallel veracity.

Discussion

The expression stability of the endogenous reference genes used as internal controls is paramount to reliable quantification of gene transcripts [22]. However, it has been reported that the reference genes are not only maintaining basic cellular functions but also participate in other cellular functions, generally resulting in theirs expressions considerable variation under different experimental conditions [23]. Therefore, it is necessary to validate the expression stability of potential reference genes in each particular experimental background prior to their use for normalization.

Siberian Apricot, belonging to the family Rosaceae, is also an important fruit and oilseed tree in China. Fast growth, easy propagation, high yield, and a high percentage of unsaturated fatty acids and a wide variety of bioactive components in the kernels make this plant one of the major resource of the formulation of protective lotion and soaps [13,14]. Also, owing to its kernel oil associated with a reduced risk of chronic diseases [24], Siberian Apricot is considered to be one of the most intensively studied tree species. However, most of the researches for Siberian Apricot kernel oil and bioactive components rest only on the level of detection [25,26]. Until now, little is known about the genetic and molecular level duo to a lack of information on reference gene stability in a variety of experimental contexts, limiting the application of RT-qPCR in the Siberian Apricot. Though 26S rRNA gene was used as a reference gene for normalization of real-time PCR data in examining carotenoid biosynthesis of apricots [15], systematic exploration and validation of stable Siberian Apricot reference genes had never been performed. In addition, the ribosomal RNA genes will not be able to participate as reference genes where cDNA synthesis is carried out using an oligo-dT primer or where only mRNA is used as template. Furthermore, they may exhibit such high levels of expression which would result in the occurrence of experimental error when normalizing weakly expressed genes [27]. All the pitfalls described above led us to not include the ribosomal RNA genes among the candidate reference genes to be evaluated in the present study.

In this present study, we used geNorm and NormFinder algorithms to select and validate the best reference genes form 11 candidates in Siberian Apricot Germplasms (including much variability) by calculating the expression stability of reference genes and determining the number of reference genes required for accurate normalization across the experimental conditions tested.

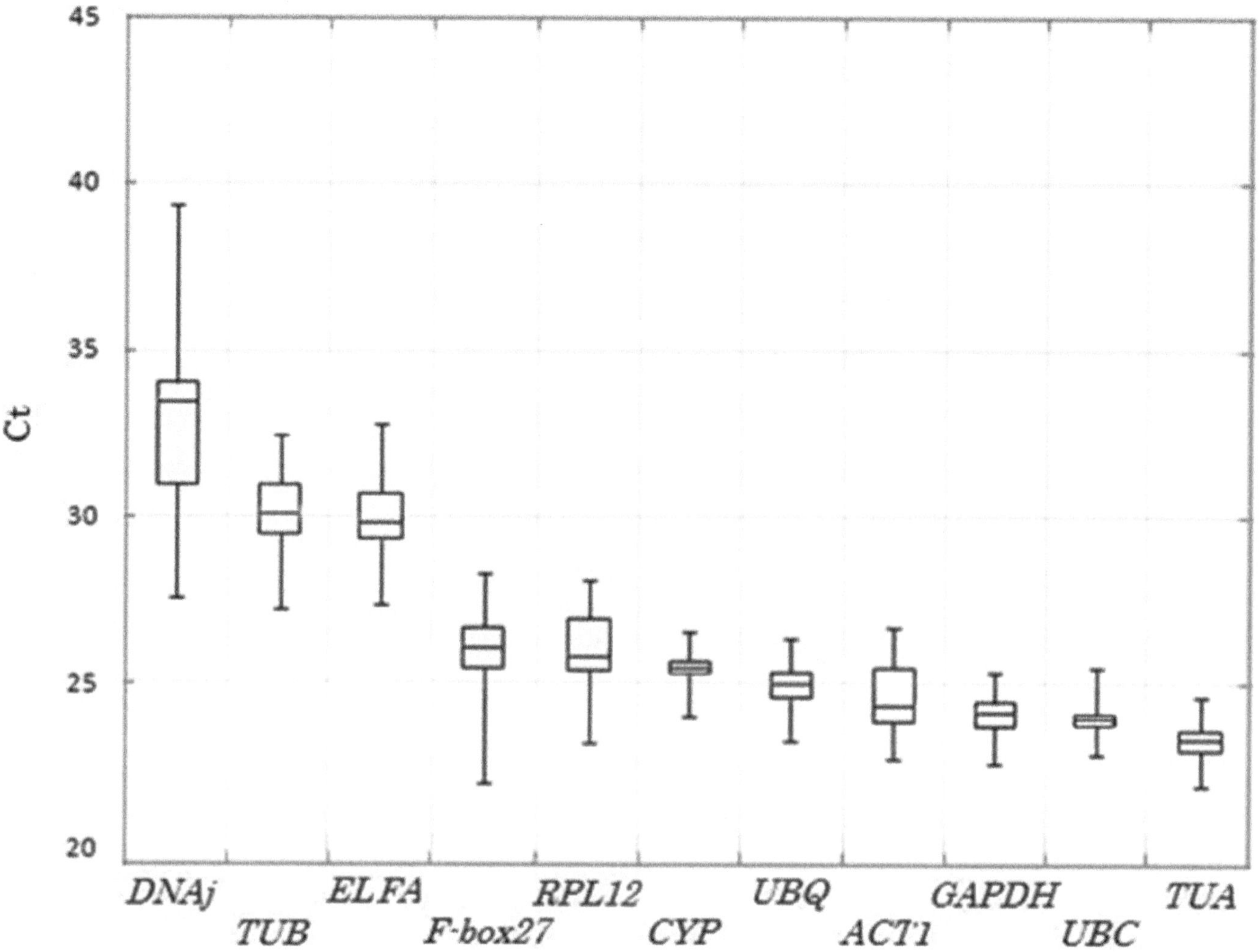

Figure 2. Cq values for 11 candidate reference genes across experimental samples. A line across the box is depicted as the median. The box indicates the 25th and 75th percentiles. Whiskers represent the maximum and minimum values.

It well known that geNorm and NormFinder relies on many of the same principles for expression variability evaluation across a panel of candidate reference genes, but NormFinder is less affected by correlated expression of the candidate genes. In contrast, one major defect in geNorm is insensitive to coregulated reference genes, leading the candidate reference genes to be preferentially selected from different pathways and functional classes [10]. However, it is sometimes difficult to avoid using coregulated genes for geNorm, especially when dealing with poorly annotated genes, or genes of unknown or hypothetical function [8]. Actually, the coregulated genes, such as *TUB* and *TUA*, were found in our experiments (Figure 3), as the case of previous studies [8,28], suggesting that error analysis must be applied to ensure reliable results when using a set of multiple reference genes that unintentionally involve coregulated genes. Therefore, when coregulated genes are absent, geNorm and NormFinder normally provide the same general ranking, with only minor differences in ordering. This was proven in our results (Figure 3C and Figure 5) in accordance with previous reports [29]. We also verified that the combination of two or more reference genes improves the stability value compared to the single most stable gene (Figure 6), *ACT*, *UBC*, *CYP*, *UBQ* and *RPL12* were ultimately identified by geNorm, *UBC* and *CYP* by NormFinder. Interestingly, *UBC* and *CYP* were ranked as the most stable by geNorm and NormFinder. In fact *UBC* was also the optimal for all developmental stages and under all stress conditions in *Platycladus orientalis* [30], but it showed less stable expression in rice grown under various environmental conditions [31]. As noted previously, *CYP* was the worst reference gene in a diverse pool of poplar [32], however the optimal stability of *CYP* expression identified in our study was in agreement with the result of *Petunia hybrid* transcriptomic analysis [33]. Additionally, the observations on the reference genes *ACT*, *UBQ* and *RPL2* with stable expression in peach (Rosaceae) [34], *Brassica oleracea* [35] and tomato [28], respectively, had been completely verified in our studies. It has been reported that *ELFA* was the most stable in potato during biotic and abiotic stress [36], but poorly ranked during light stress in tomato [28]. *ELFA* was top ranked for the developmental stage series and different times of the day, but poorly ranked for different tissues under the same developmental stage in soybean [37]. In our analysis results, *ELFA* performed poorly by geNorm and NormFinder enlightened us that *ELFA* was the most variable gene. Moreover, according to the other two genes (*DNAj* and *F-box27*) with poor rank validated in our current experimental conditions, it can be concluded that three genes (*DNAj*, *F-box27* and *ELFA*) should not be used as reference genes in Siberian Apricot Germplasms.

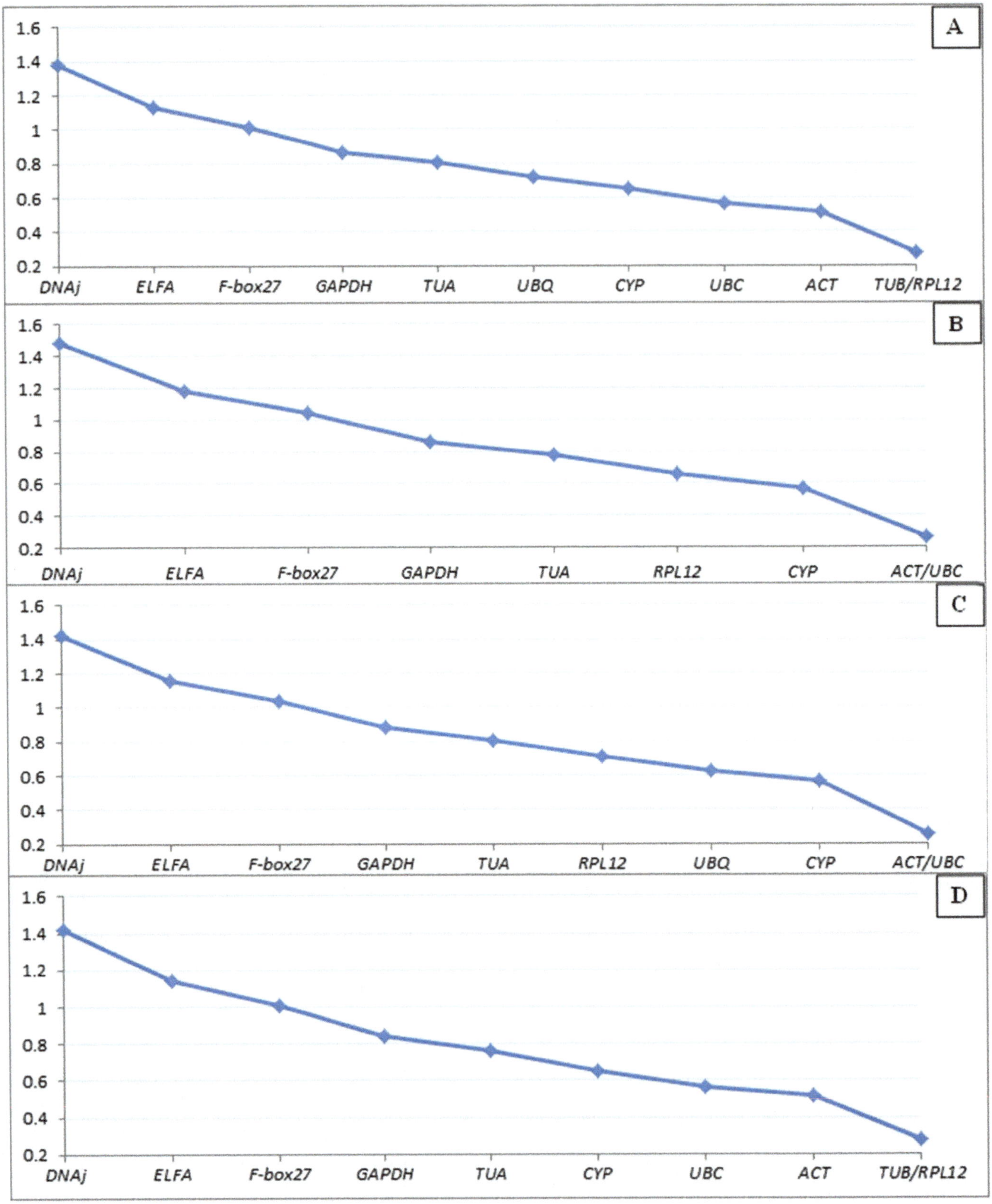

Figure 3. geNorm analysis of candidate reference genes in Siberian Apricot Germplasms. A: all reference genes. B: *TUB* and *UBQ* excluded from analysis. C: *TUB* excluded from analysis. D: *UBQ* excluded from analysis.

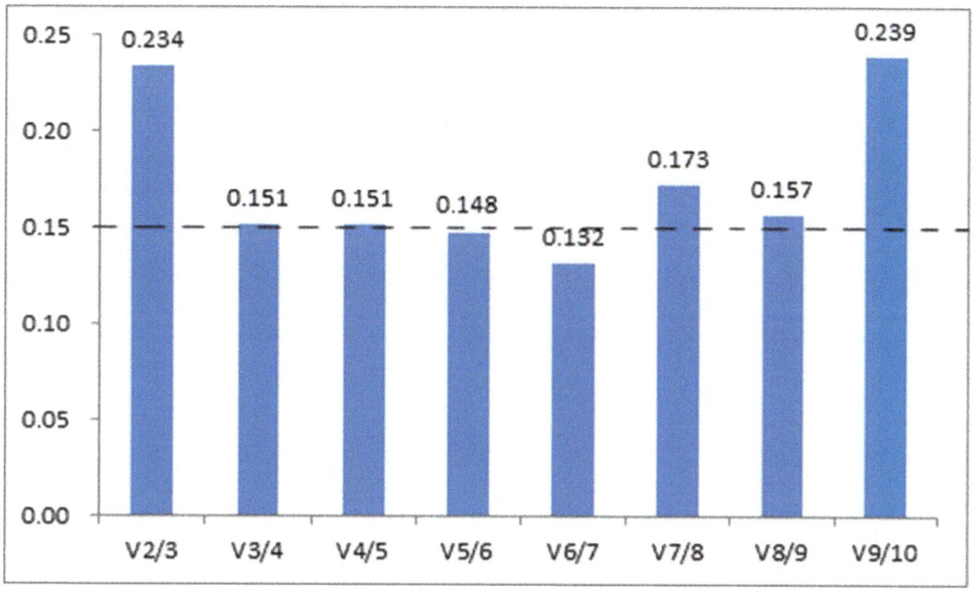

Figure 4. Pairwise variation analysis of normalization factors to determine the optimal number of reference genes. The cut-off value of 0.15, below which the inclusion of an additional reference gene is not required, is indicated by a discontinuous line.

To evaluate the effect of the choice of reference genes generated by two statistical methods on the expression profiling of genes this study, the four kinds of reference genes, including the optimal reference gene (*UBC*), the unsatisfactory one (*F-box27*), and the combination (*ACT*, *UBC*, *CYP*, *UBQ* and *RPL12*) identified by geNorm as well as another one (*UBC* and *CYP*) by NormFinder, were used to analyze the relative expression of the oleosin gene in, encoding the most abundant protein in oil bodies of plants [38]. As illustrated in Figure 6, although he kinds and amounts of reference genes differently picked by the geNorm algorithm (a set of several genes) and NormFinder (only one or two), the oleosin expression levels quantified with references selected according to either geNorm or NormFinder were conformance, as reported by Løvdal and Saha's report [29]. This was probably due to the less affected by correlated expression for NormFinder, which imply that NormFinder may be the preferred choice for stability evaluation if the experimenter is unsure about coregulation among a set of candidate references [29]. Additionally, an attempt was made to single out three reference genes (*ACT*, *CYP* and *UBC*) from candidate references by geNorm in this study, mainly due to the value of V3/4 (0.151) very close to 0.15, from which we were surprised to find that the three reference genes also could obtain satisfactory results. Thus, from a view of the economy and accuracy, it is concluded that a set of three genes (*ACT*, *CYP* and

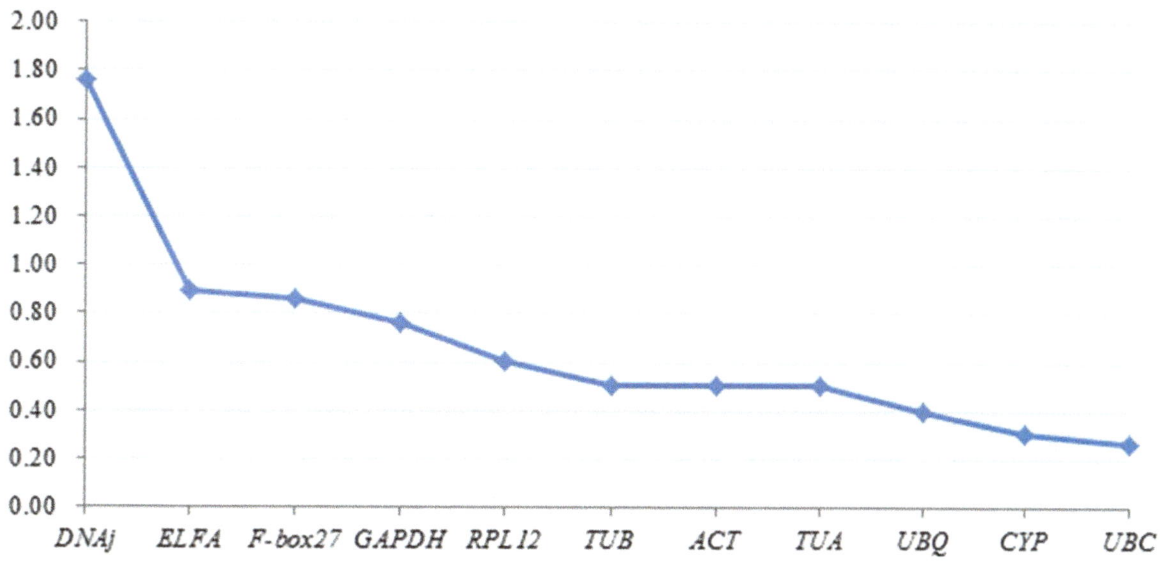

Figure 5. NormFinder analysis of candidate reference genes in Siberian Apricot Germplasms.

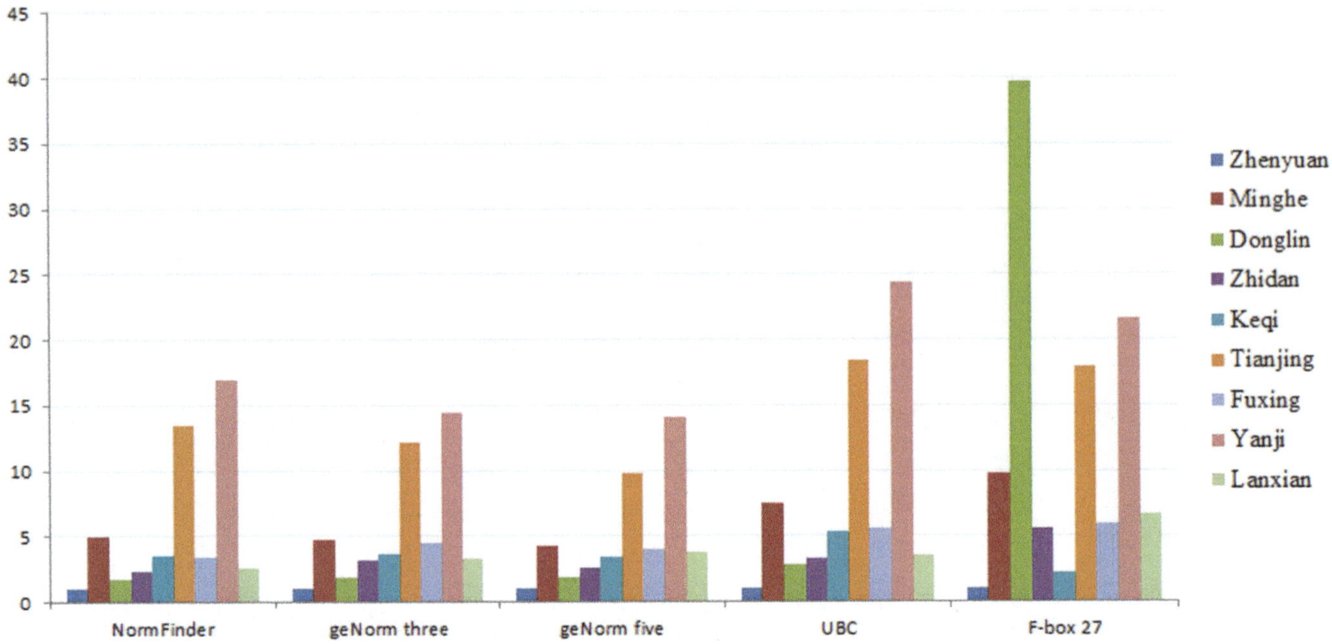

Figure 6. Expression profiles of oleosin gene in different Siberian Apricot Germplasms. Expression ratios of oleosin for the experimental calculated using (1) *CYP* and *UBC* for NormFinder; (2) *CYP, UBC* and *ACT* for geNorm; (3) *ACT, UBC, CYP, UBQ* and *RPL12* for geNorm; (4) the stable reference gene *UBC;* (5) the unsatisfactory reference gene *F-box27.*

UBC) could be used as the best control for normalization in Siberian Apricot Germplasms.

Conclusions

As far as can be ascertained, this is the first systematic study for the selection of reference genes for RT-qPCR in Siberian Apricot Germplasms. The evaluations of 11 candidate reference genes by geNorm and NormFinder indicated that the three most suitable reference genes in Siberian Apricot Germplasms are *ACT*, *UBC* and *CYP*, while the three least suitable reference genes are *DNAj*, *ELFA* and F-box27. To obtain the most reliable results from gene expression studies of Siberian Apricot, the combination of three or more stable genes should be used as internal controls for relative gene quantification.

Author Contributions

Conceived and designed the experiments: JN BZ. Performed the experiments: JN. Analyzed the data: JC. Contributed reagents/materials/analysis tools: PL LW HD LQ HY DH HZ. Wrote the paper: JN SL. Provided funding and computational guidance: ZZ SL.

References

1. Bustin S, Benes V, Nolan T, Pfaffl M (2005) Quantitative real-time RT-PCR–a perspective. J Mol Endocrinol 34: 597–601.
2. Wood SH, Clements DN, McEwan NA, Nuttall T, Carter SD (2008) Reference genes for canine skin when using quantitative real-time PCR. Vet Immunol Immunopathol 126: 392–395.
3. Huggett J, Dheda K, Bustin S, Zumla A (2005) Real-time RT-PCR normalisation; strategies and considerations. Genes Immun 6: 279–284.
4. Tricarico C, Pinzani P, Bianchi S, Paglierani M, Distante V, et al. (2002) Quantitative real-time reverse transcription polymerase chain reaction: normalization to rRNA or single housekeeping genes is inappropriate for human tissue biopsies. Anal Biochem 309: 293–300.
5. Artico S, Nardeli S, Brilhante O, Grossi-de-Sa M, Alves-Ferreira M (2010) Identification and evaluation of new reference genes in *Gossypium hirsutum* for accurate normalization of real-time quantitative RT-PCR data. BMC Plant Biol 10: 49.
6. Lee JM, Roche JR, Donaghy DJ, Thrush A, Sathish P (2010) Validation of reference genes for quantitative RT-PCR studies of gene expression in perennial ryegrass (*Lolium perenne* L.). BMC Mol Biol 11: 8.
7. Suzuki T, Higgins P, Crawford D (2000) Control selection for RNA quantitation. Biotechniques 29: 332–337.
8. Andersen CL, Jensen JL, Ørntoft TF (2004) Normalization of real-time quantitative reverse transcription-PCR data: a model-based variance estimation approach to identify genes suited for normalization, applied to bladder and colon cancer data sets. Cancer Res 64: 5245–5250.
9. Pfaffl MW, Tichopad A, Prgomet C, Neuvians TP (2004) Determination of stable housekeeping genes, differentially regulated target genes and sample integrity: BestKeeper–Excel-based tool using pair-wise correlations. Biotechnol Lett 26: 509–515.
10. Vandesompele J, De Preter K, Pattyn F, Poppe B, Van Roy N, et al. (2002) Accurate normalization of real-time quantitative RT-PCR data by geometric averaging of multiple internal control genes. Genome Biol 3: research0034–research0034.11.
11. Yin Z, Ke X, Huang D, Gao X, Voegele RT, et al. (2013) Validation of reference genes for gene expression analysis in *Valsa mali var. mali* using real-time quantitative PCR. World J Microbiol Biotechnol 29:1563–1571.
12. Du Y, Zhang L, Xu F, Huang B, Zhang G, et al. (2013) Validation of housekeeping genes as internal controls for studying gene expression during Pacific oyster (*Crassostrea gigas*) development by quantitative real-time PCR. Fish Shellfish Immunol 34: 939–945.
13. Wang L (2011) Resource investigation and distribute regular of three Armeniaca species. Forest Resour Manag 5: 65–70.

14. Li-bing W (2010) Geographic distrbution and botanical characters of 3 Armeniaca plant in China. Forest Research 3: 020.
15. Marty I, Bureau S, Sarkissian G, Gouble B, Audergon JM, et al. (2005) Ethylene regulation of carotenoid accumulation and carotenogenic gene expression in colour-contrasted apricot varieties (*Prunus armeniaca*). J Exp Bot 56: 1877–1886.
16. D'haene B, Vandesompele J, Hellemans J (2010) Accurate and objective copy number profiling using real-time quantitative PCR. Methods 50: 262–270.
17. Ramakers C, Ruijter JM, Deprez RHL, Moorman AF (2003) Assumption-free analysis of quantitative real-time polymerase chain reaction (PCR) data. Neurosci Lett 339: 62–66.
18. Bustin SA, Beaulieu J-F, Huggett J, Jaggi R, Kibenge FS, et al. (2010) MIQE precis: Practical implementation of minimum standard guidelines for fluorescence-based quantitative real-time PCR experiments. BMC Mol Biol 11: 74.
19. Bustin SA, Benes V, Garson JA, Hellemans J, Huggett J, et al. (2009) The MIQE guidelines: minimum information for publication of quantitative real-time PCR experiments. Clin Chem 55: 611–622.
20. Kumar P, Yadav P, Verma A, Singh D, De S, et al. (2012) Identification of stable reference genes for gene expression studies using quantitative real time PCR in Buffalo Oocytes and Embryos. Reprod Domest Anim 47: e88–e91.
21. Cao SN, Zhang XW, Ye NH, Fan X, Mou SL, et al. (2012) Evaluation of putative internal reference genes for gene expression normalization in *Nannochloropsis* sp. by quantitative real-time RT-PCR. Biochem Biophys Res Commun 424: 118–123.
22. Bustin S (2002) Quantification of mRNA using real-time reverse transcription PCR (RT-PCR): trends and problems. J Mol Endocrinol 29: 23–39.
23. de Vega-Bartol JJ, Santos RR, Simoes M, Miguel CM (2013) Normalizing gene expression by quantitative PCR during somatic embryogenesis in two representative conifer species: Pinus pinaster and Picea abies. Plant Cell Rep 32: 715–729.
24. Zhang J, Gu H-D, Zhang L, Tian Z-J, Zhang Z-Q, et al. (2011) Protective effects of apricot kernel oil on myocardium against ischemia–reperfusion injury in rats. Food Chem Toxicol 49: 3136–3141.
25. Manzoor M, Anwar F, Ashraf M, Alkharfy K (2012) Physico-chemical characteristics of seed oils extracted from different apricot (*Prunus armeniaca* L.) varieties from Pakistan. Grasas y Aceites 63: 193–201.
26. Dragovic-Uzelac V, Levaj B, Mrkic V, Bursac D, Boras M (2007) The content of polyphenols and carotenoids in three apricot cultivars depending on stage of maturity and geographical region. Food Chem 102: 966–975.
27. Thellin O, Zorzi W, Lakaye B, De Borman B, Coumans B, et al. (1999) Housekeeping genes as internal standards: use and limits. J Biotechnol 75: 291–295.
28. Løvdal T, Lillo C (2009) Reference gene selection for quantitative real-time PCR normalization in tomato subjected to nitrogen, cold, and light stress. Anal Biochem 387: 238–242.
29. Løvdal T, Saha A (2014) Reference gene selection in Carnobacterium maltaromaticum, Lactobacillus curvatus, and Listeria innocua subjected to temperature and salt stress. Mol Biotechnol 56: 210–222.
30. Chang EM, Shi SQ, Liu JF, Cheng TL, Xue L, et al. (2013) Selection of Reference Genes for Quantitative Gene Expression Studies in *Platycladus orientalis* (Cupressaceae) Using Real-Time PCR. Plos One 7: e33278.
31. Jain M, Nijhawan A, Tyagi AK, Khurana JP (2006) Validation of housekeeping genes as internal control for studying gene expression in rice by quantitative real-time PCR. Biochem Biophys Res Commun 345: 646–651.
32. Brunner A, Yakovlev I, Strauss S (2004) Validating internal controls for quantitative plant gene expression studies. BMC Plant Biol 4: 14.
33. Mallona I, Lischewski S, Weiss J, Hause B, Egea-Cortines M (2010) Validation of reference genes for quantitative real-time PCR during leaf and flower development in *Petunia hybrida*. BMC Plant Biol 10: 4.
34. Tong Z, Gao Z, Wang F, Zhou J, Zhang Z (2009) Selection of reliable reference genes for gene expression studies in peach using real-time PCR. BMC Mol Biol 10: 71.
35. Brulle F, Bernard F, Vandenbulcke F, Cuny D, Dumez S (2014) Identification of suitable qPCR reference genes in leaves of *Brassica oleracea* under abiotic stresses. Ecotoxicology 23: 459–471.
36. Nicot N, Hausman J-F, Hoffmann L, Evers D (2005) Housekeeping gene selection for real-time RT-PCR normalization in potato during biotic and abiotic stress. J Exp Bot 56: 2907–2914.
37. Jian B, Liu B, Bi Y, Hou W, Wu C, et al. (2008) Validation of internal control for gene expression study in soybean by quantitative real-time PCR. BMC Mol Biol 9: 59.
38. Jiang P-L, Tzen JT (2010) Caleosin serves as the major structural protein as efficient as oleosin on the surface of seed oil bodies. Plant Signal Behav 5: 447–449.

Parental Selection of Hybrid Breeding Based On Maternal and Paternal Inheritance of Traits in Rapeseed (*Brassica napus* L.)

Nailin Xing, Chuchuan Fan, Yongming Zhou*

National Key Laboratory of Crop Genetic Improvement, Huazhong Agricultural University, Wuhan, China

Abstract

Parental selection is crucial for hybrid breeding, but the methods available for such a selection are not very effective. In this study, a 6×6 incomplete diallel cross was designed using 12 rapeseed germplasms, and a total of 36 hybrids together with their parental lines were planted in 4 environments. Four yield-related traits and seed oil content (OC) were evaluated. Genetic distance (GD) was estimated with 359 simple sequence repeats (SSRs) markers. Heterosis levels, general combining ability (GCA) and specific combining ability (SCA) were evaluated. GD was found to have a significant correlation with better-parent heterosis (BPH) of thousand seed weight (TSW), SCA of seeds per silique (SS), TSW, and seed yield per plant (SY), while SCA showed a statistically significant correlation with heterosis levels of all traits at 1% significance level. Statistically significant correlations were also observed between GCA of maternal or paternal parents and heterosis levels of different traits except for SS. Interestingly, maternal (TSW, SS, and OC) and paternal (siliques per plant (SP) and SY) inheritance of traits was detected using contribution ratio of maternal and paternal GCA variance as well as correlations between GCA and heterosis levels. Phenotype and heterosis levels of all the traits except TSW of hybrids were significantly correlated with the average performance of parents. The correlations between SS and SP, SP and OC, and SY and OC were statistically significant in hybrids but not in parents. Potential applications of parental selection in hybrid breeding were discussed.

Editor: Meng-xiang Sun, Wuhan University, China

Funding: Ministry of Science and Technology of China (Grant nos. 2012AA101107 and 2012BAD49G00) http://www.most.gov.cn/eng/; Ministry of Agriculture of China (nycytx-00503) http://english.agri.gov.cn/. The funders had no role in study design, data collection and analysis, decision to publish, or preparation of the manuscript.

Competing Interests: The authors have declared that no competing interests exist.

* Email: ymzhou@mail.hzau.edu.cn

Introduction

Plant breeding currently faces the daunting task of meeting the challenges posed by global climate change and a growing need for food supply and bioenergy. Heterosis has proven to be an efficient way to improve yield and environmental adaptability, and has been used in breeding of many crops [1–3]. Rapeseed (*Brassica napus*), an important oilseed crop worldwide, is one of the most successful crops in application of heterosis [4–7]. So far, the vast majority of the hybrid cultivars has been developed through extensive selection of F_1 hybrid combinations, thus involving high cost and a lengthy turn-around period. Therefore, it is necessary to explore efficient methods that could predict hybrid performance in the parental generation.

In the past two decades, a number of studies have been reported that investigated methods for hybrid performance prediction in rapeseed. Riaz et al. and Tan et al. found a significant correlation between genetic distance (GD) and yield heterosis through molecular markers [8–9]. Ali et al. found significant correlations between GD and yield-component traits, but no significant correlation was identified for the number of branches per plant [10]. However, Yu et al. found significant correlations between some agronomic traits, but not for seed yield [11]. Diers et al. reported that GD was significantly correlated only with heterosis of seed yield in inbred diallel combinations, but not with cultivar diallel [12]. Qian et al. found no significant correlations between heterosis and GD, but significant correlations between general combining abilities (GCA) and hybrid performance [13]. Similarly, Teklewold and Becker, and Devi and Singh both found significant correlations between combining abilities and hybrid performance [14–15]. However, no effective method of parental selection has been proposed through GD and combining abilities in hybrid breeding.

A thorough understanding of the inheritance mode of traits is essential to establish an effective method for parental selection. Inheritances of traits have been analyzed widely in bi-parent populations and diallel crosses in rapeseed, and additive, dominance and epistasis effects have been identified for a wide array of traits [16–20]. In addition, maternal and paternal inheritances of traits have been studied using reciprocal crosses in many crops [21–23]. Maternal and paternal effects of heterosis have been identified only through analyzing GCA variance contributed by different parents [24]. However, no report is available about different correlations between traits in parents and hybrids, together with the correlations between heterosis levels and GCA of different parents.

The objectives of the current study are to investigate: (1) the relationships between GD, combining abilities, and heterosis levels, (2) maternal and paternal contribution to hybrid performances through analysis of GCA variance and correlations between heterosis levels and GCA of different parents, and (3) correlations between traits with parents as homozygous genotypes and hybrids as heterozygous genotypes. Our results provide novel information for parental selection in hybrid breeding of rapeseed.

Materials and Methods

Plant materials and field experiments

The experimental materials used in this study comprised 12 semi-winter rapeseed germplasms including 5 cultivars and 7 inbred lines whose characteristics was described in Table S1. With the parental lines listed in Table S1, a total of 36 hybrids were developed using a 6×6 incomplete diallel cross design (Table 1).

The hybrids along with the parents were grown in 4 natural environments at 3 different locations. They were planted in a winter rapeseed crop area at the experimental farm of Huazhong Agricultural University, Wuhan, China in the season of 2009–2010 and 2010–2011, and Huanggang Academy of Agricultural Sciences, Huanggang, China in the season of 2010–2011 [25], and in a spring rapeseed crop area in Hezheng, Gansu, China in the season of 2010. The field experiments of Huanggang were granted permission by the administrative board of the Huanggang Academy of Agricultural Sciences. The experiment location of Gansu was the experiment base of Huazhong Agricultural University, and no specific permission was required for the field trial. All parents and hybrids were used for all the environments and growing seasons and were grown for the evaluation of phenotype traits at the mature plant stage. All the field trials in this study did not involve endangered or protected species.

All the field experiments followed a randomized block design with 3 replications, and each replication comprised parents and hybrids that were planted in 4 rows with 48 plants for each plot. The crop was cultivated with standard agronomic practices. Ten mature plants in the middle of each plot were randomly selected for trait evaluation, and the mean value of each trait was used for analysis. Seed yield per plant (SY), seeds per silique (SS), siliques per plant (SP), thousand seed weight (TSW), and oil content in seeds (OC) were evaluated as described by Shi et al. [19].

Genetic diversity

Genomic DNA was isolated following the standard procedure [26]. A total of 359 simple sequence repeats (SSRs) including all the markers mapped on the linkage groups constructed by Fan et al. [27] were used to analyze the genetic diversity of parents. Genetic diversity was estimated on the basis of polymorphism generated by SSR primers. Amplification profiles of test genotypes were compared with each other and bands of DNA fragments were scored as '1' for present and '0' for absent. The genetic distance was computed using UPGMA method by NTSYS$_{PC}$ Version 2.10e [28].

Statistical analysis

The mid-parent heterosis (MPH), better-parent heterosis (BPH), general combining ability (GCA), specific combining ability (SCA) and correlation coefficients were calculated using SAS v9.3 software (SAS Institute, Cary, North Carolina, USA) using the GLM procedure [29–30].

The MPH was calculated as: MPH (%) = $(F_1 - MP)/MP \times 100$, BPH (%) = $(F_1 - BP)/BP \times 100$. where MP is the mean of the two parents, and BP the value of the better parent.

For combining ability analyses, the GCA for parents and SCA for hybrids were calculated using the following formula, respectively:

$$\hat{g}_{f.} = \frac{\sum_{m=1}^{n}(x_{fm} - \bar{x})}{n};\ \hat{g}_{.m} = \frac{\sum_{f=1}^{n}(x_{fm} - \bar{x})}{n};\ \hat{s}_{fm} = x_{fm} - \bar{x} - \hat{g}_{f.} - \hat{g}_{.m}$$

Here, $\hat{g}_{f.}$, $\hat{g}_{.m}$, and $\hat{s}_{fm}$ represent the GCA of female and male parent, and SCA of the hybrid, respectively; $\bar{x}$ represents the mean value of all the hybrids; x_{fm} represents the value of the hybrid produced with the fth female parent and the mth male parent; n represent the number of female or male parents. The variance of additive effect (A), dominant effect (D), additive by additive effect (AA), and their interaction with environment (AE, DE, and AAE) were estimated using QGA Station Version 1.0 software [31]. The broad sense of heritability (h^2b), narrow sense heritability (h^2n), variance of GCA, variance of SCA, and ANOVA analysis were conducted using DPS software [32]. Heritability was calculated using the following formula:

$$h_b^2 = \sigma_{(G)}^2 \big/ \sigma_{(P)}^2;\ h_n^2 = \sigma_{(A)}^2 \big/ \sigma_{(P)}^2$$

Table 1. Genetic distance between parental lines used for 6×6 incomplete diallel crosses.

Parental lines	P1	P2	P3	P4	P5	P6
P7	0.36	0.38	0.21	0.72	0.44	0.61
P8	0.21	0.40	0.40	0.66	0.43	0.61
P9	0.41	0.50	0.58	0.33	0.53	0.44
P10	0.37	0.41	0.44	0.53	0.40	0.46
P11	0.46	0.45	0.54	0.57	0.45	0.46
P12	0.47	0.49	0.56	0.50	0.52	0.46

P1 to P6: maternal parents; P7 to P12: paternal parents.
P1, P2, P3, P7, and P8: cultivars (P1: ZS5; P2: ZS7; P3: ZS10; P7: ZS9; P8: HS5). P4 to P6, and P9 to P12: inbred lines.
See also Table S1 for detailed characteristics of the parental lines.

Here, $\sigma^2_{(G)}$ is the genotypic variance, $\sigma^2_{(A)}$ is the additive variance, and $\sigma^2_{(P)}$ is the phenotypic variance [24]. MPH and BPH of hybrids, and GD of the parents between the hybrids, as well as GCA and SCA of the parents and hybrids were correlated with each other to assess the relationship between these genetic parameters and parental diversity.

The maternal and paternal inheritances were analyzed by calculating the correlation coefficients between the hybrids and their parents and comparing the contribution of female and male genotypic variance to total genotypic variance for each trait [24]. The correlation coefficients between the traits investigated were computed through phenotypic performance of parents and hybrids, respectively. The correlation coefficients between the hybrids and their maternal or paternal lines were computed through phenotypic performance of each trait.

Results

Estimation of variance and covariance components

Variances of genetic effect (G), environment effect (E), and G×E were statistically significant for all traits in parents, hybrids, and all the plant materials including parents and hybrids at the 1% significance level, except for the environment effect of SY in hybrids and SP in all the plant materials (Table 2). In addition, the additive, dominant and additive by additive interaction effects were statistically significant for all traits at the 1% significance level, except for the additive effect of SY, dominant effect of SS and AA of SP and TSW (Table 3). For the genetic and environment interaction effect, AE of TSW and OC, and AAE of SY, SS, and SP were not significant. For SP, the dominant effect was greater than the additive effect. On the contrary, the additive effect was higher than the dominant effect for TSW and OC. These results indicated that SY of hybrid was mainly controlled by the dominant and epistasis effects, SS was mainly controlled by the additive and epistasis effects, SP was mainly controlled by the dominant effect, and TSW and OC were mainly controlled by the additive effect. Furthermore, the SY of hybrids had stronger environment adaptability than parents did, which was consistent with prior research [33].

Heterosis performance of hybrids

The MPH ranged from −12.88% of SP for P6×P8 to 96.17% of SY for P2×P11, and SY showed the highest average MPH (45.41%) among the 5 traits (Table 4). The BPH ranged from −22.78% of SS for P5×P8 to 64.64% of SY for P2×P11, and SY also had the highest average BPH (26.07%) among the 5 traits. The average MPH and BPH showed a positive direction for all the traits, except for the average BPH of SS and TSW. This result was consistent with prior research [9].

Heritability, combining abilities and GD

For the 5 traits, h^2b and h^2n ranged from 68.27% to 93.05%, and 18.61% to 74.88% respectively (Table 5). TSW showed the highest heritability, whether h^2b or h^2n, followed by SS, while the h^2n of SY was only 18.61%.

The variance of GCA was higher than that of SCA for all traits except for SY. Additionally, TSW showed the highest GCA variance of 80.47% among the 5 traits while SY showed the lowest GCA variance of 25.71%. Furthermore, maternal genotypic variance was found to contribute more to total genotypic variance than paternal one for SS, TSW, and OC, but less for SY and SP. The maternal and paternal contributions to TSW showed the greatest difference among the 5 traits (Table 5).

Table 2. Double factors variance analysis of traits.

		DF	SY	SS	SP	TSW	OC
Parent	G	11	191.54**	110.15**	59614.20**	2.36**	34.29**
	E	3	104.47**	24.01**	32519.32**	0.78**	71.83**
	G×E	33	29.79**	65.05**	8293.86**	0.75**	14.77**
Hybrid	G	35	300.44**	98.26**	70730.35**	2.86**	30.89**
	E	3	5.68	126.44**	6482.04**	2.90**	28.71**
	G×E	105	21.50**	12.80**	7499.03**	0.22**	5.65**
Total	G	47	326.60**	107.03**	71818.80**	2.76**	48.46**
	E	3	34.52**	61.92**	923.54	1.57**	66.22**
	G×E	141	24.59**	26.64**	8335.66**	0.39**	10.17**

G: genotype; E: environment; G×E: interaction of genotype and environment; DF: degree of freedom.
SS: seeds per silique; SP: siliques per plant; SY: seed yield per plant; TSW: thousand seed weight; OC: oil content in seeds; MPH: middle parent heterosis; BPH: better parent heterosis. The same abbreviations were used below.
** Significant at P = 0.01

Table 3. Variance analysis of traits (By QGA station).

	SY	SS	SP	TSW	OC
A	0	1.76**	592.14**	0.24**	1.15**
D	0.45**	0	961.36**	0.13**	0.29**
AA	6.46**	2.59**	0	0	0.38**
AE	2.89**	3.11**	1423.52**	0	0
DE	8.66**	3.65**	2695.86**	0.02**	0.23**
AAE	0	0	0	0.03**	0.56**

A: additive effect; D: dominant effect; AA: interaction of additive and additive effect; AE: interaction between additive and environment effect; DE: interaction between dominant and environment effect; AAE: interaction of epistasis of additive and additive effect and environment effect.
** Significant at P = 0.01.

A total of 359 SSR markers were used to estimate the genetic diversity of the parents. The number of polymorphic alleles generated by the markers ranged from 1 to 11 with an average of 3. The genetic distances between both parents of the hybrids ranged from 0.21 for P3×P7 and P1×P8 to 0.72 for P4×P7 (Table 1).

Relationships of GD, combining abilities, and heterosis levels

No significant relationship was found between GD and MPH for all 5 traits (Table 6), which was consistent with prior reports [1,13]. However, a negative and statistically significant correlation was detected between GD and BPH of TSW at the 1% significance level. In addition, statistically significant correlations were detected between GD and SCA of SS, TSW, and SY at the 1% and 5% significance level, respectively. Each trait had a positive and statistically significant correlation between SCA and the heterosis levels at the 1% significance level.

The relationships between heterosis level and GCA of the maternal or paternal parent were computed (Table 6). OC and TSW showed statistically significant correlations between heterosis level and GCA of maternal parents at the 5% and 1% significance level, respectively. SP and SY displayed positive and statistically significant correlations between heterosis levels and GCA of both parents at the 5% or 1% significance level. SS showed no significant correlation between heterosis levels and GCA of both parents.

Performances of hybrids are closely related to parents

Correlation coefficients between performance of hybrids and their parents were found to be positive and statistically significant at the 1% significance level (Table 7). In addition, the correlations between performance of hybrids and the average performance of the two parents were also statistically significant at the 1% significance level. However, the correlations between heterosis levels and the average performance of their parents were negative and statistically significant at the 1% or 5% significance level, except for TSW (Figure 1, Table 8).

Correlations of phenotypes among the 5 traits were computed with parents as the homozygous genotype and hybrids as the heterozygous genotype, respectively (Table 9). Positive and statistically significant correlations were detected between SY and SP, and between SS and OC at the 1% significance level in both homozygous and heterozygous genotypes, while a negative correlation were found between SS and TSW. However, negative and statistically significant correlations were detected only in heterozygous genotypes between SY and OC, and between SP and two other traits (SS and OC) at the 5% and 1% significance level, respectively.

Discussion

In this study, all 5 traits appeared to have high heterosis levels and a wide variations (Table 4), which is consistent with previous studies [12–13]. This result suggests great potential for improving yield-related traits in rapeseed through hybrid breeding.

Though heterosis of rapeseed appears to be strong, it has not been utilized fully in hybrid breeding. One of the reasons is the difficulty in predicting heterosis level from parental generation [1,14]. To explore an effective way to heterosis prediction, relationships of GD with combining abilities and heterosis levels were extensively studied, and significant correlations were detected in some traits in the previous studies [1,8,10,34]. However, contradictory results for correlations between GD and heterosis of

Table 4. Heterosis level of incomplete diallel crosses.

Traits	MPH (%)		BPH (%)	
	Mean ±SD	Range	Mean ±SD	Range
SY	45.41±0.21	8.10~96.17	26.07±0.17	−3.35~64.64
SS	9.79±0.08	−11.29~22.59	−2.07±0.07	−22.78~10.51
SP	18.75±0.14	−12.88~45.76	7.53±0.15	−19.98~36.93
TSW	6.39±0.07	−9.00~21.94	−2.66±0.09	−23.22~16.71
OC	3.95±0.02	0.06~7.23	0.63±0.02	−5.02~3.88

Table 5. Genetic parameters of heritability and combining ability.

Traits	V_G (%)	V_S (%)	V_{gf} (%)	V_{gm} (%)	V_{gfm} (%)	h^2_b (%)	h^2_n (%)
SY	25.71	74.30	10.32	13.70	75.98	68.77	18.61
SS	68.56	31.44	42.96	26.47	30.58	89.67	61.66
SP	66.13	33.87	29.97	39.26	30.78	75.12	40.06
TSW	80.47	19.53	69.52	11.30	19.18	93.05	74.88
OC	70.39	29.61	35.80	34.83	29.37	68.27	48.18

V_G: variance of general combining ability (GCA); V_S: variance of specific combining ability (SCA); V_{gf}: contribution ratio of the female GCA variance to the total variance; V_{gm}: contribution ratio of the male GCA variance to the total variance; V_{gfm}: contribution ratio of the SCA variance to the total variance; h^2_b: broad sense heritability; h^2_n: narrow sense heritability.

Table 6. Correlation coefficients between GD, heterosis levels, and combining abilities of each trait.

Trait	GD			SCA		MPH		BPH	
	MPH	BPH	SCA	MPH	BPH	GCA_f	GCA_m	GCA_f	GCA_m
SY	0.05	0.06	0.19*	0.43**	0.49**	0.36**	0.19*	0.35**	0.27**
SS	0.13	0.08	0.43**	0.34**	0.31**	0.16	−0.09	0.15	−0.07
SP	−0.11	−0.13	0.04	0.55**	0.53**	0.37**	0.18*	0.34**	0.27**
TSW	−0.12	−0.32**	−0.35**	0.54**	0.70**	0.34**	0.05	0.59**	0.11
OC	−0.01	0.08	−0.08	0.53**	0.54**	0.05	−0.04	−0.17*	−0.12

GD: genetic distance; GCA_f: GCA of female parents; GCA_m: GCA of male parents.
* Significant at $p = 0.05$; ** significant at $p = 0.01$.

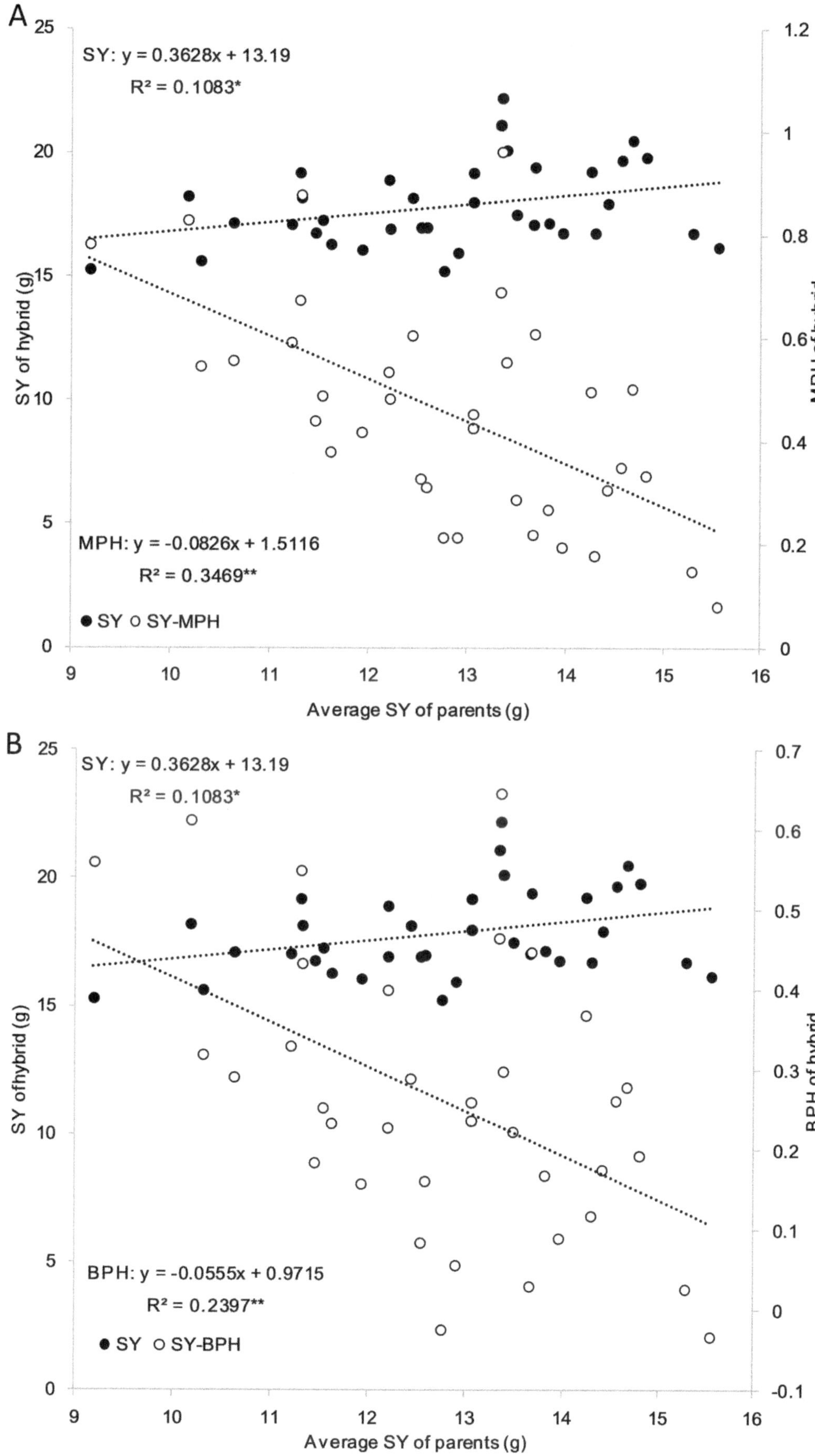

Figure 1. Relationships between average performances of parents and their hybrids for heterosis levels of seed yield. (A) relationship with MPH. (B) relationship with BPH. * Significant at p = 0.05; ** significant at p = 0.01; n = 36

Table 7. Correlation coefficients of each trait between hybrid and its parental lines.

Trait	Hybrid and female parent	Hybrid and male parent	Hybrids and two parents
SY	0.69**	0.65**	0.76**
SS	0.53**	0.62**	0.74**
SP	0.60**	0.77**	0.75**
TSW	0.76**	0.50**	0.81**
OC	0.55**	0.65**	0.77**

** Significant at p = 0.01.

seed yield and some agronomic traits were also reported [10–11]. In our study, GD was significantly correlated only with BPH of TSW, and SCA of SY, TSW, and SS (Table 6), which suggested that GD is a valuable genetic parameter for predicting heterosis levels based on the molecular markers [27,35], but should be used with cautions and with other parameters together.

Another way to hybrid prediction is to use the estimates of combining ability. Qian et al. and Rameeh found significant correlations between heterosis levels and combining abilities in rapeseed [13,36]. However, no significant correlation was found between GD and heterosis levels in prior research [13]. In maize, significant correlations were found between heterosis levels and combining abilities [15,37]. Similarly, our study found significantly positive correlations between combining abilities and heterosis levels in all 5 traits except for GCA of SS. Our results together with previous studies indicate that combining ability may be considered a more effective parameter than GD in heterosis prediction of rapeseed.

Thus far, the contributions of GCA variance of different parents together with maternal and paternal effects to hybrid performance have been rarely studied, especially in rapeseed [24]. In our study, the contribution ratios of the two parental GCA variances as well as the correlations between heterosis levels and GCA of parents showed that TSW was mainly controlled by the maternal genotype, and the maternal effect was 55.22% higher than paternal effect. Furthermore, our study showed that heterosis levels had a statistically significant and positive correlation with GCA of maternal parent (Tables 5 and 6). SS was also controlled by maternal genotype, while for OC, the contribution of maternal parent was slightly higher than that of paternal parent, which was different from Wang et al. [38]. The maternal effect of various seed traits including seed size was shown in many other crops to be correlated with parental environmental effects (e.g. availability of light, temperature, water, and nutrients) and cytoplasmic effects [39–40]. Based on the contribution ratio of GCA variance and correlation between heterosis levels and GCA of parents, paternal parent contributed more than maternal parent for SY, which provided valuable information for SY improvement in hybrid breeding of rapeseed. For SP, paternal parent contributed slightly more than maternal parent. Higher paternal effect on inheritance was also found in the leaflet area and other traits in carrot [41]. This paternal effect is likely contributed by epigenetic mechanisms [42], and plays an important role in the expression of polyphenic traits [43].

To the best of our knowledge, no report has to date has explored the relationships between traits in parental lines and their hybrids of rapeseed. Our study revealed that the correlations between SS and TSW, SS and OC, and SY and SP are statistically significant in both parents and hybrids (Table 9). In contrast, correlations between SS and SP, SP and OC, and SY and OC only appeared to be statistically significant in hybrids. The relationships between traits that are different from those obtained from homozygous genotypes or homozygous genotypes mixed with heterozygous genotypes provide valuable information for hybrid breeding.

Previously, additive, dominant, and epistasis effects were all detected for yield and yield related traits in rapeseed hybrids [19]. Zhang et al. found that SS was mainly controlled by the additive effect [44], which was similar to the results from our study (Table 2, and 5). In contrast to the result of Amiri-Oghan et al. [17], we found SY was mainly controlled by the additive by additive effect (Table 3). However, dominant effect was also detected for SY inheritance by Shen et al. and Radoev et al. [45–46]. Interestingly, traits primarily controlled by maternal parent were mainly additive inheritance, and traits primarily controlled by paternal parents were mainly dominant inheritance in our study. The reason for this phenomenon is unknown, and further studies are needed to understand the mechanism underlying this phenomenon.

In conclusion, hybrid performance in hybrid breeding might be predicted by combining ability supplemented with GD. Furthermore, the selection of maternal and paternal parents should be based on the maternal or paternal control of traits in hybrid breeding. For example, parents with a higher combining ability of TSW should be selected as a female parent, while parents with a higher combining ability of SY as a male parent. Our study

Table 8. Correlation coefficients between heterosis level and average performance of parents.

Heterosis level	Traits				
	SY	SS	SP	TSW	OC
MPH	−0.43**	−0.52**	−0.35**	−0.16	−0.46**
BPH	−0.35**	−0.30**	−0.32**	0.09	−0.42**

** Significant at p = 0.01; n = 36.

Table 9. Phenotypic correlations between traits in hybrids and parents.

Trait	SY	SS	SP	TSW	OC
SY		−0.12	0.75**	0.11	−0.18*
SS	0.16		−0.41**	−0.60**	0.36**
SP	0.73**	−0.14		−0.03	−0.44**
TSW	0.07	−0.66**	−0.08		0.06
OC	−0.00	0.47**	−0.24	−0.06	

The numbers in the upper or lower triangle are correlation coefficients between traits in hybrids, or between traits in parental lines.
* Significant at p = 0.05; ** significant at p = 0.01; n = 144.

provides useful information for parental selection in hybrid breeding of rapeseed and other crops.

Author Contributions

Conceived and designed the experiments: YMZ. Performed the experiments: NLX. Analyzed the data: NLX. Contributed reagents/materials/analysis tools: YMZ CCF NLX. Contributed to the writing of the manuscript: NLX YMZ.

References

1. Melchinger AE (1999) Genetic diversity and heterosis. J.G. . Coors, S. . Pandey (Eds.), Genetics and Exploitation of Heterosis in Crops, ASA/CSSA/SSSA, Madison, WI, pp. 99–118.
2. Sellis D, Callahan BJ, Petrov DA, Messer PW (2011) Heterozygote advantage as a natural consequence of adaptation in diploids. Proc Natl Acad Sci USA 108: 20666–20671.
3. Schnable PS, Springer NM (2013) Progress toward understanding heterosis in crop plants. Annu Rev Plant Biol 64: 71–88.
4. Thurling N (1991) Application of the ideotype concept in breeding for higher yield in the oilseed brassicas. Field Crop Res 26: 201–219.
5. Shen JX, Wang HZ, Fu TD, Tian BM (2008) Cytoplasmic male sterility with self-incompatibility, a novel approach to utilizing heterosis in rapeseed (*Brassica napus* L.). Euphytica 162: 109–115.
6. Tochigi T, Udagawa H, Li F, Kitashiba H, Nishio T (2011) The self-compatibility mechanism in *Brassica napus* L. is applicable to F_1 hybrid breeding. Theor Appl Genet 123: 475–482.
7. Chen SY, Guan CY, Wang GH, Li X, Liu ZS (2012) Breeding of Xiangzayou 763, a new double-low and high oil content rapeseed cultivar. Crop Res 26: 40–42.
8. Riaz A, Li G, Quresh Z, Swati M, Quiros C (2001) Genetic diversity of oilseed *Brassica napus* inbred lines based on sequence- related amplified polymorphism and its relation to hybrid performance. Plant Breed 120: 411–415.
9. Tan Z, Li Y, Hu Q, Mei D, Li Y, et al. (2007) Heterosis prediction based on genetic distance estimated by molecular markers in rapeseed. Chinese J. Oil Crop Sci 29: 126.
10. Ali M, Copeland L, Elias S, Kelly J (1995) Relationship between genetic distance and heterosis for yield and morphological traits in winter canola (*Brassica napus* L.). Theor Appl Genet 91: 118–121.
11. Yu C, Hu S, Zhao H, Guo A, Sun G (2005) Genetic distances revealed by morphological characters, isozymes, proteins and RAPD markers and their relationships with hybrid performance in oilseed rape (*Brassica napus* L.). Theor Appl Genet 110: 511–518.
12. Diers B, McVetty P, Osborn T (1996) Relationship between heterosis and genetic distance based on restriction fragment length polymorphism markers in oilseed rape (*Brassica napus* L.). Crop Sci 36: 79–83.
13. Qian W, Sass O, Meng J, Li M, Frauen M, et al. (2007) Heterotic patterns in rapeseed (*Brassica napus* L.): I. Crosses between spring and Chinese semi-winter lines. Theor Appl Genet 115: 27–34.
14. Teklewold A, Becker HC (2006) Comparison of phenotypic and molecular distances to predict heterosis and F1 performance in Ethiopian mustard (*Brassica carinata* A. Braun). Theor Appl Genet 112: 752–759.
15. Devi P, Singh N (2011) Heterosis, molecular diversity, combining ability and their interrelationships in short duration maize (*Zea mays* L.) across the environments. Euphytica 178: 71–81.
16. Ramsay L, Bradshaw J, Griffiths D, Kearsey M (2001) The inheritance of quantitative traits in *Brassica napus* ssp. *rapifera* (swedes): Augmented triple test cross analyses of production characters. Euphytica 121: 65–72.
17. Amiri-Oghan H, Fotokian M, Javidfar F, Alizadeh B (2009) Genetic analysis of grain yield, days to flowering and maturity in oilseed rape (*Brassica napus* L.) using diallel crosses. Int J Plant Prod 3: 19–26.
18. Chen W, Zhang Y, Yao J, Ma C, Tu J, et al. (2011) Quantitative trait loci mapping for two seed yield component traits in an oilseed rape (*Brassica napus*) cross. Plant Breed 130: 640–646.
19. Shi J, Li R, Zou J, Long Y, Meng J (2011) A dynamic and complex network regulates the heterosis of yield-correlated traits in rapeseed (*Brassica napus* L.). PloS One 6: e21645.
20. Zhang L, Yang G, Liu P, Hong D, Li S, et al. (2011) Genetic and correlation analysis of silique-traits in *Brassica napus* L. by quantitative trait locus mapping. Theor Appl Genet 122: 21–31.
21. Hua W, Li RJ, Zhan GM, Liu J, Li J, et al. (2012) Maternal control of seed oil content in *Brassica napus*: the role of silique wall photosynthesis. Plant J 69: 432–444.
22. Nodine MD, Bartel DP (2012) Maternal and paternal genomes contribute equally to the transcriptome of early plant embryos. Nature 482: 94–97.
23. Debes P, Fraser D, McBride M, Hutchings J (2013) Multigenerational hybridisation and its consequences for maternal effects in *Atlantic salmon*. Heredity 111: 238–247.
24. Lou XY, Hu QS, Bao MZ, Ye YM (2010) Analysis of combining ability of two-types of male sterile and four restorer lines of *Zinnia elegans*. Euphytica 174: 91–103.
25. Wu J, Cai G, Tu J, Li L, Liu S, et al. (2013) Identification of QTLs for resistance to *Sclerotinia* stem rot and *BnaC. IGMT5. a* as a candidate gene of the major resistant QTL *SRC6* in *Brassica napus*. PloS One 8: e67740.
26. Murray M, Thompson WF (1980) Rapid isolation of high molecular weight plant DNA. Nucleic Acids Res 8, 4321–4326.
27. Fan C, Cai G, Qin J, Li Q, Yang M, et al. (2010) Mapping of quantitative trait loci and development of allele-specific markers for seed weight in *Brassica napus*. Theor Appl Genet 121: 1289–1301.
28. Rohlf F (2000) Multivariate Analysis System, Version 2.10 e, Applied Biostatistics. Inc, New York.
29. Goffman FD, Becker HC (2001) Diallel analysis for tocopherol contents in seeds of rapeseed. Crop Sci 41: 1072–1079.
30. Valdiani A, Kadir MA, Saad MS, Talei D, Tan SG (2012) Intra-specific hybridization: Generator of genetic diversification and heterosis in *Andrographis paniculata* Nees. A bridge from extinction to survival. Gene 505: 23–36.
31. Zhu J, Weir BS (1994) Analysis of cytoplasmic and maternal effects. I. A genetic model for diploid plant seeds and animals. Theor Appl Genet 89: 153–159.
32. Tang Q, Feng M (2002) DPS data processing system for practical statistics. Science, Beijing.
33. Haussmann BIG, Obilana AB, Ayiecho PO, Blum A, Schipprack W, et al. (2000) Yield and yield stability of four population types of grain Sorghum in a semi-arid area of Kenya. Crop Sci 40: 319–329.
34. Bansal P, Banga S, Banga S (2012) Heterosis as investigated in terms of polyploidy and genetic diversity using designed *Brassica juncea* amphiploid and its progenitor diploid species. PloS One 7: e29607.
35. Shen JX, Fu TD, Yang GS, Tu JX, Ma CZ (2006) Prediction of heterosis using QTLs for yield traits in rapeseed (*Brassica napus* L.). Euphytica 151: 165–171.
36. Rameeh V (2010) Combining ability and factor analysis in F_2 diallel crosses of rapeseed varieties. Plant Breed Seed Sci 62: 73–83.
37. Balestre M, Machado J, Lima J, Souza J, Nóbrega-Filho L (2008) Genetic distance estimates among single cross hybrids and correlation with specific

combining ability and yield in corn double cross hybrids. Genet Mol Res 7: 65–73.
38. Wang X, Liu G, Yang Q, Hua W, Liu J, et al. (2010) Genetic analysis on oil content in rapeseed (*Brassica napus* L.). Euphytica 173: 17–24.
39. Xu Y, Li HN, Li GJ, Wang X, Cheng LG, et al. (2011) Mapping quantitative trait loci for seed size traits in soybean (*Glycine max* L. Merr.). Theor Appl Genet 122: 581–594.
40. Marty C, BassiriRad H (2014) Seed germination and rising atmospheric CO_2 concentration: a meta-analysis of parental and direct effects. New Phytologist 202: 401–414
41. Grebenstein C, Kos SP, Jong TJ de, Tamis WLM, Snoo GR de (2013) Morphological markers for the detection of introgression from cultivated into wild carrot (*Daucus carota* L.) reveal dominant domestication traits. Plant Biol 15: 531–540.
42. Curley JP, Mashoodh R, Champagne FA (2011) Epigenetics and the origins of paternal effects. Horm Behav 59: 306–314.
43. Buzatto BA, Simmons LW, Tomkins JL (2012) Paternal effects on the expression of a male polyphenism. Evolution 66: 3167–3178
44. Zhang LW, Liu PW, Hong DF, Huang AQ, Li SP, et al. (2010) Inheritance of seeds per silique in *Brassica napus* L. using joint segregation analysis. Field Crops Res 116: 58–67.
45. Shen JX, Fu TD, Yang GS, Ma CZ, Tu JX (2005) Genetic analysis of rapeseed self-incompatibility lines reveals significant heterosis of different patterns for yield and oil content traits. Plant Breed 124: 111–116.
46. Radoev M, Becker HC, Ecke W (2008) Genetic analysis of heterosis for yield and yield components in rapeseed (*Brassica napus* L.) by quantitative trait locus mapping. Genetics 179: 1547–1558.

11

Development of an Alfalfa SNP Array and Its Use to Evaluate Patterns of Population Structure and Linkage Disequilibrium

Xuehui Li[1], Yuanhong Han[1], Yanling Wei[1], Ananta Acharya[1], Andrew D. Farmer[2], Julie Ho[3], Maria J. Monteros[1], E. Charles Brummer[1]*

1 Forage Improvement Division, The Samuel Roberts Noble Foundation, Ardmore, Oklahoma, United States of America, 2 National Center for Genome Resources, Santa Fe, New Mexico, United States of America, 3 Forage Genetics International, Davis, California, United States of America

Abstract

A large set of genome-wide markers and a high-throughput genotyping platform can facilitate the genetic dissection of complex traits and accelerate molecular breeding applications. Previously, we identified about 0.9 million SNP markers by sequencing transcriptomes of 27 diverse alfalfa genotypes. From this SNP set, we developed an Illumina Infinium array containing 9,277 SNPs. Using this array, we genotyped 280 diverse alfalfa genotypes and several genotypes from related species. About 81% (7,476) of the SNPs met the criteria for quality control and showed polymorphisms. The alfalfa SNP array also showed a high level of transferability for several closely related *Medicago* species. Principal component analysis and model-based clustering showed clear population structure corresponding to subspecies and ploidy levels. Within cultivated tetraploid alfalfa, genotypes from dormant and nondormant cultivars were largely assigned to different clusters; genotypes from semidormant cultivars were split between the groups. The extent of linkage disequilibrium (LD) across all genotypes rapidly decayed to 26 Kbp at $r^2 = 0.2$, but the rate varied across ploidy levels and subspecies. A high level of consistency in LD was found between and within the two subpopulations of cultivated dormant and nondormant alfalfa suggesting that genome-wide association studies (GWAS) and genomic selection (GS) could be conducted using alfalfa genotypes from throughout the fall dormancy spectrum. However, the relatively low LD levels would require a large number of markers to fully saturate the genome.

Editor: Tongming Yin, Nanjing Forestry University, China

Funding: This research was funded by the USDA-DOE Plant Feedstock Genomics for Bioenergy program, award # 2009-65504-05809 to ECB and MJM. The funders had no role in study design, data collection and analysis, decision to publish, or preparation of the manuscript.

Competing Interests: JH is employed by Forage Genetics International, a commercial alfalfa breeding company. The Noble Foundation employs ECB, MJM, XL, YH, and YW and has a cooperative research agreement with FGI on alfalfa research and product development. Future cultivars developed by the Noble Foundation may be marketed by FGI. FGI has and is funding research projects at the Noble Foundation, although their funding was not involved in this project. The FGI-Noble cooperative work does not involve the project under review here. FGI does not own the germplasm evaluated in this project.

* E-mail: ecbrummer@noble.org

Introduction

Alfalfa (*M. sativa* L.) is an important forage legume throughout the world. Alfalfa provides high nutrition animal feeds and as a perennial crop, offers numerous environmental benefits. The deep root system stabilizes soil and increases soil fertility through symbiotic biological nitrogen fixation, providing all the nitrogen needed for the alfalfa crop and a subsequent grain crop.Most alfalfa cultivars are synthetic populations developed by recurrent phenotypic selection. Although important agronomic traits such as winter survival and disease resistance have been improved by phenotypic selection, multiple-year phenotypic evaluations per cycle of selection limit rate of genetic gain. Partially due to the long breeding cycle, the improvement of biomass yield in alfalfa has been much slower compared to that in annual crops [1]. Minimal genetic improvement of biomass yield in alfalfa was found when comparing cultivars released from different eras [2]. Developing cultivars with high yield that are also tolerant to abiotic and biotic stresses is crucial to enhance alfalfa productivity and competitiveness in the world market.

Molecular markers can be used to dissect the genetic control of quantitative traits and predict breeding values of individuals. Indirect selection based on molecular markers can potentially reduce time per breeding cycle and accelerate alfalfa improvement. The genetic dissection of agronomic traits in alfalfa with molecular markers began in the 1990's, and to date QTLs have been identified in biparental populations for aluminum tolerance [3], biomass yield [4], persistence [5], winter hardiness [6,7], and disease resistance [8]. However, due to large linkage blocks in these populations, the QTL were not localized to small regions of the chromosomes. Further fine mapping and validation of the alfalfa QTLs are costly and time consuming and thus have not been pursued in most research programs. Genome-wide association studies (GWAS) can be used to more precisely localize QTLs by evaluating historical recombination. Using GWAS, causal genes of agronomic traits have been successfully identified in various crops [9,10], which can be used to efficiently improve traits influenced by a few genes via marker-assisted selection (MAS) [11]. For complex traits controlled by many genes with small effects, practical application is constrained by poor predictions based on a few major loci [12]. In contrast to pyramiding a few major QTLs via marker-assisted selection (MAS), breeders can apply genomic selection (GS), which uses whole-genome markers to predict the breeding values of individuals and thus can capture the effects from both major and minor QTLs [13]. Studies based on both simulated and experimental data suggested that GS improved prediction of breeding values compared to MAS for complex traits [14–16].

A large set of genome-wide markers able to be assayed with a high-throughput genotyping platform is required to perform GWAS and GS. Advances in next generation sequencing and bioinformatics enable detection of many SNP inexpensively, even for orphan crops without a sequenced reference genome. Using 454 sequencing, approximately 11,000 SNPs were discovered by sequencing two alfalfa genotypes contrasting for lignin and cell wall composition [17] and 40,000 SNPs were identified for two genotypes with varying responses to drought stress [18]. By sequencing transcriptomes of 27 diverse alfalfa genotypes, including wild diploids and cultivated tetraploids, we identified 900,000 SNPs [19]. Some of these SNP have been developed into markers using high resolution melting (HRM) [18], and these have been used to improve the resolution of linkage maps and to map candidate genes for traits such as cold tolerance and flowering time (Li et al., unpublished data). However, a high-throughput genotyping platform for alfalfa is still not available.

The experimental design of GWAS and GS can be optimized by accounting for the patterns of population structure and linkage disequilibrium (LD) in the populations employed. Proper GWAS statistical models need to be used to minimize false positive association caused by the underlying population structure [20]. Clear population structure corresponding to the subsp. *falcata* and subsp. *sativa* has been commonly found in alfalfa germplasm [21,22], although less clear distinctions have been identified within cultivated germplasm. The number of markers required for a successful GWAS or GS project depends on the extent of LD in the population. In alfalfa, LD of 0.5 Mbp ($p<0.01$) was found in a breeding population with a relatively narrow genetic background using SSR markers, although the high mutation rate of SSR markers and the small sample size likely overestimated the LD [23]. Rapid LD decay was observed within candidate genes related to lignin biosynthesis in wild diploid alfalfa collections [24] and flowering time in ten tetraploid cultivars [25]. To facilitate GWAS and GS in alfalfa, evaluation of the genome-wide LD pattern using a large set of markers is needed.

In this study, our objectives were as follows: (a) to develop a high density SNP array that would provide a high-throughput genotyping platform for alfalfa; and (b) to assess patterns of population structure and LD in diverse alfalfa accessions that could guide appropriate experimental designs for subsequent GWAS and GS projects.

Materials and Methods

SNP detection, filtering, and selection

We previously identified about 900,000 SNPs by transcriptome sequencing 27 diverse alfalfa genotypes [19]. These ESTs and SNPs are publicly available at the Legume Information System (LIS, http://medsa.comparative-legumes.org/). From this set, we developed an Illumina Infinium array consisting of 10,441 SNPs. We identified SNP loci for the array as follows. First, only loci that had at least 10 reads in at least 20 of the 27 sequenced genotypes were retained; this resulted in about 400,000 SNPs. Second, we computed assay designing tool (ADT) scores and GoldenGate scores (http://support.illumina.com/array/array_software/assay_design_tool.ilmn) for these SNPs and about 180,000 having scores for both criteria that were greater than 0.4 were used for further selection. Third, from the literature, we identified candidate genes for aluminum tolerance, cell cycle, cell size, cell wall composition, drought tolerance, flowering time, growth rate, lignin biosynthesis, components of glandular trichomes, and winter hardiness. For sequences of the candidate genes identified in other plant species, we searched the alfalfa transcriptome sequences [19] for homologous sequences using BLAST, and the best hits with an E-value lower than 1×10^{-5} were selected. Up to three SNPs were selected from a given EST. In total, 6,468 SNPs from 2,640 candidate genes were included on the array (Table S1). Of the 6,468 SNPs, 3,241 aligned on the eight chromosomes of *M. truncatula* pseudomolecule v3.5. Lastly, in order to improve the genome coverage, we added 2,712 SNPs to fill the gaps between candidate gene SNPs. We also selected 1,261 SNPs from 552 ESTs that did not align to the eight chromosomes of *M. truncatula* pseudomolecule v3.5, assuming that some of them might be derived from genes specifically expressed in alfalfa.

SNP array evaluation

We assayed 280 genotypes including 233 cultivated tetraploid subsp. *sativa*, one tetraploid subsp. *falcata*, 25 diploid subsp. *caerulea*, 15 diploid subsp. *falcata* and six genotypes from related species (Table S2) with this SNP array. The 233 tetraploid subsp. *sativa* genotypes spanned the fall dormancy spectrum of alfalfa germplasm and included multiple genotypes from the eleven fall dormancy standard check cultivars [26] and the four winter survival standard checks cultivars [27] (Table S2). The alfalfa entries also included 11 F_1 progenies from bi-parental mapping populations (10 F_1 progenies from the DM3 × DM5 population (a biparental population with broad fall dormancy variation (X. Li and E.C. Brummer, unpub. results) and one F_1 progeny from the Altet4 × NECS-141 population [3]). In addition, we included genotypes of Bulldog 805, a cultivar adapted to the southeastern Coastal Plain of the USA and of four non-dormant populations (ND1, ND2, ND3, and ND4) selected for grazing tolerance using beef cattle in Tifton, GA. As part of our ongoing cultivar development program, we recently developed four new breeding populations by hand crossing 15 genotypes from each of the four ND populations with a different set of 15 genotypes from the cultivar Bulldog 805. The resulting populations were BR1 = ND1 × Bulldog 805, BR2 = ND2 × Bulldog 805, BR3 = ND3 × Bulldog 805, and BR4 = ND4 × Bulldog 805. We are interested in the extent of linkage disequilibrium in these populations. To estimate LD in these breeding populations, we used the genotypic data from the parental genotypes to infer the levels of LD in these populations.

For each genotype, fresh leaf tissue was used for DNA extraction using the Wizard® Genomic DNA Purification Kit (Promega, Cat # A1125) following the manufacturer's protocol. The purified DNA was quantified with the Quant-iT PicoGreen dsDNA Reagent and Kits (Invitrogen, Cat # P7589) and normalized to 55 ng/µl. Genotyping was performed on an Illumina iScan system at the National Center for Genome research (NCGR, Santa Fe, NM) following the manufacturer's protocol for the Infinium assay. The resulting intensity data were used to score SNPs with the Genotyping Module of the GenomeStudio software (Illumina Inc., San Diego CA) using a GenCall threshold of 0.15. For a given SNP marker, three clusters (AA, AB, and BB) could exist in diploid alfalfa but up to five clusters (AAAA, AAAB, AABB, ABBB, and BBBB) could be present in tetraploid alfalfa. In this study, 40 diploid alfalfa genotypes were scored at the same time as tetraploid alfalfa. The automated cluster algorithm with a cluster number of three was used to generate the initial calls. The initial SNP clusters were manually edited to refine the cluster position. For the SNPs in which five clusters could be clearly distinguished, manual editing was performed following the polyploidy protocol in GenomeStudio (Illumina Inc.). The diploid genotypes (AA, BB and AB) were useful to determine the tetraploid homozygous classes (AAAA and BBBB) and the duplex heterozygous class (AABB). To simplify analysis of population structure and LD, both diploid and tetraploid genotypes were scored with three classes (AA, AB and BB), where all three heterozygous classes (AAAB, AABB and ABBB) in tetraploid individuals were scored as heterozygous AB.

Population structure

Each allele of a SNP marker was scored as 1 (presence) or 0 (absence). For a given SNP marker, missing values were first replaced by the mean value for the marker across all individuals. Most SNP loci (92%) had less than 1% missing data (see Results section). Principal component analysis (PCA) was conducted using the R statistical software package *prcomp*. To reduce dimensionality of the data, the first five principal components (PCs) were used for model-based clustering using the R package *mclust*. Here, the choice to use the first five PCs was based on the scree plot of eigenvalues [28]. The best model, which had the optimal number of clusters, was established according to the Bayesian Information Criterion (BIC), and the use of five PCs was optimal in this case. Nei's genetic distances between clusters were calculated using GenAlex [29]. The number of polymorphic SNPs within each cluster and shared between clusters was counted. For the 21 populations including the eleven fall dormancy standard check cultivars, four winter survival standard check cultivars, and five breeding populations, Nei's genetic distances were calculated using GenAlex [29]. The pair-wise matrix was used to construct an unweighted neighbor-joining tree using PHYLIP [30].

Linkage disequilibrium

For each cluster, the LD between pairs of SNPs was estimated as r^2 and calculated using the software program TASSEL [31]. The SNPs with minor allele frequency lower than 5% were excluded in the estimates of LD. The functional relationship between LD and physical distances on the *M. truncatula* genome was evaluated by fitting a non-linear model. The expected r^2 value under drift-recombination equilibrium was $E(r^2) = 1/(1+C)$, and $C = 4ad$, where d is the physical distance in bp and a is an estimated regression coefficient [32]. With a low level of mutation and finite sample size n, the expectation becomes [33]

$$E(r^2)=\left[\frac{(10+C)}{(2+C)(11+C)}\right]\left[1+\frac{(3+C)(12+4C+C^2)}{n(2+C(11+C)}\right]$$

For the breeding populations BR1-4, we inferred linkage disequilibrium based on their parental genotypes. To evaluate consistency of LD between populations, we computed the correlation of LD (r_LD) between populations for SNP pairs located at distances less than 0.1 Mbp, 0.2 Mbp, or 1.0 Mbp and then plotted these correlations against the genetic distance between the pair of populations being analyzed.

Results

Development and validation of the SNP array

An initial set of 10,441 SNPs were selected to be included in the array, and a total of 9,277 SNPs were ultimately included. A total of 7,476 SNPs (81%) passed the quality control and showed polymorphism. The remaining SNPs either passed the quality control but were monomorphic or showed poor quality genotype clustering (Table S1). The average call rate for the 274 alfalfa genotypes was 99.6%. For the related species evaluated in this study, 97% of the 7,476 SNPs were useful for genotyping in *M. prostrata*, 91% for *M. truncatula* and *M. secundiflora*, 90% for *M. lupulina*, and 55% for *Trifolium pratense*.

Of the 7,476 SNPs, 7,433 SNPs were polymorphic among the 274 alfalfa genotypes. Of the 7,433 polymorphic SNPs, 6,826 SNPs (92%) had less than 1% missing values, 557 SNPs (7%) had 1–10% missing values, and only 50 SNPs (1%) had more than 10% missing values. Compared to diploids, up to five clusters (AAAA, AAAB, AABB, ABBB, and BBBB) could exist in tetraploid alfalfa. In total 1,025 SNPs (13.8%) showed five clearly distinguishable clusters (Table S1).

To assess the reproducibility of SNP calls, the alfalfa entry (NFAA08-258) was genotyped twice. The proportion of consistent calls for the duplicated sample was 100% (Table S3). Twenty-four samples (including 23 alfalfa genotypes and one *M. truncatula* genotype) from one SNP array chip showed a lower call rate compared to alfalfa genotypes from other chips. The four genotypes from other related species (one genotype of *M. lupulina*, one genotype of *M. secundiflora* and two genotypes of *T. pratense*) also showed lower call rates than alfalfa genotypes that were on the same chip. These 28 genotypes with lower call rates were genotyped a second time using this SNP array. The call rates for the 24 samples originating from one chip were higher in the second round of genotyping (Table S3). The average proportion of consistent calls for the 24 samples was 99.3% (Table S3). For the four samples from the closely related species, the calling rate was not significantly different and the average proportion of consistent calls was 99.9% (Table S3). Therefore, the lower calling rate for the related species was due to species differentiation rather than random technical error.

Population structure

The principal component analysis (PCA) was performed based on 7,433 polymorphic SNP among 263 alfalfa genotypes (the F_1 progenies were excluded). The first five principal components explained 18.2% of the total variation. The first principal component (PC1) explained 9.4% of the total variation and clearly separated diploid subsp. *falcata* from diploid subsp. *caerulea* and tetraploid alfalfa (Figure 1a). PC2 explained 3.5% of the total variation and separated diploid *caerulea* from tetraploid *sativa* (Figure 1a). Model-based cluster analysis with the first five PCs suggested five clusters of alfalfa accessions (Figures 1a and 1b). Cluster 1 (C1) consisted of the 15 diploid subsp. *falcata* genotypes; Cluster 2 (C2) contained 25 diploid subsp. *caerulea* genotypes; and Cluster 3 (C3) contained six tetraploid genotypes (one tetraploid subsp. *falcata* and five tetraploid subsp. *sativa*). Clusters 4 (C4) and 5 (C5) can be differentiated based on the distribution of genotypes from check cultivars with clearly defined fall dormancy scores. While both clusters contained about half of the intermediate dormancy (FD = 5–7) check cultivar genotypes, C4 included 44 of 45 genotypes from the check cultivars with FD ≤4 and C5 contained 18 of 20 genotypes from check cultivars with FD 8–11 (Table 1 and Table S2). Each of these clusters included genotypes from Bulldog and the other four, non-check populations that were

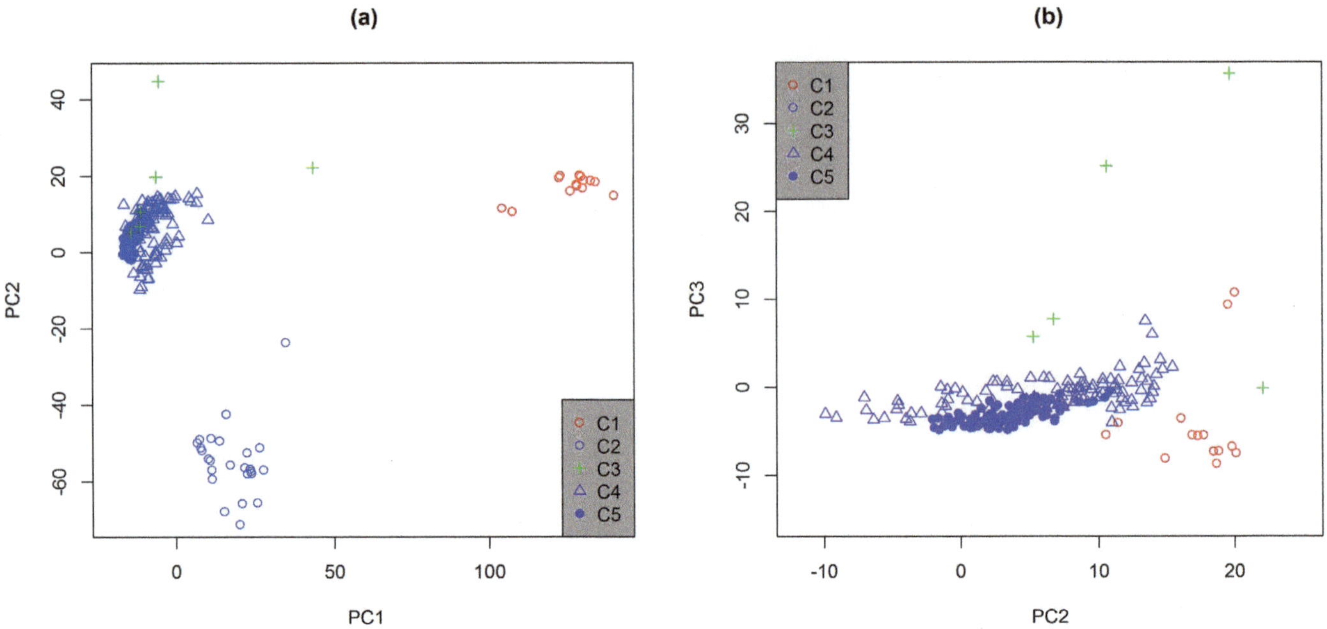

Figure 1. Plots of the first three principal components derived from a principal component analysis (PCA) of 263 alfalfa genotypes using SNP markers generated on an Infinium array, with (a) PC1 vs. PC2 and (b) PC2 vs. PC3. (C1: diploid subsp. *falcata*, C2: diploid subsp. *caerulea*, C3: miscellaneous tetraploid alfalfa, C4: dormant and semi-dormant tetraploid subsp. *sativa*, and C5: semi-dormant and non-dormant tetraploid subsp. *sativa*)

derived from non-dormant cultivars (Table 2). Among the five clusters, C1 was most distant from others, with an average distance of 0.119 (Table 3). The C4 and C5 clusters were most closely related (Table 3).

Pairwise Nei's genetic distances were calculated for the 16 tetraploid standard check cultivars and five breeding populations. The neighbor-joining trees based on the pair-wise distance matrix showed that cultivars were related to one another largely based on dormancy, with cultivars more similar in dormancy being closer together on the tree (Figure 2).

Diversity within and across sub-populations

Since C3 contained only six genotypes and its composition was not obviously related to any particular characteristic, we excluded it for the diversity analysis here. In total 7,107 SNPs were polymorphic among the remaining 257 alfalfa genotypes within C1, C2, C4, and C5. Of these, 320 SNPs (4.5%) were only polymorphic within C1 or C2 (wild diploid subsp. *caerulea* or subsp. *falcata* genotypes) but fixed in C4 and C5 (cultivated tetraploid subsp. *sativa*) (Figure 3). Of the 6,787 SNPs polymorphic within C4 or C5, most (6,191 or 91.2%) were polymorphic in both C4 and C5 (Figure 3).

Heterozygosity was calculated as the percentage of SNPs with a heterozygous call for a given genotype. The average heterozygosity was 15.8% for diploid subsp. *falcata* within the C1 cluster and 16.3% for the 25 diploid subsp. *caerulea* within C2 (Table S2). The average heterozygosity for tetraploid alfalfa genotypes was 33.8% within C4 and 33.0% within C5. For the related species, the heterozygosity level was 5.9% for *M. prostrata*, 0.8% for *M. truncatula*, 1.4% for *M. lupulina*, 1.1% for *M. secundiflora*, and 2.5% for *T. pratense* (Table S2).

Linkage disequilibrium

The functional relationship between LD (estimated as the square of the correlation coefficient (r^2) between pairs of SNPs) and physical distances on the *M. truncatula* genome was determined by fitting a non-linear model. The extent of LD rapidly decayed to 26 Kbp at $r^2 = 0.2$ across all the alfalfa genotypes, but varied among clusters and populations (Figure 4). The extent of LD was about 74 Kbp for C2, 35 Kbp for C4, and 47 Kbp for C5 at $r^2 = 0.2$ (Figure 4). However, at $r^2 = 0.2$, extensive LD was found in C1 (about 600 Kbp) (Figure 4).

For the four new breeding populations developed (BR1, BR2, BR3, and BR4), we inferred LD based on the 30 parents used to develop each of the populations. The average LD level of these

Table 1. The number of individual plants evaluated of the standard fall dormancy and winter hardiness check cultivars and the number that were assigned to either cluster 4 (C4) or cluster 5 (C5).

Fall dormancy class	Fall dormancy score	Total no. genotypes evaluated	No. genotypes in C4	No. genotypes in C5
Dormant	1–4	45	44	1
Semi-Dormant	5–7	15	7	8
Non-Dormant	8–11	20	2	18

Table 2. The number of individual plants evaluated for each of five breeding populations developed from non-dormant germplasm and the number of each population that were assigned to a particular cluster.

Entry	FD	Total no. genotypes evaluated	No. genotypes in C3	No. genotypes in C4	No. genotypes in C5
Bulldog 805	8	60	2	12	46
ND1	9	15		7	8
ND2	8	15		15	
ND3	9	15		1	14
ND4	10–11	15	1	1	13

four populations was expected to be about 170 Kbp at $r^2 = 0.2$ (Figure 4). However, the extent of LD rapidly decayed to 54 Kbp when the four populations were combined (BR1+BR2+BR3+BR4) and analyzed as a single population (Figure 4).

Correlation of LD across sub-populations

We estimated LD between pairs of SNP in different populations and then correlated these LD estimates (r_LD) between the populations. The r_LD values were negatively correlated to the genetic distance between populations (Figure 5). The LD values between pairs of SNP were more highly correlated if the SNP were physically closer together and if the populations under comparison were genetically more similar (Figure 5).

Discussion

Development and validation of SNP array

Illumina Infinium SNP arrays have been developed as a high-throughput genotyping platform for numerous plant species. Genotypes were called for about 72% of array features for apple [34], 70% for peach [35], 80% for sunflower [36], and 88% for tomato [37]. In this study, we developed and validated an Illumina Infinium alfalfa SNP array. In total 7,476 (81%) of the 9,274 SNPs on the array were verified and polymorphic. The successful call rate is comparable with those observed in other plant species. This array also showed a high level of transferability for closely related *Medicago* species including *M. prostrata*, *M. lupulina*, and *M. truncatula* and reasonable transferability to red clover (*T. pratense*). In addition, the allelic dosage of 1,025 SNPs (14%) could be determined in tetraploid alfalfa. In tetraploid potato, the Infinium array included 8,303 SNPs and 35% of those SNPs could be clustered into five genotypic groups [38]. Because allele dosage can affect gene expression and phenotypes in polyploids [39,40], integrating allele dosage information into GWAS statistical models could enhance QTL identification in tetraploid species.

Table 3. Pairwise genetic distances among five clusters of alfalfa genotypes based on SNP markers generated by the Infinium array.

Clusters	C2	C3	C4	C5
C1	0.107	0.124	0.112	0.132
C2		0.063	0.036	0.041
C3			0.022	0.024
C4				0.005

Population structure and diversity

Population structure has implications for experimental design and data analysis in GWAS and GS. As in previous studies [21,22], population structure was observed between the two subspecies of subsp. *sativa* and subsp. *falcata* in this study. In a previous experiment, no differentiation of chloroplast haplotypes was found between tetraploid subsp. *sativa* and diploid subsp. *caerulea*, which suggested that the tetraploid subsp. *sativa* simply originated from diploid subsp. *caerulea* through genome doubling [22]. Based on a small set of SSR markers, diploid subsp. *caerulea* and tetraploid subsp. *sativa* could not be distinguished from each other (Li et al., unpublished data). However, clear differentiation between them was observed using a large set of SNP markers, but only a few genotypes were analyzed [19]. In this study, we provided further supportive evidence using 263 genotypes that included wild diploid subsp. *caerulea* and cultivated tetraploid subsp. *sativa* that the two subspecies are clearly differentiated genetically. Niche differentiation may have occurred following polyploidization, early domestication, and/or modern selection in tetraploid subsp. *sativa*, and this may have caused modern alfalfa to diverge from the wild diploid progenitor subsp. *caerulea*.

Fall dormancy (FD) in alfalfa refers to the reduction of growth and the development of decumbent shoot orientation that occurs in autumn due to shortening photoperiods and/or lowering temperatures [41]. Fall dormancy is positively correlated with winter survival [42], and the alfalfa community generally uses the fall dormancy level as one criterion for choosing appropriate cultivars for planting zones. Based on crown traits, discriminant analysis of 110 cultivars showed that three classes of cultivars were identified – (1) highly dormant cultivars (FD1), (2) dormant cultivars (FD2, 3, 4, and 5), and (3) non-dormant cultivars (FD7, 8, 9); cultivars with FD = 6 overlapped with both dormant and non-dormant classes [43]. The model-based cluster analysis generated in this study showed that C5 contained mostly genotypes from semi- and non-dormant check cultivars, and that C4 contained mostly genotypes from dormant and semi-dormant cultivars. The neighbor-joining tree of standard check cultivars suggested a gradation of relatedness based largely on dormancy level. Interestingly, the non-dormant breeding populations ND1-4 were the surviving plants selected from non-dormant germplasm following beef cattle grazing. Based on our experience, intercrossing genotypes surviving grazing pressure, as was done here, typically results in a population that is more dormant than the source population. The results here suggest that the ND1 and ND2 populations, especially, may be more dormant than their progenitor populations. Bulldog 805, itself a FD = 8 cultivar, was selected under grazing, and some genotypes within the population fall into the more dormant cluster. Alfalfa winterhardiness has been improved by the incorporation of subsp. *falcata* germplasm,

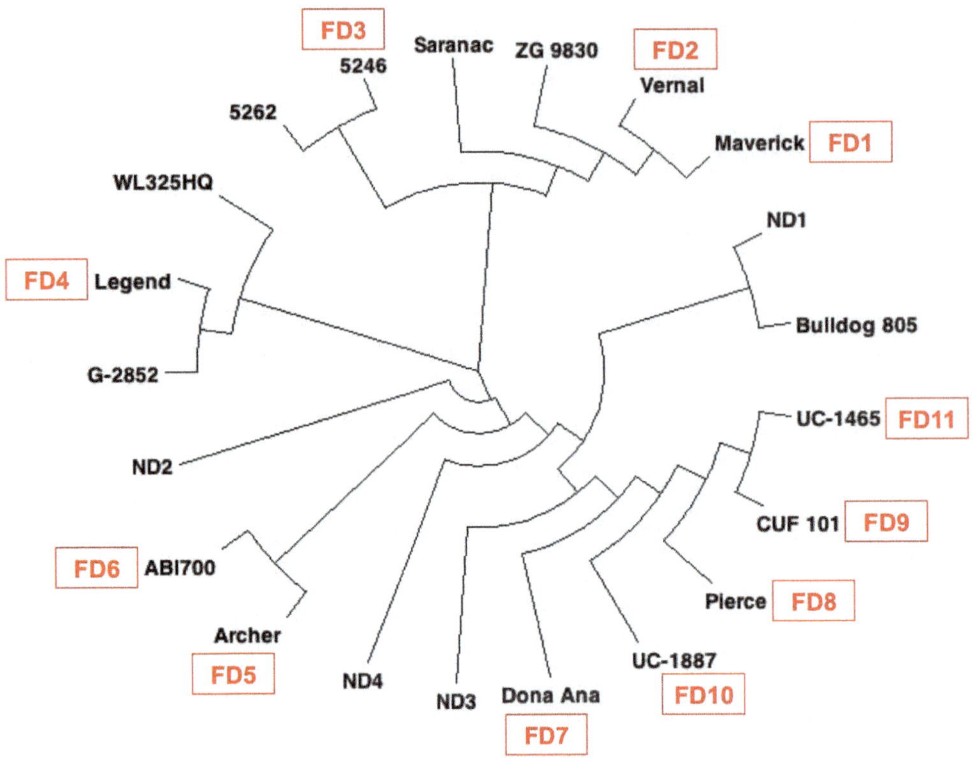

Figure 2. Neighbor-joining tree among 21 alfalfa cultivars and breeding populations based on SNP markers generated from an Infinium array. The fall dormancy (FD) classifications of the cultivars, which include standard check cultivars for fall dormancy and winter hardiness, are indicated on a scale of FD = 1 as very dormant to FD = 11 as very non-dormant.

which is highly grazing tolerant, and the relationships seen among germplasm sources may reflect the amount of historical *falcata* introgression into the germplasm.

Breeding history and population diversity

High levels of heterozygosity were observed in alfalfa, a perennial outcrossing species, while low levels of heterozygosity

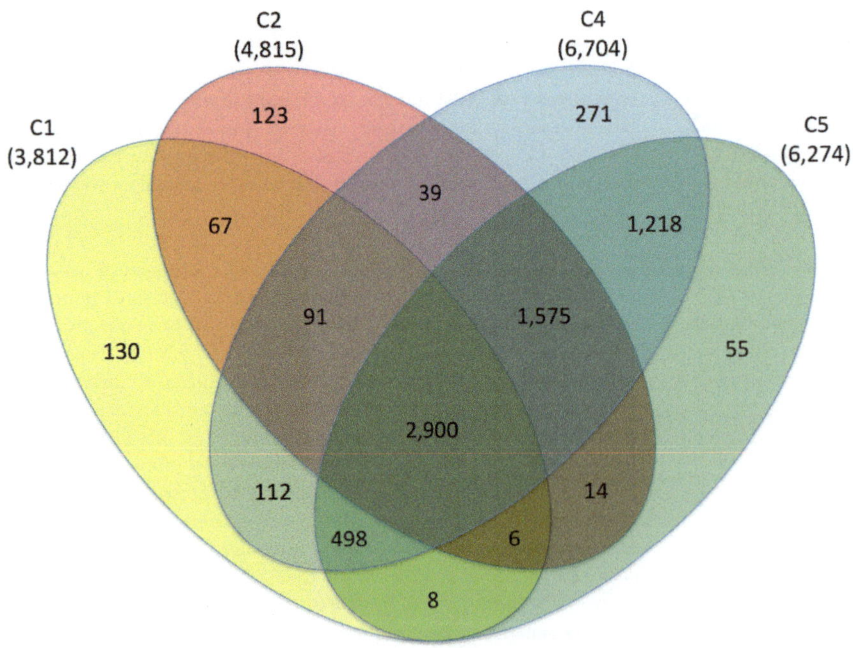

Figure 3. The number of polymorphic SNPs within and between four clusters developed based on a model-based clustering analysis. Cluster 3, which included only six genotypes, was omitted from the analysis.

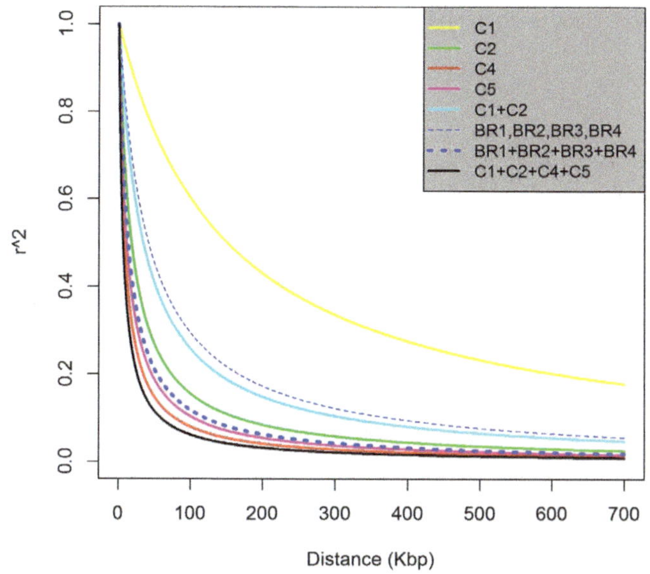

Figure 4. Genome-wide linkage disequilibrium patterns within clusters and populations. (C1: diploid subsp. *falcata*, C2: diploid subsp. *caerulea*, C4: dormant and semi-dormant tetraploid subsp. *sativa*, and C5: semi-dormant and non-dormant tetraploid subsp. *sativa*)

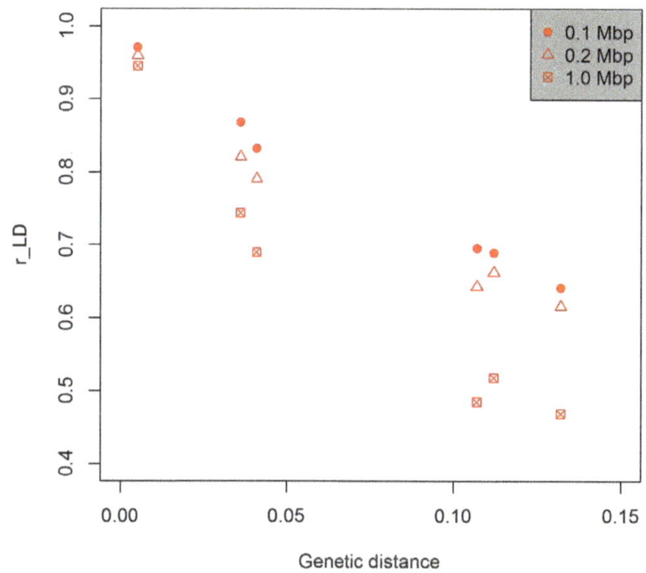

Figure 5. Correlation of LD values between pairs of SNP markers grouped based on three different physical distances apart and plotted as a function of genetic distances between clusters and breeding populations being correlated.

were found in the annual inbreeding species *M. truncatula*, or the autogamous short-lived perennials *M. lupulina* and *M. secundiflora*. The heterozygosity level of *M. prostrata* was 5.9%, which fell between that of the annual self-pollinated species and perennial outcrossing alfalfa. The perennial *Medicago* species are known to be outcrossing, although some evidence (such as low pollen to ovule ratio [44] and small flowers) suggests that *M. prostrata* may be an exception. Our results indicate that *M. prostrata* likely outbreeds to some extent in the wild. *M. prostrata* can easily be crossed with alfalfa and used to improve pest resistance [45].

Minimal diversity appeared to have been lost in cultivated tetraploid alfalfa compared to wild diploid alfalfa based on a transcriptome analysis of 27 genotypes [19]. In this study, we genotyped a larger population of individuals with the SNP array and found a similar result. Our results are in contrast to a previous experiment that showed that domesticated alfalfa contained on average 31% less diversity than wild alfalfa [46]. The different results could be due to the different populations being evaluated, the marker types used, and the number of markers. Nine major alfalfa germplasms have been introduced into the US historically [47]. However, many new landraces and plant introductions, including even wild diploid germplasms, have been subsequently continuously integrated into elite breeding germplasms [48]. This germplasm introduction process, together with a relatively short breeding history and the outcrossing and polyploid nature of alfalfa, could explain the high diversity retained in cultivated tetraploid alfalfa. However, ascertainment bias could partially explain the results, because the SNPs used for the array were primarily derived from cultivated tetraploid alfalfa sequences.

More than 90% of the SNPs polymorphic in tetraploid alfalfa were polymorphic in both cultivated alfalfa sub-populations. Multiple germplasms have been mixed in most breeding programs since the 1950's [47] and both dormant and non-dormant cultivars have often included germplasm from many of the historical introduction groups in their parentage [49]. However, the major historically important germplasms contributing to the dormant and non-dormant alfalfa gene pools are different. For example, the top five germplasms contributing to current dormant cultivars are Flemish (39%), *varia* (22%), Turkistan (12%), Ladak (8%), and Chilean (7%) [49]. The top four major germplasm sources for non-dormant cultivars are African (34%), Indian (17%), Turkistan (14%), and Chilean (9%) [49]. Therefore, although a broad differentiation in the background of dormant and nondormant germplasm exists, recombination during breeding has made this distinction less clear, and consequently, our clustering results reflect this pattern.

Patterns of linkage disequilibrium

Linkage disequilibrium is a key factor in determining the number of markers needed for GWAS and GS. Using the Infinium alfalfa SNP array, we evaluated the genome wide LD pattern for the first time using a large set of markers for alfalfa. Overall, LD decayed to an $r^2 = 0.20$ within 26 Kbp, but this varied greatly among sub-populations. We estimated LD to be about 170 Kbp at $r^2 = 0.2$ within four newly developed breeding populations (BR1, BR2, BR3, and BR4), each generated from 30 parents. However, LD rapidly decayed within 54 Kbp when we evaluated all 120 parents together. Populations with fewer founders will require a lower marker density for GWAS or GS, but using populations of this sort may result in inbreeding depression, which can affect results and needs to be carefully considered [50]. Nevertheless, we argue that alfalfa breeding programs will need to move to smaller populations to fully realize efficiency gains from genetic marker resources.

Many factors such as recombination, mutation, genetic drift, selection, and population admixture affect the extent of LD. If we analyzed diploid subsp. *caerulea* and subsp. *falcata* together, LD increased, possibly due to the divergent genomes generating extensive admixture LD. Similarly, extensive admixture LD was found in barley when divergent materials were mixed for LD evaluation [51]. The more extensive LD that we found within the BR1/2/3/4 breeding populations (of 30 parental genotypes each) compared to the cluster of 25 wild subsp. *caerulea* genotypes indicates that by selecting subsets of genotypes from modern cultivars, breeders can increase LD to acceptable levels for genome-wide marker-based breeding projects.

Successful GWAS and GS experiments also rely on the consistency of LD within the employed population. Natural and/or artificial selection that occurred in wild and breeding populations may result in different allelic combinations and in phasing across loci. Marker-trait association in GWAS and estimated marker effects in GS could be biased or obscured when divergent populations are combined [51]. We identified a high level of consistency in LD between two large subpopulations of cultivated alfalfa in this study. However, wild diploid alfalfa, especially diploid subsp. *falcata*, is divergent from cultivated tetraploid alfalfa and showed a low level of LD consistency at a distance of 0.1 Mbp. Therefore, it is not advisable to combine those germplasms for GWAS and GS.

This experiment reports on the development and testing of an Illumina Infinium array for alfalfa. The array can assay nearly 8,000 SNP markers simultaneously, providing the community with the most advanced genotyping platform currently available. It should prove useful not only for genetic diversity experiments but also for genetic mapping, GWAS, and possibly GS. The array is available to any researcher and can be obtained commercially from Illumina, Inc. (www.illumina.com).

Acknowledgments

The authors appreciate assistance from Kazuyo Ueda and Christy Motes for sample preparation and DNA extraction.

Author Contributions

Conceived and designed the experiments: ECB MJM. Performed the experiments: XL YH YW AA ADF JH. Analyzed the data: XL YH YW AA ADF. Contributed reagents/materials/analysis tools: ECB ADF MJM. Wrote the paper: XL ECB MJM YW YH AA JH ADF.

References

1. Brummer EC (1999) Capturing heterosis in forage crop cultivar development. Crop Sci 39: 943–954.
2. Lamb JFS, Sheaffer CC, Rhodes LH, Sulc RM, Undersander DJ, et al. (2006) Five decades of alfalfa cultivar improvement: Impact on forage yield, persistence, and nutritive value. Crop Sci 46: 902–909.
3. Khu D-M, Reyno R, Han Y, Zhao PX, Bouton JH, et al. (2013) Identification of aluminum tolerance quantitative trait loci in tetraploid alfalfa. Crop Sci 53: 148–163.
4. Robins JG, Luth D, Campbell IA, Bauchan GR, He CL, et al. (2007) Genetic mapping of biomass production in tetraploid alfalfa. Crop Sci 47: 1–10.
5. Robins JG, Hansen JL, Viands DR, Brummer EC (2008) Genetic mapping of persistence in tetraploid alfalfa. Crop Sci 48: 1780–1786.
6. Alarcon Zuniga B, Scott P, Moore K, Luth D, Brummer E (2004) Quantitative trait locus mapping of winter hardiness metabolites in autotetraploid alfalfa (*M. sativa*). In: Hopkins A, Wang ZY, Mian R, Sledge M, Barker RE, editors. Molecular breeding of forage and turf. Kluwer, Dordrecht, The Netherlands. pp. 97–104.
7. Brouwer DJ, Duke SH, Osborn TC (2000) Mapping genetic factors associated with winter hardiness, fall growth, and freezing injury in autotetraploid alfalfa. Crop Sci 40: 1387–1396.
8. Musial JM, Mackie JM, Armour DJ, Phan HT, Ellwood SE, et al. (2007) Identification of QTL for resistance and susceptibility to *Stagonospora meliloti* in autotetraploid lucerne. Theor Appl Genet 114: 1427–1435.
9. Famoso AN, Zhao K, Clark RT, Tung CW, Wright MH, et al. (2011) Genetic architecture of aluminum tolerance in rice (*Oryza sativa*) determined through genome-wide association analysis and QTL mapping. PLoS Genet 7: e1002221.
10. Kump KL, Bradbury PJ, Wisser RJ, Buckler ES, Belcher AR, et al. (2011) Genome-wide association study of quantitative resistance to southern leaf blight in the maize nested association mapping population. Nat Genet 43: 163–168.
11. Dekkers JCM, Hospital F (2002) The use of molecular genetics in the improvement of agricultural populations. Nat Rev Genet 3: 22–32.
12. Jannink JL, Lorenz AJ, Iwata H (2010) Genomic selection in plant breeding: from theory to practice. Brief Funct Genomics 9: 166–177.
13. Meuwissen TH, Hayes BJ, Goddard ME (2001) Prediction of total genetic value using genome-wide dense marker maps. Genetics 157: 1819–1829.
14. Heffner EL, Lorenz AJ, Jannink JL, Sorrels ME (2010) Plant breeding with genomic selection: gain per unit time and cost. Crop Sci 50: 1681–1690.
15. Lorenzana RE, Bernardo R (2009) Accuracy of genotypic value predictions for marker-based selection in biparental plant populations. Theor Appl Genet 120: 151–161.
16. Zhong S, Dekkers JC, Fernando RL, Jannink JL (2009) Factors affecting accuracy from genomic selection in populations derived from multiple inbred lines: a Barley case study. Genetics 182: 355–364.
17. Yang SS, Tu ZJ, Cheung F, Xu WW, Lamb JF, et al. (2011) Using RNA-Seq for gene identification, polymorphism detection and transcript profiling in two alfalfa genotypes with divergent cell wall composition in stems. BMC Genomics 12: 199.
18. Han Y, Kang Y, Torres-Jerez I, Cheung F, Town CD, et al. (2011) Genome-wide SNP discovery in tetraploid alfalfa using 454 sequencing and high resolution melting analysis. BMC Genomics 12: 350.
19. Li X, Acharya A, Farmer AD, Crow JA, Bharti AK, et al. (2012) Prevalence of single nucleotide polymorphism among 27 diverse alfalfa genotypes as assessed by transcriptome sequencing. BMC Genomics 13: 568.
20. Yu J, Pressoir G, Briggs WH, Vroh Bi I, Yamasaki M, et al. (2006) A unified mixed-model method for association mapping that accounts for multiple levels of relatedness. Nat Genet 38: 203–208.
21. Sakiroglu M, Doyle JJ, Charles Brummer E (2010) Inferring population structure and genetic diversity of broad range of wild diploid alfalfa (*Medicago sativa* L.) accessions using SSR markers. Theor Appl Genet 121: 403–415.
22. Havananda T, Brummer EC, Doyle JJ (2011) Complex patterns of autopolyploid evolution in alfalfa and allies (*Medicago sativa*; Leguminosae). Am J Bot 98: 1633–1646.
23. Li X, Wei Y, Moore KJ, Michaud R, Viands DR, et al. (2011b) Association mapping of biomass yield and stem composition in a tetraploid alfalfa breeding population. Plant Genome 4: 24–35.
24. Sakiroglu M, Sherman-Broyles S, Story A, Doyle JJ, Moore KJ, et al. (2012) Paterns of linkage diequilibium and association mapping in diploid alfalfa (*M. sativa* L.) Theor Appl Genet 125: 577–590.
25. Herrmann D, Barre P, Santoni S, Julier B (2010) Association of a CONSTANS-LIKE gene to flowering and height in autotetraploid alfalfa. Theor Appl Genet 121: 865–876.
26. Teuber LR, Taggard KL, Gibbs LK, McCaslin MH, Peterson MA, et al. (1998) Fall dormancy. In Fox CC, Berberet L, Gray FA, Grau CR, Jessen DL, Peterson MA, editors, Standard tests to characterize alfalfa cultivars. North America Alfalfa Improvement Conference, Beltsville, MD. (Revised 2002). Available at http://www.naaic.org/stdtests/Dormancy2.html (verified 20 May 2013).
27. McCaslin MH, Woodward T, Undersander D (2003) Winter survival. In Fox CC, Berberet L, Gray FA, Grau CR, Jessen DL, Peterson MA, editors, Standard tests to characterize alfalfa cultivars. North America Alfalfa Improvement Conference, Beltsville, MD. Available at http://www.naaic.org/stdtests/wintersurvivalnew.htm (verified 20 May 2013).
28. Cattell RB (1966) The scree test for the number of factors. Multivar Behav Res 1:245–276
29. Peakall R, Smouse PE (2012) GenAlEx 6.5: genetic analysis in Excel. Population genetic software for teaching and research—an update. Bioinformatics 28: 2537–2539.
30. Felsenstein J (1989) PHYLIP: phylogeny inference package (version 3.2). Cladistics 5: 164–166.
31. Bradbury PJ, Zhang Z, Kroon DE, Casstevens TM, Ramdoss Y, et al. (2007) TASSEL: software for association mapping of complex traits in diverse samples. Bioinformatics 23: 2633–2635.
32. Sved JA (1971) Linkage disequilibrium and homozygosity of chromosome segments in finite populations. Theor Popul Biol 2:125–141
33. Hill WG, Weir BS (1988) Variances and covariances of squared linkage disequilibria in finite populations. Theor Popul Biol 33: 54–78.
34. Chagne D, Crowhurst RN, Troggio M, Davey MW, Gilmore B, et al. (2012) Genome-wide SNP detection, validation, and development of an 8K SNP array for apple. PLoS One 7: e31745.
35. Verde I, Bassil N, Scalabrin S, Gilmore B, Lawley CT, et al. (2012) Development and evaluation of a 9K SNP array for peach by internationally

coordinated SNP detection and validation in breeding germplasm. PLoS One 7: e35668.
36. Bachlava E, Taylor CA, Tang S, Bowers JE, Mandel JR, et al. (2012) SNP discovery and development of a high-density genotyping array for sunflower. PLoS One 7: e29814.
37. Sim SC, Van Deynze A, Stoffel K, Douches DS, Zarka D, et al. (2012) High-density SNP genotyping of tomato (*Solanum lycopersicum* L.) reveals patterns of genetic variation due to breeding. PLoS One 7: e45520.
38. Hirsch CN, Hirsch CD, Felcher K, Coombs J, Zarka D, et al. (2013) Retrospective view of North American potato (*Solanum tuberosum* L.) breeding in the 20th and 21st centuries. G3: Genes|Genomes|Genetics 3: 1003–1013.
39. Stupar RM, Bhaskar PB, Yandell BS, Rensink WA, Hart AL, et al. (2007) Phenotypic and transcriptomic changes associated with potato autopolyploidization. Genetics 176: 2055–2067.
40. Jackson S, Chen ZJ (2010) Genomic and expression plasticity of polyploidy. Curr Opin Plant Biol 13: 153–159.
41. Castonguay Y, Laberge S, Brummer EC, Volenec JJ (2006) Alfalfa winter hardiness: A research retrospective and integrated perspective. Adv Agron 90: 203–265.
42. Schwab PM, Barnes DK, Sheaffer CC (1996) The relationship between field winter survival and fall growth score for 251 alfalfa cultivars. Crop Sci 36: 418–426.
43. Marquez-Ortiz JJ, Johnson LD, Barnes DK, Basigalup DH (1996) Crown morphology relationships among alfalfa plant introductions and cultivars. Crop Sci 36: 766–770.
44. Small E (1988) Pollen-ovule patterns in tribe *Trifolieae* (Leguminosae). Plant Syst Evol 160: 195–205.
45. Sorensen EL, Stuteville DL, Horber EK, Skinner DZ (1994) Registration of KS224 Glandular-haired alfalfa germplasm with multiple pest resistance. Crop Sci 34: 544–544.
46. Muller MH, Poncet C, Prosperi JM, Santoni S, Ronfort J (2006) Domestication history in the Medicago *sativa* species complex: inferences from nuclear sequence polymorphism. Mol Ecol 15: 1589–1602.
47. Barnes DK, Bingham ET, Murphy RP, Hunt OJ, Beard DF, et al. (1977) Alfalfa germplasm in the United States: Genetic vulnerability, use, improvement, and maintenance. Washington, D.C.: U.S. Gov. Print. Office.
48. Li X, Brummer EC (2012) Applied genetics and genomics in alfalfa breeding. Agronomy 2: 40–61.
49. Volenec J, Teuber L, Stoner A, Bauchan G, Berberet R, et al. (2000) Report on the status of Medicago germplasm in the United States. Available at http://www.ars-grin.gov/npgs/cgc_reports/alfalfa/alfalfacgc2000.htm (verified 20 May 2013).
50. Hayes BJ, Cogan NOI, Pembleton LW, Goddard ME, Wang JP, et al. (2013) Prospects for genomic selection in forage plant species. Plant Breeding 132: 133–143.
51. Hamblin MT, Close TJ, Bhat PR, Chao SM, Kling JG, et al. (2010) Population structure and linkage disequilibrium in US Barley germplasm: implications for association mapping. Crop Sci 50: 556–566.

12

Development and Validation of a 20K Single Nucleotide Polymorphism (SNP) Whole Genome Genotyping Array for Apple (*Malus* × *domestica* Borkh)

Luca Bianco[1], Alessandro Cestaro[1], Daniel James Sargent[1], Elisa Banchi[1], Sophia Derdak[2], Mario Di Guardo[1,3], Silvio Salvi[4], Johannes Jansen[5], Roberto Viola[1], Ivo Gut[2], Francois Laurens[6], David Chagné[7], Riccardo Velasco[1], Eric van de Weg[3], Michela Troggio[1]*

1 Research and Innovation Centre, Fondazione Edmund Mach, San Michele all'Adige, Trento, Italy, **2** CNAG – Centro Nacional de Análisis Genómico, Parc Científic de Barcelona, Barcelona, Spain, **3** Wageningen UR Plant Breeding, Wageningen University and Research Centre, Wageningen, The Netherlands, **4** DipSA - University of Bologna, Bologna, Italy, **5** Biometris, Wageningen University and Research Centre, Wageningen, The Netherlands, **6** INRA, UMR1345 Institut de Recherche en Horticulture and Semences, Beaucouzé, France, **7** Plant & Food Research, Palmerston North Research Centre, Palmerston North, New Zealand

Abstract

High-density SNP arrays for genome-wide assessment of allelic variation have made high resolution genetic characterization of crop germplasm feasible. A medium density array for apple, the IRSC 8 K SNP array, has been successfully developed and used for screens of bi-parental populations. However, the number of robust and well-distributed markers contained on this array was not sufficient to perform genome-wide association analyses in wider germplasm sets, or Pedigree-Based Analysis at high precision, because of rapid decay of linkage disequilibrium. We describe the development of an Illumina Infinium array targeting 20 K SNPs. The SNPs were predicted from re-sequencing data derived from the genomes of 13 *Malus* × *domestica* apple cultivars and one accession belonging to a crab apple species (*M. micromalus*). A pipeline for SNP selection was devised that avoided the pitfalls associated with the inclusion of paralogous sequence variants, supported the construction of robust multi-allelic SNP haploblocks and selected up to 11 entries within narrow genomic regions of ±5 kb, termed focal points (FPs). Broad genome coverage was attained by placing FPs at 1 cM intervals on a consensus genetic map, complementing them with FPs to enrich the ends of each of the chromosomes, and by bridging physical intervals greater than 400 Kbps. The selection also included ~3.7 K validated SNPs from the IRSC 8 K array. The array has already been used in other studies where ~15.8 K SNP markers were mapped with an average of ~6.8 K SNPs per full-sib family. The newly developed array with its high density of polymorphic validated SNPs is expected to be of great utility for Pedigree-Based Analysis and Genomic Selection. It will also be a valuable tool to help dissect the genetic mechanisms controlling important fruit quality traits, and to aid the identification of marker-trait associations suitable for the application of Marker Assisted Selection in apple breeding programs.

Editor: Pankaj Jaiswal, Oregon State University, United States of America

Funding: This work has been funded under the EU seventh Framework Programme by the FruitBreedomics project N°. 265582: Integrated Approach for increasing breeding efficiency in fruit tree crops. The views expressed in this work are the sole responsibility of the authors and do not necessary reflect the views of the European Commission. The funders had no role in study design, data collection and analysis, decision to publish, or preparation of the manuscript.

Competing Interests: The authors have declared that no competing interests exist.

* Email: michela.troggio@fmach.it

Introduction

Cultivated apple (*Malus* × *domestica*) is the most economically important deciduous fruit tree crop worldwide [1]. Breeding of novel cultivars with superior fruit quality characteristics is a slow and costly process, because of the extended juvenility period of the species. Typically, the selection process takes on average more than 20 years from seed to introduction, even when rootstocks conferring precocious flowering are exploited. Breeding for disease resistance takes even longer, requiring introgression of novel resistance genes from wild germplasm, followed by successive generations of back-crossing to restore fruit quality to a commercially acceptable level. The use of molecular markers could accelerate and enhance the breeding process, particularly for traits that are difficult to select for phenotypically, such as pyramided disease resistances, or for traits that are expressed only in mature trees, such as fruit characteristics.

Recent advances in genomics technologies have enabled the sequencing of the 'Golden Delicious' genome [2] and the subsequent development of a whole genome genotyping (WGG) (micro-) array for the species [3]. The International RosBREED Single nucleotide polymorphism (SNP) Consortium (IRSC) WGG array contains 7,867 *Malus* SNP markers as well as 921 SNPs derived from *Pyrus* [4]. The IRSC array greatly facilitated the development of high density linkage maps for segregating apple progenies [5–7]. It was used in a Genome-Wide Association Study (GWAS) of the genetic control of several significant fruit traits [8], for the implementation of Genomic Selection (GS) [9] and for resolving pedigrees [10]. However, these 8,788 potential genetic markers are not sufficient to perform GWAS in wider germplasm sets, or to perform Pedigree-Based Analysis (PBA) [11] with high levels of precision. This is due to the rapid linkage disequilibrium decay in apple [12], the limited proportion of robust, easy to score makers included on the IRSC SNP array [6,10], and their uneven distribution across the genome. A higher density array, with robust genome-wide markers is therefore required to perform such studies successfully.

In this investigation, we describe the development of such a high density WGG array for apple, using a focal point approach and stringent selection criteria built from experience of the analysis of the apple IRSC array [3,5,6]. A haplotype-targeting strategy similar to that adopted to design the IRSC array was implemented for the design of the array to combine information from individual SNPs into haploblocks and provide fully informative multi-allelic markers. We also summarize available metadata on the application of the array to a separate genetic mapping study.

Materials and Methods

SNP Discovery Panel and Re-sequencing

To enable the identification of SNPs, a discovery panel comprising the following 13 apple cultivars, including some of the core European apple breeding founder varieties [13–15], was re-sequenced using short-read sequencing technology: 'Braeburn', 'Cox's Orange Pippin', 'Common Antonovka', 'Delicious', 'Dr. Oldenburg', F2-26829-2-2, 'Fuji', 'Jonathan', 'Lady Williams', 'McIntosh', 'Macoun', 'Priscilla-NL' and 'Worcester Pearmain-USA'. Additionally, a scab-resistant accession of *M. micromalus* and two *M.* × *domestica* double haploid (DH) accessions, X9273 and X9748, which were derived from 'Golden Delicious', were included [16]. Leaf material was procured from various institutions (Table 1). For 'Priscilla-NL' and 'Worcester Pearmain-USA', the country of origin of the leaf material is included, to distinguish them from other genotypes with the same cultivar name [17]. The DHs were included to help identify pseudo-SNPs created from paralogous sequences of the apple genome that are erroneously assembled into a single locus, or that are located at different segments/chromosomes but that are targeted by the same Illumina probes. Since the DH lines are homozygous across their entire genomes, any heterozygous calls in these genotypes were thus considered evidence of paralogous sequences. DNA was extracted from freeze-dried, newly emerged leaf material using a phenol-chloroform isoamyl alcohol extraction method [10] and quantified with a Qubit fluorometer (Life Technologies). Sequencing libraries were constructed according to the TruSeq DNA sample preparation protocol (Illumina) with minor modifications, in particular employing double size selection steps. Two micrograms of genomic DNA were fragmented with a Covaris E210 and size selected to 300–600 bps. The resulting fragments were end-repaired, adenylated and ligated to Illumina paired-end adaptors. The size of the library was confirmed on the BioAnalyzer 2100 (Agilent) and the library was sequenced on an Illumina HiSeq 2000 platform with paired end runs of 2×101+7 bps. Base calling and quality control were performed using the Illumina RTA sequence analysis pipeline.

Read Alignment, Variant Detection and Quality Filtering

A schematic representation of all the steps in the pipeline for array development is given in Figure 1. As the first step, reads were sequentially aligned to the primary assembly and three alternative assemblies of the 'Golden Delicious' genome v2.0 (http://www.rosaceae.org/species/malus/malus_x_domestica/genome_v2.0) allowing up to seven mismatches in a four-step exhaustive alignment procedure using GEM [18] and BFAST [19]. Version 2.0 of the apple genome was created by removing 34,882 problematic contigs from the previous version [2]. The primary assembly, representing about 80% of the assembled and anchored genome, and the three alternative assemblies were produced following the NCBI AGP v2.0 specifications (http://www.ncbi.nlm.nih.gov/projects/genome/assembly/agp/AGP_Specification_v2.0.shtml).

Read pairs that always mapped to a single genomic location were kept in the alignment file for that reference at each step, while unaligned read pairs were submitted to the next mapping round. Pairs that did not align uniquely to any of the references were discarded (Figure 1). As a result, four .bam files were generated for each sample from the four alternative assemblies. Variant calling was performed for each of these .bam files separately, applying samtools and bcftools (version 0.1.18) [20] using default parameters. Positions with significant strand bias (p-value<0.001), significant tail distance bias (p-value<0.05) or within regions of low mapping confidence [21] were excluded. The variant-calling pipeline was set to produce genotype calls for each variant for all the re-sequenced cultivars. Calls with a support value of at least 10 reads were retained in the .vcf file, while the others were considered unreliable and therefore set to empty. To avoid the inclusion of potential paralogy-related SNPs, variants with a heterozygous genotype in the DHs were filtered out. Finally, known ambiguous bases in the primary reference were annotated.

A quality filter was then applied to remove potentially unreliable variants. Custom scripts written in Python (www.python.org) were developed to remove variants with low phred-scaled quality scores (i.e. below 20); a high combined read depth (i.e. higher than 2,000); and a low single-cultivar read depth (i.e. lower than 10) in more than 50% of the cultivars. A minimum phred-scaled quality score of 20 was chosen to ensure that only SNPs with a probability of less than 1% for the alternative allele being called wrongly were selected, and the maximum read depth value was used to ensure the removal of SNPs derived from paralogous regions rather than true heterozygous regions as done previously in the development of the IRSC apple array [3] and the 9 K peach array [22]. Additionally, a cut-off of at least ten reads per single cultivar in more than 50% of the cultivars was used to distinguish real variants from potential sequencing errors. The 50% cut-off was chosen since some cultivars, such as 'Common Antonovka' and F2-26829-2-2, and the accession of *M. micromalus*, were derived from different genetic backgrounds compared to the other accessions, and absence of sequence coverage in those regions could have been due to genomic variation. Finally, all SNPs with an allele frequency (AF) = 1 were discarded since this equated to all the re-sequenced varieties carrying an allele that was different from the 'Golden Delicious' reference genome. Such SNPs were discarded as they were likely to represent potential false SNPs resulting from sequencing errors in the 'Golden Delicious' reference sequence, or rare alleles derived from 'Golden Delicious'. For the same reason, rare alleles derived from other cultivars were discarded during the SNP selection phase described below.

Illumina Specific Filtering and SNP Selection

Quality filtered variants from the pipeline described above were then processed to meet Illumina Infinium II array design requirements (http://res.illumina.com/documents/products/technotes/technote_iselect_design.pdf). This third step removed

Table 1. Origin of samples and re-sequencing statistics for the 14 genotypes of the discovery panel and the two double haploids.

Sample name	Source of leaf material*	Tot read pairs	Mean percent unique reads	Mean coverage
'Dr Oldenburg'	JKI	134,605,846	55.98	43.13
'Fuji'	JKI	116,541,162	56.82	37.95
'Lady Williams'	JKI	115,667,659	56.17	38.12
F2-26829-2-2	UNIBO	110,254,636	55.14	35.52
'Macoun'	JKI	95,279,675	55.35	30.77
'Cox's Orange Pippin'	WUR	98,094,673	56.53	32.61
'Worchester Pearmain-USA' PI 206035	PGRU-Geneva	104,600,260	54.67	34.30
'Jonathan'	WUR	137,126,235	56.00	45.45
Malus micromalus	WUR	117,046,241	50.11	34.65
'McIntosh'	INRA	135,946,435	51.10	44.33
'Delicious'	INRA	139,815,864	57.49	45.88
'Braeburn'	KUL	122,735,693	50.48	39.37
Common Antonovka	VNIIISPK	133,423,835	50.73	43.32
'Priscilla-NL'	WUR	127,017,433	55.02	41.17
X9748	INRA	197,011,579	59.17	66.84
X9273	INRA	207,626,734	58.81	69.70

The 16 genomes were sequenced on 8 Illumina HiSeq 2000 lanes. Double haploids (DHs, acc. X9748 and X9273) feature almost a double coverage compared with the other genomes; the total number of read pairs ranges from 95 Million to 207 Million. The percentage of uniquely aligning reads is evenly spread over each genome and ranges from 50 to 60% of the total. The mean coverage similarly ranges from 30 to 45X, while it is higher for the DHs.
*(JKI) Julius Kühn-Institut; (UNIBO) Università di Bologna; (WUR) Wageningen University and Research centre; (PGRU-Geneva) Plant Genetic Resources Unit; (INRA) Institut National de la Recherche Agronomique; (KUL) Katholieke Universiteit Leuven; (VNIIISPK) The All Russian Research Institute of Horticultural Breeding.

variants that were either indels, A/T or C/G transversions, or tri-allelic SNPs. Additionally, variants located at sites containing additional high quality SNPs in either the up/down-stream 50 bp; or that contained both up-stream and down-stream sequence that appeared multiple times across the genome in the directly flanking 24 bp were removed [6]. Approximately one million SNPs obtained in this way were subsequently submitted to the Illumina Assay Design Tool (http://support.illumina.com/array/array_software/assay_design_tool.ilmn) for a preliminary estimation of the conversion rate (the Illumina SNP_Score). Following Illumina guidelines, markers with a SNP_Score below the threshold of 0.7 were discarded, while others were kept as high quality SNPs for the final selection of the 20,000 targets to be included on the 20 K Infinium array. In addition to these newly identified and validated SNPs from the discovery panel re-sequencing, 3,670 validated SNPs from the previously developed apple IRSC SNP WGG array were added to this selection. The SNPs selected from the IRSC array were chosen as they were robustly positioned in different genetic linkage maps, including those reported in [3] and [5].

The focal point strategy adopted for the design of this array followed that of the IRSC array design [3], with some modifications (Figure 2). The size of focal point intervals was reduced from 50 K bp, to 5 K bp up- or down-stream of the focal point itself to provide greater robustness in haplotype building (fewer recombination events within a focal point), and the number of SNPs within each FP was increased where feasible. Focal points evenly spread across the genome were first identified on the basis of the genetic positions available from previous mapping studies [3,5], which resulted in a total of 718 FPs. A further 1,184 FPs were then added to cover physical intervals of an average of 400 Kbps to reach an average genetic distance of one FP per centiMorgan (cM) (using an expected Kbp/cM ratio of 440 Kbps based on a previous estimate by [2]). Finally, an additional 120 FPs were selected to enrich the ends of each of the 17 *Malus* chromosomes, making the number of FPs selected 2,022. Through a SNP short-listing process, up to 11 markers for each FP were selected. Entries featuring the same genotyping profile across the 14 re-sequenced accessions within a given FP were removed to avoid having too many SNPs at a single point potentially derived from the same haplotype. Additionally, SNPs displaying a percentage of heterozygous genotypes in the re-sequencing panel greater than 65% were also excluded as they were most likely to be the result of paralogy instead of true heterozygosity. In cases where 11 or fewer SNPs remained within a given FP, these were all selected for inclusion on the array. Where more than 11 SNPs remained, a step-wise inclusion procedure was followed. Firstly, up to five SNPs were selected that were polymorphic in 35–60% of the panel members, which generally corresponds to 5–8 members. Their heterozygous profiles had to be complementary as far as possible, and panel members had to be more or less equally represented, thus balancing the number of heterozygous SNPs across all (diploid) panel members. Secondly, this first set of highly polymorphic SNPs was complemented with SNPs that were polymorphic in 10–35% of the panel members, usually corresponding to 2–4 members, following the same principles on complementarity and representation. Thirdly, the remaining positions were filled up with SNPs that were polymorphic in just one panel member. This meant that complementarity within any given FP was accounted for as far as possible. This selection

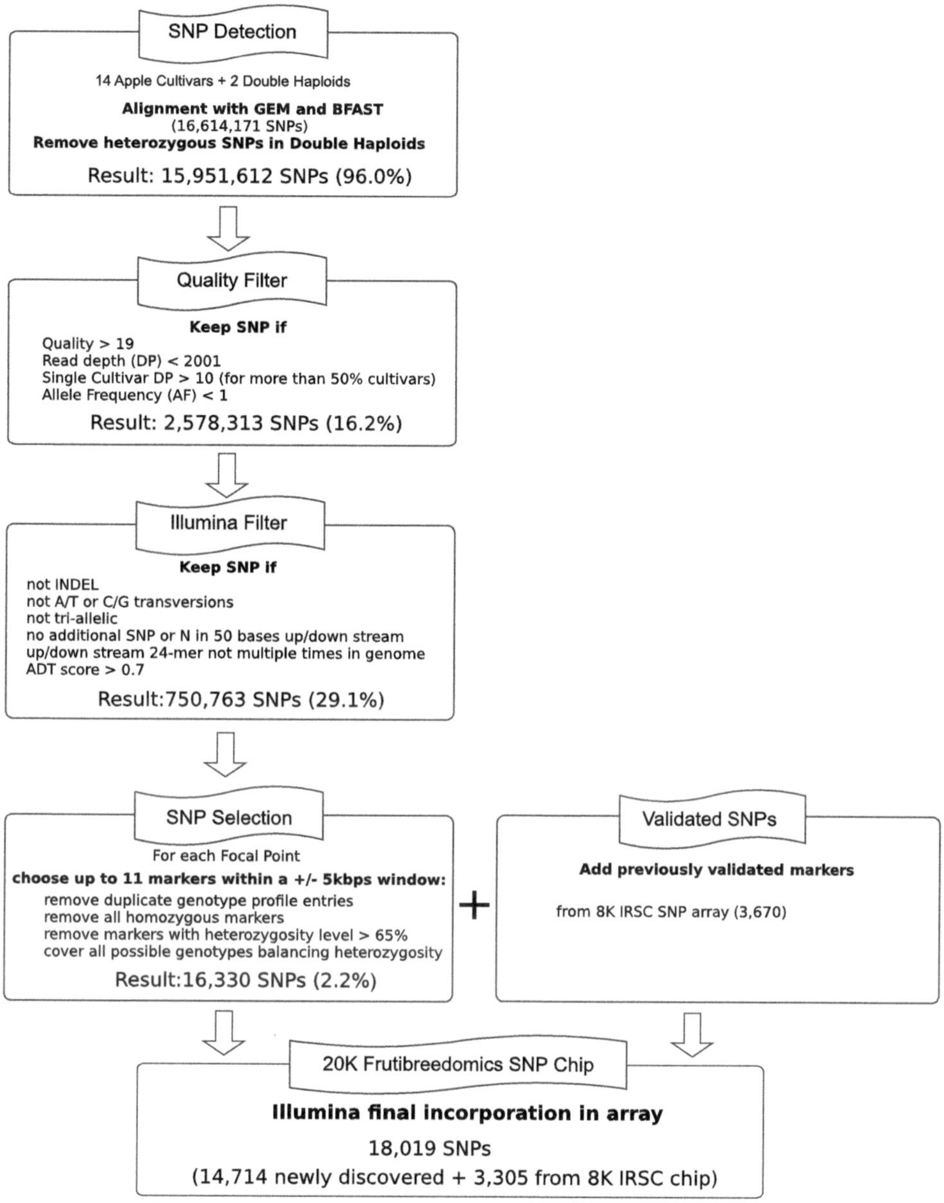

Figure 1. The SNP selection workflow. The SNP selection process was a cascade involving four steps: Detection, Quality Filtering, Illumina specific Filtering and SNP Selection, with the number of SNPs reduced after each step. The specific filtering criteria have been reported for each filtering stage, as well as the number of resulting SNPs, with the corresponding percentage survival, relative to the total number of SNP markers after the previous step. From the 20,000 selected, Illumina successfully incorporated a total of 18,019 SNP probes into the new array, 14,714 of which were newly discovered and 3,305 previously validated.

process was automated through a customized script that is available upon request from the corresponding author. Finally, the single member heterozygotes were manually scrutinized across all FPs, to ensure that FPs in the same region did not have a single heterozygous SNP for the same panel member. Also, *M. micromalus* specific SNPs were maintained at reduced frequency because of the currently limited use of this species in breeding (Figure 2).

SNP Validation

The discovery panel was genotyped with the 20 K SNP array following the standard Illumina protocol detailed in the IRSC apple array and its application papers [3,5]. In the framework of an ongoing PBA-based QTL mapping study of the FruitBreedomics project, over 1,600 seedlings from 21 full-sib families and their direct parents and additional progenitors were also genotyped and used for the construction of genetic maps. Although this work is not part of the current paper, we will present some of its outcomes, as they relate to the performance of the array. The full-sib families on which these data are based are listed in Table S1.

Genotyping data from the discovery panel and the full-sib families were analyzed using the GenomeStudio software (Illumina) with a GenCall threshold of 0.15 and a SNP filtering pipeline [23]. This employs the multipoint maximum likelihood mapping algorithm approach for cross pollinators in JoinMap 4.1 ([24,25] E. Van de Weg, unpublished data).

Genotype calls from the 14 accessions of the discovery panel were compared with those obtained through high-coverage

SNP #	LG	Position (bps)	HET	Calls of diploid discovery panel members: Braeburn'	Common Antonovka'	Cox's Orange Pippin'	Delicious'	Dr. Oldenburg'	F2-26829-2-2	Fuji'	Jonathan'	Lady Williams'	Macoun'	*M.Micromalus*	McIntosh'	Priscilla'	Worcester Pearmain-USA'	Di-haploids: X9273	X9748
1	1	633,002	7	0/1	1/1	0/0	0/1	0/0	0/0	0/1	0/0	0/1	1/1	0/1	0/1	0/0	0/1	0/0	0/0
2	1	***629,347***	6	0/1	1/1	0/0	0/1	0/0	0/0	0/1	0/0	0/1	1/1	0/0	0/1	0/0	0/1	0/0	0/0
3	1	**633,660**	6	1/2.	1/1	2/2	1/1	1/2.	2/2	1/2.	2/2	1/2.	1/1	1/1	1/2.	2/2	1/2.	1/1	1/1
4	1	632,163	3	1/1	1/1	1/1	1/1	1/1	1/2.	1/2.	2/2	1/2.	1/1		1/1	2/2	1/1	1/1	1/1
5	1	632,821	1	0/0	0/0	0/0	0/0	0/0	0/0	0/0	0/0	0/0	0/0	0/1	0/0	0/0	0/0	0/0	0/0
6	1	629,904	6	0/1	1/1	0/0	0/1	0/0	0/0	0/1	0/0	0/1	1/1	1/1	0/1	0/0	0/1	0/0	0/0
		Representation heterozygotes	*All*	4			3	1	1	5		5		2	4		4		
			Selected	3			2	1	1	4		4		2	3		3		
7	9	39,002,373	7	1/2.	1/1	1/1	1/2.	1/1	1/2.	1/1	1/1	1/2.	1/2.	1/1	1/2.	1/1	1/2.	2/2	1/1
8	9	39,000,404	5	0/0	1/1		0/1	1/1	0/0		0/1	0/1	0/0	0/0	0/1	1/1	0/1	0/0	0/0
9	9	39,002,203	5	0/0	0/0	0/0	0/1	0/1	0/1	0/1	0/1	0/0	0/0	0/0	0/0	0/0	0/0	0/0	0/0
10	9	39,002,529	5	1/1	2/2	1/1	1/1	1/2.	1/1	1/2.	2/2	1/2.	1/2.	1/1	1/1	1/2.	1/1	1/1	2/2
11	9	39,004,315	5	0/1	0/1	1/1	0/0	0/0	0/1	0/0	1/1	0/0	0/1	0/0	0/0	0/0	0/1	0/0	0/0
12	9	38,999,475	4	0/1	0/0	0/1	0/1	0/1	0/0		0/0	0/0	0/0		0/0	0/0	0/0	0/0	
13	9	39,002,125	4	0/0	0/0	0/1	0/0	0/0	0/0	0/0	0/1	0/0	0/0	0/0	0/1	0/1		0/0	0/0
14	9	39,002,242	4	0/0	0/1	0/0	0/0	0/1	0/0	0/0	0/0	0/1	0/0	0/0	0/0	0/1	0/0	0/0	0/0
15	9	**39,004,972**	4	0/1	0/1	0/1	0/0	0/0	0/0	0/0	1/1	0/0	0/1	0/0	0/0	0/0	0/0	0/0	0/0
16	9	38,999,410	3	0/0	0/0	0/0	0/0	0/0	0/0	0/0	0/0	0/0	0/0		0/1	0/1	0/1	0/0	
17	9	39,001,465	2	1/1	1/1	1/1	1/1	1/1	1/1	1/2.	1/2.	1/1	1/1	2/2	1/1	1/1	1/1	1/1	2/2
18	9	38,999,308	8	0/1	0/1	1/1	0/1	0/1	0/0	1/1	1/1	0/0	0/1		0/1	0/1	0/1	0/0	
19	9	***38,998,668***	6	0/1	1/1	1/1	0/1	1/1	0/0		1/1	0/1	0/1		0/1	1/1	0/1	0/0	
20	9	38,998,717	6	0/1	0/1	1/1	0/1	1/1	0/0	1/1	0/0	0/1	0/0		0/1	1/1	0/1	0/0	
21	9	39,002,445	6	1/2.	1/1	1/1	1/2.	1/1	1/2.	1/1	1/1	1/2.	1/2.	1/1	2/2	1/2.	2/2	2/2	1/1
22	9	39,003,048	6	1/2.	1/2.	1/1	1/2.	1/1	1/2.	1/1	2/2	1/2.	2/2	1/1	1/2.	1/1	1/1	2/2	1/1
23	9	38,999,843	5	0/0	0/1	0/0	0/0	0/1	0/0	0/0	0/0	0/1	0/0	0/0	0/1	1/1	0/1	0/0	
24	9	38,999,981	5	0/1	1/1	1/1	0/1	1/1	0/0	1/1	1/1	0/0	0/1		0/1	1/1	0/1	0/0	1/1
25	9	39,003,120	4	1/2.	1/2.	1/1	1/2.	1/1	2/2	1/1	2/2	1/2.	2/2	1/1	2/2	1/1		2/2	1/1
26	9	39,000,665	2	0/0	0/0		0/0	0/0	0/1		0/0	0/1	0/0		0/0	0/0	0/0	0/0	
27	9	39,004,669	2	0/0	0/0	0/0	0/0	0/0	0/0	0/0	0/0	0/0	0/0	0/0	0/1	0/1	1/1	0/0	0/0
28	9	39,000,300	1	0/0	0/0	0/1	0/0	0/0	0/0	0/0	0/0	0/0	0/0	0/0	0/0	0/0	0/0	0/0	0/0
		Representation heterozygotes	*All*	11	8	4	11	6	6	3	4	11	8		11	7	9		
			Selected	4	3	3	4	4	3	3	4	4	4		4	4	4		

Figure 2. SNP selection within a single focal point (FP) of chromosomes (Chrs) 1 and 9. Genotype calls are presented for the discovery panel members: 0/1 and 1/2 indicates heterozygosity, all other codes indicate absence of polymorphism within an individual. Green, blue and orange filled cells relate to selected SNPs, having 5–8, 2–4, or just 1 heterozygous panel member/s, respectively. The physical positions at the extremities of the FP are in bold and italics (most proximal) or underlined (most distal). Within each FP, SNPs are sorted firstly by being selected or not, and secondly according to the number of heterozygous members (HET). In the Chr1-FP example, SNP-6 was not selected, as its heterozygosity pattern is identical to that of SNP-2. The single-member-heterozygote SNP-5 was included because no other more polymorphic SNP markers were available to reach the target of 11 SNPs and no nearby FPs had a specific SNP for the same panel member. In the Chr9-FP example all highly specific SNPs were ignored, as there were sufficient markers of higher heterozygosity available. The selected SNPs show a homogeneous representation of the diploid panel members.

Illumina re-sequencing. Genotyping data were exported from GenomeStudio genotyping report with the option "Forward Strand" for consistency with the re-sequencing strand. Regarding re-sequencing data, only calls with a support of at least 10 reads were extracted from the .vcf file and used for the comparison.

Results

SNP Detection and Filtering

Sequencing resulted in 3,376 million reads from the 14 accessions (13 *M.* × *domestica* apple cultivars and one accession of *M. micromalus*), and an additional 809 million reads from the two DH accessions (http://bioinformatics.tecnoparco.org/fruitbreedomics/node/2). Table 1 details the depth of sequencing and the estimated genome coverage obtained for each of the 14 genotypes of the discovery panel. Variant calling resulted in a variant call format (.vcf) file containing 16,614,171 variants including indels, derived from all the re-sequenced accessions (Figure 1). A total of 662,559 variants were found to be heterozygous in one or both of the two DH accessions re-sequenced and were removed from further analysis. Thus 15,951,612 variants were retained after this filtering step. Following quality filtering described in the paragraph "Read Alignment, Variant Detection and Quality Filtering" of the Methods section, 13,373,299 variants were removed from further analysis, leaving 2,578,313 variants.

SNP Selection

When quality filtered variants were processed to meet the Illumina Infinium II array requirements, a further 1,719,293 variants were removed, leaving 859,020 SNPs that were submitted to the Illumina Assay Design Tool pipeline. Following the Illumina recommendations and the parameters used by [3], 108,257 SNPs with SNP_Score below 0.7 were discarded leaving 750,763 high quality SNPs from the discovery panel for the selection of targets to be included on the array.

Using the focal point approach detailed in the methods, 16,330 SNPs were identified from the re-sequencing of 14 genotypes of the discovery panel, as well as 3,670 validated markers from a previous IRSC array [3] and submitted to Illumina for array production. The Infinium array manufacturing produced a total of 18,019 SNP probes (14,714 newly identified SNPs and 3,305 from the IRSC array) to be incorporated in the final array. A total of 15,669 SNPs were located in 2,019 focal points (Figure 3). This included 955 SNPs from the IRSC array, which meant that some FPs contained more than 11 SNPs. The remaining 2,350 SNPs from the IRSC array did not fall in a FP and their physical position has been highlighted in Figure 3. The number of SNPs per FP ranged from 1 to 15, with a mean of 7.7. The mean distance between focal points was 311 Kbps, with two regions larger than 1 Mbps at the distal end of chromosomes 9 and 13. Overall, most regions that were not effectively covered by the IRSC array are now well represented by newly designed markers.

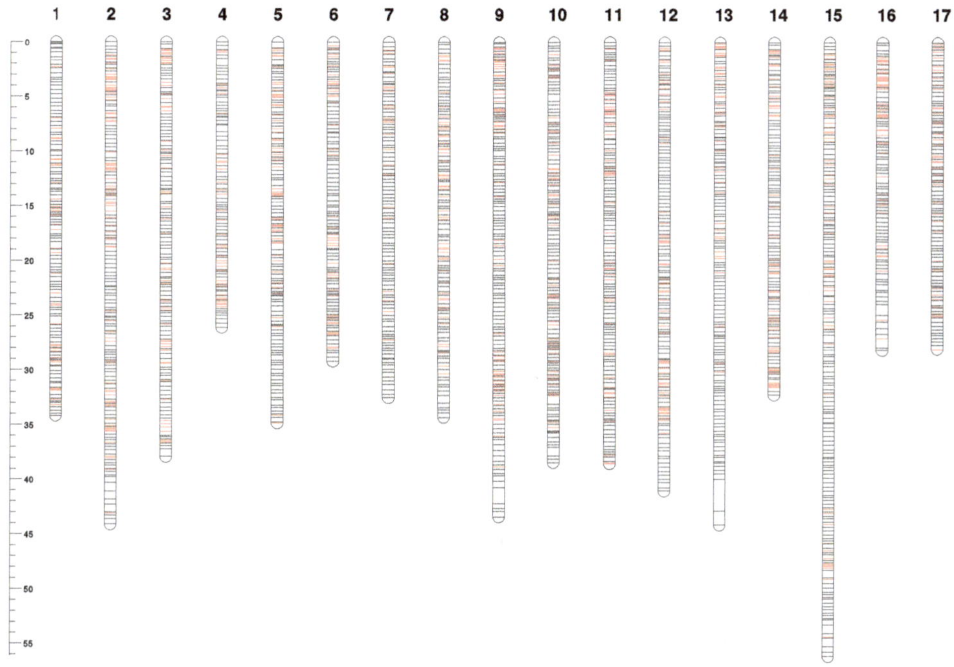

Figure 3. Distribution of focal points (FPs) (black lines) and IRSC SNPs (red lines) in the apple physical map v2. All positions are in Mbps. The average distance between FPs is 311 Kbps. Only two regions longer than 1 Mbps located at the distal end of Chrs 9 and 13 are not covered by FPs and SNPs from the 8 K IRSC SNP array. Scale bars = 5 Mbps.

SNP Mapping and Validation

Within the framework of the FruitBreedomics project, 21 full sib families were SNP genotyped, using the 20 K array, and were used for the generation of linkage maps. This resulted in the genetic mapping of 15.8 K SNP markers, and included 12,611 (success rate 86%) newly developed SNPs and 3,160 (success rate 96%) from the IRSC array that were informative in this new array and were mapped in at least one of the full-sib families screened (Table S2, Dataset S1). Of the non-mapped markers, 271 currently true monomorphic SNP markers may be mapped once screened over a wider germplasm set, as they are based on a single polymorphism in one of the more specific discovery panel members (e.g. *M. micromalus* or 'Common Antonovka'). Additionally, 747 SNPs that featured complex cluster patterns (suggesting the probes annealed to paralogous genomic regions or contained additional SNPs within the probe binding sites) were found informative by visual inspection but would have required additional elaboration on the calling pipeline or manual annotation as previously reported in [6].

The number of SNPs successfully mapped for each of the 21 full-sib families varied from ~5.2 K to ~8.5 K. Examples of a GenomeStudio cluster for a robust mapped SNP and a failed/difficult to score SNP are shown in Figure 4.

Genotype Call Comparison

The concordance of genotype calls made through re-sequencing with those obtained from genotyping with the 20 K array was assessed for the set of 15,771 validated SNPs. Excluding missing data, on average 12,347 comparisons were made for each of the 14 accessions (min: 10,402 for *M. micromalus*; max: 12,647 for 'Fuji') and the average concordance was 95%. In the majority of cases, the genotype call discordance identified was for low frequency SNPs observed in just one or two accessions of the discovery panel. This could have been the result of unreliable genotype calls in the re-sequencing analysis, owing to a lower read depth in those specific regions, or it might have been caused by unreliable genotype calls made in the array analysis owing to the presence of additional SNPs in probe binding sequences.

Discussion

The outcome of our present work, a 20 K WGG array for *M.* × *domestica* using Illumina Infinium technology, comprising a total of 18,019 SNPs, has greatly increased the genotyping and analysis opportunities for apple researchers and breeders. High throughput array-based genotyping has revolutionized the study of genome-wide genetic variation, reducing costs and increasing the reliability and efficiency of data produced, as well as significantly reducing the time spent on genotyping itself. Medium or high density arrays are available for a range of other crop plant species, including cherry [26], grapevine [27], maize [28], peach [22], potato [29], soybean [30], tomato [31], sorghum [32], white spruce [33], alfalfa [34], and rice [35]. Of our 18,019 apple SNPs, 88% were validated in 21 full-sib families, with a further reservoir of a thousand of potentially informative SNPs if a wider germplasm base had been surveyed or if the remaining markers had been called manually. An efficiency rate of 88% corresponds well with efficiencies calculated for arrays developed for other plant species such as peach (84.3%; [22]), and is significantly higher than that of the previous *Malus* IRSC array (72%) [3]. Taking into account the polyploid origin of apple, the efficiency rate of 88% is the highest reached to date for a species that has undergone a recent whole genome duplication (WGD). This increased array efficiency was achieved through the implementation of a novel SNP detection and filtering pipeline in the design of the array, together with a new SNP calling pipeline for the use of the array [23] and the screening of a relatively large number of mapping populations. The new SNP calling pipeline successfully avoided the problem of extensive genome paralogy in apple [2], which caused calling

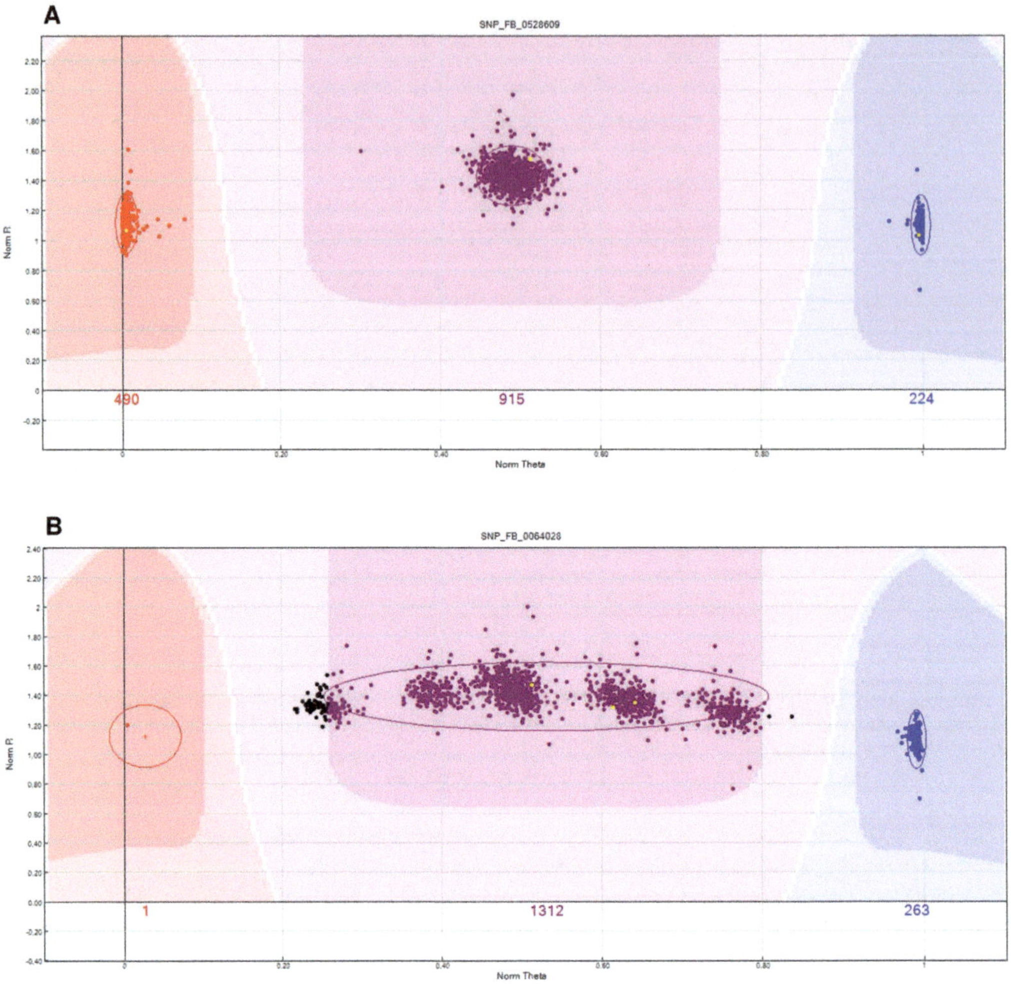

Figure 4. GenomeStudio cluster plot for the 21 F_1 full-sib families used for the construction of genetic maps. Yellow dots indicate panel members that were re-sequenced and used as parents. Genotypes are called for each sample (dot) by their signal intensity (norm R) and Allele Frequency (Norm Theta) relative to canonical cluster positions (dark shading) for a given SNP marker (red = AA, purple = AB, blue = BB) a) GenomeStudio cluster plot of a newly developed robust SNP marker. b) GenomeStudio cluster plot of a failed/difficult to score SNP marker.

difficulties using the IRSC apple array [6]. We employed a more stringent probe design by removing regions that showed multiple matches in the genome of the 24 nucleotides from either side of the SNP. Moreover, more stringent SNP filtering was achieved by including two DH accessions as discovery panel members and removing heterozygous SNPs that were identified in either of those genotypes during the SNP selection process. Finally, the high coverage re-sequencing strategy applied in this work allowed genotypes to be called for most panel members, and a minimum of ten reads per variant was imposed to successfully call a genotype. The comparison of sequenced reads for a potential variant site across multiple samples has the potential to differentiate systematic sequencing errors from real SNPs [36]. Such an approach permitted the detection of variant carriers without any need for complicated bioinformatics algorithms such as the one reported by [37], and reduced the number of false positive SNPs included in the array design. Comparison of read alignments across multiple samples also has the potential to filter out SNPs that are an artifact of inaccurate read alignments [36]. Moreover, it allowed the selection of sets of both more widely and more narrowly polymorphic markers that together represented the diversity of the discovery panel. This combination is expected to be useful for the tracing of markers (SNP haploblocks) along pedigrees for several successive generations. We did not include many SNPs of low heterozygosity (MAF) as they would have low probability of heterozygosity, and thus a low probability of being informative. As a result of these measures, the number of true monomorphic SNPs contained on the final array was very low, approximately 1% of the total. The use of this array to screen the FruitBreedomics families (Table S1) demonstrated the robust performance of the array, as the maps constructed had an average of 6.8 K SNPs uniformly distributed along the genome of each parent.

The final set of 2,578,313 high quality SNPs detected using the 14 re-sequenced accessions of the discovery panel corresponds to an average number of 4.8 SNPs/1,000 bp, in the 530 Mbps assembled and anchored 'Golden Delicious' reference genome sequence. This value was similar to the value reported in the apple genome paper [2] (4.4 SNPs/1,000 bp) but is somewhat higher than that found when surveying the polymorphism rate within a set of *M.* × *domestica* cultivars [38] (3.8 SNPs/1,000 bp), where analyses were limited to genic regions. Since, in contrast to what was done for the IRSC apple array [3], SNPs in this study were not selected exclusively from coding regions of the 'Golden Delicious' reference genome, the SNP heterozygosity rate reported here probably more closely represents the real genome-wide heterozygosity rate of *M.* × *domestica*. However, the actual SNP density may still be higher, as some of the filtered-out SNPs may actually be true polymorphisms.

Other methods have been developed recently for high-throughput genotyping of eukaryotic genomes using short-read sequencing technologies [39–41], and these techniques have been employed in the development of linkage maps and the identification of markers linked to agronomic traits in plants. Such studies include, for instance, genotyping by sequencing (GBS) based genetic maps of *Rubus idaeus* [42], *Hordeum vulgare* [43,44], and restriction-site associated DNA (RAD) sequencing in *Lolium perenne* [41], *Hordeum vulgare* [45] and *Lupinus angustifolius* [46]. These techniques employ reduced genome representation achieved through restriction enzyme digestion and subsequent PCR analysis from adapted linker sequences, and require no *a priori* knowledge of the SNPs being interrogated, making them useful for genetic analysis in species where no reference sequence is currently available. In addition, since up to 96 samples can be multiplexed in a single lane of Illumina HiSeq sequencing, the genotyping cost per sample can be as low as USD12.00 at the time of writing (http://www.igd.cornell.edu/index.cfm/page/GBS/GBSpricing.htm). However, despite clear advantages to the use of GBS under certain experimental conditions for under-resourced species, GBS datasets contain a large proportion of missing values and false homozygote calls, due to low, and uneven genome coverage among individuals (36% in the case of *Rubus*; [42]), and hence data imputation strategies are required for effective data analysis. Moreover sampling chromosomes based on restriction digestion may introduce a bias in allele frequency estimation due to polymorphisms in restriction sites [47]. Whilst the cost of consumables for genotyping per individual is higher employing WGG arrays than for GBS, and only previously characterized SNPs that are present on the array can be interrogated, the data produced are robust and reliable, typically containing almost no missing values. Moreover when a robust and efficient SNP calling pipeline is available, there is less need for bioinformatic capacity, reducing the time and cost associated with data analysis. Coupled with the cost-effective production of arrays such as the one described in this investigation, containing nearly 16,000 validated SNPs, genetic analysis using WGG arrays provides tangible advantages over GBS, particularly for PBA, where genotyping is employed in related germplasm, and variation in hybridization efficiency and indel related null-alleles can be accounted for.

Concluding remarks

We have developed a high-throughput WGG array for apple containing over 16 K validated SNP markers, spanning the apple genome at over 2,000 focal points evenly distributed throughout the 17 chromosomes of *M.* × *domestica*. Through the use of a novel SNP selection strategy informed through the design of the IRSC WGG array [3] and validation of data generated with that array [5,6,10], we have increased the robustness of the markers contained on the array, and reduced the occurrence of monomorphic SNPs and those that display cluster patterns indicating binding to paralogous loci or binding sites containing additional SNPs. The newly developed array, with its high density of polymorphic, validated SNPs, and its suitability for building multi-allelic SNP haploblocks, is expected to be of great utility for pedigree-based analysis, genomic selection and population genetics studies in *M.* × *domestica*. Moreover, the approach in SNP filtering and array design may be of use for the development of much higher density arrays for genome-wide association studies.

Supporting Information

Table S1 Full-sib families screened with the 20 K SNP array. Parents and number of seedlings are listed. Pedigrees of the X-numbered accessions are reported by [11].

Table S2 List of the 18,019 SNPs included in the 20 K SNP array. For each SNP, the physical location (Chromosome, Position), minor allele frequency (MAF), validation in 21 full-sib families, SNP genotype detected by the Illumina array, Illumina strand, source strand, source sequence, probe sequence, genotype calls (gt) made through re-sequencing and no. of high-quality bases (dp) for each discovery panel member, are reported.

Acknowledgments

We wish to thank: INRA-Angers for providing us with the DNA of the two double haploids, Yolanda Noordijk, Andreas Peil (JKI), Anna Pikunova (VNIIISPK), Gennaro Fazio (GPRU), Mark Davey (KUL), and Stefano Tartarini (UNIBO) for providing leaf material for the 14 genotypes of the discovery panel, Yolanda Noordijk (DLO) for preparing the DNA samples, Diego Micheletti, Paolo Fontana, Pietro Franceschi and Ross Crowhurst for the useful discussions on data analysis, Sue Gardiner for manuscript editing, and the International Doctoral Programme in the Genomics and Molecular Physiology of Fruit Plants (GMPF) and the Graduate School of Experimental Plant Sciences of Wageningen University (The Netherlands) for a fellowship awarded to Mario di Guardo.

Author Contributions

Conceived and designed the experiments: R. Velasco SS MT LB EVDW FL. Performed the experiments: SD IG EB. Analyzed the data: AC LB MT EVDW MDG DJS JJ. Contributed reagents/materials/analysis tools: IG SD. Contributed to the writing of the manuscript: MT LB DJS EVDW DC R. Viola.

References

1. Shulaev V, Korban SS, Sosinski B, Abbott AG, Aldwinckle HS, et al. (2008) Multiple models for Rosaceae genomics. Plant Physiol 47(3): 985–1003.
2. Velasco R, Zharkikh A, Affourtit J, Dhingra A, Cestaro A, et al. (2010) The genome of the domesticated apple (*Malus x domestica* Borkh.). Nat Genet 42(10): 833–839.
3. Chagne D, Crowhurst RN, Troggio M, Davey MW, Gilmore B, et al. (2012) Genome-Wide SNP Detection, Validation, and Development of an 8K SNP Array for Apple. PLoS One 7(2): e31745.
4. Montanari S, Saeed M, Knaebel M, Kim Y, Troggio M, et al (2013) Identification of *Pyrus* Single Nucleotide Polymorphisms (SNPs) and Evaluation for Genetic Mapping in European Pear and Interspecific Pyrus Hybrids. PLoS One 8(10): e77022.
5. Antanaviciute L, Fernandez-Fernandez F, Jansen J, Banchi E, Evans KM, et al. (2012): Development of a dense SNP-based linkage map of an apple rootstock progeny using the *Malus* Infinium whole genome genotyping array. BMC Genomics 13: 203.
6. Troggio M, Surbanovski N, Bianco L, Moretto M, Giongo L, et al. (2013) Evaluation of SNP Data from the Malus Infinium Array Identifies Challenges for Genetic Analysis of Complex Genomes of Polyploid Origin. PLoS One 8(6): e67407.
7. Clark MD, Schmitz CA, Rosyara UR, Luby JJ, Bradeen JM (2014) A consensus'Honeycrisp' apple (*Malus* × *domestica*) genetic linkage map from three full-sib progeny populations. Tree Genet Genomes DOI 10.1007/s11295-014-0709-1.

8. Kumar S, Garrick DJ, Bink MCAM, Whitworth C, Chagne D, Volz RK (2013) Novel genomic approaches unravel genetic architecture of complex traits in apple. BMC Genomics 14: 393.
9. Kumar S, Chagne D, Bink MCAM, Volz RK, Whitworth C, et al. (2012) Genomic Selection for Fruit Quality Traits in Apple (*Malus x domestica* Borkh.). PLoS One 7(5): e36674.
10. Pikunova A, Madduri M, Sedov E, Noordijk Y, Peil A, et al. (2014) 'Schmidt's Antonovka' is identical to 'Common Antonovka', an apple cultivar widely used in Russia in the breeding for biotic and abiotic stresses. Tree Genet Genomes 10: 261–271.
11. Bink M, J J, Madduri M, Voorrips R, Durel C, Kouassi A, et al. (2014) Bayesian QTL analyses using pedigreed families of an outcrossing species, with application to fruit firmness in apple. Theor Appl Genet 127: 1073–1090.
12. Khan MA, Korban SS (2012) Association mapping in forest trees and fruit crops. J Exp Bot 63(11): 4045–4060.
13. Noiton DAM, Alspach PA (1996) Founding clones, inbreeding, coancestry, and status number of modern apple cultivars. J Am Socr Hortic Sci 121(5): 773–782.
14. Durel CE, Laurens F, Fouillet A, Lespinasse Y (1998) Utilization of pedigree information to estimate genetic parameters from large unbalanced data sets in apple. Theor Appl Genet 96(8): 1077–1085.
15. Salvi S, Micheletti D, Magnago P, Fontanari M, Viola R, et al. (2014) One-step reconstruction of multigeneration pedigree networks in apple (*Malus* × *domestica* Borkh.) and the parentage of Golden Delicious. Mol Breed DOI 10.1007/s11032-014-0054-y.
16. Lespinasse Y, Bouvier L, Djulbic M, Chevreau E (1999) Haploidy in apple and pear. Acta Horticult 484: 49–54.
17. Evans KM, Patocchi A, Rezzonico F, Mathis F, Durel CE, et al. (2011) Genotyping of pedigreed apple breeding material with a genome-covering set of SSRs: trueness-to-type of cultivars and their parentages. Mol Breed 28(4): 535–547.
18. Marco-Sola S, Sammeth M, Guigo R, Ribeca P, (2012) The GEM mapper: fast, accurate and versatile alignment by filtration. Nat Meth 9(12): 1185–1188.
19. Homer N, Merriman B, Nelson SF (2009) BFAST: An Alignment Tool for Large Scale Genome Resequencing. PLoS One 4(11): e7767.
20. Li H, Handsaker B, Wysoker A, Fennell T, Ruan J, et al. (2009) Genome Project Data P: The Sequence Alignment/Map format and SAMtools. Bioinformatics 25(16): 2078–2079.
21. Derrien T, Estelle J, Marco Sola S, Knowles DG, Raineri E, et al. (2012). Fast Computation and Applications of Genome Mappability. PLoS One 7(1): e30377.
22. Verde I, Bassil N, Scalabrin S, Gilmore B, Lawley CT, et al. (2012) Development and Evaluation of a 9K SNP Array for Peach by Internationally Coordinated SNP Detection and Validation in Breeding Germplasm. PLoS One 7(4): e35668.
23. Van de Weg W, Di Guardo M, Koehorst-Putten H, Longhi S, Noordijk Y, et al. (2014) A Pipeline for Robust Marker Calling from Infinium SNP Arrays for Diploid Crops. In: International Plant & Animal Genome XXII, San Diego, California, U.S.A.
24. Van Ooijen J (2006) JoinMap® 4, Software for the calculation of genetic linkage maps in experimental populations. In. Wageningen, Netherlands: Kyazma B.V.
25. Van Ooijen JW (2011) Multipoint maximum likelihood mapping in a full-sib family of an outbreeding species. Genet Res 93(5): 343–349.
26. Peace C, Bassil N, Main D, Ficklin S, Rosyara UR, et al. (2012) Development and Evaluation of a Genome-Wide 6K SNP Array for Diploid Sweet Cherry and Tetraploid Sour Cherry. PLoS One 7(12): e48305.
27. Myles S, Chia J-M, Hurwitz B, Simon C, Zhong GY, et al. (2010) Rapid Genomic Characterization of the Genus *Vitis*. PLoS One 5(1): e8219.
28. Ganal MW, Durstewitz G, Polley A, Berard A, Buckler ES, et al. (2011) A Large Maize (Zea mays L.) SNP Genotyping Array: Development and Germplasm Genotyping, and Genetic Mapping to Compare with the B73 Reference Genome. PLoS One 6(12): e28334.
29. Felcher KJ, Coombs JJ, Massa AN, Hansey CN, Hamilton JP, et al. (2012) Integration of Two Diploid Potato Linkage Maps with the Potato Genome Sequence. PLoS One 7(4): e36347.
30. Song Q, Hyten DL, Jia G, Quigley CV, Fickus EW, et al. (2013) Development and Evaluation of SoySNP50K, a High-Density Genotyping Array for Soybean. PLoS One 8(1): e54985.
31. Sim S-C, Durstewitz G, Plieske J, Wieseke R, Ganal MW, et al. (2012) Development of a Large SNP Genotyping Array and Generation of High-Density Genetic Maps in Tomato. PLoS One 7(7): e40563.
32. Bekele WA, Wieckhorst S, Friedt W, Snowdon RJ (2013) High-throughput genomics in sorghum: from whole-genome resequencing to a SNP screening array. Plant Biotech J 11(9): 1112–1125.
33. Pavy N, Gagnon F, Rigault P, Blais S, Deschenes A, et al. (2013) Development of high-density SNP genotyping arrays for white spruce (*Picea glauca*) and transferability to subtropical and nordic congeners. Mol Ecol Resour 13(2): 324–336.
34. Li X, Han Y, Wei Y, Acharya A, Farmer AD, et al. (2014) Development of an Alfalfa SNP Array and Its Use to Evaluate Patterns of Population Structure and Linkage Disequilibrium. PLoS One 9(1): e84329.
35. Chen H, Xie W, He H, Yu H, Chen W, et al. (2014) A high-density SNP genotyping array for rice biology and molecular breeding. Mol Plant 7(3): 541–553.
36. Bansal V, Harismendy O, Tewhey R, Murray SS, Schork NJ, et al. (2010) Accurate detection and genotyping of SNPs utilizing population sequencing data. Genome Res 20(4): 537–545.
37. Wang Z, Cai H, Bai X, Ji W, Li Y, et al. (2012) Isolation of GsGST19 from Glycine soja and analysis of saline-alkaline tolerance for transgenic *Medicago sativa*. Acta Agron Sin 38(6): 971–979.
38. Micheletti D, Troggio M, Zharkikh A, Costa F, Malnoy M, et al. (2011) Genetic diversity of the genus *Malus* and implications for linkage mapping with SNPs. Tree Genet Genomes 7(4): 857–868.
39. Baird NA, Etter PD, Atwood TS, Currey MC, Shiver AL, et al. (2008) Rapid SNP Discovery and Genetic Mapping Using Sequenced RAD Markers. PLoS One 3(10): e3376.
40. Elshire RJ, Glaubitz JC, Sun Q, Poland JA, Kawamoto K, et al. (2011) A Robust, Simple Genotyping-by-Sequencing (GBS) Approach for High Diversity Species. PLoS One 6(5): e19379.
41. Pfender WF, Saha MC, Johnson EA, Slabaugh MB (2011) Mapping with RAD (restriction-site associated DNA) markers to rapidly identify QTL for stem rust resistance in *Lolium perenne*. Theor Appl Genet 122(8): 1467–1480.
42. Ward JA, Bhangoo J, Fernández-Fernández F, Moore P, Swanson JD, et al. (2013) Saturated linkage map construction in *Rubus idaeus* using genotyping by sequencing and genome-independent imputation. BMC Genomics 14: 2.
43. Liu H, Bayer M, Druka A, Russell JR, Hackett CA, et al. (2014) An evaluation of genotyping by sequencing (GBS) to map the Breviaristatum-e (ari-e) locus in cultivated barley. BMC Genomics 15: 104.
44. Poland JA, Brown PJ, Sorrells ME, Jannink J-L (2012) Development of High-Density Genetic Maps for Barley and Wheat Using a Novel Two-Enzyme Genotyping-by-Sequencing Approach. PLoS One 7(2): e32253.
45. Chutimanitsakun Y, Nipper RW, Cuesta-Marcos A, Cistue L, Corey A, et al. (2011) Construction and application for QTL analysis of a Restriction Site Associated DNA (RAD) linkage map in barley. BMC Genomics 12: 4.
46. Yang H, Tao Y, Zheng Z, Zhang Q, Zhou G, et al. (2013) Draft Genome Sequence, and a Sequence-Defined Genetic Linkage Map of the Legume Crop Species *Lupinus angustifolius* L. PLoS One 8(5): e64799.
47. Arnold B, Corbett-Detig RB, Hartl D, Bomblies K (2013) RADseq underestimates diversity and introduces genealogical biases due to nonrandom haplotype sampling. Mol Ecol 22(11): 3179–3190.

Genetic Analysis and QTL Detection for Resistance to White Tip Disease in Rice

Tong Zhou[1], Cunyi Gao[1,2], Linlin Du[1], Hui Feng[1], Lijiao Wang[1], Ying Lan[1], Feng Sun[1], Lihui Wei[1], Yongjian Fan[1], Wenbiao Shen[2]*, Yijun Zhou[1]*

1 Institute of Plant Protection, Jiangsu Academy of Agricultural Sciences, Nanjing, China, **2** College of Life Science, Nanjing Agricultural University, Nanjing, China

Abstract

The inheritance of resistance to white tip disease (WTDR) in rice (*Oryza sativa* L.) was analyzed with an artificial inoculation test in a segregating population derived from the cross between Tetep, a highly resistant variety that was identified in a previous study, and a susceptible cultivar. Three resistance-associated traits, including the number of *Aphelenchoides besseyi* (*A. besseyi*) individuals in 100 grains (NA), the loss rate of panicle weight (LRPW) and the loss rate of the total grains per panicle (LRGPP) were analyzed for the detection of the quantitative trait locus (QTL) in the population after construction of a genetic map. Six QTLs distributed on chromosomes 3, 5 and 9 were mapped. *qNA3* and *qNA9*, conferring reproduction number of *A. besseyi* in the panicle, accounted for 16.91% and 12.54% of the total phenotypic variance, respectively. *qDRPW5a* and *qDRPW5b*, associated with yield loss, were located at two adjacent marker intervals on chromosome 5 and explained 14.15% and 14.59% of the total phenotypic variation and possessed LOD values of 3.40 and 3.39, respectively. *qDRPW9* was considered as a minor QTL and only explained 1.02% of the phenotypic variation. *qLRGPP5* contributed to the loss in the number of grains and explained 10.91% of the phenotypic variation. This study provides useful information for the breeding of resistant cultivars against white tip disease in rice.

Editor: Tai Wang, Institute of Botany, Chinese Academy of Sciences, China

Funding: This research was financially supported by grants from the National Key Basic Research and Development Program (973 Program) of China (2013CBA01403), the Special Fund for Agro-Scientific Research in the Public Interest of China (201303021, 201003031), the Jiangsu Province Science and Technology Support Project (BE2012303) and the Jiangsu Agricultural Science and Technology Independent Innovation Fund (cx[12]1003). The funders had no role in study design, data collection and analysis, decision to publish, or preparation of the manuscript.

Competing Interests: The authors have declared that no competing interests exist.

* Email: yjzhou@jaas.ac.cn (YZ); wbshenh@njau.edu.cn (WS)

Introduction

White tip disease of rice (WTDR), which is caused by the rice white tip nematode (*Aphelenchoides besseyi*, *A. besseyi*), is one of the most serious nematode diseases affecting rice worldwide [1–5]. Plants infected with *A. besseyi* exhibit whitening and withering at the tip of the leaf, and the symptoms also include small grains and erect panicles in later growth stages, which can result in large losses in production. The traditional methods used to control *A. besseyi* (including insecticide application and crop rotation) are expensive and cause serious environmental problems [4]. In addition, the wide-scale use of direct-sowing technology for rice, which is not compatible with the treatment of seeds, has recently made this problem more acute in China. Thus, the development of resistant rice cultivars has been considered as the primary strategy for controlling white tip disease [6,7].

Rice varieties that are resistant against WTDR have been reported by researchers, e.g., *cvs Arkansas Fortuna*, *Nira 43*, *Asa-Hi*, *Binam* and *Domsiah*, which were screened from a large number of rice varieties [4,8,9]. However, the genetic mechanism of WTDR resistance is still poorly understood. In addition, some of the resistant varieties exhibited resistance to WTDR only in particular regions or were highly susceptible to other pathogens [10], which hindered projects aimed at breeding for resistance against WTDR. In addition, many nematode resistance loci, such as *H1*, *GroV1*, *Cre* and *Mi3*, have been identified in tomato, potato, soybean and other crops [11–14], and some have even been cloned and functionally analyzed [15–18]. In addition, a gene (*Has-1*^{Og}) resistant to the cyst nematode (*Heterodera sacchari*) has been identified in rice [19]. These results suggested that host plants, including rice, harbored defense mechanisms to fight against nematodes, and these mechanisms have developed over long-term co-evolution. The lack of information on the inheritance of the resistance to WTDR may slow the progress of breeding programs. Therefore, elucidating the resistance mechanisms involved will contribute to a better understanding of nematode-plant interactions and assist with the breeding of nematode-resistant cultivars [20].

In our previous study [21], a collection of germplasm resources was screened using an inoculation test, and an *indica* variety, Tetep, showed high resistance to WTDR. Herein, we present the inheritance mode of resistance to WTDR in rice and the quantitative trait loci (QTLs) related to resistance against WTDR.

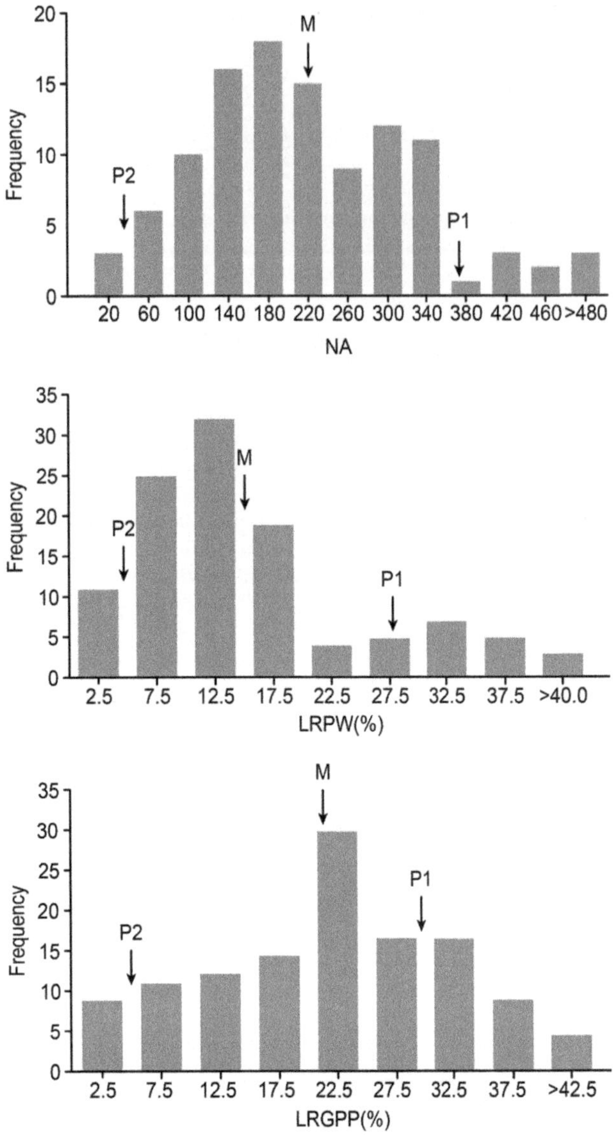

Figure 1. The frequency distributions of three resistance-associated traits in the $F_{2:3}$ lines. P_1 represents Huaidao No.5, P_2 represents Tetep, and M represents the mean value of the $F_{2:3}$ lines.

Materials and Methods

Plant materials

In 2008, Huaidao No.5, a *japonica* cultivar highly susceptible to WTDR, and Tetep (an *indica* variety) were grown at the experimental station in Jiangsu Academy of Agricultural Sciences. In the same year, F_1 was developed from a cross between Huaidao No.5 and Tetep. In 2009, the F_1 was grown and self-pollinated at the experimental station to generate F_2 lines that were used as mapping populations in Nanjing, Jiangsu Province. In 2009–2010, the F_2 lines were grown at the experimental station in Lingshui, Hainan Province, and the leaves of the F_2 lines were collected, numbered and stored at −70°C. A total of 138 F_2 individuals were selected and self-pollinated to generate 138 $F_{2:3}$ families. The resistance of the two parents, the F_1 generation and the $F_{2:3}$ lines were evaluated at the experimental station in Jiangsu Academy of Agricultural Sciences in 2012. Seeds were pretreated by soaking them in water at 55°C for 15 minutes to ensure there were no live *A. besseyi* before the seeds were sown [4].

Nematode preparation

The seed-borne ectoparasitic *A. besseyi* was initially isolated from infected rice seeds using the Baermann funnel technique and surface-sterilized with 3% H_2O_2 for 10 minutes. After isolation, *A. besseyi* was cultured on *Botrytis cinerea*, which were grown on potato dextrose agar (PDA) medium for approximately 3 weeks at 25°C [22]. Then, the nematodes were rinsed thoroughly with distilled water and used as inoculation material [23].

Evaluation of resistance

40-μl nematode suspension with 400 juveniles of *A. besseyi* was inoculated between the leaf sheath and the culm with a pipette at the top tillering stage. To keep the micro-environment moist so that the nematodes were able to move and feed easily, a small wad of absorbent cotton was placed in the infection spot [7,21,24]. The mature seeds of each inoculated panicle were harvested. The number of grains per panicle and the weight of the panicle were determined. The number of nematodes in 100 grains from each inoculated panicle was counted using the following protocol. The peeled seed and the shell were soaked in distilled water for 24 hours, after which the free *A. besseyi* were collected from the solution using the Baermann funnel technique, and a microscope was used to count the number of *A. besseyi* [21]. At least 2 panicles of 25 plants from each $F_{2:3}$ line were used for the inoculation. Distilled water without *A. besseyi* was used as the control.

The number of *A. besseyi* individuals in 100 grains (NA), the loss rate of panicle weight (LRPW) and the loss rate of the total grains per panicle (LRGPP) were calculated using the following formulas: NA = (the total number of *A. besseyi* in the counted grains/the total number of counted grains) × 100, LRPW = (the panicle weight of the control–the panicle weight of the inoculated plant)/the panicle weight of the control and LRGPP = (the total number of grains per panicle in the control–the total number of grains per panicle in the inoculated plant)/the total number of grains per panicle in the control.

Genotyping, linkage map construction and QTL analysis

The DNA of the two parents, the F_1 generation and the 138 F_2 lines was extracted using the CTAB method described by Rogers and Bendich [25]. SSR markers (842 pairs), obtained from Gramene (http://www.gramene.org), and polymorphism markers for the two parents were tested. PCR amplification was performed as described by Septiningsih et al. [26], and PCR products were detected using 8% denaturing polyacrylamide gel electrophoresis. The polymorphic markers were used for genotype analysis of the F_2 lines to assemble the linkage map. The genetic linkage map was constructed by employing MAPMAKER/EXP 3.0 software [27,28]. Marker distances in centimorgans (cM) were calculated using the Kosambi function.

Composite interval mapping (CIM) was applied to analyze the phenotypic and genotypic data for detecting the QTLs responsible for resistance to *A. besseyi* by using Windows QTL Cartographer 2.5 software [29]. Experiment-wide significance ($P<0.05$) threshold values of the LOD scores for putative QTL detection were determined with 1000 permutations. The threshold value of the LOD was 2.5 at a significance level of $P = 0.05$ for CIM. The additive effect and the explanation of the phenotypic variance for each QTL were also acquired using this software.

Table 1. Three performance traits of the two parents and the $F_{2:3}$ lines.

Traits	Huaidao No.5	Tetep	$\|P_1-P_2\|$	$F_{2:3}$ lines	
				Mean±SD	Range
NA	472±80	51±19	421	222±110	24–514
LRPW	0.28±0.19	0.04±0.01	0.24	0.15±0.10	0.01–0.44
LRGPP	0.29±0.09	0.05±0.01	0.24	0.22±0.13	0.00–0.64

P_1 and P_2 represent Huaidao No.5 and Tetep, respectively.

Results

Phenotypic variation

A dramatic difference in NA, LRPW and LRGPP was observed between Huaidao No.5 and Tetep (Table 1). The values for NA, LRPW and LRGPP of Huaidao No.5 were 472, 0.28 and 0.29, respectively. In comparison, the values for Tetep were much lower, at 51, 0.04 and 0.05, respectively. These data supported the hypothesis that Huaidao No.5 was highly susceptible to WTDR, whereas Tetep was highly resistant. The values for NA, LRPW and LRGPP of the $F_{2:3}$ lines ranged from 24 to 514, from 0.01 to 0.44 and from 0 to 0.64, respectively. Each of the frequency distributions for NA, LRPW and LRGPP showed a continuous distribution, which indicated that these traits behaved as quantitative variables (Figure 1).

Correlation coefficients among the three resistance-associated traits

The correlation analysis among the three resistance-associated traits was performed using SPSS ver. 20.0 software (Table 2). The results showed that NA was not significantly correlated with LRPW or LRGPP. However, a significant positive correlation was observed between LRPW and LRGPP at the $P<0.01$ level.

Genetic linkage map

A total of 160 polymorphic SSR markers were found between the two parents of 842 total markers, for a ratio of 19.01%. A linkage map was constructed, which included 12 linkage groups and spanned a total of 2179.6 cM in genetic distance with an average of 17.16 cM among 127 SSR polymorphic markers. Because these SSR markers were evenly distributed on 12 chromosomes, the linkage map was suitable for QTL detection.

QTL analysis

QTL identification using CIM indicated that a total of six QTLs for resistance to WTDR were found on chromosomes 3, 5 and 9 (Table 3, Figure 2). Two important QTLs (*qNA3* and *qNA9*) responsible for the reproduction numbers of *A. besseyi* in the panicles were detected at the marker intervals of RM5626–RM7097 and RM5526–RM3912 on chromosomes 3 and 9, respectively. The LOD values of *qNA3* and *qNA9* were 3.04 and 2.62, which accounted for 16.91% and 12.54% of the total phenotypic variation, respectively. Three QTLs, i.e., *qLRPW5a*, *qLRPW5b* and *qLRPW9*, which conferred yield loss, were mapped at the marker intervals of RM163–RM18620, RM440–RM161 and RM5526–RM3912, respectively, on chromosomes 5 and 9 (Table 3). *qLRPW5a* and *qLRPW5b*, which were located at two adjacent marker intervals on chromosome 5, explained 14.15% and 14.59% of the phenotypic variation and possessed LOD values 3.40 and 3.39, respectively. *qLRPW9* was considered as a minor QTL because it only explained 1.02% of the phenotypic variation and had an LOD value of 2.55. We also identified a QTL (*qLRGPP5*) responsible for the loss in the number of the grains at the marker interval RM18632–RM163 on chromosome 5 (Table 3), which explained 10.91% of the phenotypic variation. Among these QTLs, four (*qNA3*, *qLRPW5a*, *qLRPW5b* and *qLRGPP5*) were derived from Tetep, and the others (*qNA9* and *qLRPW9*) originated from Huaidao No.5.

Discussion

It has been reported that not all the susceptible rice varieties exhibited symptoms of both white leaf tips and small grains after infection with *A. besseyi*, although plants without symptoms still showed large losses in production [21]. Therefore, the relative yield loss was preferentially used in this study to evaluate the resistance level of plants against WTDR, which was described by the LRPW. The final number of nematodes in the mature grains was considered as an important assessment index by most researchers because of the potential transmission of the nematodes by seeds and because nematodes can contribute to the loss of the grain, which directly reduces the harvest. The number of *A. besseyi* in 100 grains was used as another assessment index based on a previous study [21]. We also used the LRGPP to measure the loss in the number of grains, which caused the small grain symptoms. Another symptom, i.e., white leaf tips, was barely seen in our study, which was reported by Feng [21] and Fortuner [30], and

Table 2. Correlation coefficients among the three resistance-associated traits.

Traits	NA	LRPW	LRGPP
NA	1		
LRPW	0.022	1	
LRGPP	−0.059	0.424**	1

**Significant at the 0.01 level.

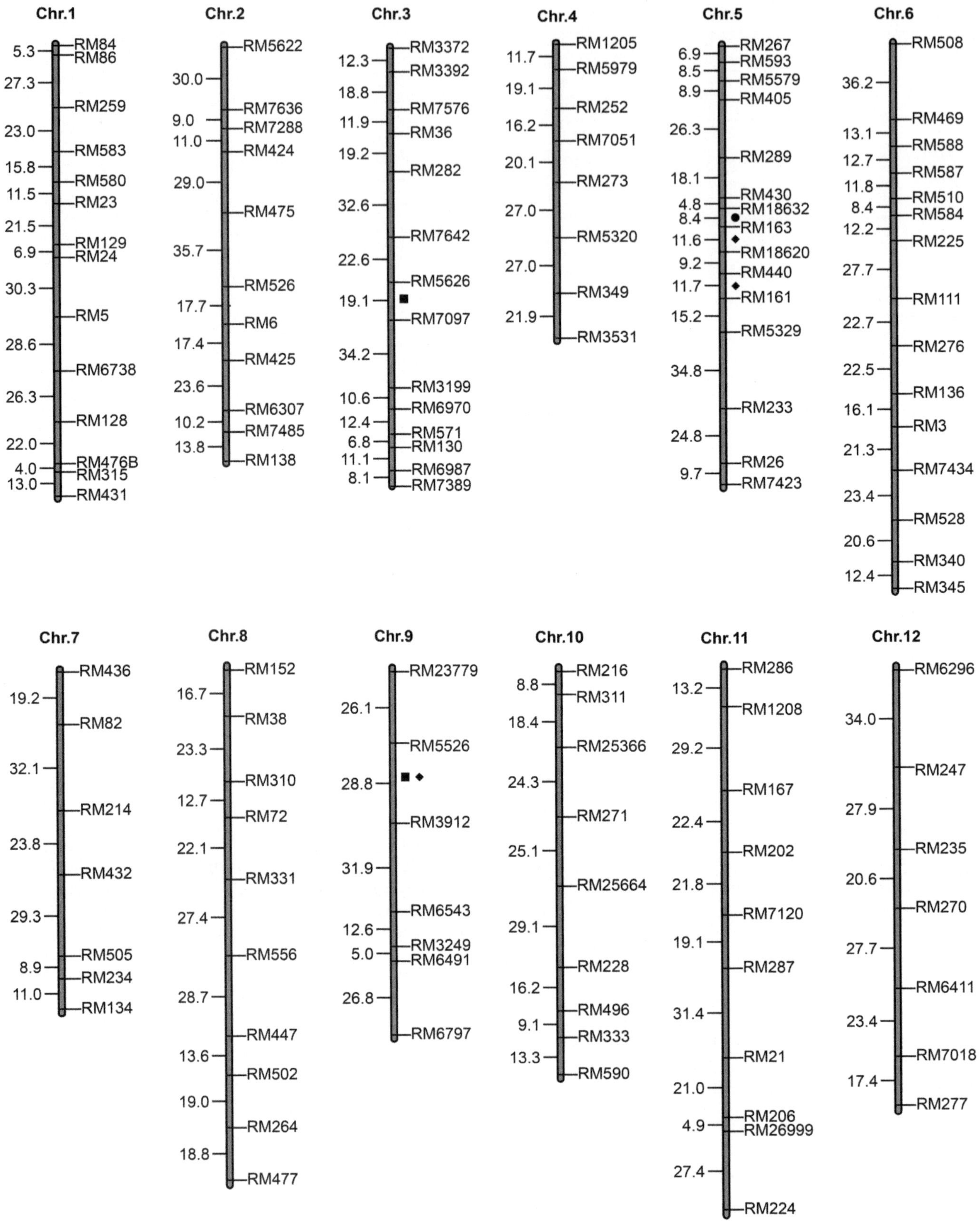

Figure 2. The SSR linkage map and chromosomal distributions of the putative loci responsible for resistance to WTDR. ■, ◆ and ● indicate the resistance QTLs detected using the resistance-associated traits of NA, LRPW and LRGPP, respectively.

Table 3. QTLs for resistance to WTDR detected in the F_2 lines developed from crossing Huaidao No.5 and Tetep.

Traits	QTL	Chr.	Interval	R^2 (%)	A (%)	LOD
NA	*qNA3*	3	RM5626-RM7097	16.91	71.18	3.04
	qNA9	9	RM5526-RM3912	12.54	−61.08	2.62
LRPW	*qLRPW5a*	5	RM163-RM18620	14.15	5.43	3.40
	qLRPW5b	5	RM440-RM161	14.59	5.51	3.39
	qLRPW9	9	RM5526-RM3912	1.02	−1.5	2.55
LRGPP	*qLRGPP5*	5	RM18632-RM163	10.91	1.94	2.83

QTL represents the putative QTL at an LOD level ≥2.5.
Chr. represents Chromosome.
R^2 represents the phenotypic variance of each QTL.
A is an additive effect.

was not investigated. Correlation analysis showed that there was no relationship between NA and LRPW or LRGPP. This finding suggested that the reproduction level of *A. besseyi* in the grains was not entirely responsible for the loss in production and that there may be unknown factors involved in the resistance process. Roberts [31] believed there was tolerance factor involved in reducing the yield loss caused by WTDR. In contrast, LRPW was significantly correlated with LRGPP, which indicated that the decrease in the number of grains in the mature panicles infected by *A. besseyi* caused a direct loss in yield. The mapping results confirmed this hypothesis, i.e., one of the loci for LRPW, *qLRPW5a*, was close to *qLRGPP5*, a locus for LRGPP. *qLRPW9* and *qNA9* were close to each other, but there was no relationship between these two traits. This lack of a relationship might stem from the fact that *qLRPW9* was a minor QTL and only accounted for 1.02% the phenotypic variation.

In this study, we identified 6 QTLs related to resistance against WTDR. Three important QTLs (*qLRPW5a*, *qLRPW5b* and *qLRGPP5*) located on chromosome 5 contributed to a decrease in the yield loss and grain number loss caused by WTDR. There were several genes related to the number, yield and weight of the grain, including *gw-5*, *gw5-1*, *Gwt5a*, *tgwt5* and *qYI-5* in the region of *qLRPW5a* and *qLRGPP5*, which were very close to each other [32–36]. The region containing another QTL (*qLRPW5b*) also included yield-related genes such as *gp5*, *QGwp5*, *gwt5*, *Pdw5*, *ssp-5* and *tgwt5b* [37–41]. This pattern suggests that these genes may be involved in reducing the damage in the grains after infection by *A. besseyi*, i.e., although these genes do not contribute directly to resistance against WTDR, they are involved in yield protection and are useful for resistance breeding projects. The resistance loci responsible for different types of pathogens are thought to be clustered in a chromosomal region in many organisms [42]. We found several resistance genes (e.g., *qBB3-1* and *rbr3*) against pathogens (rice bacterial blight and rice blast) by comparative mapping using a common marker [43,44]. Additionally, the additive effect of *qNA9* and *qLRPW9* were −61.8% and −1.5%, respectively, i.e. the resistance of these two QTLs originated from Huaidao No.5. It suggested that Huaidao No.5, a susceptible variety, also possessed the resistance loci. Lorieux et al. [19] mapped a resistance gene (*Has-1*^{Og}) to the cyst nematode on rice chromosome 11 between two markers, i.e., RM206 and RM254. In our study, however, no gene was identified on chromosome 11, most likely because different resistance mechanisms occurring in rice depended on the different types of nematodes.

Recently, the *A. besseyi* infestation has caused increasing losses in rice yield in China [21,45–47]. Breeding and utilizing resistant cultivars is considered the most effective strategy for resolving this urgent problem. In this study, we identified the QTLs responsible for resistance to WTDR, which can be further improved by fine mapping to identify the tightly linked markers. Based on these findings, molecular-level breeding efforts target at WTDR can be more efficient, especially for locating both the nematode-resistance genes and the yield-protection genes.

It was reported previously that some resistance genes against nematodes had been cloned. Functional analysis has shown that these resistance genes can be divided into two types. The first type, which possessed a nucleotide binding site-leucine-rich repeat (NBS-LRR) structure, was located in the cytoplasm but lacked a signal sequence, e.g., *Mi-1*, *HeroA*, *Gpa2* and *Gro1-4* [15,17,18,20]. The second type has a transmembrane structure and a signal sequence, e.g., *Hs1*$^{pro\text{-}1}$, *Rhg1*, and *Rhg4* [16,48]. Hosts have evolved various defense pathways to protect themselves against invading nematode diseases. The vast amount of available information can help to rapidly and easily identify the set of resistance genes in this cultivar.

Acknowledgments

We gratefully acknowledge Ying Wang (Nanjing Agricultural University, China) for constructing the segregating population used in this study.

Author Contributions

Conceived and designed the experiments: TZ YZ WS. Performed the experiments: TZ CG LD L. Wang. Analyzed the data: TZ CG L. Wang LD. Contributed reagents/materials/analysis tools: YL HF FS L. Wei YF. Contributed to the writing of the manuscript: TZ CG LD YZ.

References

1. Yoshii H, Yamamoto S (1950) A rice nematode disease, "Senchu Shingare Byo." Symptom and pathogenic nematode. J Fac Agric Kyushu Univ 9: 309–333.
2. Huang CS, Chiang YC (1975) The influence of temperature on the ability of *Aphelenchoides besseyi* to survive dehydration. Nematologica 21: 351–357.
3. Jamali S, Pourjam E, Alizadeh A, Alinia F (2006) Incidence and distribution of *Aphelenchoides besseyi* in rice areas in Iran. J Agric Technol 2: 337–344.
4. Jamali S, Pourjam E, Safai N (2011) Determining the relationship between population density of white tip nematode and rice yield. J Agric Sci Technol 14: 195–203.

5. Nicol JM, Turner SJ, Coyne DL, den Nijs L, Hockland S, et al. (2011) Current nematode threats to world agriculture. Genomics Mol Genet Plant-Nematode Interact 21–43.
6. Hulbert SH, Webb CA, Smith SM, Sun Q (2001) Resistance gene complexes: Evolution and utilization. Annu Rev Phytopathol 39: 285–312.
7. De Waele D (2002) Foliar nematodes: *Aphelenchoides* species. Plant Resist Parasitic Nematodes 141–151.
8. Cralley EM (1949) White tip of rice. Phytopathology 39: 5.
9. Nishizawa T (1953) Studies on the varietal resistance of rice plant to the rice nematode disease "senchu shingare byo" (VI). Bull Kyushu Agric Experiment Station 1: 339–349.
10. Bridge J, Plowright RA, Peng D (2005) Nematode parasites of rice. In: Luc M, Sikora RA, Bridge J, editors. Plant parasitic nematodes in subtropical and tropical agriculture. Wallingford, UK: CAB International. pp. 87–130.
11. Pineda O, Bonierbale MW, Plaisted RL, Brodie BB, Tanksley SD (1993) Identification of RFLP markers linked to the H1 gene conferring resistance to the potato cyst nematode *Globodera rostochiensis*. Genome 36: 152–156.
12. Jacobs JME, van Eck HJ, Horsman K, Arens PFP, Verkerk-Bakker B, et al. (1996) Mapping of resistance to the potato cyst nematode *Globodera rostochiensis* from the wild potato species *Solanum vernei*. Mol Breed 2: 51–60.
13. Williams KJ, Fisher JM, Langridge P (1994) Identification of RFLP markers linked to the cereal cyst nematode resistance gene (*Cre*) in wheat. Theor Appl Genet 89: 927–930.
14. Yaghoobi J, Kaloshian I, Wen Y, Williamson VM (1995) Mapping a new nematode resistance locus in *Lycopersicon peruvianum*. Theor Appl Genet 91: 457–464.
15. Ernst K, Kumar A, Kriseleit D, Kloos DU, Phillips MS, et al. (2002) The broad-spectrum potato cyst nematode resistance gene (*Hero*) from tomato is the only member of a large gene family of NBS-LRR genes with an unusual amino acid repeat in the LRR region. Plant J 31: 127–136.
16. Cai D, Kleine M, Kifle S, Harloff HJ, Sandal NN, et al. (1997) Positional cloning of a gene for nematode resistance in sugar beet. Science 275: 832–834.
17. Vos P, Simons G, Jesse T, Wijbrandi J, Heinen L, et al. (1998) The tomato *Mi-1* gene confers resistance to both root-knot nematodes and potato aphids. Nat Biotechnol 16: 1365–1369.
18. Van Der Vossen EAG, Der Voort V, Rouppe JNAM, Bendahmane A, Sandbrink H, et al. (2000) Homologues of a single resistance-gene cluster in potato confer resistance to distinct pathogens: a virus and a nematode. Plant J 23: 567–576.
19. Lorieux M, Reversat G, Diaz SXG, Denance C, Jouvenet N, et al. (2003) Linkage mapping of *Hsa-1*^{Og}, a resistance gene of African rice to the cyst nematode, *Heterodera sacchari*. Theor Appl Genet 107: 691–696.
20. Paal J, Henselewski H, Muth J, Meksem K, Menéndez CM, et al. (2004) Molecular cloning of the potato *Gro1–4* gene conferring resistance to pathotype Ro1 of the root cyst nematode *Globodera rostochiensis*, based on a candidate gene approach. Plant J 38: 285–297.
21. Feng H, Wei L, Lin M, Zhou Y (2013) Assessment of rice cultivars in China for field resistance to *Aphelenchoides besseyi*. J Integr Agric doi:10.1016/S2095-3119(13)60608-5.
22. Yoshida K, Hasegawa K, Mochiji N, Miwa J (2009) Early embryogenesis and anterior-posterior axis formation in the white-tip nematode *Aphelenchoides besseyi* (Nematoda: Aphelenchoididae). J Nematol 41: 17–22.
23. Sun MJ, Liu WH, Lin MS (2009) Effects of temperature, humidity and different rice growth stages on vertical migration of *Aphelenchoides besseyi*. Rice Sci 16: 301–306.
24. McGawley EC, Rush MC, Hollis JP (1984) Occurrence of *Aphelenchoides besseyi* in Louisiana rice seed and its interaction with *Sclerotium oryzae* in selected cultivars. J Nematol 16: 65–68.
25. Rogers SO, Bendich AJ (1985) Extraction of DNA from milligram amounts of fresh, herbarium and mummified plant tissues. Plant Mol Biol 5: 69–76.
26. Septiningsih EM, Sanchez DL, Singh N, Sendon PM, Pamplona AM, et al. (2012) Identifying novel QTLs for submergence tolerance in rice cultivars IR72 and Madabaru. Theor Appl Genet 124: 867–874.
27. Lander ES, Green P, Abrahamson J, Barlow A, Daly MJ, et al. (1987) MAPMAKER: an interactive computer package for constructing primary genetic linkage maps of experimental and natural populations. Genomics 1: 174–181.
28. Lincoln SE, Lander SL (1993) Mapmaker/exp 3.0 and mapmaker. qtl 1.1. technical report. Cambridge: Whitehead Institute of Medical Research.
29. Wang S, Basten CJ, Zeng ZB (2005) Windows QTL cartographer, version 2.5. Statistical genetics. Raleigh: North Carolina State University.
30. Fortuner R, Williams KJO (1975) Review of the literature on *Aphelenchoides besseyi* hristie, 1942, the nematode causing "white tip" disease in rice. Helminthological Abstracts Series B, Plant Nematology 44: 1–40.
31. Roberts PA (1982) Plant resistance in nematode pest management. J Nematol 14: 24–33.
32. Lu C, Shen L, He P, Chen Y, Zhu L, et al. (1997) Comparative mapping of QTLs for agronomic traits of rice across environments by using a doubled-haploid population. Theor Appl Genet 94: 145–150.
33. Thomson MJ, Tai TH, McClung AM, Lai XH, Hinga ME, et al. (2003) Mapping quantitative trait loci for yield, yield components and morphological traits in an advanced backcross population between *Oryza rufipogon* and the *Oryza sativa* cultivar Jefferson. Theor Appl Genet 107: 479–493.
34. Tan Z, Shen L, Yuan Z, Lu CF, Chen Y, et al. (1996) Identification of QTLs for ratooning ability and grain yield traits of rice and analysis of their genetic effects. Zuo Wu Xue Bao 23: 289–295.
35. Lin HX, Qian HR, Zhuang JY, Lu J, Min SK, et al. (1996) RFLP mapping of QTLs for yield and related characters in rice (*Oryza sativa* L.). Theor Appl Genet 92: 920–927.
36. Cho YC, Suh JP, Choi IS, Hong HC, Baek MK, et al. (2003) QTLs analysis of yield and its related traits in wild rice relative *Oryza rufipogon*. Treat Crop Res 4: 19–29.
37. Yu SB, Li JX, Xu CG, Tan YF, Gao YJ, et al. (1997) Importance of epistasis as the genetic basis of heterosis in an elite rice hybrid. Proc Natl Acad Sci USA 94: 9226–9231.
38. Li Z, Pinson SRM, Park WD, Paterson AH, Stansel JW (1997) Epistasis for three grain yield components in rice (*Oryxa sativa* L.). Genetics 145: 453–465.
39. Zhuang JY, Lin HX, Lu J, Qian HR, Hittalmani S, et al. (1997) Analysis of QTL× environment interaction for yield components and plant height in rice. Theor Appl Genet 95: 799–808.
40. CiXin H, Jun Z, JuQiang Y, Benmoussa M, Ping W (2000) QTL mapping for developmental behaviour of panicle dry weight in rice. Zhongguo Nong Ye Ke Xue 33: 24–32.
41. Lu C, Shen L, Tan Z, Xu Y, He P, et al. (1996) Comparative mapping of QTLs for agronomic traits of rice across environments using a doubled haploid population. Theor Appl Genet 93: 1211–1217.
42. Gebhardt C, Mugniery D, Ritter E, Salamini F, Bonnel E (1993) Identification of RFLP markers closely linked to the H1 gene conferring resistance to *Globodera rostochiensis* in potato. Theor Appl Genet 85: 541–544.
43. Ramalingam J, Vera Cruz CM, Kukreja K, Chittoor JM, Wu JL, et al. (2003) Candidate defense genes from rice, barley, and maize and their association with qualitative and quantitative resistance in rice. Mol Plant Microbe Interact 16: 14–24.
44. Chen H, Wang S, Xing Y, Xu C, Hayes PM, et al. (2003) Comparative analyses of genomic locations and race specificities of loci for quantitative resistance to *Pyricularia grisea* in rice and barley. Proc Natl Acad Sci USA 100: 2544–2549.
45. Lin MS, Ding XF, Wang ZM, Zhou FM, Lin N (2004) Description of *Aphelenchoides besseyi* from abnormal rice with "small grains and erect panicles" symptom in China. Rice Sci 12: 289–294.
46. Wang ZY, Yang HF, Ji MX (2006) The small grains and erect panicle of rice caused by nematode and control the disease in Jiangsu Province. Nanjing Nong Ye Da Xue Xue Bao 29: 54–56.
47. Liu W, Lin M, Li H, Sun MJ (2007) Study on dynamic development of *Aphelenchoides besseyi* on rice plant by artificial inoculations in greenhouse. Zhongguo Nong Ye Ke Xue 12: 14.
48. Hauge BM, Wang ML, Parsons JD, Wang M (2006) Nucleic acid molecules and other molecules associated with soybean cyst nematode resistance: U.S. Patent 7,154,021[P].

14

Association Mapping of Seed Oil and Protein Content in *Sesamum indicum* L. Using SSR Markers

Chun Li, Hongmei Miao, Libin Wei, Tide Zhang, Xiuhua Han, Haiyang Zhang*

Henan Sesame Research Center, Henan Academy of Agricultural Sciences, Zhengzhou, Henan, People's Republic of China

Abstract

Sesame is an important oil crop for the high oil content and quality. The seed oil and protein contents are two important traits in sesame. To identify the molecular markers associated with the seed oil and protein contents in sesame, we systematically performed the association mapping among 369 worldwide germplasm accessions under 5 environments using 112 polymorphic SSR markers. The general linear model (GLM) was applied with the criteria of $\log P \geq 3.0$ and high stability under all 5 environments. Among the 369 sesame accessions, the oil content ranged from 27.89%–58.73% and the protein content ranged from 16.72%–27.79%. A significant negative correlation of the oil content with the protein content was found in the population. A total of 19 markers for oil content were detected with a R^2 value range from 4% to 29%; 24 markers for protein content were detected with a R^2 value range from 3% to 29%, of which 19 markers were associated with both traits. Moreover, partial markers were confirmed using mixed linear model (MLM) method, which suggested that the oil and protein contents are controlled mostly by major genes. Allele effect analysis showed that the allele associated with high oil content was always associated with low protein content, and vice versa. Of the 19 markers associated with oil content, 17 presented near the locations of the plant lipid pathway genes and 2 were located just next to a fatty acid elongation gene and a gene encoding Stearoyl-ACP Desaturase, respectively. The findings provided a valuable foundation for oil synthesis gene identification and molecular marker assistant selection (MAS) breeding in sesame.

Editor: Tianzhen Zhang, Nanjing Agricultural University, China

Funding: This work was supported by China National "973" Project (2011CB109304) and the earmarked fund for the China Agriculture Research System (CARS-15). The funders had no role in study design, data collection and analysis, decision to publish, or preparation of the manuscript.

Competing Interests: The authors have declared that no competing interests exist.

* Email: zhy@hnagri.org.cn

Introduction

Sesame (*Sesamum indicum* L.) is an ancient and important oilseed crop and is cultivated mainly in the tropical and subtropical regions of Asia, Africa and Southern America. The harvested area of world sesame reaches to 7.3×10^7 hm^2, and the total product per year is roughly 3.7×10^7 ton from 2001 to 2010 (FAO data). Compared with other main oil crops, e.g., soybean (18% of average oil content) [1], oilseed rape (41%) [2], sunflower (40–44%) [3] and peanut (51%) [4], sesame is one of the few oil crops with the highest oil content and quality. Sesame seeds contain 55–58% oil and almost 18% proteins. Among the fatty acid compositions in sesame seeds, oleic acid (18:1) (39.6%) and linoleic acid (18:2) (46.0%) are the two main components with the ideal ratio of almost 1:1 [5,6]. Apart from the seed yield, the content of seed storage oil and proteins is the highlight agronomic trait in sesame breeding [7].

In the past two decades, in order to clarify the high quality of sesame oil and protein, many researchers focused on exploring seed development and fatty acid and storage protein synthesis processes, as well as identifying the lipid synthesis related genes and molecular markers in sesame [8–12]. Of all the three available cDNA libraries, two libraries are constructed for seed development analysis [13,14]. However, oil and protein contents are complex quantitative traits and always are affected by genotype and environment [15]. At present, the mechanism of high oil content and quality in sesame seeds is still unclear. No loci of oil and protein content traits have ever been found in the sesame linkage maps. Even though Wei et al. [12] performed the association analysis of seed oil and protein content and fatty acid composition within 216 Chinese sesame accessions using 79 molecular primer pairs (including SSRs, SRAPs and AFLPs), only one association marker (M15E10-3) was identified under two environments. Therefore, in order to precisely detect the genes or markers associated with oil and protein content traits and to improve the sesame breeding, more efficient markers and germplasm resources with larger phenotypic variation need to be applied [16,17].

Currently, linkage analysis (QTL mapping) and association mapping are two main and common analysis tools for dissecting complex phenotypic variation. Compared with the traditional linkage analysis based on mapping populations, association mapping offers higher precision for locating QTLs and selecting molecular markers [16,18]. Till now, association mapping has been extensively used for analyzing important agronomic and quantity traits in wheat, maize, cotton, oilseed rape and other crops [18–22]. In the past several years, vast simple sequence repeat (SSR) or microsatellite markers with the high polymorphism are developed in sesame [7,23,24]. Accordingly, the association mapping is getting reliable and powerful for detecting the genes or markers associated with key traits and improving the molecular marker-assisted selection (MAS) in sesame breeding programs.

The objectives of this study are: (1) to perform the association mapping of seed oil content (OC) and protein content (PC) traits in worldwide sesame accessions using the GLM and MLM models, (2) to reflect the characteristics of sesame oil and protein contents under various environments, and (3) to determine the key SSR markers associated with seed quality. In this report, a natural population covering 369 worldwide accessions from China and other 15 countries and 112 pairs of polymorphic SSR markers were applied. The results give a foundation for investigating the seed development-related genes and seed quality in sesame.

Results

Seed oil and protein content variation in the natural population

Both the seed oil content (OC) and the protein content (PC) are often influenced by environment. To decrease the environmental effect, we collected the phenotypic data of the 369 sesame accessions under 5 environments of Pingyu and Yuanyang locations in 2011, and Pingyu, Yuanyang and Xinyang locations in 2012, The descriptive parameters under each environment were calculated (Table 1). The results showed that the OC and PC significantly varied among the 369 accessions. In total, the OC of the natural population ranged from 27.89%–58.73%, with an average of 49.59%–53.14%; meanwhile, the PC ranged from 16.72%–27.79% with an average of 20.28%–22.51%. All the datasets showed a normal or nearly normal distribution. To determine the heritability of the phenotypes, we performed the variance analysis of oil and protein contents (Table 2). Results indicated that the OC and PC traits were significantly influenced by genotype and environments (i.e., year and location). No significant interactions between variety and environmental factors (year and location) were detected. Moreover, the OC and PC traits presented the significant negative correlation under the 5 environments, as the correlation coefficient (r) between OC and PC varied from $-0.66\sim-0.72$ ($P<0.01$) in 2011 and from $-0.52\sim-0.74$ ($P<0.01$) in 2012, respectively (Data not listed).

Linkage disequilibrium

Linkage disequilibrium (LD) refers to the non-random association of alleles between the genetic loci. A total of 112 SSR markers were used for estimating the LD level among the 369 sesame germplasm accessions (Table S2). These SSR markers distributed in 33 contigs/scaffolds with a total length of 180.86 Mb, which approximately represented 67 percentage of the assembly genome size (270 Mb) and 50 percentage of the estimated genome size (360Mb). The average SSR density was 1 SSR per 1.6 Mb. To reflect the associations between the polymorphic loci of the 112 SSR markers, LD P-values were determined among the 6,216 locus pairs (i.e., 112*(112-1)/2) using Fisher's exact test and two indexes of D' and r^2 (Figure 1). The average values of D' and r^2 for the 6,216 pairs were 0.1649 and 0.0173, respectively. Of the 6216 pairs, 2584 pairs (41.57%) showed a significant linkage disequilibrium ($P<0.01$), 363 pairs (5.84%) showed a higher linkage disequilibrium of $D'>0.5$, and 33 pairs (0.5%) gave a D' value of 1.0 (i.e., complete linkage). The data indicated that linkage disequilibrium existed among the sesame accessions.

SSR marker diversity and population structure

Before analyzing the association, the polymorphism of the 112 SSR markers within the 369 germplasms was investigated (Table S2). Results showed that the number of alleles ranged from 2–5, with an average of 2.47 per locus. The PIC values of the markers

Table 1. Phenotypic variation of the seed oil and protein contents in the 369 accessions under 5 environments.

Trait	Year	Place	Min. (%)	Max. (%)	Mean (%)	Std. ev.	Skewness	Kurtosis
OC	2011	Pingyu	27.89	55.03	49.59	4.61	−2.19	1.69
	2011	Yuanyang	34.28	57.53	51.94	3.72	−2.26	2.85
	2012	Pingyu	31.95	54.80	49.82	3.43	−2.54	5.13
	2012	Yuanyang	30.77	57.88	52.23	4.22	−2.53	3.98
	2012	Xinyang	32.74	58.73	53.14	4.59	−2.32	2.67
PC	2011	Pingyu	18.74	26.49	21.65	1.30	0.90	1.28
	2011	Yuanyang	17.60	25.10	20.28	1.16	0.97	1.55
	2012	Pingyu	20.22	27.79	22.51	1.04	0.78	1.68
	2012	Yuanyang	16.72	27.44	20.94	1.25	1.05	5.27
	2012	Xinyang	17.24	26.84	20.59	1.57	0.85	1.30

Table 2. Analysis of Variance of the oil and protein contents in the population under 5 environments.

Source of variation	DF	OC		PC	
		MS	F value	MS	F value
Year	1	27.33	7.46**	211.90	412.83**
Location	2	1564.58	427.02**	551.56	1074.55**
Variety	368	62.78	17.13**	5.23	10.19**
Year×Variety	368	3.19	0.87	0.55	1.07
Location×Variety	736	2.94	0.80	0.65	1.27
Residual	367	3.66		0.51	

** The significance at $P<0.01$.
DF denotes degree of freedom; MS denotes mean square.

ranged from 0.028 (Hs373) to 0.669 (Y1994) with an average value of 0.302. The percentage of heterozygotes per marker varied from 0.27% (Hs373) to 23.12% (Hs4325), with an average of 10.14%. The data indicated that the natural population had the high heterozygosity and was suitable for association mapping analysis. Subsequently, we estimated the population structure, as admixture of subpopulation could result in LD and produce the false-positive results. The most likely number (K) of subgroups in the 369 germplasm accessions was estimated using the 112 SSR markers (Figure 2). As K values increased from 1 to 10, the value of LnP(D) elevated directly; meanwhile, the ΔK reduced straightly with a clear peak value of 643.5 at $K=2$. The results indicated that the population was roughly composed of two divergent subgroups according to the Bayesian posterior probability analysis (Figure 3).

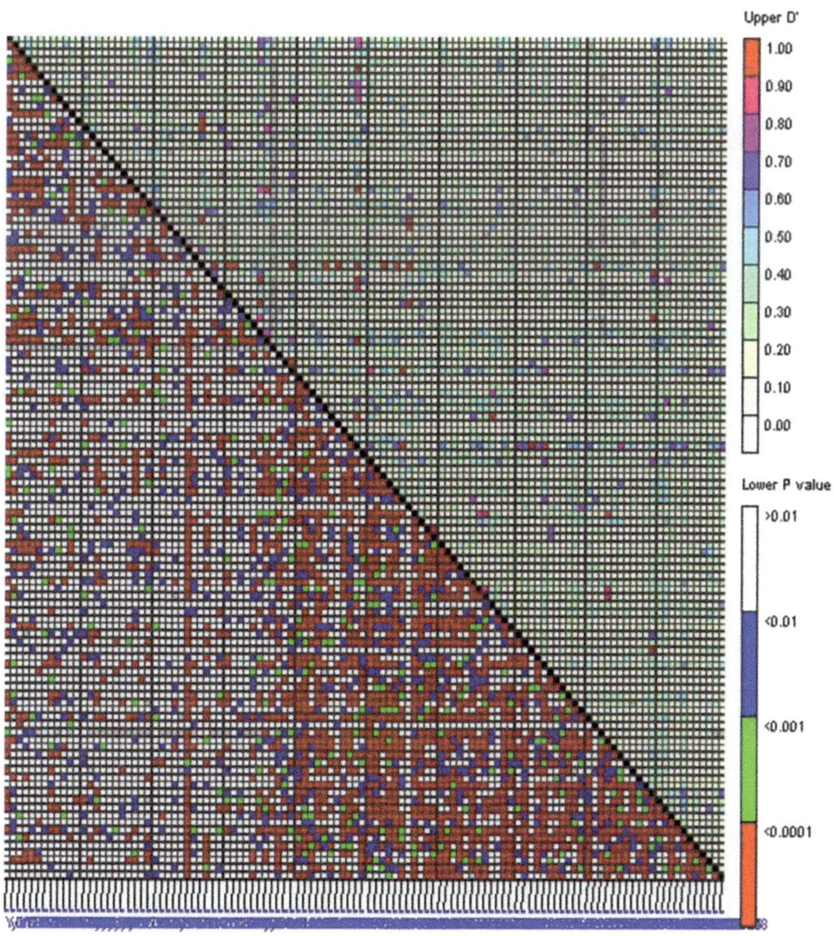

Figure 1. Disequilibrium matrix of 112 SSR polymorphic sites with both the X-axis and Y-axis. The matrix is divided into two regions by the diagonal line. The upper right region indicates the D′ value of each SSR couples, and the corresponding blocks in the lower left region indicates the significance of D′. Various value intervals of the D′ or P values are shown in different colors according to the right color columns.

Of the 369 accessions, 126 accessions were grouped into subgroup 1 (the green ones in Figure 3), 243 accessions into subgroup 2 (the red ones in Figure 3). Most (47 out of 51) of the foreign lines were located in subgroup 2. The two subgroups in the population were considered having the complex 'admixture' relationship, and no significant correlation of geographic origin with the subgroups in the Chinese lines were found.

Marker-trait associations

In this study, the association analysis was performed using general liner model (GLM) method. The stringent criterion of log$P\geq3.0$ under 5 environments was used for determining the association significance between the OC and PC traits. For the OC trait, 19 markers were detected and the R^2 values ranged from 4%–29% (Table 3 & Table S2). According to the primer locations in the sesame genome, 19 markers were mapped in 11 contigs/scaffolds. Of the 19 markers, Hs485 and Hs586 are located in scaffold00001, Hs4381, Hs02 and Hs4082 are located in C01, and Hs4061, Hs345 and Hs19563 are in C04. The distance between the markers that located in the same contig/scaffold was short than 2 Mb (Table S2). For the PC trait, 24 markers were detected with the R^2 value range of 3%–29% and mapped in 15 contigs/scaffolds (Table 4 & Table S2). Comparison results indicated that 19 of the above 24 markers were associated with OC trait at the same time; the other 5 markers of Hs425 (in C08), Hs560 (in C11), Hs4265 (in C12), and Hs4089 and Hs1514 (in C19) were unique to PC trait. Simultaneously, in order to assay the stability, we performed association mapping using mixed liner model (MLM) method, with the criterion of log$P\geq3.0$ under at least 3 environments.

For the OC trait, 9 markers were detected in MLM model and 8 markers of Hs345, Hs4381, Hs485, Hs1036, Hs4061, Hs635, Hs376 and Hs586 were confirmed using GLM and MLM methods. Especially, 4 markers of Hs345, Hs4381, Hs485 and Hs1036 had high R^2 values ($\geq$10%) under 5 environments (Table 3 and Table S3). For the PC trait, 9 markers were found and confirmed in both models, of which 7 markers had high R^2 values ($\geq$10%) under 5 environments (Table 4 and Table S3). These results suggested that the OC and PC traits are controlled mostly by major genes in sesame.

Marker effect on the phenotypic variation

To explore the association between the above markers and phenotypic variation and the utility potential in sesame MAS breeding program, we performed the allelic effects of the five markers associated with both traits (Table 5). For each marker, the effects estimated were in accordance with the allele character under the 5 environments. Hs345 had the largest effect on the variation of seed oil and protein content. The Hs345-1:1 and Hs345-2:2 of Hs345 showed the different variation effects on OC trait, as the average oil content in the accessions ranged from 52.05%–42.82%. In the genotypes carrying the Hs4381-1:1 allele, oil and protein contents were 43.89% and 23.25%, respectively, while the samples with the Hs485-2:2 contained 52.00% oil and 21.00% proteins. Furthermore, the specific allele indicated the negative or positive effect to a large extent on the OC or PC trait. For example, the allele effect of Hs345-2:2 on OC trait ranged from −8.13 to −12.18, whereas the effect on PC trait ranged from 2.05 to 3.82. Comparison results suggested that the alleles of all 5 markers give the opposite effects on OC and PC traits, respectively. The allele that increased the oil content certainly gave the negative effect on protein content, and vice versa. Therefore, these markers could be used for screening sesame lines with either high oil or protein content.

Comparative genome analysis

To clarify the distribution and more information of the above associated markers, we performed the comparative genome analysis of the 19 SSR markers associated with the OC trait (Table 6). All genes closest to the markers were explored. Of the 19 markers, 11 are located in gene regions and 8 are in intergenic regions. The flanking genes had various functions, such as ligase (C01.560), transcription factor (C04.26, C13.438) and kinase (C14.22). Moreover, we found that the markers of Hs4082 and Hs345 were just located next to C01.883 (ABC transporter G family member 3 gene) and C04.786 (Stearoyl-ACP Desaturase gene), respectively, which were proved involved in plant lipid biosynthesis. We also screened the upstream and downstream sequences of 500 Kb far from each marker. 17 (out of 19) markers were proved close to plant lipid pathway genes. These result further confirmed our association mapping conclusions.

Discussion

To clarify the genetic mechanisms of fatty acid and protein synthesis in sesame seeds, we herein performed the association mapping analysis of the OC and PC traits among 369 sesame accessions using 112 genic-SSR markers. These accessions were collected from 19 provinces of China and 15 other countries, and represented the genetic diversity of sesame germplasm for association mapping study. These accessions included many released Chinese and foreign cultivars. Compared with the traditional linkage analysis (QTL mapping), the association

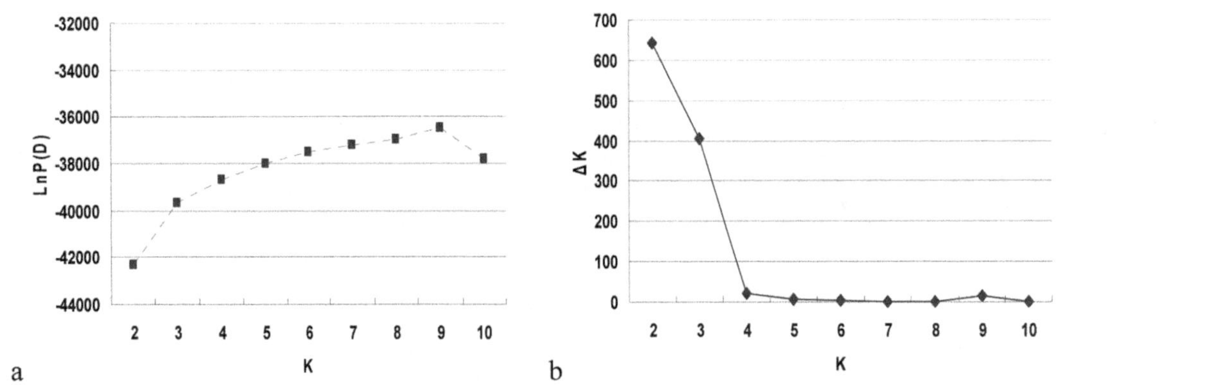

Figure 2. Estimated (a) Ln*P*(D) and (b) Δ*K* values for a given *K*.

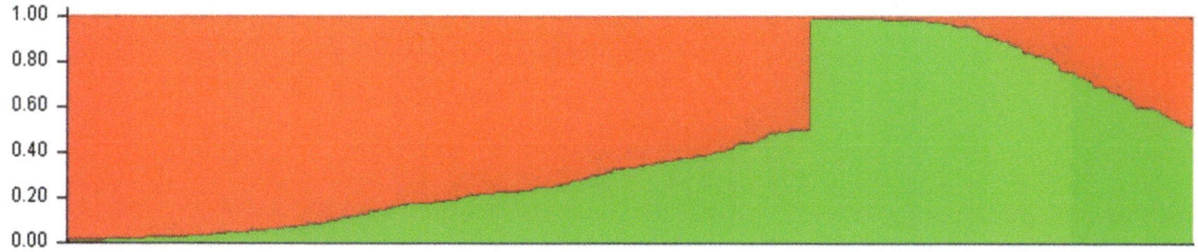

Figure 3. Genetic composition of individuals based on Bayesian posterior probability. According to the Bayesian posterior probability, the natural population of 369 worldwide sesame lines is divided into two groups in green and red colors, respectively.

analysis based on linkage disequilibrium (LD) has been applied for the quantitative trait loci (QTLs) detection and location in many crops. Meanwhile, GLM and MLM models are applied individually for evaluating the marker association. In this article, 19 SSR markers associated with the OC trait were detected under each 5 environments in GLM model, while 24 markers were determined and associated with the protein content.

Sesame genetic diversity and the population structure

Sesame is a diploid and self-pollinated oil crop with the karyotype of 2n = 2x = 26. As all cultivars are originated from the sole cultivated species, *Sesamum indicum* L, the narrow genetic diversity presents in regional sesame resources to a large extent [24,49–51]. In addition, many reports reflect that there is no clear association between genotype and geographical origin, as many sesame accessions from the different geographic locations are clustered into the same group in the dendrogram [51–53]. Apart from the natural history of predomesticated ancestors, the diversity pattern of domestic species could be influenced by the breeding practice, germplasm collection and human activity [53–55]. In this article, in order to guarantee the broad genetic variation, we selected the 369 worldwide accessions for seed nutrition genetic analysis according to the geographical origin, molecular clustering and the morphologic diversity (Table S1). The heterozygosity ranged from 0.27% (Hs373)-23.12% (Hs4325). The result proved that the natural population could be used as the core germplasm for association mapping (Table S2). Population structure analysis showed that many sesame accessions collected from the same geographic locations were not grouped together, which further proved the unclear association between genotype and geographical origin in sesame germplasm resource [51–53].

During performing the association mapping in a population, LD patterns between the functional loci and markers should be analyzed at first [56]. We analyzed the *P*-values of linkage disequilibrium between the polymorphisms of the 112 SSR marker loci using Fisher's exact test (Figure 1). As 363 (5.84%) pairwise comparisons had the high LDs ($D'>0.5$), the linkage disequilibrium existed in 369 sesame accessions. Accordingly, we believed that the natural population is suitable for association analysis of the oil and protein contents.

Oil and protein content variation and associated SSR markers

Among large-scale sesame germplasm resources, the oil and protein contents varied significantly. Yermanos et al. [57] evaluated 721 sesame samples collected from more than 19 countries, and found that the oil content varied from 40.4–59.8% with the mean of 53.1%. The protein content ranged from 19–31% with an average of about 25% [58]. In our population, the oil content varied from 27.89%–58.73% with an average of 51.34%, while the protein content varied from 16.72%–27.79% with an average of 21.19% (Table 1). The data reflected the great variation of sesame seed compositions in germplasm [57,59,60].

Comparison analysis proved that there is a strong negative correlation of the oil content with protein content. Interestingly, the stringency relationship was also exhibited in the association analysis. As shown in the GLM analysis in Table 3 and 5, all the 19 markers associated with OC were detected for PC trait; and the alleles exhibited the opposite effects on OC and PC at the same time. These phenomena also present in other oil crops, such as oilseed rape, cotton, soybean and peanut [1,61–63]. Zhao et al. [62] found 6 QTLs with pleiotropic effects on both oil content and protein content in oilseed rape. In the cotton backcross inbred population, of 17 QTLs for oil content and 20 QTLs for protein content, 8 QTLs co-localized in the same chromosome regions and controlled oil and protein contents with opposite additive effects [63]. Hwang et al. (2014) detected 25 SNPs associated with seed oil in 13 different genomic regions through GWAS (genome wide association study), and 7 SNPs were significantly associated with both protein and oil traits. For the six of seven marker loci, a negative relationship existed between the protein effect versus that on oil [64]. Meanwhile, QTLs or markers associated with increased protein and oil contents were also found. For example, Zhao et al. (2006) found that 2 QTLs that controlled oil content were independent from protein content by conditional QTL mapping [65]; Hwang et al. (2014) found a SNP at the 4.92 Mb position on Chr 9 was associated with increased protein and oil contents [64].

In many crops, the seed oil content seems to be controlled mostly by major genes [66–69]. In this study, we detected 19 markers associated with OC trait in sesame using GLM method, and the R^2 values ranged from 4.0–29.0%. Chen et al. (2010) screened 27 QTLs related to oil content in oilseed rape and the individual explanation was high with the range of 4.20–30.20% [66]. In *Arabidopsis thaliana*, a single QTL or marker could give an explanation of 17–19% for the seed oil content [68]. In soybean, the explanation reached to 14.3–45.6% [69]. In this report, 112 polymorphic SSR markers were used for association mapping. Compared with other common molecular markers, such as SRAP and AFLP, SSR marker is more suitable for sesame diversity analysis due to the narrow genetic basis [24,49–51]. The marker distribution and density were analyzed using the sesame genome assembly data. The contigs/scaffolds carrying the 112 markers approximately covered 180 Mb of the sesame genome and occupied ~67% of the assembly size (270 Mb) and 50% of the estimated genome size (~360 Mb, in which 90 Mb was believed to be repeat sequences) (Table S2) [70,71]. Therefore, the association mapping using 112 markers is meaningful and reliable, even though some QTLs might be missed. To detect more QTLs, new SSR or SNP markers could be applied in further research.

Table 3. Association mapping of the OC trait using GLM method under the 5 environments.

Marker	2011						2012								
	Pingyu			Yuanyang			Pingyu			Yuanyang			Xinyang		
	F_{maker}	P_{marker}	R^2	F_{maker}	P_{marker}	R^2	F_{maker}	P_{marker}	R^2	F_{maker}	P_{marker}	R^2	F_{maker}	P_{marker}	R^2
Hs345	35.93	3.05E-25	0.28	26.22	4.37E-19	0.22	13.41	3.41E-10	0.13	26.44	3.11E-19	0.22	37.34	4.08E-26	0.29
Hs4381	61.72	8.30E-24	0.25	43.17	1.48E-17	0.19	40.31	1.53E-16	0.18	46.18	1.32E-18	0.20	75.46	3.73E-28	0.29
Hs1956	47.51	7.46E-19	0.22	39.98	2.83E-16	0.19	16.15	2.05E-07	0.09	24.86	8.80E-11	0.13	40.25	2.29E-16	0.19
Hs485	31.50	4.06E-18	0.20	20.63	2.29E-12	0.14	21.10	1.27E-12	0.15	21.48	7.82E-13	0.15	36.07	1.98E-20	0.23
Hs1036	41.73	6.75E-17	0.20	25.69	4.19E-11	0.13	17.87	4.24E-08	0.10	26.00	3.19E-11	0.13	43.84	1.24E-17	0.20
Hs205	17.18	3.11E-15	0.19	10.41	2.42E-09	0.12	10.37	2.68E-09	0.13	12.46	3.65E-11	0.15	16.29	1.75E-14	0.18
Hs4209	14.91	2.67E-13	0.17	8.15	2.64E-07	0.10	9.28	2.55E-08	0.12	12.37	4.41E-11	0.14	23.92	9.21E-21	0.25
Hs672	16.81	1.19E-12	0.16	11.35	1.09E-08	0.11	16.82	1.16E-12	0.16	20.27	4.38E-15	0.18	26.53	2.58E-19	0.22
Hs393	17.90	2.01E-13	0.16	10.10	9.18E-08	0.10	12.19	2.61E-09	0.12	16.74	1.32E-12	0.15	29.18	4.76E-21	0.24
Hs4061	30.93	3.93E-13	0.14	19.32	1.06E-08	0.09	21.42	1.60E-09	0.11	23.28	3.05E-10	0.11	30.72	4.66E-13	0.14
Y2129	22.99	4.73E-10	0.13	13.89	1.64E-06	0.08	15.29	4.57E-07	0.09	24.86	9.25E-11	0.13	30.65	6.69E-13	0.16
Hs618	25.22	5.54E-11	0.12	12.90	3.86E-06	0.06	7.54	6.19E-04	0.04	17.66	4.78E-08	0.09	22.91	4.26E-10	0.11
Hs377	4.14	2.01E-05	0.11	3.16	6.87E-04	0.08	3.92	4.49E-05	0.10	6.26	7.83E-09	0.15	7.09	3.63E-10	0.17
Hs635	22.15	8.38E-10	0.11	17.23	7.11E-08	0.08	10.34	4.30E-05	0.05	18.79	1.71E-08	0.09	31.73	1.98E-13	0.15
Hs02	20.98	2.38E-09	0.10	17.19	7.34E-08	0.08	10.31	4.43E-05	0.05	12.41	6.09E-06	0.06	36.20	4.55E-15	0.16
Hs376	19.25	1.15E-08	0.10	11.49	1.45E-05	0.06	12.53	5.52E-06	0.07	16.27	1.72E-07	0.08	21.26	1.89E-09	0.10
Hs1638	17.45	6.23E-08	0.09	7.29	7.96E-04	0.04	18.27	2.97E-08	0.10	19.25	1.23E-08	0.10	24.86	8.66E-11	0.13
Hs586	16.84	1.01E-07	0.08	13.48	2.25E-06	0.07	7.30	7.82E-04	0.04	10.90	2.52E-05	0.06	20.14	5.06E-09	0.10
Hs4082	8.18	3.36E-04	0.04	7.56	6.08E-04	0.04	18.30	2.69E-08	0.09	15.51	3.44E-07	0.08	20.61	3.33E-09	0.10

The markers associated with the OC trait under 5 environments are listed in the table.

Table 4. Association mapping of PC trait using GLM method under the 5 environments.

Marker	2011						2012								
	Pingyu			Yuanyang			Pingyu			Yuanyang			Xinyang		
	F_{maker}	P_{marker}	R^2	F_{maker}	P_{marker}	R^2	F_{maker}	P_{marker}	R^2	F_{maker}	P_{marker}	R^2	F_{maker}	P_{marker}	R^2
Hs345	25.46	1.41E-18	0.21	21.52	6.24E-16	0.19	13.41	3.41E-10	0.13	26.44	3.11E-19	0.22	37.34	4.08E-26	0.29
Hs4381	50.30	5.14E-20	0.21	33.86	3.25E-14	0.15	40.31	1.53E-16	0.18	46.18	1.32E-18	0.20	75.46	3.73E-28	0.29
Hs485	26.93	9.51E-16	0.18	21.07	1.32E-12	0.14	21.10	1.27E-12	0.15	21.48	7.82E-13	0.15	36.07	1.98E-20	0.23
Hs205	16.43	1.33E-14	0.18	7.90	4.52E-07	0.10	10.37	2.68E-09	0.13	12.46	3.65E-11	0.15	16.29	1.75E-14	0.18
Hs1036	31.71	2.48E-13	0.16	17.49	5.98E-08	0.09	17.87	4.24E-08	0.10	26.00	3.19E-11	0.13	43.84	1.24E-17	0.20
Hs393	16.01	4.43E-12	0.15	13.34	3.73E-10	0.12	12.19	2.61E-09	0.12	16.74	1.32E-12	0.15	29.18	4.76E-21	0.24
Hs672	17.02	8.43E-13	0.15	11.63	6.73E-09	0.11	16.82	1.16E-12	0.16	20.27	4.38E-15	0.18	26.53	2.58E-19	0.22
Hs1956	26.91	1.50E-11	0.14	22.88	4.93E-10	0.12	16.15	2.05E-07	0.09	24.86	8.80E-11	0.13	40.25	2.29E-16	0.19
Hs377	5.73	5.69E-08	0.14	4.46	6.21E-06	0.11	3.92	4.49E-05	0.10	6.26	7.83E-09	0.15	7.09	3.63E-10	0.17
Y2129	26.65	1.99E-11	0.14	16.79	1.17E-07	0.09	15.29	4.57E-07	0.09	24.86	9.25E-11	0.13	30.65	6.69E-13	0.16
Hs4209	11.18	5.03E-10	0.13	7.07	2.56E-06	0.09	9.28	2.55E-08	0.12	12.37	4.41E-11	0.14	23.92	9.21E-21	0.25
Hs4061	25.55	4.14E-11	0.12	17.49	5.59E-08	0.08	21.42	1.60E-09	0.11	23.28	3.05E-10	0.11	30.72	4.66E-13	0.14
Hs1638	20.72	3.30E-09	0.11	14.13	1.29E-06	0.08	18.27	2.97E-08	0.10	19.25	1.23E-08	0.10	24.86	8.66E-11	0.13
Hs618	22.58	5.71E-10	0.11	12.16	7.75E-06	0.06	7.54	6.19E-04	0.04	17.66	4.78E-08	0.09	22.91	4.26E-10	0.11
Hs1514	19.02	1.53E-08	0.10	15.50	3.69E-07	0.08	6.23	2.20E-03	0.04	14.59	8.48E-07	0.08	14.77	7.19E-07	0.08
Hs4089	19.68	7.67E-09	0.09	18.49	2.25E-08	0.09	6.44	1.80E-03	0.03	17.68	4.71E-08	0.09	13.84	1.61E-06	0.07
Hs425	17.07	8.21E-08	0.08	8.38	2.75E-04	0.04	7.90	4.39E-04	0.04	17.02	8.55E-08	0.08	18.86	1.61E-08	0.09
Hs560	4.53	1.96E-04	0.07	4.98	6.48E-05	0.08	4.69	1.33E-04	0.07	6.43	1.91E-06	0.10	7.36	1.97E-07	0.11
Hs635	13.54	2.12E-06	0.07	13.28	2.72E-06	0.07	10.34	4.30E-05	0.05	18.79	1.71E-08	0.09	31.73	1.98E-13	0.15
Hs02	14.67	7.44E-07	0.07	10.16	5.08E-05	0.05	10.31	4.43E-05	0.05	12.41	6.09E-06	0.06	36.20	4.55E-15	0.16
Hs4265	5.17	1.34E-04	0.06	6.58	7.07E-06	0.08	6.70	5.50E-06	0.08	8.54	1.16E-07	0.10	11.90	1.13E-10	0.14
Hs376	11.29	1.76E-05	0.06	11.04	2.22E-05	0.06	12.53	5.52E-06	0.07	16.27	1.72E-07	0.08	21.26	1.89E-09	0.10
Hs586	11.07	2.15E-05	0.06	11.38	1.60E-05	0.06	7.30	7.82E-04	0.04	10.90	2.52E-05	0.06	20.14	5.06E-09	0.10
Hs4082	12.36	6.41E-06	0.06	8.72	2.00E-04	0.04	18.30	2.69E-08	0.09	15.51	3.44E-07	0.08	20.61	3.33E-09	0.10

The markers associated with the PC trait under 5 environments are listed in the table.

Table 5. Allele effects of 5 markers simutaneously associated with OC and PC traits under 5 environments.

Marker	Allele[a]	2011				2012					
		Pingyu		Yuanyang		Pingyu		Yuanyang		Xinyang	
		OC	PC	OC	PC	OC	PC	OC	PC	OC	PC
Hs345	1:1	−0.04	−0.17	−0.59	0.25	−1.17	0.44	−1.61	0.57	−0.93	0.39
	2:2	−10.74	2.30	−8.13	2.12	−8.87	2.05	−12.18	2.96	−11.73	3.82
	1:2	−5.53	1.38	−4.59	1.67	−5.13	1.29	−5.84	2.08	−6.66	2.58
Hs4061	1:1	0.00	0.00	0.00	0.00	0.00	0.00	0.00	0.00	0.00	0.00
	2:2	8.78	−2.41	4.46	−1.95	6.37	−1.84	8.54	−2.33	7.94	−2.96
	1:2	2.58	−0.98	−0.18	−1.06	1.78	−0.80	2.58	−1.14	1.60	−0.86
Hs1956	1:1	1.12	−0.36	1.01	−0.49	0.85	−0.32	1.18	−0.39	1.67	−0.81
	2:2	−5.89	1.21	−4.13	0.79	−4.03	0.63	−5.03	0.95	−5.15	1.34
	1:2	−0.74	0.15	0.11	−0.28	−0.86	0.26	−0.60	0.49	−0.85	0.02
Hs4381	1:1	0.00	0.00	0.00	0.00	0.00	0.00	0.00	0.00	0.00	0.00
	2:2	9.75	−2.45	6.33	−1.73	6.97	−1.82	9.06	−2.17	9.80	−3.41
	1:2	5.02	−1.05	2.34	−0.49	3.16	−0.90	5.12	−0.75	4.58	−1.51
Hs485	1:1	−4.80	0.47	−3.23	0.25	−3.87	0.51	−6.88	1.62	−7.19	2.24
	2:2	2.04	−1.28	0.82	−1.21	1.71	−0.96	−0.02	0.02	0.20	−0.29
	1:2	−3.99	0.33	−3.85	−0.03	−2.59	−0.12	−4.64	1.38	−5.76	1.74

Note: [a] two of the most common alleles are listed.

Table 6. Information of candidate genes related to the markers associated with OC traits.

Marker	Related gene[a]	Annotation	Nearby lipid genes[b]	*Arabidopsis thaliana* homologous genes		Ref.
				Loci[c]	Annotation	
Hs4381	C01.460	Thioredoxin M3	C01.526	At1G55260	Lipid Transfer Protein type 5	[25]
Hs02	C01.560	E3 ubiquitin-protei	C01.548	At3G25110	Acyl-ACP Thioesterase A	[26]
		ligase RHF2A	C01.575	At3G07400	Lipid Acylhydrolase-like	
	C01.561	Unknown	C01.601	At2G30720	Acyl-CoA Thioesterase	[27]
Hs4082	C01.882	Cell cycle protein	C01.873	At1G53390	ABC Transporter	[28]
	C01.883	ABC transporter G	C01.883	At2G28070	ABC Transporter	[28]
		family member 3	C01.928	At3G44830	Diacylglycerol Acyltransferase	[29]
Hs635	C02.696	RAS-related protein Rab11A	C02.739	At1G47620	Midchain Alkane Hydroxylase	30]
Hs4209	C04.25	Transmembrane domain	C04.38	At1G31770	ABC Transporter	[31]
		containing protein	C04.56	At1G71010	Phosphatidylinositol-Phosphate Kinase	[32]
	C04.26	TCP transcription factor 12	C04.81	At1G10900	Phosphatidylinositol-Phosphate Kinase	[32]
			C04.96	At1G77660	Phosphatidylinositol-Phosphate Kinase	[32]
Hs4061	C04.681	Inositol trisphosphate 5-phosphatase 1	—			
Hs345	C04.785	Unknown	C04.767	At1G05630	Phosphoinositide 5-Phosphatase	—
	C04.786	Stearoyl-ACP desaturase	C04.786	At2G43710	Stearoyl-ACP Desaturase	[33]
Hs1956	C04.845	FAR1-related sequence	The same as Hs345			
Hs393	C13.438	bZIP transcription	C13.388	At1G10900	Phosphatidylinositol-Phosphate Kinase	[32]
		factor 40	C13.471	At5G10160	Hydroxyacyl-ACP Dehydrase	[34]
			C13.504	At1G17840	ABC Transporter	[35]
			C13.514	At4G36480	Subunit of Serine Palmitoyltransferase	[36]
Hs205	C14.21	Thylakoid membrane	C14.49	At4G33550	Lipid Transfer Protein type 3	[37]
		phosphoprotein	C14.66	At1G49430	Long-Chain Acyl-CoA Synthetase	[38]
	C14.22	Serine/threonine-protein	C14.111	At2G46210	Sphingobase-D8 Desaturase	[39]
		kinase WNK4	C14.132	At2G46090	Diacylglycerol Kinase	—
Hs672	C14.367	Cyclin-P3.1 F-box	C14.359	At2G45150	CDP-DAG Synthase	[40]
		protein	C14.413	At4G33355	Lipid Transfer Protein type 1	—
	C14.368	Unknown	C14.428	At2G44810	Acylhydrolase (DAD1-like)	—
Y2129	C15.825	Unknown	C15.92	At2G29980	Linoleate Desaturase	[41]
Hs1638	C15.840	Unknown	The same as Y2129			
Hs618	C25.41	IAA-alanine resistance	C25.69	At5G08415	Lipoate Synthase	[42]
		protein 1				
	C25.42	Beta-D-xylosidase				
Hs376	C26.474	ADP-ribosylation factor	C26.417	At1G15110	Base-Exchange-type Phosphatidylserine	[43]
					Synthase	
			C26.454	At1G71010	Phosphatidylinositol-Phosphate Kinase	—
			C26.515	At1G31812	Acyl CoA Binding Protein	[44]
Hs377	C26.475	Unknown	The same as Hs376			
Hs485	sf00001.885	Unknown	sf00001.95	At4G04930	Dihydrosphingosine Delta-4 Desaturase	[45]
	sf00001.886	Unknown				
Hs586	sf00001.754	ADP-ribosylation factor	—			
Hs1036	sf00044.12	ABC transporter I	sf00044.12	At3G20320	Acid-Binding Protein	[46]
		family member 15	sf00044.41	At2G25170	Chromatin remodeling factor	[47]

Table 6. Cont.

Marker	Related gene[a]	Annotation	Nearby lipid genes[b]	*Arabidopsis thaliana* homologous genes		Ref.
				Loci[c]	Annotation	
			sf00044.61	At4G19860	Acyl acceptor Acyltransferase	—
			sf00044.90	At1G13210	Translocase	—
			sf00044.113	At1G74320	Choline Kinase	[48]

Note: [a]Related genes are the genes containing or close to the screened markers.
[b]Nearby lipid genes refer to the genes located in the upstream and downstream of 500 Kb far from the marker.
[c]Only one of the homologous genes is listed in the table.
— refers to no known lipid genes in the location.

Candidate genes and oil components

To clarify the marker location and more genome information for OC and PC traits, we screened the genes that were close to the associated markers using the sesame genome assembly data. As a result, 36 candidate genes related to lipid pathway were identified (Table 6) [25–48]. We found that most candidate genes were involved in three pathways: (1) fatty acid and TAG (triacylglycerol) synthesis and elongation, e.g., C01.526, C01.548 and C01.928; (2) TAG degradation, e.g., C01.601; and (3) fatty acid dehydrogenation, e.g., C04.786 (Stearoyl-ACP desaturase, determining the ratio of saturated and unsaturated fatty acids) (Table S4). Therefore, the seed oil content in the sesame accessions could be regulated by three factors, i.e., oil synthesis ability, oil degradation ability and oil component ratio (e.g., 18:1 and 18:2 fatty acids). In various accessions, any alleles of the genes involved in fatty acid and TAG synthesis, TAG degradation or dehydrogenase genes could give effects on oil content. To confirm this hypothesis, further studies of seed oil synthesis should be performed in the future.

Conclusion

We systematically explored the association mapping of seed oil and protein content traits in 369 worldwide sesame accessions using the 112 SSR markers. A significant negative correlationof the oil content with the protein content existed in the population. 19 SSR markers were associated with the oil content trait with high phenotypic variation explanation, and 24 SSR markers were associated with the protein content trait using GLM method. This association results would provide an efficient platform for seed development research and MAS breeding in sesame.

Methods

Plant materials

A population of 369 core sesame germplasm resources was chosen according to the genetic diversity analysis results and phenotypic variation [51]. These core genotypes comprised 318 lines from 19 provinces of China and 51 worldwide lines from the 15 countries, which were reserved at the Sesame Germplasm Bank of Henan Sesame Research Center (HSRC), HAAS (Henan Academy of Agricultural Sciences) (Table S1). All accessions were grown with three replications at three different experimental stations of Yuanyang (113.96°E, 35.05°N), Pingyu (114.62°E, 32.97°N), and Xinyang (114.08°E, 32.13°N) in 2011 and 2012. Five or six young leaves of individual accessions were collected and reserved at −70°C for DNA extraction.

Oil and protein content analysis

After harvested, ~20 g of seeds were collected from five plants per line, and the seed OC and PC were measured on infrared determination equipment (Perten DA7250, Sweden) according to the manufactures' instructions. The standard curve for measuring the sesame oil and protein contents had been established according to the chemical analysis results of 300 sesame accessions (Unpublished data, HSRC). Three replications of each samples were assayed for phenotypic analysis. Mean value, broad-sense heritability and correlation coefficient of the phenotypic data were analyzed using the statistical analysis system software (SAS Institute Inc. 2002) [72].

SSR genotyping

The 112 polymorphic SSR pairs were selected from our SSR marker bank [24,51,73] (Table S2). DNA extraction, PCR amplification, electrophoresis and SSR genotyping analysis were performed according to the methods described by Zhang et al. [24]. The total number of polymorphic alleles at each SSR locus was calculated according to the results of all 369 lines. The polymorphic SSR alleles presented only within 4 (1%) or fewer accessions were recorded as rare alleles.

Linkage disequilibrium (LD)

As the population structure could result in the spurious associations between phenotypes and marker loci, we analyzed the extent and structure of LD within the population before selecting the appropriate association mapping strategy. To assay whether the 112 polymorphic SSR markers were segregated independently or not, LD analysis was conducted according to the dedicated procedure of the TASSEL software [49]. Both D' and r^2 were used for quantifying LD values [74,75]. Significance (P values) of D' for each SSR pairs was determined with 100,000 permutations.

Population structure and relatedness analysis

The population structure was determined using STRUCTURE 2.2 [76]. The mixture model and the independent allele frequency model were used to analyze the population dataset. Five runs of STRUCTURE were carried out for each number of populations (K) (from 1–10), and each run started with 10,000 burn-ins, followed by 100,000 iterations. While performing the STRUCTURE, we assumed that the inferred population accord with Hardy Weinberg equilibrium (HWE) and the loci are unlinked. To correct the relatedness of individuals in further analyses, the relatedness between individuals (relative kinship) was evaluated using SPAGeDi 1.2 software [77]. The matrix with the relative kinship coefficients (*K* matrix) was applied for association analysis using the Mixed Linear Model(MLM, Q+K)method.

Association mapping and marker distribution in sesame genome

Associations between the SSR markers and the oil and protein content traits were investigated using both methods of the general linear model (GLM, Q) and the mixed linear model (MLM, Q+*K*) in TASSEL 2.1 described by Bradbury et al. [49]. The mean value of the markers at $P<0.005$ was used for determining the significance of marker-trait associations.

To determine distributions of the associated markers in sesame genome, we carried out the alignment of SSR markers and transcripts with the updated sesame genome data [70,71]. In the present genome assembly, the number of N50 scaffold was 14, and the number of N90 scaffold was 64. 29,798 gene models were identified. Among related scaffolds or contigs, the sesame lipid synthesis related genes were identified according to the homologous comparison using the genes of *A. thaliana* from the Acyl Lipids pathway database [78] as queries.

Supporting Information

Table S1 Origins of 369 sesame accessions.

Table S2 Diversity statistics of 112 SSR markers in 369 sesame accessions and locations in the sesame draft genome

Table S3 Association mapping of the OC and PC traits using MLM method.

Table S4 Functions and involved pathways of the candidate genes.

Author Contributions

Conceived and designed the experiments: HZ. Performed the experiments: CL HM LW. Analyzed the data: CL HM LW. Contributed reagents/materials/analysis tools: TZ XH. Wrote the paper: CL HM LW.

References

1. Brummer EC, Graef GL, Orf J, Wilcox JR, Shoemaker RC (1997) Mapping QTL for seed protein and oil content in eight soybean populations. Crop Sci 37: 370–378.
2. Guo SX (2011) Genetic and QTL analysis of oil content and other important quality traits on *Brassica Napus* L.). PhD thesis. Sichuan Agricultural University, Rice Institute.
3. Turhan H, Citak N, Pehlivanoglu H, Mengul Z (2010) Effects of ecological and topographic conditions on oil content and fatty acid composition in sunflower. Bulgarian J Agri Sci 16:553–558.
4. Chen SL, Li YR, Cheng ZS, Liao BS, Lei Y, et al (2009) Heterosis and Genetic Analysis of Oil Content in Peanut Using Mixed Model of Major Gene and Polygene. Scientia Agricultura Sinic 42:3048–3057.
5. Anilakumar KR, Pal A, Khanum F, Bawa AS (2010) Nutritional, medicinal and industrial uses of sesame (*Sesamum indicum* L.) seeds- an overview. Agric Conspec Sci 75:159–168.
6. Bedigian D, Harlan JR (1986) Evidence for cultivation of sesame in the ancient world. Econ Bot 40: 137–154.
7. Zhang HY (2013) Sesame genetic and breeding status and development strategy in China. In: Chinese Journal of Oil Crop Sciences Editorial Office, editors. VII China National Oil Crops Conference altar Paper compilations. Wenchang, China. pp. 29–36.
8. Jin UH, Lee JW, Chung YS, Lee JH, Yi YB, et al (2001) Characterization and temporal expression of a ω-6 fatty acid desaturase cDNA from sesame (*Sesamum indicum* L.) seeds. Plant Sci 161:935–941.
9. Chun JA, Jin UH, Lee JW, Yi YB, Hyung NI, et al (2003) Isolation and characterization of a myo-inositol 1-phosphate synthase cDNA from developing sesame (*Sesamum indicum* L.) seeds: functional and differential expression, and salt–induced transcription during germination. Planta 216:874–880.
10. Leduc V, Moneret-Vautrin DA, Tzen JTC, Morisset M, Guerin L, et al (2006) Identification of oleosins as major allergens in sesame seed allergic patients. Allergy 61:349–356.
11. Magni C, Ballabio C, Restani P, Fuggetta D. Alessandri C, et al (2010) Molecular insight into IgE-mediated reactions to sesame (*Sesamum indicum* L.) seed proteins. Ann Allerg Asthma Im 105:458–464.
12. Wei W, Zhang Y, Lv H, Li D, Wang L, Zhang X (2013) Association analysis for quality traits in a diverse panel of Chinese sesame (*Sesamum indicum* L.) germplasm. J Integr Plant Biol 55:745–758.
13. Suh MC, Kim MJ, Hur CG, Bae JM, Park YI, et al (2003) Comparative analysis of expressed sequence tags from *Sesamum indicum* and *Arabidopsis thaliana* developing seeds. Plant Mol Biol 52:1107–1123.
14. Ke T, Dong C, Han M, Zhao Y, Chen H, et al (2011) Analysis of expression sequence tags from a full-length-enriched cDNA library of developing sesame seeds (*Sesamum indicum*). BMC Plant Biol 11:180.
15. Singkham N, Jogloy S, Kesmala T, Swatsitang P, Jaisil P, et al (2010) Genotypic variability and genotype by environment interactions in oil and fatty acids in high, intermediate, and low oleic acid peanut genotypes. J Agr Food chem 58:6257–6263.
16. Zhao J, Artemyeva A, Del Carpio DP, Basnet RK, Zhang N, et al (2010) Design of a *Brassica rapa* core collection for association mapping studies. Genome 53:884–898.
17. Zou J, Jiang C, Cao Z, Li R, Long Y, et al (2010) Association mapping of seed oil content in Brassica napus and comparison with quantitative trait loci identified from linkage mapping. Genome 53:908–916.
18. Abdurakhmonov IY, Saha S, Jenkins JN, Buriev ZT, Shermatov SE, et al (2009) Linkage disequilibrium based association mapping of fiber quality traits in *G. hirsutum* L. variety germplasm. Genetica 136:401–417.
19. Ecke W, Clemens R, Honsdorf N, Becker HC (2010) Extent and structure of linkage disequilibrium in canola quality winter rapeseed (*Brassica napus* L.). Theor Appl Genet 2010, 120:921–931.
20. Eleuch L, Jilal A, Grando S, Ceccarelli S, Schmising MVK, et al (2008) Genetic diversity and association analysis for salinity tolerance, heading date and plant height of barley germplasm using simple sequence repeat markers. J Integr Plant Biol 50:1004–1014.
21. Famoso AN, Zhao K, Clark RT, Tung CW, Wright MH, et al (2011) Genetic architecture of aluminum tolerance in rice (*Oryza sativa*) determined through genome-wide association analysis and QTL mapping. PLoS Genet 7:e1002221.
22. He D, Xing H, Zhao J, Li T, Tang Y, et al (2011) Population structure inferring and association analysis of fiber quality in cultivated cotton. J Northwest A & F Univ (Nat Sci Ed) 39:103–112.
23. Wei W, Qi X, Wang L, Zhang Y, Wei H, et al (2011) Characterization of the sesame (*Sesamum indicum* L.) global transcriptome using Illumina paired–end sequencing and development of EST–SSR markers. BMC genomics 12:451.
24. Zhang H, Wei L, Miao H, Zhang T, Wang C (2012) Development and validation of genic–SSR markers in sesame by RNA-seq. BMC genomics 13:316.
25. Pagnussat GC, Yu HJ, Ngo QA, Rajani S, Mayalagu S., et al (2005) Genetic and molecular identification of genes required for female gametophyte development and function in Arabidopsis. Development 132:603–614.
26. Salas JJ, Ohlrogge JB (2002) Characterization of substrate specificity of plant FatA and FatB acyl-ACP thioesterases. Arch Biochem Biophys 403:25–34.
27. Tilton G, Shockey J, Browse J (2000) Two families of acyl-CoA thioesterases in Arabidopsis. Biochem Soc Trans 28:946–947.
28. Quilichini TD, Friedmann MC, Samuels AL, Douglas CJ (2010) ATP-Binding cassette transporter G26 is required for male fertility and pollen exine formation in Arabidopsis. Plant Physiol 154:678–690.
29. Zhang M, Fan JL, Taylor DC, Ohlrogge JB (2009) DGAT1 and PDAT1 Acyltransferases have overlapping functions in Arabidopsis triacylglycerol biosynthesis and are essential for normal pollen and seed development. Plant Cell 21:3885–3901.
30. Cheng D, Nelson TC, Chen J, Walker SG, Wardwell-Swanson J, et al (2003) Identification of acyl coenzyme A:Monoacylglycerol acyltransferase 3, an intestinal specific enzyme implicated in dietary fat absorption. J Biol Chem 278:13611–13614.
31. Panikashvili D, Shi JX, Schreiber L, Aharoni A (2011) The Arabidopsis ABCG13 transporter is required for flower cuticle secretion and patterning of the petal epidermis. New Phytolt 190:113–124.
32. Xue HW, Hosaka K, Plesch G, Mueller-Roeber B (2000) Cloning of *Arabidopsis thaliana* phosphatidylinositol synthase and functional expression in the yeast pis mutant. Plant Mol Biol 42:757–764.
33. Kachroo A, Shanklin J, Whittle E, Lapchyk L, Hildebrand D, Kachroo P (2007) The Arabidopsis stearoyl-acyl carrier protein-desaturase family and the contribution of leaf isoforms to oleic acid synthesis. Plant Mol Biol 63:257–271.
34. Brown A, Affleck V, Kroon J, Slabas A (2009) Proof of function of a putative 3-hydroxyacyl-acyl carrier protein dehydratase from higher plants by mass spectrometry of product formation. FEBS lett 583:363–368.

35. Bird D, Beisson F, Brigham A, Shin J, Greer S, et al (2007). Characterization of Arabidopsis ABCG11/WBC11, an ATP binding cassette (ABC) transporter that is required for cuticular lipid secretion. Plant J 52:485–498.
36. Chen M, Han GS, Dietrich CR, Dunn TM, Cahoon EB (2006) The essential nature of sphingolipids in plants as revealed by the functional identification and characterization of the Arabidopsis LCB1 subunit of serine palmitoyltransferase. Plant Cell 18:3576–3593.
37. Xing S, Zachgo S (2008) ROXY1 and ROXY2, two Arabidopsis glutaredoxin genes, are required for anther development. Plant J 53:790–801.
38. Schnurr J, Shockey J, Browse J (2004). The acyl-CoA synthetase encoded by LACS2 is essential for normal cuticle development in Arabidopsis. Plant Cell 16:629–642.
39. Chen M, Markham JE, Cahoon EB (2012) Sphingolipid delta 8 unsaturation is important for glucosylceramide biosynthesis and low-temperature performance in Arabidopsis. Plant J 69:769–781.
40. Haselier A, Akbari H, Weth A, Baumgartner W, Frentzen M (2010) Two closely related genes of Arabidopsis encode plastidial cytidinediphosphate diacylglycerol synthases essential for photoautotrophic growth. Plant Physiol 153:1372–1384.
41. Arondel V, Lemieux B, Hwang I, Gibson S, Goodman HM, et al (1992) Map-based cloning of a gene controlling omega-3fatty-acid desaturation in Arabidopis. Science 258:1353–1355.
42. Yasuno R, Wada H (2002) The biosynthetic pathway for lipoic acid is present in plastids and mitochondria in Arabidopsis thaliana. FEBS Lett 517:110–114.
43. Yamaoka Y, Yu Y, Mizoi J, Fujiki Y, Saito K, et al (2011) PHOSPHATIDYL-SERINE SYNTHASE1 is required for microspore development in *Arabidopsis thaliana*. Plant J 67:648–661.
44. Engeseth NJ, Pacovsky RS, Newman T, Ohlrogge JB (1996) Characterization of an Acyl-CoA-binding protein from *Arabidopsis thaliana*. *Arch Biochem Biophys* 331:55–62.
45. Michaelson LV, Zauner S, Markham JE, Haslam RP, Dësikan R, et al (2009) Functional characterization of a higher plant sphingolipid Delta4-desaturase: Defining the role of sphingosine and sphingosine-1-phosphate in Arabidopsis. Plant Physiol 149:487–498.
46. Awai K, Xu CC, Tamot B, Benning C (2006) A phosphatidic acid-binding protein of the chloroplast inner envelope membrane involved in lipid trafficking. Proc Natl Acad Sci 103:10817–10822.
47. Ogas J, Cheng JC, Sung ZR, Somerville C (1997) Cellular differentiation regulated by gibberellin in the *Arabidopsis thaliana* pickle mutant. Science 277:91–94.
48. Monks DE, Goode JH, Dewey RE (1996) Characterization of soybean choline kinase cDNAs and their expression in yeast and Escherichia coli. Plant Physiol 11: 1197–1205.
49. Laurentin HE, Karlovsky P (2006) Genetic relationship and diversity in a sesame (*Sesamum indicum* L.) germplasm collection using amplified fragment length polymorphism (AFLP). BMC genetics 7:10.
50. Zhang P, Zhang HY, Guo WZ, Zheng YZ, Wei LB, et al (2007) Genetic diversity analysis of *Sesamum indicum* L. germplasms using SRAP and EST-SSR markers. Acta Agro Sin 33:1696–1702.
51. Yue WD, Wei L, Zhang T, Li C, Miao H, et al (2012) Genetic diversity and population structure of germplasm resources in sesame (*Sesamum indicum* L.) by SSR markers. Acta Agro Sin 38:2286–2296.
52. Kim DH, Zur G, Danin-Poleg Y, Lee SW, Shim KB, et al (2002) Genetic relationships of sesame germplasm collection as revealed by inter-simple sequence repeats. Plant Breed 121:259–262.
53. Kumar H, Kaur G, Banga S (2012) Molecular characterization and assessment of genetic diversity in sesame (*Sesamum indicum* L.) germplasm collection using ISSR markers. J Crop Improve 26:540–557.
54. Stankiewicz M, Gadamski G, Gawronski SW (2001) Genetic variation and phylogenetic relationships of triazine resistant and triazine susceptible biotypes of *Solanum nigrum*-analysis using RAPD markers. Weed Res 41:287–300.
55. Garris AJ, Tai TH, Coburn J, Kresovich S, McCOUCH S (2005) Genetic structure and diversity in *Oryza sativa* L. Genetics 169:1631–1638.
56. Ersoz ES, Yu J, Buckler ES (2007) Applications of linkage disequilibrium and association mapping in crop plants. *In*: Varshney RK, Tuberosa R, editors. Genomics-assisted crop improvement. Volume 1. Netherlands: Springer. pp. 97–119.
57. Yermanos DM, Hemstreet S, Saleeb W, Huszar CK (1972) Oil content and composition of the seed in the world collection of sesame introductions. J Am Oil Chem Soc 49:20–23.
58. Johnson LA, Suleiman TM, Lusas EW (1979) Sesame protein: A review and prospectus. J Am Oil Chem Soc 56:463–468.
59. El Tinay AH, Khattab AH, Khidir MO (1976) Protein and oil compositions of sesame seed. J Am Oil Chem Soc 53:648–653.
60. Uzun B, Arslan Ç, Furat Ş (2008) Variation in fatty acid compositions, oil content and oil yield in a germplasm collection of sesame (*Sesamum indicum* L.). J Am Oil Chem Soc 85:1135–1142.
61. Sarvamangala C, Gowda MVC, Varshney RK (2011) Identification of quantitative trait loci for protein content, oil content and oil quality for groundnut (*Arachis hypogaea* L.). Field Crops Res 122:49–59.
62. Zhao J, Becker HC, Zhang D, Zhang Y, Ecke W (2006) Conditional QTL mapping of oil content in rapeseed with respect to protein content and traits related to plant development and grain yield. Theor Appl Genet 113:33–38.
63. Yu J, Yu S, Fan S, Song M, Zhai H, et al (2012) Mapping quantitative trait loci for cottonseed oil, protein and gossypol content in a *Gossypium hirsutum*×*Gossypium barbadense* backcross inbred line population. Euphytica 187:191–201.
64. Hwang EY, Song Q, Jia G, Specht JE, Hyten DL, et al (2014) A genome-wide association study of seed protein and oil content in soybean. BMC Genomics 15:1.
65. Zhao JY, Becker HC, Zhang D, Zhang Y, Ecke W (2006) Conditional QTL mapping of oil content in rapeseed with respect to protein content and traits related to plant development and grain yield. Theor Appl Genet 113:33–38.
66. Chen G, Geng J, Rahman M, Liu X, Tu J (2010) Identification of QTL for oil content, seed yield, and flowering time in oilseed rape (*Brassica napus*). Euphytica 175:161–174.
67. Zheng P, Allen WB, Roesler K, Williams ME, Zhang S, et al (2008) A phenylalanine in DGAT is a key determinant of oil content and composition in maize. Nat Genet 40:367–372.
68. Hobbs DH1, Flintham JE, Hills MJ (2004) Genetic control of storage oil synthesis in seeds of Arabidopsis. Plant Physiol 136:3341–3349.
69. Shi A, Chen P, Zhang B, Hou A (2000) Genetic diversity and association analysis of protein and oil content in food-grade soybeans from Asia and the United States. Plant Breeding 129:250–256.
70. Zhang H, Miao HM, Wang L, Qu L, Liu H, et al (2013) Genome sequencing of the important oilseed crop *Sesamum indicum* L. Genome Biol 14:401.
71. Miao HM. The Sesame Genome Project and Sesame Genome Sequencing. Plant and Animal Genome XXII Conference. Plant and Animal Genome, 2014.
72. Holland JB, Nyquist WE, Cervantes-Martnez CT (2003) Estimating and interpreting heritability for plant breeding: An update. Plant breed reviews 22:9–112.
73. Wei LB, Zhang HY, Zheng YZ,Miao HM, Zhang TZ, et al (2009) A Genetic Linkage Map Construction for Sesame (*Sesamum indicum* L.). Genes & genomics 31:199–208.
74. Hedrick PW (1987) Gametic disequilibrium measures: proceed with caution. Genetics 117:331–341.
75. Weir BS (1996) Genetic data analysis 2: methods for discrete population genetic data. Sunderland: Sinauer Associates Inc Publishers. 445 p.
76. Pritchard JK, Stephens M, Donnelly P (2000) Inference of population structure using multilocus genotype data. Genetics 155:945–959.
77. Hardy OJ, Vekemans X (2002) SPAGeDi: a versatile computer program to analyse spatial genetic structure at the individual or population levels. Mol Ecol Notes 2:618–620.
78. Acyl Lipid: pathways Available at: http://aralip.plantbiology.msu.edu

15

Appraisal of Artificial Screening Techniques of Tomato to Accurately Reflect Field Performance of the Late Blight Resistance

Marzena Nowakowska[1], Marcin Nowicki[1], Urszula Kłosińska[1], Robert Maciorowski[2], Elżbieta U. Kozik[1]*

1 Department of Genetics, Breeding, and Biotechnology of Vegetable Crops, Research Institute of Horticulture, Skierniewice, Poland, **2** Unit of Economics and Statistics, Research Institute of Horticulture, Skierniewice, Poland

Abstract

Late blight (LB) caused by the oomycete *Phytophthora infestans* continues to thwart global tomato production, while only few resistant cultivars have been introduced locally. In order to gain from the released tomato germplasm with LB resistance, we compared the 5-year field performance of LB resistance in several tomato cultigens, with the results of controlled conditions testing (i.e., detached leaflet/leaf, whole plant). In case of these artificial screening techniques, the effects of plant age and inoculum concentration were additionally considered. In the field trials, LA 1033, L 3707, L 3708 displayed the highest LB resistance, and could be used for cultivar development under Polish conditions. Of the three methods using controlled conditions, the detached leaf and the whole plant tests had the highest correlation with thefield experiments. The plant age effect on LB resistance in tomato reported here, irrespective of the cultigen tested or inoculum concentration used, makes it important to standardize the test parameters when screening for resistance. Our results help show why other reports disagree on LB resistance in tomato.

Editor: Mark Gijzen, Agriculture and Agri-Food Canada, Canada

Funding: This research was supported by The Polish Ministry of Agriculture and Rural Development (Grant # HOR hn-801-15/13; "Searching for new sources of resistance to tomato late blight with regard to pathogenicity changes in Phytophthora infestans populations and attempts of markers identification linked with resistance genes"), carried out in the period 2008–2013. The funders had no role in study design, data collection and analysis, decision to publish, or preparation of the manuscript.

Competing Interests: The authors have declared that no competing interests exist.

* Email: elzbieta.kozik@inhort.pl

Introduction

Tomato (*Solanum lycopersicum* L.) is the fourth most economically important crop in the world: after rice, wheat, and soybean [1,2]. Of the 200 pathogens affecting tomato production worldwide, *Phytophthora infestans* (Mont.) de Bary, the oomycete causing late blight (LB), is the primary cause of tomato crop loss [2]. Losses are more frequent and more severe in areas where tomato is grown near potato [3].

Research on plant age-dependent expression of *P. infestans* resistance in tomato is needed, especially for the method of inoculation, plant evaluation, and inoculum concentration. In contrast, numerous studies have been done on the methods for testing *P. infestans* resistance in potato [4–14]. Tomato germplasm exist with resistance against *P. infestans* [15–20]; however, little information is available on standardized methods for evaluating LB resistance in tomato. Information on the correlation between seedlings and mature plant resistance is also limited [21–23]. Moreover, the last systematic comparison of methods for testing tomato LB resistance was reported over 40 years ago [24]. Lack of standardization already causes problems. For example, the wild tomato accession *S. pimpinellifolium* L 3708 was reported to have LB resistance conferred by one incompletely dominant gene [15,25]. That was amended a mere seven years later, when other studies reported complex inheritance of the trait [26–29]. Similarly, the accession *S. habrochaites* LA 1033 was designated *Ph - 4* and used as a LB resistance standard by several research centers [30–32], despite evidence of multiple QTLs conferring the trait [17,33–35]. Such discrepancies make it difficult for tomato breeders to use resistant germplasms.

The objective of this study was to (i) systematically compare the *P. infestans* resistance in the previously and recently reported resistant tomatoes using an array of methods, both in the field (naturally occurring infections) and with artificial inoculation (employing whole plant and detached leaf/leaflet bio-assays); (ii) determine the importance of the age-dependent expression of LB resistance; and (iii) investigate the level and stability of LB resistance of four sources of this trait in a multi-year field experiment.

Materials and Methods

Plant material

Tomato germplasm used in this study included commercial cultivars, breeding lines, landraces, and wild tomato accessions from the tomato germplasm collection at the Research Institute of Horticulture (RIH; Skierniewice, Poland), collectively referred to as 'cultigens'. Cultigens included three accessions of *S. pimpinelli-*

Table 1. Tomato cultigens used in the study, their origin, and LB resistance status (if known).

Cultigen[a]	Origin			Seed source
	country	province	collection site	
S. pimpinellifolium				
West Virginia 700 (WVa 700)[b]	-	-	-	INRA, Montfavet, France
L 3707 (PI 365951)[c]	Peru	-	-	Bar-Ilan University, Ramat-Gan, Israel
L 3708 (PI 365957)[d]	Peru	Lima	Pisiquillo	Bar-Ilan University, Ramat-Gan, Israel
S. habrochaites				
LA 1033 (6326A)[e]	Peru	Lambayeque	Hacienda Tanlis	NCSU, Raleigh, USA
LA 1353 (365934)	Peru	Cajamarca	Contumasa	TGRC, Davis, USA
LA 2552 (PE 36)	Peru	Cajamarca	Las Flores	TGRC, Davis, USA
LA 2650 (PI 503514)	Peru	Prura	Ayabaca	TGRC, Davis, USA
LA 407	Ecuador			TGRC, Davis, USA
S. huaylasense				
LA 1360 (PI 365952)	Peru	Ancash	Apricot	TGRC, Davis, USA
LA 1365 (PI 365953)	Peru	Ancash	Carnaquilloc	TGRC, Davis, USA
LA 1983	Peru	Ancash	Rio Manta	TGRC, Davis, USA
S. corneliomuelleri				
LA 1395 (PI 379014)	Peru	Amazonas	Chachapoyas	TGRC, Davis, USA
LA 1910	Peru	Huancavelica	Tambillo	TGRC, Davis, USA
S. peruvianum				
LA 1929	Peru	Ica	La Yapana	TGRC, Davis, USA
LA 2581	Chile	Arica i Parinacota	Chacarilla	TGRC, Davis, USA
LA 2744	Chile	Arica i Parinacota	Sobraya	TGRC, Davis, USA
S. lycopersicum				
LA 1673	Peru	Lima	Nana	TGRC, Davis, USA
LA 2416	-	-	-	TGRC, Davis, USA
LA 3845 (NC EBR-5)	-	-	-	NCSU, Raleigh, USA
LA 3846 (NC EBR-6)	-	-	-	NCSU, Raleigh, USA
'New Yorker' ('NY')[f]	-	-	-	INRA, Montfavet, France
LB susceptible control				
'Rumba'	-	-	-	PNOS Ożarów, Poland

[a]Plant introductions numbers (PI or PE) were added where available.
[b]Source of *Ph-2*; single incomplete-dominant gene mapped to the long arm of chromosome 10 [19,24,38].
[c]Source of race-non-specific LB resistance conferred by two independent genes: A partially-dominant gene and a dominant epistatic gene, both mapped to chromosome 9 [46,47].
[d]Source of race-specific LB resistance dubbed *Ph-3*; originally reported as conferred by single incomplete-dominant gene, corrected by subsequent research; resistance conferred by at least two QTLs [15,25–29,31–33,46,67].
[e]Source of polygenic LB resistance, conferred by several QTLs [16,17,31–35].
[f]Source of LB resistance dubbed *Ph-1*, conferred by a single completely dominant gene mapped to the distal end of chromosome 7 with morphological markers [36–39].

folium, five of *S. habrochaites*, three of *S. huaylasense*, two of *S. corneliomuelleri*, three of *S. peruvianum*, and four of *S. lycopersicum*. 'Rumba' (*S. lycopersicum*; PNOS Ożarów, Poland) served as the LB susceptible control, and it was readily infected by *P. infestans* under field conditions. The cultigens used in this study, their origin, and LB resistance background (if known) are listed in Table 1.

Plants were grown from seeds, transplanted at first true leaf stage into ∅ 10 cm plastic pots containing Classman Potgrond substrate (Lasland, Grady, Poland), and placed in a greenhouse. Growing plants were kept at approximately 24/18°C day/night for the tests under controlled conditions, or grown for 4 to 5 weeks in the greenhouse, and then transplanted to the field in the second half of May each year.

Field evaluations

In 2007 and 2008, all cultigens were evaluated for resistance against *P. infestans* under epiphytotic conditions at the Department of Genetics, Breeding, and Biotechnology experimental field area (RIH, Skierniewice, Poland). Four- to five-week old plants were transplanted to the field in a randomized complete block design, with three replications. Each plot consisted of ten *S. lycopersicum* plants or five plants of the wild tomato cultigens, spaced 50×100 cm (within rows × between rows, respectively). Additionally, plants of the LB susceptible 'Rumba' were planted around the border of the experimental field to ensure high and uniform pathogen pressure. We also included a *S. lycopersicum* cv. New Yorker ('NY') which carries the LB resistance gene *Ph-1* [36–

39]. No fungicide control was applied throughout the growing seasons in any of the field trials.

Field plots were inspected weekly throughout the season. When the susceptible control 'Rumba' foliage developed maximal area symptomatic for LB in a given year, disease ratings for each cultigen were collected. Observations were made by the same individual throughout the experiment to avoid that source of variation. LB lesions on the leaves and stems of each plant were rated using the modified scale of Zarzycka [40]:

- 1 – infection over 97.1 to 100% of the investigated plant organ
- 2 – infection over 87.1 to 97% of the investigated plant organ
- 3 – infection over 75.1 to 87% of the investigated plant organ
- 4 – infection over 50.1 to 75% of the investigated plant organ
- 5 – infection over 30.1 to 50% of the investigated plant organ
- 6 – infection over 18.1 to 30% of the investigated plant organ
- 7 – infection over 10.1 to 18% of the investigated plant organ
- 8 – infection over 3.1 to 10% of the investigated plant organ
- 9 – no infections or small lesions.

At the beginning of this study, such arithmetically biased methods of LB assessment were employed predominantly [5,8,15,19,20,24,25,31,35], and keep on being used until this day [9,16,21,26], although more accurate methods were developed [7,11,29].

Subsequently, the disease severity index (DSI) was calculated for each cultigen, respectively, as a mean of ratings for the plants, similar to other studies of this pathosystem [26,40].

In the initial field screenings performed in 2007 and 2008, four cultigens (WVa 700, L 3708, L 3707, LA 1033) exhibited high *P. infestans* resistance. In order to assess the stability of resistance in these cultigens, tests were conducted in 2009 to 2013, under natural LB infestations. Field trials were run in two locations: RIH, Skierniewice (Central Poland), and Boguchwała (Southern Poland), which are 300 km apart from one another. In these experiments, tomato plants were evaluated for resistance against *P. infestans* as described earlier.

Pathogen cultures and inoculum preparation

P. infestans isolates used in this study, collected from tomato plants grown in different regions of Poland and tested in a pilot study (Fig. 1), were deposited at RIH (Skierniewice; n = 19) or obtained from IHAR (the Plant Breeding and Acclimatization Institute - National Research Institute, Młochów, Poland; n = 27 isolates). Isolates were transferred from rye agar onto leaflets of cv. Rumba and cultured for at least two generations, each 7 to 8 days long, with incubations in darkness at 100% RH and 16°C [40]. Inoculum consisted of a sporangial suspension that was washed off the sporulating lesions on the 'Rumba' leaflets using distilled water. Sporangia counting was done with a haemocytometer, and final inoculum concentrations (inocula loads) were adjusted according to the assay protocol described below. Prior to dilution, the suspension was chilled for 2 h at 4°C, and then incubated at RT for 30 min.

The isolate IWP 13, collected in 2008 from our experimental field (RIH Skierniewice), was used in all subsequent tests. Detailed isolate characteristics are as follows: race according to Black [3,4,7,10,11], mating type A1, mtDNA haplotype Ia, and intermediate resistance to metalaxyl [41]. This isolate was chosen based on the results of three independent disease severity tests. These tests were performed with 46 isolates of *P. infestans* on the detached leaflets of cv. Rumba, where it induced the highest disease severity (DSI = 1.7±0.1). Moreover, it produced the smallest variation in symptoms among the isolates tested in a pilot study on 'Rumba' (Fig. 1).

Seedling tests

Resistance screening experiments under controlled conditions were conducted in growth chambers and in greenhouse at the RIH, Skierniewice, Poland. These assays included detached leaflet, detached leaf, and whole plant bio-assays, and are described in detail in the paragraphs that follow. In all test methods we evaluated the influence of plant developmental stage (age in weeks) and inoculum concentration on the disease severity in four tomato cultigens with various levels of *P. infestans* resistance/susceptibility: WVa 700, L 3708, LA 1033, and 'Rumba'. All tests were conducted in the spring and summer (April to September) of 2011 to 2013.

The detached leaflet assay

We determined the effects of the plant age (4- to 8-week old plant, in one week increments) and the inoculum concentration (5×10^3, 10^4, 5×10^4 sporangia/ml) on disease symptoms development. Each treatment combination (inoculum concentration vs. plant age) was tested in a series of three independent trials with 25 leaflets per cultigen. The third to fifth fully expanded leaves, counting from the plant's top, were collected. This criterion was established by our earlier experiments as well as by previous studies [42] on the relationship between the intensity of LB symptoms and leaf position on plants. Side leaflets were detached with scissors and immediately placed on wet cellulose wadding in a plastic tray. Sporangial suspension (40 µl) was placed on the center of the abaxial side of each leaflet. Trays with leaflets were covered with glass to maintain 90 to 100% RH and incubated at 16°C. After an initial 24 h incubation in the dark, the leaflets were turned abaxial side down and incubated at the same temperature and RH for 6 days under 12 h photoperiod and a light intensity (PPFD) of 650 $\mu mol/m^2/s$. Symptoms of LB were assessed 7 days after inoculation, using the DSI scale described earlier.

In our preliminary tests with 20 local isolates of the pathogen, individual leaflets of the resistant cultigens displayed high variability in LB reaction. Additionally, the size of necrotic lesions did not correspond with the intensity of sporulation. Within susceptible cultigens (i.e., compatible reactions) that was not observed, and in agreement with other studies [43] the sporulation often preceded the necrosis. These differences caused difficulties in classifying cultigens using this method. Therefore, at the time of assessment by the DSI scale described above, the inoculated leaflets were tested for sporulation intensity by shaking off the sporangia (vortexing each sample for 30 sec in 1 ml of water) and counting the sporangia with a haemocytometer.

The detached leaf assay

The severity of LB symptoms on detached tomato leaves was studied in 11 stages of development (5- to 15-week old plants, in one week increments), using three concentrations of inoculum (5×10^3, 10^4, 5×10^4 sporangia/ml). The third to fifth fully expanded leaves (counting from the plant's top) were excised from greenhouse-grown plants. The petiole of each leaf was immediately placed into distilled water (100 ml) in a plastic container with a ∅ 1 cm hole in the lid center. The leaves in plastic containers were then placed in a plastic box and hand-sprayed with sporangial suspension, until the upper surface of each excised tomato leaf was completely covered. The inoculated leaves were incubated in darkness for 24 h at 18°C, and 95 to 100% RH. Following this initial incubation, samples were then incubated for 6 days under conditions similar to the detached leaflet method.

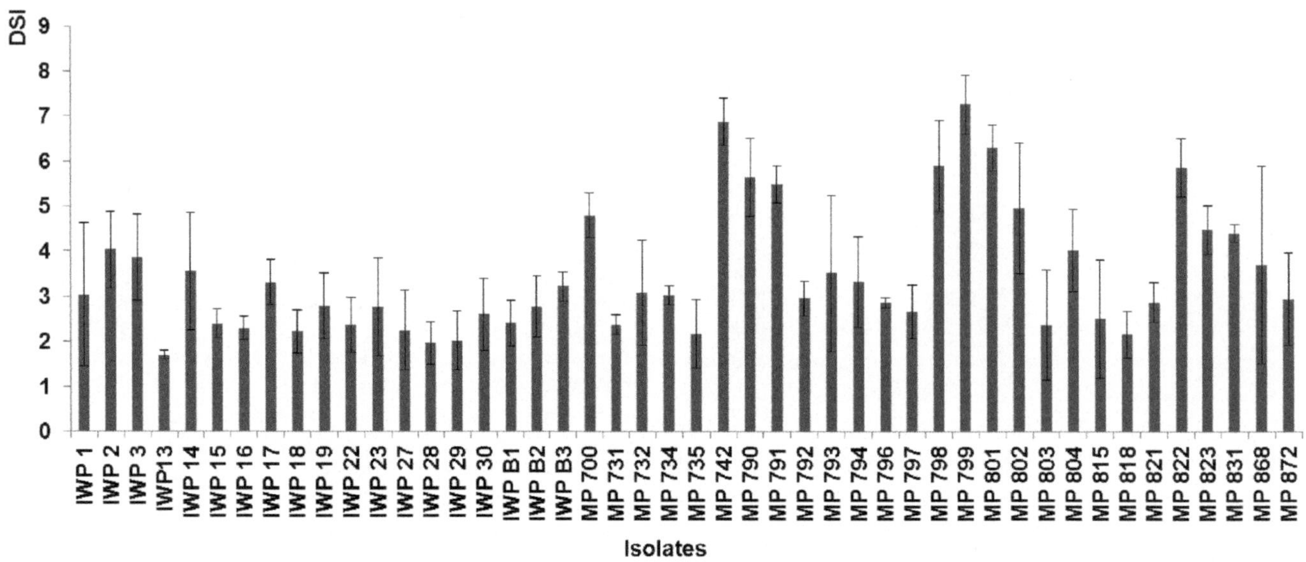

Figure 1. DSI from artificial *P. infestans* inoculations of 'Rumba' detached leaflets. A total of 46 local isolates were used. Each isolate was tested on 10 leaflets, in a series of three independent assays. Inoculum concentration of 5×10^4 sporangia/ml was used. Assay evaluations were performed on the 7th day after inoculation. Disease assessment scale was based on the % of leaflet area being infected; 1 = 100% area infected; 9 = lack of disease symptoms or few and small necrotic spots. Error bars indicate standard deviation (SD).

Leaves of each cultigen were evaluated using the scale described earlier (1 = 97.1–100% leaf area covered with lesions or dead; to 9 = no lesions). For each of the four cultigens undergoing evaluation, and for each treatment combination (inoculum concentration vs. plant age), a different number of leaves (12 to 24) per cultigen was used due to availability of plant materials. In addition, each treatment combination was examined in three independent sets of experiments.

The whole plant assay

LB severity on whole plants was tested at six developmental stages (3- to 8-weeks of age, in one week increments) of greenhouse-grown plants, using three concentrations of inoculum (5×10^3, 10^4, 5×10^4 sporangia/ml). Plants were hand-sprayed with the sporangial suspension until complete leaf coverage and excess run-off was observed. The inoculated seedlings were incubated in the dark at growth chamber for 24 h, at 16°C and 100% RH. After this initial incubation, the inoculated seedlings were grown at 16°C with 12 h of light. The plants were rated individually seven days after inoculation. The symptomatic area of both leaves and stems were evaluated using the DSI scale described earlier. Each treatment combination (developmental stage vs. inoculum concentration) was repeated three to seven times, with each cultigen represented by 12 to 72 plants depending on the experiment.

Statistical analyses

Data from all experiments were analyzed by means of the general linear model with year, location, cultigen, plant age, inoculum concentration, and their interaction as the tested variables. Means were separated with the Tukey multiple comparison procedure at significance level of 0.05. Regression and correlation analyses were used to compare results from the different testing methods. All calculations were done with the statistical software STATISTICA 8.0 (StatSoft, Inc. 2009).

Results

Field evaluations

Based on the results from the initial field experiments in 2007 and 2008 (Table 2), four tomato cultigens were chosen (LA 1033, L 3708, L 3707, WVa 700) from the original experimental group of 20 cultigens, to verify their *P. infestans* resistance in the field at two separate locations (Skierniewice, Boguchwała).

Comparative analysis of the LB resistance levels among the tested tomato cultigens indicated significant differences across the five years of study (2009 to 2013), at both locations (Table 3). The highest and most stable level of LB resistance in the field, comparable with the baseline years 2007 and 2008, was found in LA 1033 *S. habrochaites* (DSI ranging from 7.2 to 9.0, depending on the year). Plants of this cultigen were either free from any LB symptoms, or developed only slight infection symptoms (classes 6 to 8). Indeed, this cultigen demonstrated high levels of resistance in the field, even under strong *P. infestans* pressure. For example, in 2011, both locations experienced conditions particularly conducive to LB, as confirmed by disease intensities recorded for both susceptible tested tomato cultigens ('Rumba', 'NY'), while LA 1033 showed no LB symptoms (Table 3). Moreover, LA 1033 plants displayed a moderate infection intensity on leaves and stems only in the first week of September 2012 (DSI = 7.2±0.8) and 2013 (DSI = 7.8±0.7), in the Skierniewice experimental field.

Intensity of LB symptoms on the *S. pimpinellifolium* accessions L 3708 and L 3707 which ranged from DSI = 7.4 to 8.8, was generally comparable with *S. habrochaites* LA 1033, and depended on the year and the location. The WVa 700 plants displayed significant differences in LB intensity levels, depending on the location and the year (Table 3). In Skierniewice, we recorded partial infection of this cultigen in 2011 and 2013 (DSI = 4.9±1.0 and 4.5±1.1, respectively). In the remaining years of the study, the WVa 700 plants exhibited very low LB intensities (DSI = 8.0 to 8.4). In the Boguchwała study, we observed consistently high levels of disease severity in this cultigen in all years of testing (DSI ranging from 4.3 to 6.1). The 'NY' (*Ph-1*)

Table 2. Tomato cultigens' responses to *P. infestans* under natural infection in the Skierniewice experimental field.

Cultigen	DSI[a]		Mean
	2007	2008	
WV 700	8.0±0.5	8.3±0.3	8.2
L 3707	8.0±0.4	9.0±0.0	8.5
L 3708	8.1±0.6	9.0±0.0	8.5
LA 1033	8.8±0.1	9.0±0.0	8.9
LA 1353	6.3±1.1	6.9±0.7	6.6
LA 2552	7.0±0.2	5.5±0.3	6.3
LA 2650	5.0±1.1	5.8±0.8	5.4
LA 407	4.7±1.2	5.3±1.1	5.0
LA 1360	3.1±1.2	2.7±1.1	2.9
LA 1365	3.8±1.5	2.8±0.8	3.3
LA 1983	1.9±0.2	1.5±0.2	1.7
LA 1395	2.6±1.2	2.8±0.7	2.7
LA 1910	1.5±0.2	1.6±0.3	1.6
LA 1929	1.4±0.2	1.8±0.2	1.6
LA 2581	1.0±0.0	1.0±0.0	1.0
LA 2744	1.5±0.2	1.4±0.3	1.5
LA 1673	2.6±1.5	3.0±1.1	2.8
LA 2416	2.1±1.0	2.4±0.7	2.3
LA 3845	1.2±0.2	1.1±0.2	1.2
LA 3846	1.0±0.0	1.0±0.0	1.0
'NY'	2.6±1.1	1.4±0.2	2.0
'Rumba'	2.1±1.6	1.2±0.2	1.7

[a]The disease assessment scale is based on the % area of leaf and stem infection (1 = 100% percent area infected; 9 = lack of symptoms or few small necrotic spots). Disease symptoms were scored yearly, when LB susceptible control 'Rumba' turned fully symptomatic for the infection. Data presented is the mean ± SD from DSI recorded on leaves and stems. $LSD_{0.05}$ calculated according to Tukey procedure for comparing the cultigens in each year: 1.59, and for comparing the years for each cultigen: 0.99.

developed intensive LB symptoms, at levels comparable with the susceptibility control 'Rumba', irrespective of the location or year of study (DSI = 1.0 to 4.3).

The detached leaflet assay

Experimental results indicated significant effects of the plant age on the LB intensity levels across all tested cultigens. Leaflets from 4- to 5-week old plants of both *S. pimpinellifolium* cultigens tested (WVa 700, L 3708) exhibited higher degrees of infection than those from 7- to 8-week old plants, irrespective of the inoculum concentration used (Fig. 2). The LA 1033 leaflets showed the highest intensity of LB symptoms when inoculated with the 5×10^4 sporangia/ml (DSI = 4.8 to 5.8, dependent on plant age). Comparatively, at the two lower inoculum concentrations used, the disease symptoms proved significantly decreased at all developmental stages tested.

We found a broad range of intra-cultigen variation in the lesion area of individual leaflets of resistant cultigens (WVa 700, L 3708, LA 1033) inoculated with *P. infestans*. This variation was observed even in the oldest specimens (8-week old) tested with the lowest inoculum load (Fig. 2). The observed high variability of this assay makes it impossible to unequivocally distinguish genotypes exhibiting specific LB reactions. Regardless of the differences in the necrotic area, the LB resistant cultigens showed a low intensity of sporulation and were not significantly different from each other. 'Rumba' displayed sporulation intensity of 23,4±5.3 thousands of sporangia/mm^2, exceeding those of the resistant cultigens approximately 50-fold (WV 700: 0.7±0.6; L 3708: 0.5±0.5; LA 1033: 0.5±0.5 thousands of sporangia/mm^2, respectively). Also, only in case of 'Rumba', did the size of necrotic spots correspond well with the intensity of sporulation. Finally, in contrast to the resistant cultigens tested, 'Rumba' showed consistently high and uniform disease symptoms in all detached leaflet assays (Fig. 2), regardless of plant age or inoculum load.

The detached leaf assay

Significant effects of all variables tested (cultigen, plant age, and inoculum concentration) on the LB intensity were observed. The highest variation was observed for the interaction between cultigen and plant age, and the lowest was between plant age and inoculum concentration.

'Rumba' leaves displayed the highest degree of LB infection (DSI = 1.0 to 1.1), regardless of the plant age, and lacked significant differences in their respective DSI scores (Fig. 3). Relationships between plant age and degree of disease symptoms were observed for all three remaining cultigens in the study (WVa 700, L 3708, LA 1033). Additionally, we noted a different trend in each of the three resistant cultigens. The WVa 700 leaves showed low levels of LB intensity, irrespective of plant age or inoculum concentration. The lowest LB symptoms levels displayed by L 3708 were similar to those of WVa 700, but proved comparatively more dependent on plant age and inoculum concentration (Fig. 3).

Table 3. Late blight symptoms severity on chosen tomato cultigens under natural *P. infestans* field infections.

Year[a]	Cultigen	DSI[b]	
		Skierniewice	Boguchwała
2009	LA 1033	8.9±0.1	8.7±0.4
2010		8.8±0.2	8.8±0.2
2011		9.0±0.0	8.0±1.1
2012		7.2±0.8	9.0±0.0
2013		7.8±0.7	8.2±0.6
2009	WVa 700	8.0±0.4	6.1±0.9
2010		8.4±0.3	5.7±1.0
2011		4.9±1.0	5.0±0.7
2012		8.0±0.4	5.7±0.5
2013		4.5±1.1	4.3±1.3
2009	L 3708	7.9±0.4	7.9±0.3
2010		8.6±0.2	8.5±0.2
2011		8.7±0.0	8.0±0.3
2012		8.3±0.4	8.8±0.2
2013		8.2±0.4	7.9±0.7
2009	L 3707	7.8±0.4	7.4±0.6
2010		8.5±0.3	7.5±0.8
2011		8.3±0.7	8.0±0.3
2012		8.3±0.2	8.2±0.2
2013		7.4±0.5	8.3±0.7
2009	'NY'	4.3±1.4	1.0±0.0
2010		2.4±1.2	1.2±0.2
2011		1.0±0.0	1.2±0.2
2012		1.0±0.0	2.6±0.3
2013		1.7±1.3	1.0±0.0
2009	'Rumba'	3.6±1.4	1.0±0.0
2010		2.7±1.7	1.4±0.2
2011		1.0±0.0	1.1±0.2
2012		1.0±0.0	1.0±0.0
2013		1.8±0.9	1.0±0.0

[a]Data were collected from 2009 to 2013 in two locations (Skierniewice, Boguchwała).
[b]The disease assessment scale is based on the % area of leaf and stem infection (1 = 100% percent area infected; 9 = lack of symptoms or few small necrotic spots). Disease symptoms were scored yearly, when LB susceptible control 'Rumba' reached maximal intensity of LB symptoms. Data presented is the mean ± SD from DSI recorded on leaves and stems. $LSD_{0.05}$ calculated according to Tukey procedure for comparing cultigens in each year and in either location: 0.76, and for comparing cultigens in each year and between locations: 0.55.

In L 3708, at the lowest inoculum concentration tested, the impact of plant age on LB intensity was evident, as response dropped below the significance level in 10-week old and older plants. Cultigen LA 1033 also showed varied reactions to *P. infestans* inoculation, dependent on plant age; the observed differences greatly surpassed those noted for WVa 700 or L 3708 (Fig. 3). The lowest degree of disease symptoms in LA 1033 was observed in the leaves of 15-week old plants (DSI = 8.8±0.4), after application of the lowest concentration of inoculum.

The whole plant assay

As in the detached leaf experiments, also in the whole plant assays we found significant effects of all variables tested (cultigen, plant age, inoculum concentration) on the intensity of LB symptoms. This necessitated an independent analysis for each variable.

In 'Rumba', we observed a significant interaction between the disease severity and the inoculum concentration (Fig. 4). Plants inoculated with the highest inoculum concentration exhibited high disease severity levels (DSI = 1.0 to 1.6, depending on plant age) in all tested stages of development (3 to 8 weeks of age). In contrast, plants treated with the lowest inoculum concentration showed lower intensity of disease symptoms (DSI = 2.6 to 5.4, depending on plant age). This wide DSI range was dependent on plant age, and for each developmental stage tested was distributed over at least two of the nine severity rating classes (reaching max. 7 classes of spread). A low degree of disease symptoms coupled with large variation in the LB susceptible 'Rumba' under 5×10^3 sporangia/

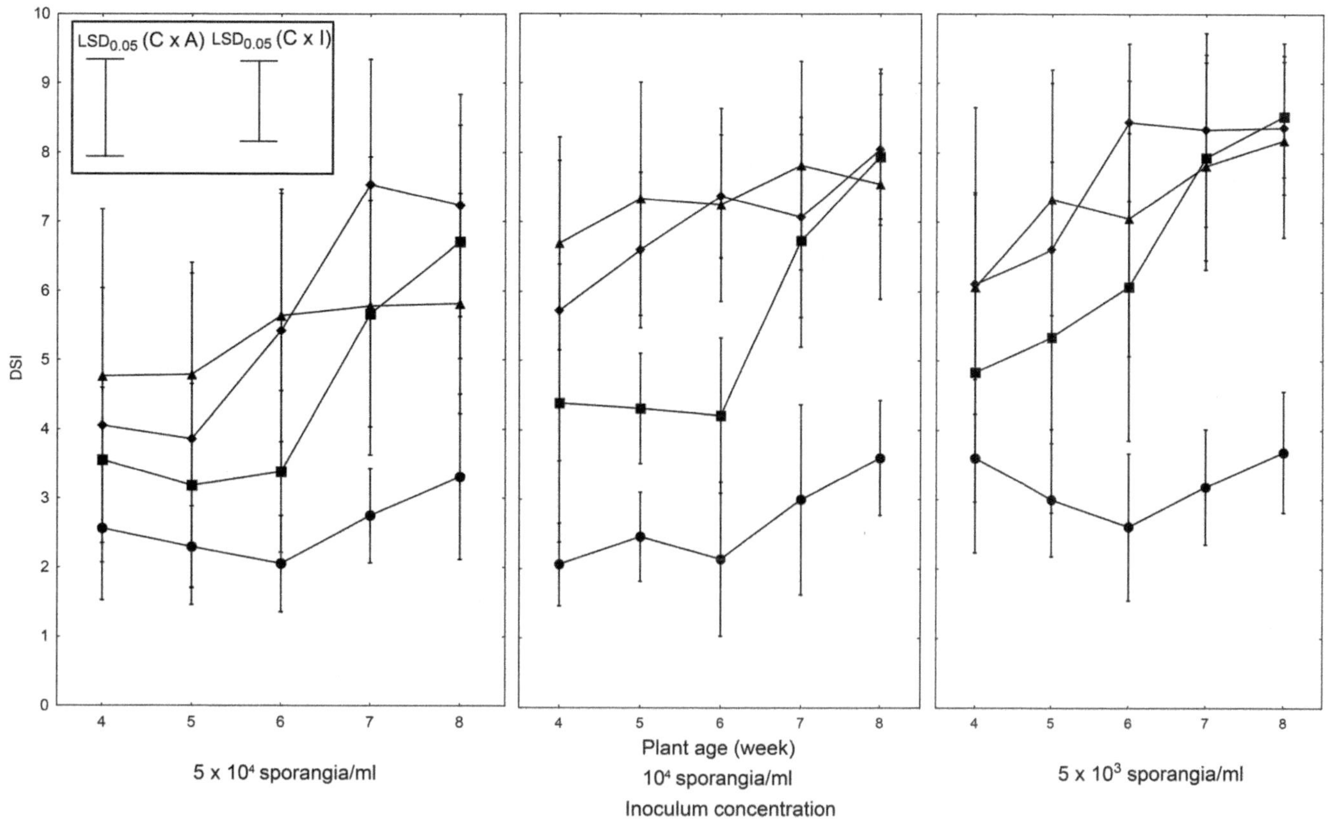

Figure 2. Severity of LB symptoms on tomato cultigens in the detached leaflet assay. Detached leaflets of tomato cultigens (● 'Rumba'; ■ WVa 700; ♦ L 3708; ▲ LA 1033) were inoculated with the *P. infestans* isolate IWP 13, at indicated age (weeks), and concentration. Assay evaluations were performed on the 7th day after inoculation. Data shown are the means of ratings from at least three independent experiment sets for each combination; 25 leaflets per cultigen were inoculated per experiment. Vertical bars at each data point signify the standard deviations (SD). $LSD_{0.05}$ calculated according to Tukey procedure (inset) for comparing cultigens at each plant age and inoculum concentration (C×I): 1.34, and for comparing inoculum concentration for each cultigen and plant age (C×A): 1.43.

ml, rendered this inoculum concentration unsuitable to correctly differentiate between the LB resistant and LB susceptible plants.

Of the resistant cultigens, WVa 700 had the lowest variation in LB symptoms at different phases of development. Indeed, WVa 700 exhibited a low degree of infection (DSI>7) at all stages of development, regardless of inoculum concentration (Fig. 4), with one exception: the 3-week old plants had increased disease severity with DSI ratings spread over all nine rating classes at the highest inoculum concentration used. We found the lowest, most uniform levels of LB intensity in 7-week old and older WVa 700 plants, regardless of inoculum concentration.

In L 3708, we observed variable levels of LB symptoms intensity, dependent on plant age and inoculum concentration (Fig. 4). Generally, younger plants (3- and 4-week old) exhibited distinctly higher levels of LB sysmptoms than older plants (7- and 8-week old). We also observed higher variation in the range of severity ratings in younger plants compared with older plants. Moreover, the range of variation in severity ratings depended on the inoculum concentration used and correspondingly increased. In addition, we observed an inverse relationship between plant age and inoculum concentration on LB intensity levels. Older plants showed a more uniform intensity of disease symptoms across inoculum loads tested, than did the younger plants. Indeed, among the L 3708 developmental stages tested, the lowest and most uniform levels of LB symptoms were observed in 7- and 8-week old plants at all inoculum concentrations tested.

LA 1033 displayed a more diverse response to *P. infestans* challenges than WVa 700 or L 3708 plants. This reaction, however, remained dependent on plant age at all tested inoculum concentrations (Fig. 4). A comparative analysis of DSI values for all inoculum concentrations at all developmental stages examined in this cultigen showed greater severity of LB symptoms, compared with WVa 700 or L 3708. The lowest level of LB intensity (DSI = 8.2±0.6) was found in 8-week old plants inoculated with the lowest concentration of inoculum. Higher inoculum concentrations caused an increase in the severity of symptoms, with only sporadic significant increases in the variation of severity ratings (Fig. 4).

In general, all control conditions assays indicated an impact of plant age on the intensity of LB symptoms at levels specific for a given tomato cultigen. Symptoms of LB tended to decrease with plant age. High variability noted for the detached leaflet assay makes this method unreliable for standard testing of the LB resistance, or requires additional assessments (e.g., sporulation intensity), but the assays may be used for efficient identification of the LB susceptible individuals, such as during the early stages of selection and breeding.

Cross-test comparison

Pooled data from the controlled condition tests (detached leaflet, detached leaf, whole plant) from all treatment combinations (cultigens, inoculum concentrations, plant ages), were individually

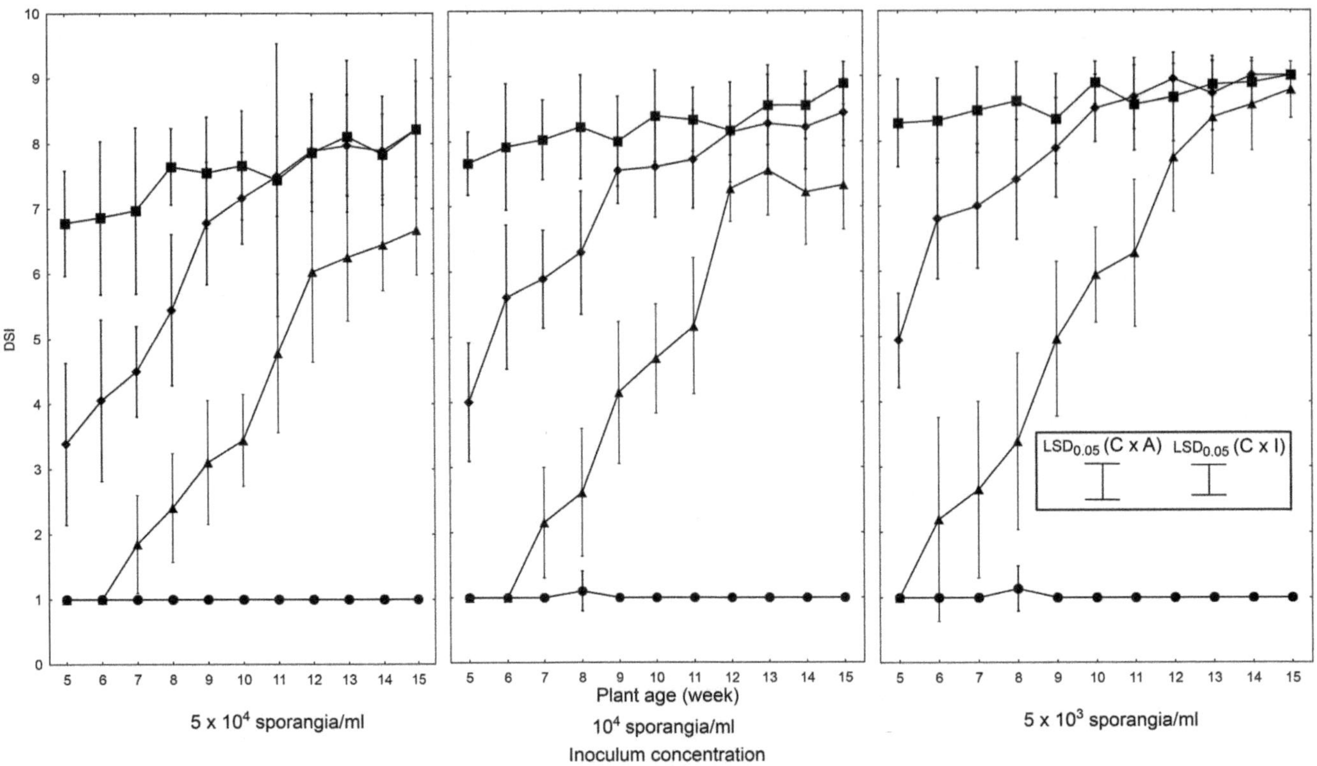

Figure 3. Severity of LB symptoms on tomato cultigens in the detached leaves assay. Third to fifth fully expanded leaves were collected for testing from the plants (● 'Rumba'; ■ WVa 700; ◆ L 3708; ▲ LA 1033) at indicated ages (weeks). Detached leaves of the tested tomato cultigens (12 to 24 leaves per cultigen and per developmental stage) were inoculated with suspension of the *P. infestans* isolate IWP 13, by spraying at indicated concentration. Presented data, for each treatment combination, are the means of ratings from three independent experiment sets. Vertical bars at each data point signify the standard deviations (SD). $LSD_{0.05}$ calculated according to Tukey procedure (inset) for comparing cultigens at each plant age and inoculum concentration (C×I): 0.58, and for comparing inoculum concentration for each cultigen and plant age (C×A): 0.74.

compared with the data of field experiments. All tested laboratory techniques showed significant linear relationships with the field results. But, the detached leaf assay and whole plant assay correlated better with the field assay than did the detached leaflet assay. The determination coefficients were: 0.94, 0.83, and 0.41, respectively (Fig. 5). The stronger relationship observed for the detached leaf assay and whole plant assay with reference to the field data resulted from lower variation within the plant materials, and the domination of extreme DSI values from the field observations.

In summary, this indicates that the detached leaf and the whole plant assays may serve as reliable tools for testing tomato LB resistance. These assays could thus replace the field trials at the early stages of testing. Additionally, it is noteworthy that at high disease intensity (DSI = 1 to 3) the detached leaflet assay showed good correlation with the field results, which further supports our suggestion to use this method for initial testing in tomato LB resistance breeding.

Discussion

The choice of methods for accurate testing of genotypes for LB resistance is an important area for plant breeding programs, including the tomato-*P. infestans* pathosystem. Lack of agreement among published reports for LB resistance in several tomato cultigens (e.g., L 3708, LA 1033) prompted the current study. Here, our aim was to standardize and compare several methods for testing the *P. infestans* resistance under controlled conditions using four tomato cultigens, over several plant ages and inoculum concentrations. This would allow us to establish a uniform approach to assess the benefit of each cultigen for use in breeding for LB resistance. We also attempted to optimize the methods and to relate the results to those from natural field infection experiments.

Results of the multi-year (2008 to 2013) field experiments, from two locations (approx. 300 km apart), show that *S. habrochaites* LA 1033 had the lowest and most stable *P. infestans* infection levels of the cultigens tested. These findings are in agreement with other studies for LA 1033 reporting high resistance under natural field infection against a diverse set of *P. infestans* isolates, in the USA [17,33,35]. In other studies, isolates from Taiwan succeeded in infecting LA 1033 [16,31,34]. Our results of LA 1033 showing modest LB symptoms in the field could not be confirmed with detached leaf assay using *P. infestans* isolates derived from symptomatic LA 1033 plants (Nowakowska et al., unpublished data). The latter observation is in agreement with potato-LB studies, where *P. infestans* isolates derived from LB-symptomatic potato plants, carrying the *Rpi-phu1* gene conferring high levels of potato LB resistance, and grown in the field [44], failed to induce the disease symptoms in the laboratory [45]. This demonstrates the complexity of the pathosystem under field conditions.

Other tomatoes with outstanding LB resistance in the field, including the *S. pimpinellifolium* cultigens L 3707 and L 3708 [46,47], performed insignificantly lower than LA 1033. We observed, however, their stable low LB intensities in the field at all study years and in both locations, even with high *P. infestans*

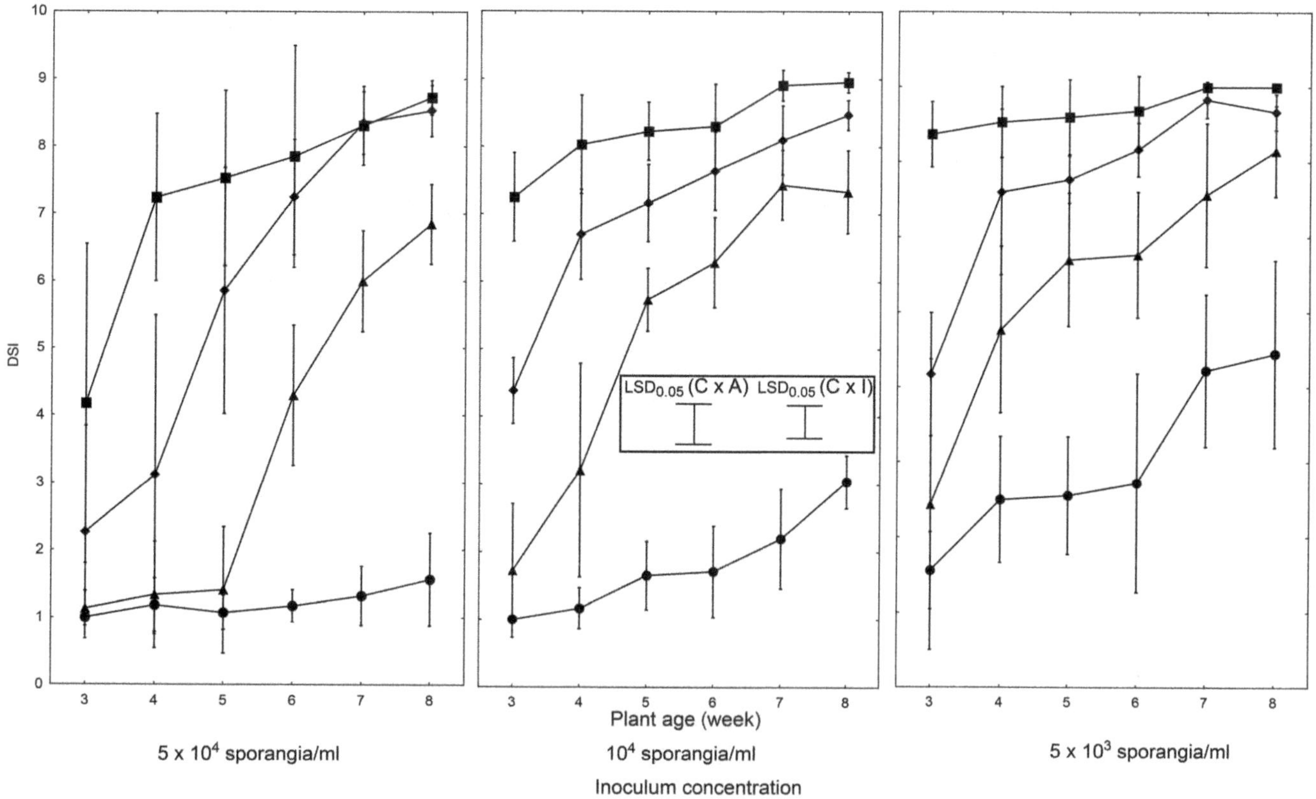

Figure 4. Severity of LB symptoms on tomato cultigens in the whole plant assay. Plants (● 'Rumba'; ■ WVa 700; ◆ L 3708; ▲ LA 1033) at indicated ages (3- to 8-weeks) were inoculated with suspension of the *P. infestans* isolate IWP 13, by spraying at indicated concentration (12 to 72 plants per cultigen and per developmental stage). Each combination was tested in three independent experiment sets. Vertical bars at each data point (means of the ratings) signify the standard deviations (SD). $LSD_{0.05}$ calculated according to Tukey procedure (inset) for comparing cultigens at each plant age and inoculum concentration (C×I): 0.55 and for comparing inoculum concentration for each cultigen and plant age (C×A): 0.61.

incidence. High LB resistance has been reported in these cultigens [15,46,47], despite a lack of clear elucidation of the trait's genetic background [3,26–29]. These cultigens have been successfully used as sources for pyramiding *Ph-3* with *Ph-2* LB resistance into new tomato cultivars [48–51]. Only few research groups reported rare instances, when high pathogen pressure, under highly conducive conditions, led to overcoming the resistance of L 3708, particularly in controlled condition assays [15,30,31,52].

In contrast to the aforementioned LB resistant cultigens, WVa 700 [19,38,53] exhibited varying levels of LB symptoms in the field, depending on the study year and location. These findings indicate the differences in the local pathogen populations. Consequently, such unstable expression of LB resistance in WVa 700 may occur due to a simpler genetic background for this trait, compared with those found in the other resistant cultigens. Similarly to our findings, this cultigen has failed to display stable LB resistance in other studies [19,30,31,54,55]. Finally, the *Ph-1* gene present in 'NY' [36–39] provides no reliable protection against *P. infestans* for field tomatoes grown in Poland, as well as in other locations [2,24,30,32,54,56,57].

Collectively, the field studies indicated that under Polish conditions, LA 1033, L 3707, and L 3708 could be considered promising sources for breeding tomato for LB resistance. This has implications for Central Europe, with field production of both tomatoes and potatoes [1].

In the controlled conditions testing methods used, LA 1033 showed the largest variability in age-dependent reaction to *P. infestans* inoculation. Furthermore, in contrast to its consistently superior performance in the field assays, LA 1033 proved inferior to both WVa 700 and L 3708 cultigens in the detached leaf and whole plant tests. This observation underscores that the assays under controlled conditions may differ from the field tests (weather conditions, plant age, heterogenic isolate mixture, constant pathogen pressure, presence of other (a)biotic stresses). These results also suggest that full expression of LB resistance in LA 1033 occurs later than the oldest developmental stages investigated (3- to 8-week old). Our observation of lower LB intensities in the detached leaves of the 13- to 15-week old plants of this cultigen further supports this hypothesis. The abundant growth of this cultigen may pose problems with generation of appropriate plant materials in the greenhouse for large scale bio-assays. This problem, however, can be solved using the detached leaf tests.

In our study, L 3708 showed higher symptoms variability relative to plant age compared with the WVa 700 cultigen. For L 3708, the trait reached stable expression in 7- to 8-week old plants in the whole plant assays, or in 10-week old leaves. This is in contrast to other studies on L 3708 [15,25], which described high levels of LB resistance when testing only 5-week old plants. Possible reasons for these discordant results include differences in the assumed methodology (isolates used, inoculum preparation and load, conditions of the assays) or in the climatic conditions related to location (e.g., intensity of sun exposure, length of day).

In the detached leaf and whole plant assays, we recorded the lowest variability for WVa 700, with low LB symptoms levels in plants of this cultigen older than 3 weeks. Moderate LB intensities were seen in whole plant assays. Turkensteen [24] reported age-

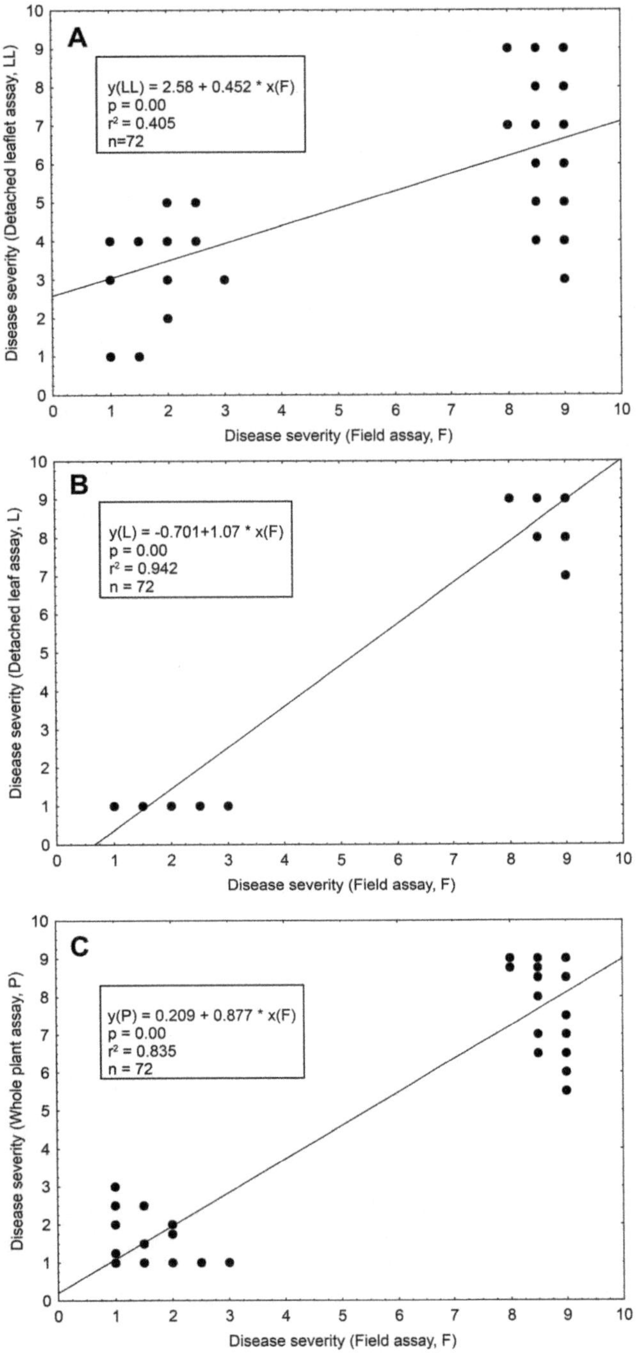

Figure. 5. Cross-comparison of the methods for testing the tomato LB resistance. Data from the field experiments (F) were pairwise compared with the results of each of the controlled-conditions method used, under the conditions optimized towards the maximal LB resistance expression (LL: Detached leaflets tested on 8-week old plants, under 5×10^4 sporangia/ml; L: Detached leaves tested on 15-week old plants, under 5×10^3 sporangia/ml; P: Whole plants tested on 8-week old plants, under 5×10^4 sporangia/ml). Calculated trend lines, with respective determination coefficients (r^2) and *P*-values are indicated. A: Comparison of detached leaflet assays with field assays (LL and F); B: Comparison of detached leaf assays with field assays (L and F); C: Comparison of whole plant assays with field assays (P and F).

dependent and progressively increasing LB resistance in the detached leaf assays of 6- to 8-week old WVa 700. Similarly, the 6-week old seedlings of this cultigen exhibited high resistance against both pathogen isolates tested by Moreau et al. [19]. In contrast to our results showing high LB infection in 3- to 4-week old WVa 700 plants, under 5×10^4 sporangia/ml, previous studies reported low LB intensity in this cultigen, under comparable developmental stages and inoculum concentrations [58,59]. The most likely reasons for the observed discrepancies are the pathogen isolates used or the assumed methodology, including the double isolate activation employed in our study.

Apart from the significant influence of plant developmental stage and inoculum concentration on the intensity of LB symptoms in the detached leaflet assay, we observed that LB intensity decreases with plant age. Variability of lesion size on leaflets observed using this assay, particularly in the resistant cultigens, indicated the need for improved methods for testing LB resistance (e.g., sporulation intensity). Thus, the detached leaflet assays were an inefficient testing method, although it is fast and easy. The detached leaflet assays can be used only for identification of susceptible genotypes in initial stages of breeding. Our results of tomato LB intensities are in line with the studies of LB in potato [5,9,60,61], where reported problems regarded high variability, especially for potato genotypes with moderate resistance. Although several tomato LB studies used the detached leaflet method to evaluate LB intensity [16,33,34,42,43,62], they generally included additional evaluations, such as sporulation intensity. In contrast to the detached leaflet assays, other methods used in this study (detached leaf and whole plant assays) proved more reliable for testing tomato LB resistance under controlled conditions. The observed variability among cultigens tested with these methods was significantly lower than this observed for the detached-leaflet assays. These two most effective methods, however, required different inoculum concentrations for successful separation of resistant genotypes. Both detached leaf and whole plant assays suggested an age-related LB intensity. Our results are similar to those using detached leaflet assays, except for more accurate distinction of the resistant genotypes.

Inoculation and incubation conditions influenced the reproducibility of our results. Here, infection and subsequent development of LB symptoms clearly depended on changes in temperature or RH, in accordance with the biology of the pathogen [5,8,33,57,60,63]. Standardizing the test methods (choice of isolate, preparation of inoculum, concentration of inoculum, and conditions of incubation after inoculation) resulted in greater precision in distinguishing LB resistant plants. In the light of our findings, we postulate it very useful to study the plant LB resistance in an age-dependent manner and under the controlled conditions, for each cultigen being reported, in order to better reflect field performance. This might provide an explanation for the differences in mapping of the genes or QTLs controlling the LB resistance trait [15,25,26,28,64–67], and might be of help in detailed analyses of the emerging cultigens reported as LB resistant [16,18,20,34,50].

Of the methods studied, the detached leaf and the whole plant tests resulted in the lowest discrepancies, when compared with the field experiments. This may be due to inoculation of larger plant surface area, which may generate lower assessment errors and permit a more accurate evaluation of LB resistance. Both leaf and leaflet assays may additionally exhibit reactions to *P. infestans* inoculation different from those of whole plants. These may be due to differences in environmental conditions or influences on the physiological and/or biochemical processes, as a result of detachment from the plant. Thus, we propose the whole plant assays as the most reliable method of testing the tomato LB resistance under controlled conditions. This, in the case of some cultigens (notably LA 1033), may pose other challenges, to be circumvented by using alternative methods, such as the detached leaf assays. We agree with previous reports on potato-LB [4–12,14] and tomato-LB pathosystems [21–23], that the ultimate assessment of LB resistance should be performed with field tests.

Conclusions

Our five-year study under natural field infection, in two distinct Polish locations, confirmed low and stable levels of LB symptoms in LA 1033, L 3708, and L 3707 tomato cultigens. These cultigens are useful for resistance breeding programs for Central Europe. Based on field results, we consider the cultigens carrying the *Ph-1* gene (e.g., 'New Yorker') an unsuitable source of LB resistance in Poland. The same is true for cultigens carrying *Ph-2* (e.g., WVa 700) if they are used as the only source of LB resistance. Our comparison of three methods for assessing tomato resistance against *P. infestans* under controlled conditions indicated that each method may be used for different purposes in the resistance breeding. The detached leaflet assay proved useful only to separate the LB susceptible genotypes (such as within the segregating populations) and to maintain the pathogen isolates, but was unreliable for systematic screens due to high variability. Tests on detached leaves and whole plants (greenhouse) generated lower variability than those performed on the detached leaflets and were also found to be highly correlated with field tests. Our results indicate congruent trends for age-dependent expression of LB resistance in all tested tomato cultigens, irrespective of the testing method. The plant age-related LB resistance in tomato reported here, shows the need to optimize and standardize the testing parameters when reporting new sources of resistance. As documented in this study, the reliable comprehensive evaluation of a given cultigen, by means of the optimized and well-suited assays, remains crucial to maximize the benefits from the best performing crops.

Acknowledgments

The Authors are thankful to Krystyna Szewczyk and Marzena Czajka for skillful technical assistance. Dr. Dorothy M. Tappenden PhD (Michigan State University) and Dr. Todd C. Wehner (North Carolina State University) are gratefully recognized for copy editing and critical reading of this manuscript.

Author Contributions

Conceived and designed the experiments: M. Nowakowska EUK. Performed the experiments: M. Nowakowska EUK. Analyzed the data: M. Nowakowska M. Nowicki RM UK EUK. Contributed reagents/materials/analysis tools: M. Nowakowska M. Nowicki EUK. Wrote the paper: M. Nowakowska M. Nowicki EUK.

References

1. Anonymous (2011) FAOSTAT final 2011 data. pp. http://faostat.fao.org/site/339/default.aspx.
2. Nowicki M, Kozik EU, Foolad MR (2013) Late blight of tomato. In: Varshney RK, Tuberosa R, editors. Translational genomics for crop breeding: John Wiley & Sons Ltd. pp. 241–265.
3. Nowicki M, Foolad MR, Nowakowska M, Kozik EU (2012) Potato and tomato late blight caused by *Phytophthora infestans*: An overview of pathology and resistance breeding. Plant Disease 96: 4–17.
4. Darsow U (2004) The use of four different assessment methods to establish relative potato tuber blight resistance for breeding. Potato Research 47: 163–174.
5. Dorrance AE, Inglis DA (1997) Assessment of greenhouse and laboratory screening methods for evaluating potato foliage for resistance to late blight. Plant Disease 81: 1206–1213.
6. Dorrance AE, Inglis DA (1998) Assessment of laboratory methods for evaluating potato tubers for resistance to late blight. Plant Disease 82: 442–446.
7. Douches DS, Kirk WW, Bertram MA, Coombs JJ, Niemira BA (2002) Foliar and tuber assessment of late blight (*Phytophthora infestans* (Mont.) de Bary) reaction in cultivated potato (*Solanum tuberosum* L.). Potato Research 45: 215–224.
8. Huang S, Vleeshouwers VGAA, Visser RGF, Jacobsen E (2005) An accurate in vitro assay for high-throughput disease testing of *Phytophthora infestans* in potato. Plant Disease 89: 1263–1267.
9. Michalska AM, Zimnoch-Guzowska E, Sobkowiak S, Plich J (2011) Resistance of potato to stem infection by *Phytophthora infestans* and a comparison to detached leaflet and field resistance assessments. American Journal of Potato Research 88: 367–373.
10. Park T-H, Vleeshouwers VGAA, Kim J-B, Hutten RCB, Visser RGF (2005) Dissection of foliage and tuber late blight resistance in mapping populations of potato. Euphytica 143: 75–83.
11. Sharma BP, Forbes GA, Manandhar HK, Shrestha SM, Thapa RB (2013) Determination of resistance to *Phytophthora infestans* on potato plants in field, laboratory and greenhouse conditions. Journal of Agricultural Science 5: p148.
12. Tai G (1998) Relationship between resistance to late blight in potato foliage and tubers of cultivars and breeding selections with different resistance levels. American Journal of Potato Research 75: 173–178.
13. Visker M (2005) Association between late blight resistance and foliage maturity type in potato: Physiological and genetic studies. 160 pp. p.
14. Vleeshouwers VGAA, van Dooijeweert W, Keizer LCP, Sijpkes L, Govers F, et al. (1999) A laboratory assay for *Phytophthora infestans* resistance in various *Solanum* species reflects the field situation. European Journal of Plant Pathology 105: 241–250.
15. Chunwongse J, Chunwongse C, Black L, Hanson P (2002) Molecular mapping of the *Ph-3* gene for late blight resistance in tomato. Journal of Horticultural Science & Biotechnology 77: 281–286.
16. Li J, Liu L, Bai Y, Finkers R, Wang F, et al. (2011) Identification and mapping of quantitative resistance to late blight (*Phytophthora infestans*) in *Solanum habrochaites* LA1777. Euphytica 179: 427–438.
17. Lough RC, Gardner RG (2000) Inheritance of tomato late blight resistance derived from *Lycopersicon hirsutum* LA1033 and identification of molecular markers. Hortscience 35: 490.
18. Merk HL, Ashrafi H, Foolad MR (2012) Selective genotyping to identify late blight resistance genes in an accession of the tomato wild species *Solanum pimpinellifolium*. Euphytica 187: 63–75.
19. Moreau P, Thoquet P, Olivier J, Laterrot H, Grimsley N (1998) Genetic mapping of *Ph-2*, a single locus controlling partial resistance to *Phytophthora infestans* in tomato. Molecular Plant-Microbe Interactions 11: 259–269.
20. Smart CD, Tanksley SD, Mayton H, Fry WE (2007) Resistance to *Phytophthora infestans* in *Lycopersicon pennellii*. Plant Disease 91: 1045–1049.
21. Akhtar KP, Saleem MY, Asghar M, Ali S, Sarwar N, et al. (2012) Resistance of *Solanum* species to *Phytophthora infestans* evaluated in the detached-leaf and whole-plant assays. Pak J Bot 44: 1141–1146.
22. Kim BS (2012) Evaluation of tomato genetic resources for the development of resistance breeding lines against late blight. Research in Plant Disease 18(1): 342–350.
23. Luo D, Zhang XC, Wen XH (2013) Identification and screening of introduced tomato varieties resistant to late blight. Advanced Materials Research 610: 3472–3477.
24. Turkensteen LJ (1973) Partial resistance of tomatoes against *Phytophthora infestans*, the late blight fungus. Wageningen, The Netherlands: Wageningen University.
25. Chunwongse J, Chunwongse C, Black LL, Hanson P (1998) Mapping of the *Ph-3* gene for late blight from *L. pimpinellifolium* L3708. Rep Tomato Genet Cooperative 48: 13–16.
26. Chen A-L, Liu C-Y, Chen C-H, Wang J-F, Liao Y-C, et al. (2014) Reassessment of QTLs for late blight resistance in the tomato accession L3708 using a restriction site associated DNA (RAD) linkage map and highly aggressive isolates of *Phytophthora infestans*. PLoS One 9: e96417.
27. Frary A, Graham E, Jacobs J, Chetelat RT, Tanksley SD (1998) Identification of QTL for late blight resistance from *L. pimpinellifolium* L3708. Tomato Genetics Cooperative Report 48: 19–21.
28. Kim MJ, Mutschler MA (2005) Transfer to processing tomato and characterization of late blight resistance derived from *Solanum pimpinellifolium* L. L3708. Journal of the American Society for Horticultural Science 130: 877–884.
29. Kim M-J, Mutschler MA (2006) Characterization of late blight resistance derived from *Solanum pimpinellifolium* L3708 against multiple isolates of the pathogen *Phytophthora infestans*. Journal of the American Society for Horticultural Science 131: 637–645.
30. AVDRC (2005) 2005 Progress Report. Shanhua, Tainan, Taiwan: Asian Vegetable Research and Development Center. 49–51 p.

31. Chen C-H, Sheu Z-M, Wang T-C (2008) Host specificity and tomato-related race composition of *Phytophthora infestans* isolates in Taiwan during 2004 and 2005. Plant Disease 92: 751–755.
32. Wang TC, Chen CH (2005) The variation of race composition of *Phytophthora infestans* in Taiwan during 1991–2004. 2005 APS Annual Meeting. Austin, TX: APS: Phytopathology. pp. S109.
33. Kim M, Mutschler M (2000) Differential response of resistant lines derived from the *L. pimpinellifolium* accession L3708 and *L. hirsutum* accession LA 1033 against different isolates of *Phytophthora infestans* in detached leaf lab assays. Tomato Genetics Cooperative Report 540: 23–25.
34. Li J (2010) Exploration of wild relatives of tomato for enhanced stress tolerance. Wageningen, The Netherlands: Wageningen University. 158 p.
35. Lough RC (2003) Inheritance of tomato late blight resistance in *Lycopersicum hirsutum* LA1033. Raleigh: North Carolina State university.
36. Bonde R, Murphy EF (1952) Resistance of certain tomato varieties and crosses to late blight. Bull Me Agric Exp Sta 497: 15.
37. Gallegly ME (1952) Sources of resistance to two races of the tomato late blight fungus. Phytopathology 42.
38. Gallegly ME, Marvel ME (1955) Inheritance of resistance to tomato Race-0 of *Phytophthora infestans*. Phytopathology 45: 103–109.
39. Peirce LC (1971) Linkage tests with *Ph* conditioning resistance to race 0, *Phytophthora infestans*. Rep Tomato Genet Coop 21: 30.
40. Zarzycka H (2001) Evaluation of resistance to *Phytophthora infestans* in detached leaflet assay. Preparation of the inoculum. The methods of evaluation and selection applied in potato research and breeding Monografie i rozprawy naukowe IHAR, 10a [in Polish]: 75–77.
41. Nowicki M, Lichocka M, Nowakowska M, Kłosińska U, Kozik EU (2012) A simple dual stain for detailed investigations of plant-fungal pathogen interactions. Vegetable Crops Research Bulletin 77: 61–74.
42. Nelson HE (2006) Bioassay to detect small differences in resistance of tomato to late blight according to leaf age, leaf and leaflet position, and plant age. Australasian Plant Pathology 35: 297–301.
43. Smart C, Myers K, Restrepo S, Martin G, Fry W (2003) Partial resistance of tomato to *Phytophthora infestans* is not dependent upon ethylene, jasmonic acid, or salicylic acid signaling pathways. Molecular Plant-Microbe Interactions 16: 141–148.
44. Śliwka J, Jakuczun H, Lebecka R, Marczewski W, Gebhardt C, et al. (2006) The novel, major locus *Rpi-phu1* for late blight resistance maps to potato chromosome IX and is not correlated with long vegetation period. TAG Theoretical and Applied Genetics 113: 685–695.
45. Stefańczyk E, Świątek M, Chmielarz M, Tomczyńska I, Śliwka J (2013) Influence of chosen factors on the expression of the *Rpi-phu1* gene, conferring the potato resistance against *Phytophthora infestans*. In: Śmiałowski T, Strzembicka A, Szecówka PS, editors. Ogólnopolska konferencja naukowa - Nauka dla hodowli i nasiennictwa roślin uprawnych Zakopane Poland. Plant Breeding and Acclimatization Institute (IHAR) - National Research Institute in Radzików and IHAR Cereals Department in Cracow.pp. 67–70.
46. Black LL, Wang TC, Hanson PM, Chen JT (1996) Late blight resistance in four wild tomato accessions: Effectiveness in diverse locations and inheritance of resistance. Phytopathology 86: S24.
47. Irzhansky I, Cohen Y (2006) Inheritance of resistance against *Phytophthora infestans* in *Lycopersicon pimpenellifolium* L3707. Euphytica 149: 309–316.
48. Gardner RG, Panthee DR (2010) NC 1 CELBR and NC 2 CELBR: Early blight and late blight-resistant fresh market tomato breeding lines. Hortscience 45: 975–976.
49. Gardner RG, Panthee DR (2010) 'Plum Regal' Fresh-market plum tomato hybrid and its parents, NC 25P and NC 30P. HortScience 45: 824–825.
50. Ojiewo C, Swai I, Oluoch M, Silué D, Nono-Womdim R, et al. (2010) Development and release of late blight-resistant tomato varieties 'Meru' and 'Kiboko'. International Journal of Vegetable Science 16: 134–147.
51. Panthee DR, Gardner RG (2010) 'Mountain Merit': A late blight-resistant large-fruited tomato hybrid. Hortscience 45: 1547–1548.
52. Scott J, Gardner R (2007) Breeding for resistance to fungal pathogens. In: Razdan MK, Mattoo AK, editors. Genetic improvement of Solanaceous crops: Taylor & Francis Science Publishers. pp. 421–456.
53. Gallegly ME (1960) Resistance to the late blight fungus in tomato. Proceedings of Plant Science Seminar, Camden, New Jersey 1960. Camden, New Jersey pp. 113–135.
54. Foolad MR, Merk H, Ashrafi H, Kinkade M (2006) Identification of new sources of late blight resistance in tomato and mapping of a new resistance gene. In: Moyer J, editor. 22nd Annual Tomato Disease Workshop. Mountain Horticultural Crops Research & Extension Center Fletcher, NC, USA. North Carolina State University. pp. 4–8.
55. Laterrot H (1975) Selection for the resistance to *Phytophthora infestans* in tomato. Annales de l'Amelioration des Plantes 25: 129–150.
56. Cohen Y (2002) Populations of *Phytophthora infestans* in Israel underwent three major genetic changes during 1983 to 2000. Phytopathology 92: 300–307.
57. Klarfeld S, Rubin A, Cohen Y (2009) Pathogenic fitness of oosporic progeny isolates of *Phytophthora infestans* on late-blight-resistant tomato lines. Plant Disease 93: 947–953.
58. Michalska A, Pazio M (2002) A new method for evaluating tomato leaf resistance to *Phytophthora infestans* using seedling test. Plant Breeding and Seed Science 46: 3–21.
59. Michalska AM, Pazio M (2005) Inheritance of tomato leaf resistance to *Phytophthora infestans* - new information based on laboratory tests on seedlings. Plant Breeding and Seed Science 51: 31–42.
60. Mizubuti ES, Fry WE (1998) Temperature effects on developmental stages of isolates from three clonal lineages of *Phytophthora infestans*. Phytopathology 88: 837–843.
61. Sobkowiak S, Zarzycka H, Lebecka R, Zimnoch-Guzowska E (2009) Reaction of potato standards to *Phytophthora infestans* in resistance tests. Biuletyn Instytutu Hodowli i Aklimatyzacji Roślin: 253–268.
62. Legard DE, Fry WE (1996) Evaluation of field experiments by direct allozyme analysis of late blight lesions caused by *Phytophthora infestans*. Mycologia: 608–612.
63. Hardham AR, Shan W (2009) Cellular and molecular biology of *Phytophthora*–plant interactions. In: Deising HB, editor. The Mycota. Berlin Heidelberg: Springer Verlag. pp. 4–27.
64. Huang X-M, Xu X-Y, Li J-F, Chen X-L, Xu Y-H (2009) Construction of tomato molecular genetic map and QTL analysis of resistance of gene cluster *ph-3* to tomato *Phytophthora infestans*. Scientia Agricultura Sinica 42: 3571–3580.
65. Park HP, Chae Y, Kim H-R, Chung K-H, Oh D-G, et al. (2010) Development of a SCAR marker linked to *Ph-3* in *Solanum* ssp. Korean J Breed Sci 42: 139–143.
66. Park Y, Hwang J, Kim K, Kang J, Kim B, et al. (2013) Development of the gene-based SCARs for the *Ph-3* locus, which confers late blight resistance in tomato. Scientia Horticulturae 164: 9–16.
67. Zhang C, Liu L, Wang X, Vossen J, Li G, et al. (2014) The *Ph-3* gene from *Solanum pimpinellifolium* encodes CC-NBS-LRR protein conferring resistance to *Phytophthora infestans*. TAG Theoretical and applied genetics 127: 1353.

The FIGS (Focused Identification of Germplasm Strategy) Approach Identifies Traits Related to Drought Adaptation in *Vicia faba* Genetic Resources

Hamid Khazaei[1]*, Kenneth Street[2], Abdallah Bari[2], Michael Mackay[3], Frederick L. Stoddard[1]

1 Department of Agricultural Sciences, University of Helsinki, Helsinki, Finland, **2** International Centre for Agricultural Research in the Dry Areas, Aleppo, Syria, **3** Queensland Alliance for Agricultural and Food Innovation, The University of Queensland, St Lucia, Queensland, Australia

Abstract

Efficient methods to explore plant agro-biodiversity for climate change adaptive traits are urgently required. The focused identification of germplasm strategy (FIGS) is one such approach. FIGS works on the premise that germplasm is likely to reflect the selection pressures of the environment in which it developed. Environmental parameters describing plant germplasm collection sites are used as selection criteria to improve the probability of uncovering useful variation. This study was designed to test the effectiveness of FIGS to search a large faba bean (*Vicia faba* L.) collection for traits related to drought adaptation. Two sets of faba bean accessions were created, one from moisture-limited environments, and the other from wetter sites. The two sets were grown under well watered conditions and leaf morpho-physiological traits related to plant water use were measured. Machine-learning algorithms split the accessions into two groups based on the evaluation data and the groups created by this process were compared to the original climate-based FIGS sets. The sets defined by trait data were in almost perfect agreement to the FIGS sets, demonstrating that ecotypic differentiation driven by moisture availability has occurred within the faba bean genepool. Leaflet and canopy temperature as well as relative water content contributed more than other traits to the discrimination between sets, indicating that their utility as drought-tolerance selection criteria for faba bean germplasm. This study supports the assertion that FIGS could be an effective tool to enhance the discovery of new genes for abiotic stress adaptation.

Editor: Jean-Marc Lacape, CIRAD, France

Funding: HK was supported by CIMO (Centre for International Mobility) and is currently supported by Emil Aaltonen Foundation (Emil Aaltosen Säätiö) fellowship. The project was also supported by the University of Helsinki and Legume Futures (KBBE 2009-1-2-01, contract 245216 CP-FP 'Legume Futures: Legume supported cropping systems for Europe'). The funders had no role in study design, data collection and analysis, decision to publish, or preparation of the manuscript.

Competing Interests: The authors have declared that no competing interests exist.

* E-mail: hamid.khazaei@helsinki.fi

Introduction

Drought coupled with heat stress, expected to increase in frequency and intensity is likely to expand due to climate change [1,2]. Faba bean (*Vicia faba* L.) is an important source of protein, often referred to as poor man's meat, in those dry areas of developing countries most likely to be impacted by climate change [3,4]. This has significant food security implications because faba bean is relatively sensitive to terminal moisture stress when compared to other temperate-season grain legumes [5–7] so drought is a major constraint to its production and yield stability. Therefore it is imperative that natural variation for traits related to drought adaptation be identified from the faba bean genepool and introduced into improved cultivars. Economic analysis of cultivar development showed that the identification of a desirable trait is of equal importance to the process of transferring it into improved backgrounds because it reduces the time taken to develop improved cultivars [8].

Genetic resource collections conserved in genebanks are the most obvious place to look for useful traits, but given the size of these collections, searching for specific and often rare traits has been likened to searching for a needle in a haystack. Further, evaluating large collections for some parameters can be extremely costly. For example, the International Center for Agricultural Research in the Dry Areas (ICARDA) houses a globally important collection of over 9500 faba bean accessions. It would be beyond the resources of most research programs to evaluate this entire collection for variation in leaf morpho-physiological traits related to plant moisture stress. What is needed therefore is a means of wisely selecting an economically feasible set size that has a better probability of capturing useful variation than if material was selected at random or through the use of other techniques that do not focus on the sought-after trait.

The core collection was proposed as a way to work with fewer accessions that would represent, "with a minimum of repetitiveness, the genetic diversity of a crop species and its relatives" [9]. There are numerous examples of methodologies to develop core collections (*see* Hodgkin et al. [10] for examples), which in practice tend towards limiting the size of the sub-set to around 10% [11,12] of the original collection size. Although one of the stated purposes of core collections is to improve utilization, the vast majority of reported research seems to focus more on methods (or sampling strategies) to establish core collections [13–16] and the analysis of the diversity held within core collections [17–20]. A number of references suggest alternative types of collections, or sets of collections, to enhance the efficiency of capturing diversity or addressing utilization, including specialized core collections [21], mini core sets [22], nested core collections [23] and composite collections [24]. Despite this diversity of core collection methodology, there seems to be a lack of literature that demonstrates that core collections have had a significant impact on the utilization of genetic resources. Rare and adaptive alleles, most of which are thought to be functional, may even be missed from a core collections [21,25–29].

The Focused Identification of Germplasm Strategy (FIGS) was designed to improve the efficiency with which specific adaptive traits are identified from genetic resource collections. FIGS is based on the premise that adaptive traits displayed by an accession will reflect the selection pressures of the environment from which it was originally sampled [30–33]. The FIGS approach uses both trait and environmental (climate) data to develop *a priori* information or specialized knowledge as per Gollin et al. [8] based on a quantification of the trait-environment relationship [32,34,35]. This *a priori* information is then used to define a set of accessions with a high probability of containing the desired traits.

Many adaptive traits can be linked to agro-climatic parameters. For example, using monthly values for a range of climatic variables, FIGS detected sources of resistance in wheat (*Triticum* ssp.) for biotic stresses such as powdery mildew (*Blumeria graminis* (DC) Speer f.sp. *tritici*) [36,37], Sunn pest (*Eurygaster intergriceps* Put.) [38], Russian wheat aphid (*Diuraphis noxia* Kurdj.) [39] and stem rust (*Puccinia graminis* Pers.) [34,35] as well as net blotch (*Pyrenophora teres* Drechs.) in barley (*Hordeum vulgare* L.) [34]. Further, Endresen et al. [40] demonstrated how eco-geographic data from the collection sites of 14 Nordic barley landraces (*Hordeum vulgare* L.) was successfully correlated to morphological traits using multilinear data modelling techniques and conclude that the FIGS approach can be used efficiently as a targeted germplasm selection method.

However, so far studies are few on the effectiveness of FIGS to detect traits that impart tolerance to abiotic stresses such as moisture availability, and there are certainly none for faba bean.

The aim of this study was to compare the leaf morpho-physiology and phenology of two sets of faba bean accessions originating from environments with contrasting seasonal moisture availabilities. The underlying hypothesis is that ecotypic (climatic) differentiation occurred so traits associated with plant moisture regulation and lifecycle will differ between the two sets. From this, we would further expect that set membership based on collection site environmental descriptors would be the same when the accessions are classified using trait measurements.

Materials and Methods

Construction of FIGS Sets

Two sets containing landrace accessions of faba bean were selected from the collection conserved by ICARDA that contains 9545 entries, representing 21% of the worldwide germplasm collection [41]. One set was chosen to maximize the probability of having drought-related adaptive traits, the "dry set", and the other was constructed as a control from accessions originating from environments with higher moisture profiles - the "wet set". The origin of the selected accessions is presented in Figure 1, and the ICARDA accession numbers are given in Table S1.

The dry set (201) was constructed as follows. Accessions from collection sites where the annual rainfall was below 300 mm/year or greater than 550 mm/year were not considered. Of the remaining accessions, one accession per collection site was chosen at random. A hierarchical cluster analysis was performed using the following collection site agro-climatic parameters: *precyr, ariyr, tminyr, tmaxyr, bio4, bio15, bio16, and bio19* extracted from ICARDA and Worldclim-databases and Hijmans et al. [42] (Table 1). The climate variables were chosen to combine temperature and precipitation factors that would influence the length of growing season and seasonal moisture availability. The between-groups linkage option was set as the clustering algorithm, using squared euclidian distances as the distance measure. The procedure created 20 clusters. Accessions contained in 6 clusters were dropped because the average aridity index for the cluster was above 0.6 or below 0.1 (indicating irrigated sites). For each of the remaining clusters the accessions were sorted according to the *bio15* climate variable (a measure of the variation in seasonal moisture availability) for their respective collection sites. Any accession with a score of 50 or lower was discarded. The remaining accessions within each cluster were ranked based on collection site long-term yearly precipitation. A set of 201 accessions was chosen by selecting the lowest ranked accession in each cluster and repeating the process until the set size was achieved.

The wet set was chosen from sites that receive over 800 mm/year of rainfall (long-term average). One accession per site was chosen at random. The remaining accessions were sorted according to collection site yearly average aridity index and 201 accessions were chosen from sites with the highest aridity indices.

Growth Conditions

Accessions were planted in a randomized complete block design (RCBD) with 4 replicates in a climate-controlled greenhouse of the University of Helsinki, Finland during 2010–2011, giving a total of 1608 pots. Before sowing, seeds were inoculated with *Rhizobium leguminosarum* biovar. *viciae* (faba bean strain, Elomestari Oy, Tornio, Finland). Three seeds were sown per 2 L plastic pot, which held a mixture of sand and peat (White 420 W, Kekkilä Oy, Vantaa, Finland) (3:1 *v/v*). After 10 days, the seedlings were thinned to one per pot. Soil moisture levels were maintained at field capacity with an automatic irrigation system to ensure that each plant received the same amount of water during the experiment. At three and five weeks after sowing, 70 ml of fertilizer solution (equivalent to 20 kg of P and 24 kg of K per hectare) was added to each pot. The photoperiod was adjusted to 14 h light and 10 h dark, and the temperature was set to 21°C day/15°C night ±2°C. Photosynthetic photon flux density (PPFD) was about 300 µmol m^{-2} s^{-1} at the canopy level. The relative humidity was maintained at 60±5%.

Pest control. Thrips were controlled biologically using *Amblyseius cucumeris*, especially at seedling and flowering stages.

Morphological Measurements

Stomatal density and morphology. Stomatal density (SD), length (SL) and width (SW) were measured on the middle part of the abaxial surface of the youngest, fully expanded leaflet of 8-week-old plants using the impression method [43]. The number of stomata was counted from ten different microscopic fields of view at 250× magnification. To estimate SD, the number of stomata per field of view was converted to the number of stomata per mm^2 of leaf using a standard scale. SL and SW were measured on ten stomata from the impressions using a scaled 500× eyepiece of microscope and converted to µm. Stomatal area (SA) was calculated as SA = SL × SW. Stomatal area per unit area of leaflet (SAAL) was calculated as the product of SA and SD.

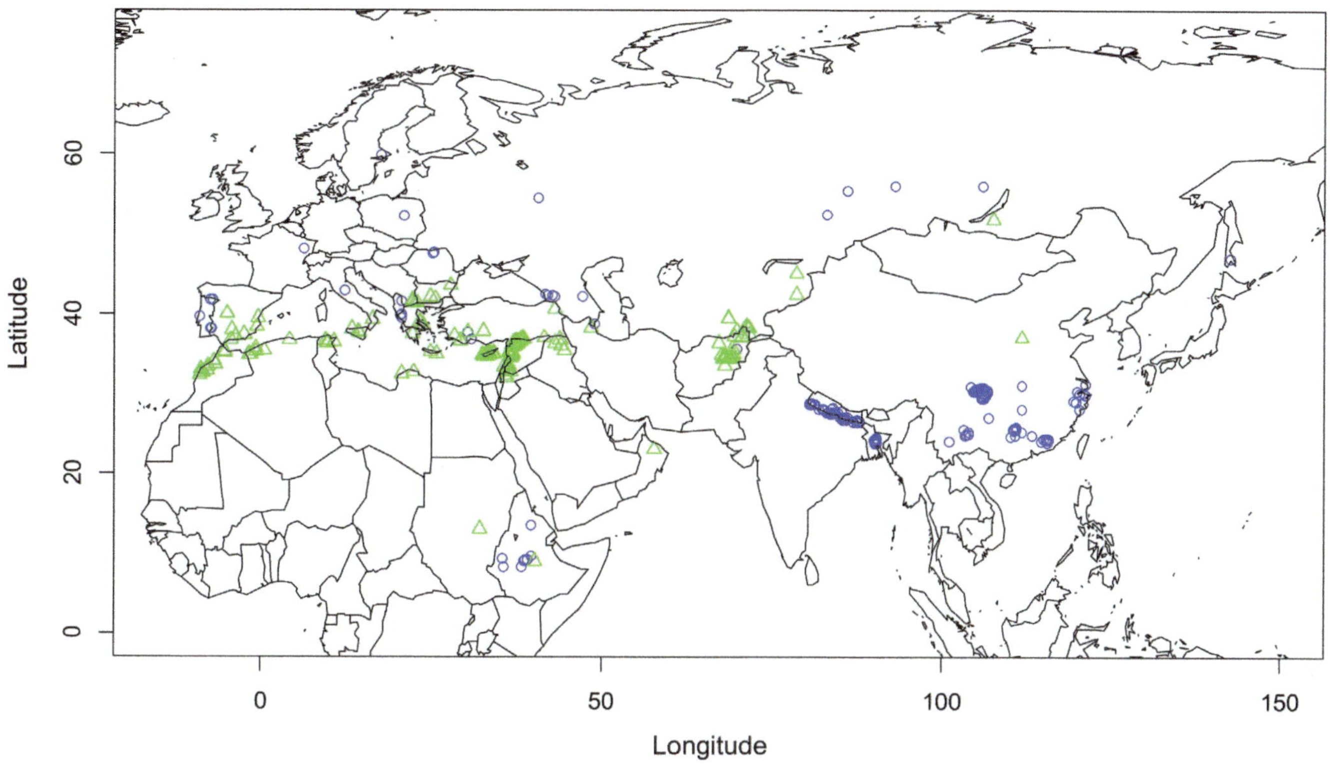

Figure 1. Geographical distribution of the two sets (wet set, blue circle and dry set, green triangle).

Leaflet area. Leaflet area was measured using a LI-6200 leaf area meter (LI-COR, Inc., Lincoln, NE, USA). Means of four leaflets per plant were used for analysis.

Fertile tillers. The number of fertile tillers was counted at 16 weeks after sowing.

Seed size. Ten seeds from each accession were measured in order to classify them to the traditional seed size class, *minor*, *equina* and *major* according to seed length and mass [44,45].

Physiological Measurements

Gas exchange traits. Gas exchange was measured on each plant at 6 weeks and 8 weeks after sowing, using a LI-6400 portable photosynthesis system (LI-COR, Inc.) equipped with a 2×3 cm leaf chamber with a LED light source (6400-02B, 90% red and 10% blue). Photosynthesis photon flux density (PPFD) was 1000 μmol m^{-2} s^{-1}. A CO_2-injecting cartridge was attached to the system to control reference CO_2 concentration at 400 μmol mol^{-1}, a value close to that during plant growth. The flow rate was 400 μmol s^{-1}. All the gas exchange measurements were done between 9 and 11 am using the youngest, fully expanded leaflet which was also used for stomatal morphology and leaflet area measurements. Measurements were logged only when the stability criteria were met, according to the manufacturer's instructions. For logistical reason, each replicate was measured on a separate day. The gas exchange measurements taken were: photosynthetic rate (A_{net}), stomatal conductance (g_s), transpiration rate (E), and intercellular CO_2 (C_i). Intrinsic water use efficiency (WUE) was calculated as gas exchange rate divided by stomatal conductance (A_{net}/g_s) [46].

Table 1. The climatic variables used in the selection of FIGS sets.

Code	Description
precyr	Long term yearly precipitation
ariyr	Long term yearly aridity index
tminyr	Long term yearly minimum temperature
tmaxyr	Long term yearly maximum temperature
Bio4	Temperature seasonality (standard deviation × 100)
Bio15	Precipitation seasonality (coefficient of variation)
Bio16	Precipitation of wettest quarter
Bio19	Precipitation of coldest quarter

Source: ICARDA GIS (Geographic information system) unit and world global climate data (http://www.worldclim.org/bioclim).

Leaflet and canopy temperatures. Leaflet temperature was measured along with gas exchange on the LI-6400. Canopy temperature was measured using a FLUKE® 574 thermometer gun (FLUKE, Everett, WA, USA) from the fully expanded leaves used for the other measurements. Canopy temperate was measured at 6 weeks and 8 weeks after sowing. Air temperature was recorded at the time of measuring leaf temperature. Leaflet temperature is presented as: Leaflet temperature – air temperature and canopy temperature as: canopy temperature – air temperature.

Relative water content. Five leaflets were used for determining leaf relative water content (RWC%) according to the initial principles by Barrs and Weatherley [47]. First, fresh weight (FW) was determined. Turgid weight (TW) was measured after floating the sample on distilled water in Petri dishes in darkness at 4̊C for 24 h. Dry weight (DW) was calculated by putting the samples for 48h in a 60̊C oven. RWC (%) = (FW–DW)/(TW–DW) × 100.

Phenological Measurement

The number of days to the onset of flowering was recorded.

Statistical Analysis

The membership of the two contrasting FIGS sets was based on *a priori* information, namely the long-term climatic conditions of the sites from which the accessions were collected. The underlying assumption was that morpho-physiological traits related to moisture stress adaptation would differ between two sets of selected germplasm. Two methods were used to determine whether the two sets are different in terms of morpho-physiological phenotypic expression.

To determine if there were differences between the sets, they were subjected to a t-test, using means across replicates for each accession, with the R statistical package [48] after testing for normality.

Multivariate analysis was employed for deeper investigation because the relationships between the collection site agro-climatic conditions and trait expression are likely to be non-linear and multi-dimensional and thus not captured in a linear framework. When trait expression differs between the two sets, this should be reflected in how the classification algorithms discriminate between accessions. Thus, we would expect the algorithms to correctly assign accessions into the sets created on climatic descriptors. Three models (Table 2) were used to classify accessions, discriminate between sets and to highlight those traits that contributed most to the discrimination. The algorithms used a learning-based approach, in which they were "trained" on a set of accessions whose set membership (wet or dry) was made "known" to the algorithm. The trained algorithm was then used to classify the accessions whose set membership was "unknown" to the algorithm into two sets (wet or dry). This is an iterative process where the model that is finally chosen by the algorithm is based on the "best" values for accuracy parameters that measure the model's ability to classify the unknown accessions into their respective climate-based sets. These learning-based techniques need fewer assumptions and thus are more suitable when highly complex non-linear relationships are expected among input variables. They were used to overcome the problem of restrictive parametric paradigms on one hand and the prerequisite distribution assumptions on the other [56,57].

The parameters used to measure the accuracy of these models are the AUC and Kappa values. The AUC refers to the area under the curve (AUC) of the Receiver Operating Characteristics (ROC) [58,59], which is a plot of true positive rate versus false positive [60]. An AUC value of 0.5 represents randomness and would indicate that the FIGS sets are no different from randomly chosen sets. An AUC value of 0.7 and above represents high model performance [59] indicating that the wet and dry sets are highly distinguishable and that the dry set is more prone to harbor traits that favor drought adaptation. Similar to the AUC, Kappa is a measure of agreement, where a value of 0.4 and above is an indication of good agreement between the model's prediction and the trait measurements [61].

The datasets were presented to the algorithms as follows: the mean value for each variable was calculated over the replicates for each accession. This accession level data was combined (wet+dry sets) and standardized so that the dataset mean was zero with standard deviation of 1. The algorithms split the combined data into 2 datasets containing 2/3 and 1/3 of the accessions on a random basis. The larger dataset was used to "train" the models and quantify the association between the membership (wet/dry) and the drought-related attributes. The association was then used in turn (in reverse) to classify the "unknown" accessions of the smaller dataset. This process was performed 10 times and the results were averaged.

Selection of important parameters. Some of the parameters used to differentiate between two sets are expected to have more influence on the classification defined by the algorithms (Table 2). The importance of each variable was calculated based on the Gini, or impurity index, where a split node that has a mixture of both tolerance and susceptible membership (wet and dry set) is less pure.

Results

Eleven of the 16 parameters measured differed between the sets. The members of the dry set had 21% fewer fertile tillers, flowered 2.4 days earlier, had longer stomata (4%), greater stomatal area (4%), more stomatal area per unit of leaflet (3%), 48% more leaflet area, 5% higher transpiration rate, 5% higher RWC, and cooler leaves than the wet set. The transpiration rate was 9% higher in the wet set while leaflet and canopy temperatures were lower in the dry set (Table 3). Furthermore, three quarters of the material from the dry set were large-seeded (*major* type) compared to only 20% in the wet set, whose remaining seeds were distributed equally between the *minor* and *equina* classes (Figure S1). The two sets thus contained accessions that, on average, differed morphologically and physiologically.

This assertion is supported by all 3 models used to classify the accessions based on the trait data; the accessions were placed into sets that agreed with the original climate-based classifications. The Kappa scores were all close to one, which demonstrates a high degree of accuracy given that an acceptable score is above 0.4. Likewise the AUC values were well in excess of the acceptable value of 0.7. Thus the models classified the accessions into their climate-based sets with accuracies approaching 100% (Table 4).

Table 2. Models used in the study to test the difference between the two sets and to select the best splitters.

Model	Tuning parameters	Library (R language)	References
Classification and Regression Training (CARET)*		caret	[49,50]
Random Forests (RF)*	Number of trees (n.tree) Number of predictors chosen at each node (mtry)	randomForest	[51–53]
Support Vector Machines (SVM)	gamma/sigma, cost (C)	svm (e1071) ksvm (kernalab)	[49,54,55]

*Variable importance is available for these models.

Table 3. The mean (± standard deviation) of morphological, physiological and phenological measurements on sets of 201 wet adapted and 201 dry adapted faba bean accessions, along with the difference between the set means and the value of the t-test.

Parameters	means		difference	$t_{(df=400)}$
	Wet	Dry		
Morphology				
Stomatal density *(No. mm^{-2})*	**49.6**±12.4	**48.3**±6.1	−1.3	1.29^{ns}
Stomatal length *(μm)*	**53.2**±3.9	**55.4**±2.6	+2.2	6.63***
Stomatal width *(μm)*	**30.4**±1.61	**30.3**±0.92	−0.1	0.20^{ns}
Stomatal area *(μm^2)*	**1622**±184	**1685**±119	+63	4.07***
Stomatal area per unit area of leaflet × 10^{-3} *(μm^2.mm^{-2})*	**77.9**±12.4	**80.6**±6.8	+2.7	2.67**
Leaflet area *(cm^2)*	**11.4**±5.9	**16.9**±5.2	+5.5	9.86***
Number of fertile tillers	**2.85**±1.52	**2.25**±1.1	−0.60	4.54***
Physiology				
Photosynthetic rate *(μmol m^{-2} s^{-1})*	**7.8**±2.0	**8.2**±1.1	+0.4	2.25*
Intercellular CO_2 *(ppm)*	**322.4**±22.3	**325.2**±10.5	+2.8	1.64^{ns}
Stomatal conductance *(mol m^{-2} s^{-1})*	**0.329**±0.119	**0.320**±0.072	−0.009	0.92^{ns}
Water use efficiency *(μmol mol^{-1})*	**25.6**±7.8	**26.4**±4.2	+0.8	1.24^{ns}
Transpiration rate *(mmol m^{-2} s^{-1})*	**3.26**±0.9	**3.42**±0.5	+0.16	2.15*
Leaflet temperature *(°C)*	**−0.17**±0.31	**−0.72**±0.19	−0.55	34.98***
Canopy temperature *(°C)*	**−0.80**±1.21	**−1.59**±0.48	−0.79	13.77***
Relative water content *(%)*	**82.3**±3.9	**86.3**±2.4	+4.0	12.59***
Phenology				
Days to flowering	**50.7**±13.1	**48.3**±7.4	−2.4	2.24*

df: degrees of freedom; ns: non significant; *, $P<0.05$; **, $P<0.01$; ***, $P<0.001$.

The accuracy of the models is also illustrated by the ROC plots (Figure 2), where displacement above the diagonal indicates non-random assignment of accessions to the correct subset. In the rpat-caret plot, there is some overlap between the sets, but both RF and SVM show mutual exclusivity of the two sets. The prediction density plots to the right of the ROC plots demonstrate that the wet and dry sets include accessions which, in a multivariate sense, are different and that the basis for the difference will be related to the selection criteria, in this case the seasonal moisture availability at collection sites.

Table 4. Model accuracy values for learning-based techniques used on test data (1/3) of faba bean over 10 runs of the algorithms.

Model		AUC	omission rate	sensitivity	specificity	correct classification rate	Kappa
caret-rpart	**Mean**	**0.97**	**0.03**	**0.97**	**0.97**	**0.97**	**0.93**
	Lower	0.96	0.01	0.95	0.95	0.96	0.92
	Upper	0.98	0.05	0.99	0.98	0.97	0.95
RF	**Mean**	**0.99**	**0.01**	**0.99**	**1.00**	**0.99**	**0.99**
	Lower	0.99	0.00	0.98	1.00	0.99	0.98
	Upper	1.00	0.02	1.00	1.00	1.00	1.00
SVM	**Mean**	**0.99**	**0.00**	**1.00**	**0.98**	**0.99**	**0.98**
	Lower	0.99	0.00	1.00	0.97	0.99	0.97
	Upper	1.00	0.00	1.00	0.99	1.00	0.99

AUC: Area under the ROC curve.
RF: Random Forest.
caret-rpart: Classification and Regression Training.
SVM: Support vector machine.
Correct classification rate: the overall classification of both wet and dry accessions to their respective membership group. It is the total of both correctly classified accessions as either wet or dry divided by the total of all the accessions (402).
Omission rates: the opposite of correctly classified accessions with drought-related traits in this case, which is the number of accessions lacking the traits yet they have been classified (incorrectly) as having the.

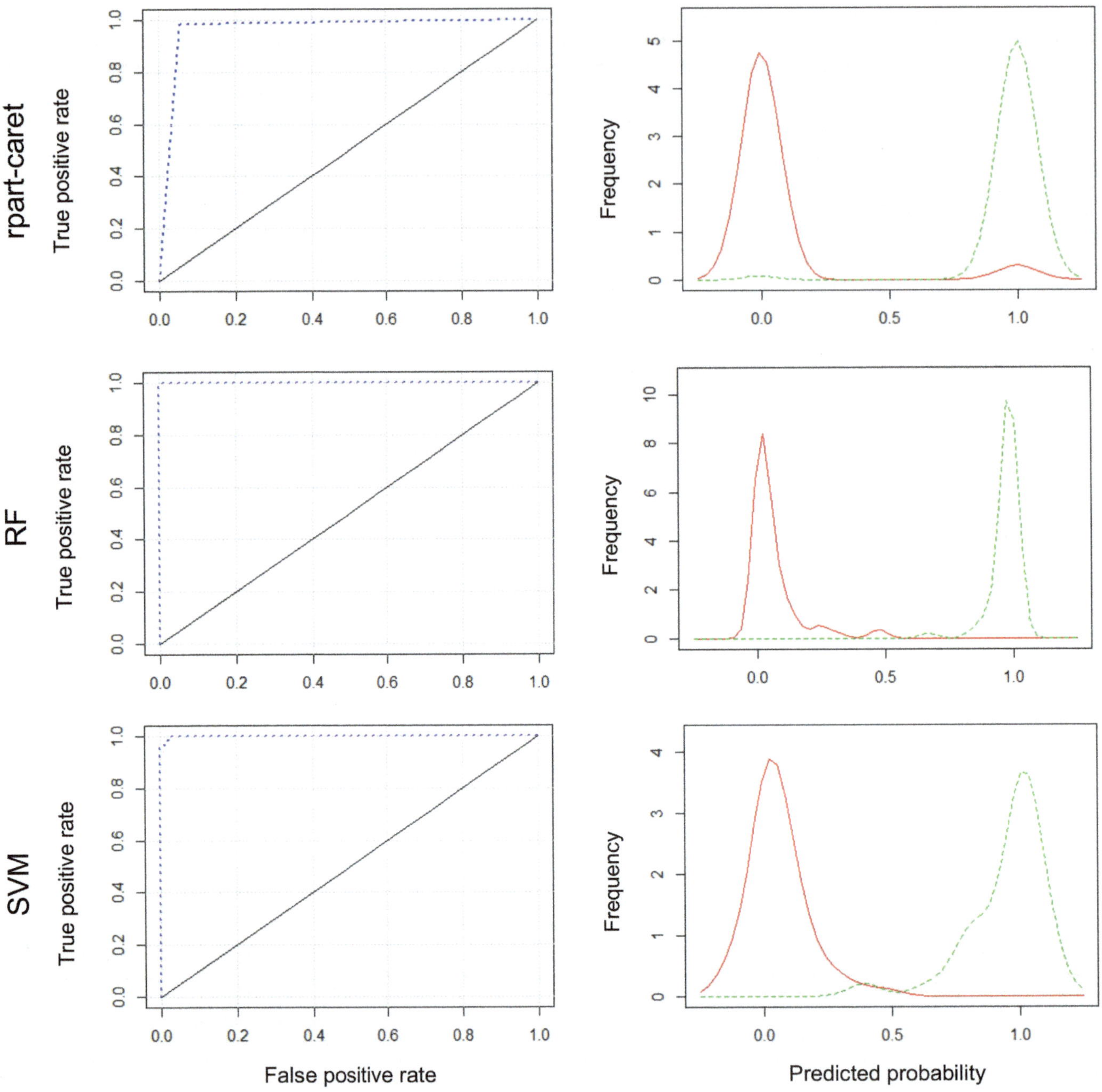

Figure 2. ROC plots (left) and density plots class prediction (right) for dry and wet sets using the three models; The class predictions fall out of range (0, 1) as a result of linearity/interpolation in some of the models.

Variable Importance

Of the 16 variables, leaflet temperature depression was the most informative, followed by canopy temperature, RWC, leaflet area and stomatal length (Table 5). The relative importance of the other variables differed between the three assessment methods, with transpiration rate being the third most important in RF mean decrease accuracy and fourth in RF mean decrease Gini, for example.

Discussion

While other studies have shown that the FIGS approach was effective when employed in the search for resistance to pests and diseases (*e.g.* [34,35,37,38]), this study demonstrates its effectiveness as a method to search for adaptive traits associated with abiotic constraints. The set selection process, based on indicators of moisture availability, yielded sets whose morpho-physiology and phenology were significantly different.

This result is not all that surprising, since it has been comprehensively shown that the environment strongly influences gene flow, natural selection and thus spatial/geographic differen-

Table 5. Potential climate predictors based on caret R and RF packages.

Rank	drought related parameter	model		
		rpart-caret	RF	
			mean decrease accuracy	mean decrease Gini
1	**Leaflet temperature**	**34.91**	**0.26**	**55.41**
2	**Canopy temperature**	**13.68**	**0.10**	**31.64**
3	**Relative water content**	**12.46**	**0.02**	**9.64**
4	Leaflet area	9.95	0.01	6.39
5	Stomatal length	6.70	0.01	2.30
6	Fertile tillers	4.72	0.00	0.47
7	stomatal area	4.13	0.00	1.01
8	Transpiration rate	3.61	0.03	6.94
9	Stomatal area per unit area of leaflet	2.75	0.01	3.52
10	Photosynthetic rate	2.34	0.01	4.25
11	Days to flowering	2.21	0.00	1.95
12	Intercellular CO_2	1.64	0.00	0.93
13	Stomatal density	1.26	0.01	2.42
14	Water use efficiency	1.21	0.01	1.89
15	Stomatal conductance	0.86	0.01	1.83
16	Stomatal width	0.14	0.00	0.82

tiation [62–64]. Numerous studies have documented eco-geographic variation for drought-related traits linked to environmental parameters such as phenology and carbon isotope discrimination in *Triticum turgidum* spp. *dicoccoides* (Körn.) Thell [65], as well as leaf area, electrolyte linkage and RWC in *Arabidopsis thaliana* [66]. In this context, FIGS represents a logical extension of N. I. Vavilov's work that by the 1920s had developed and illustrated the concept of centres of diversity that established the association between diversity and eco-geographic distribution [67].

Despite the above, using an eco-geographic approach to select germplasm for utilization has not been industry standard in genetic resource conservation circles. Rather, there has been a focus on the core collection concept (*e.g.* [21]). In fact, the Food and Agriculture Organization of the United Nations (FAO), in its global strategy for plant genetic resources (PGR) conservation, called for and financially supported the development of core collections as a standard and recommended practice.

However, the authors of this paper have determined that a large percentage of germplasm requests from the ICARDA genetic resources database are for specific adaptive traits. Thus it is argued that, in contrast to core collections, FIGS represents a dynamic, direct and practical approach that focuses on specific adaptive traits rather than on generalized measures of diversity, and as such could be of considerable value to the genebank user community if deployed on a regular basis. It is further suggested that as the plant breeding community prepare to tackle climate change, the efficient utilization of genetic resource collections will become increasingly important [68]. In this context, it is argued that the FIGS approach can reduce the cost and effectiveness of evaluation by reducing the number of accessions screened while providing a higher probability of identifying sought-after traits.

While this study supports the assertion that FIGS is an effective way to search for adaptive traits, there is considerable room for improvement in the approach. Since FIGS is still in its infancy, it is acknowledged that the procedure used to select the sets in this study was more a common sense process rather one based on previous research. The rationale behind the selection of the dry set was to select material from environments that were most likely to impose relatively dry conditions during the growing season whilst not so dry that a crop would need irrigation. Faba bean is unlikely to be planted to rain fed conditions much below the 300 mm/year limit. Further the criteria on narrow range in rainfall and low aridity index were selected to favour environments where there is more likely to be higher seasonal variation for moisture availability, low rainfall tending to be coupled with high variability. The rationale here is that higher seasonal moisture variation is likely to push populations towards physiological adaptation to dry conditions rather than drought avoidance strategies (earliness, for example). The *bio15* parameter, a measure of seasonal variation in rainfall, was then used to select high variation environments. The *tminyr, tmaxyr, bio4, bio16,* and *bio19* parameters were included in the clustering procedure because they all represent factors that influence growth conditions and it was desirable to include a range of different low-moisture environments. The approach outlined above to select the dry set could have been done in different ways and further experimentation is needed to determine the optimal strategy.

Different approaches could also have been used to define the set of material originating in environments with higher seasonal moisture profiles. In this case it was considered desirable to include a wide range of environments provided they received over 800 mm of precipitation, which is considered to be favourable for faba bean cultivation.

Both sets were chosen by applying selection criteria to long-term average yearly data. However, these data do not necessarily reflect the conditions within the growing season. A more effective approach would be to use climatic data presented on the basis of growing season or different crop development phases rather than calendar year. To do this effectively there is a need for accurate continuous surface maps detailing the onset of the growing season for different crop species. Additionally, the machine-learning algorithms used in this study could be used to create the FIGS sets using climatic variables as the input data.

While this study demonstrates that there is a difference in leaf morphology and physiology associated with water use between the two sets, it was performed under well watered conditions and thus we cannot firmly conclude that the dry set is in fact more drought tolerant. Nevertheless, the existence of a difference indicates that eco-typic differentiation has occurred in faba bean accessions from dryer environments, so we can infer that differentiation is in some way associated with adaptation to dryer seasonal moisture profiles. Indeed, eco-geographic differentiation has been found for leaf morphology in other species. For example leaf area was found to be negatively correlated with altitude (and by inference the probability of chilling stress) for *Dodonaea viscosa* subsp. *angustissima* [69]. It would appear that the same holds true for faba bean, since leaf width in this study was linked to maximum temperature regionally (latitude gradient) and leaf area to minimum temperature locally (altitude gradient).

While leaf area and RWC were positively correlated in *Quercus acutissima* [70], as found in this study ($R^2 = 0.29$, $P<0.001$, $n = 402$), leaf area and size diminished with declining water availability, in contrast to this study. The present results may be seen as somewhat counter-intuitive if one expects reduced leaf areas to present less evaporative surface, thus favouring tighter control on water use, which is certainly the case in xerophytic perennials. However, large leaf areas cover the soil surface more effectively, minimizing unproductive evaporation. Furthermore, 75% of the dry set accessions belong to the *major* seed type of faba bean (Figure S1) and these larger seeds tend to produce bigger seedlings with larger leaflets and more extensive root systems, which bestow the adaptive advantage of rapidly exploiting available soil moisture earlier in the season. In *Panicum virgatum* L., for example, larger seeds were linked to higher seedling vigour and better root establishment in dry environments [71], while in oat (*Avena sativa* L.) larger seeds lead to better germination under osmotic stress [72], and in faba bean larger seeds were related to higher transpiration efficiency and lower transpiration rates [73]. Furthermore, in some legume species seed size was found to be an indicator of abiotic adaptation [74].

RWC has been recognized as a reliable indicator of plant water status, and thus has been widely used as a screening parameter for drought adaptation in crop plants [47,75]. Nevertheless, screening large quantities of germplasm using RWC measurements is costly and time consuming. Since lower RWC in this study was associated with lower canopy temperatures ($R^2 = 0.54$, $P<0.001$, $n = 402$), it supports the assertion of Blum [76] that leaf temperature can be used as a rapid and economical phenotyping method to screen germplasm for drought adaptation. The slightly earlier flowering in the dry set is in line with expectations that earlier flowering is part of drought escape in faba bean as in many other species [6].

The current work involved the aerial part of the plant. Nevertheless, for drought adaptation, root morphology and function also play a significant role [75,77]. For example, the roots of sorghum genotypes from dry African environments were found to be deeper and more highly branched than US-derived genotypes [78]. Variation for root traits linked to drought adaptation is of particular interest, especially if they can be linked to more easily evaluated above-ground marker. A logical extension of the work reported here would be to assess differences in root morphology between the two sets.

Many genetic diversity studies still use linear based approaches such as principal component analysis (PCA). The machine learning/recursive algorithms used here represent a novel approach deserving some comment. This study demonstrates that the RF and SMV approaches are suited to studies such as this, since they can detect patterns or relationships between a dependent variable (trait data) and a set of independent variables (climate data) in large datasets [79]. They can also identify parameters that have the greatest impact on the discrimination. Used in this context, the algorithms can point to which trait or combination of traits confers the adaptation.

Further, the use of recursive partitioning is gaining momentum in areas where the data are too highly dimensional for standard regression methods such as PCA in which the decomposition of variables into reduced components leads to the loss of their individual effects, thus rendering the important variable unidentifiable in the interpretation [80]. In the present algorithms, the variables that have a strong relationship to the trait would be those that split the accessions correctly [81]. At the split, the variable that produces less entropy measured using either information theory (Shannon index) or Gini index (known as impurity measure) is ranked first. A reduction in the impurity is a prerequisite for the variable ranking/importance which can be best visualised in the graphs generated by these algorithms [82].

A further advantage of the algorithms used here is that the input data does not have to be normally distributed or conform to other assumptions related to linear models and thus do not require the tedious and time consuming pre-analysis required for linear models to ensure that the assumptions are not violated.

Conclusions

The methods used were effective at creating sets that were different in terms of leaf morphology, physiology and phenology. This demonstrates that eco-geographic differentiation in faba beans has occurred and is related, in part, to moisture availability. Thus the underlying premise upon which FIGS is based was supported, indicating that it can be an effective tool to enhance the discovery and deployment of new genes, although the FIGS process can be improved to select for drought-adapted genetic resources. Further, the use of machine-learning algorithms was demonstrated here as an effective tool to investigate datasets that are complex and highly dimensional, so it is suggested that they are particularly suited to eco-geographic diversity studies. The results also indicate that leaf and canopy temperature could be an economical way to screen for potentially drought-adapted material as has been suggested by other authors.

Acknowledgments

We would like to thank Markku Tykkyläinen, Sini Lindstrom and Sanna Peltola, technical assistants of the glasshouse of the University of Helsinki for their kind assistance during the experiments.

Author Contributions

Conceived and designed the experiments: HK KS FLS. Performed the experiments: HK. Analyzed the data: AB HK. Contributed reagents/materials/analysis tools: HK KS AB MM FLS. Wrote the paper: HK KS AB FLS.

References

1. Barnabás B, Jäger K, Fehér A (2008) The effect of drought and heat stress on reproductive processes in cereals. Plant Cell Environ 31: 11–38.
2. IPCC (2012) Managing the risks of extreme events and disasters to advance climate change adaptation. A special report of working groups I and II of the intergovernmental panel on climate change [Field CB, Barros V, Stocker TF, Qin D, Dokken DG, et al. editors]. Cambridge University Press, Cambridge, UK, and New York, NY, USA, 582 pp.
3. Duc G (1997) Faba bean (*Vicia faba* L.). Field Crops Res 53: 99–109.
4. Crépon K, Marget P, Peyronnet C, Carrouée B, Arese P, et al. (2010) Nutritional value of faba bean (*Vicia faba* L.) seeds for feed and food. Field Crops Res 115: 329–339.
5. Khan HR, Link W, Hocking TJH, Stoddard FL (2007) Evaluation of physiological traits for improving drought tolerance in faba bean (*Vicia faba* L.). Plant Soil 292: 205–217.
6. Khan HR, Paull JG, Siddique KHM, Stoddard FL (2010) Faba bean breeding for drought-affected environments: A physiological and agronomic perspective. Field Crops Res 115: 279–286.
7. Duc G, Link W, Marget P, Redden RJ, Stoddard FL, et al. (2011) Genetic adjustment to changing climates: faba bean. In: Yadav SS, Redden RJ, Hatfield JL, Lotze-Campen H, Hall AE, editors. Crop adaption to climate change, 1rd edn. John Wiley & Sons, 269–286.
8. Gollin D, Smale M, Skovmand B (2000) Searching an *ex situ* collection of wheat genetic resources. Am J Agric Econ 82: 812–827.
9. Frankel OH (1984) Genetic perspectives of germplasm conservation. In: Arber W, Llimensee K, Peacock WJ, Starlinger P, editors. Genetic manipulation: Impact on man and society, Cambridge University Press, Cambridge, 161–170.
10. Hodgkin T, Brown ADH, van Hintum Th JL, Morales EAV (1995) Core collections of plant genetic resources, John Wiley & Sons, Chichester UK, 265 p.
11. Brown AHD (1989) Core collections: a practical approach to genetic resources management. Genome 31: 818–824.
12. Brown AHD (1989) The case for core collections. In: Brown AHD, Frankel OH, Marshall DR, Williams JT, editors. The use of plant genetic resources. Cambridge, Cambridge University Press, 136–156.
13. Holbrook CC, Anderson WF, Pittman RN (1993) Selection of a core collection from the United-States germplasm collection of peanut. Crop Sci 33: 859–861.
14. Ortiz R, Ruiz-Tapia EN, Mujica-Sanchez A (1998) Sampling strategy for a core collection of Peruvian quinoa germplasm. Theor Appl Genet 96: 475–483.
15. Hu J, Zhu J, Xu HM (2000) Methods of constructing core collections by stepwise clustering with three sampling strategies based on the genotypic values of crops. Theor Appl Genet 101: 264–268.
16. Malosetti M, Abadie T (2001) Sampling strategy to develop a core collection of Uruguayan maize landraces based on morphological traits. Genet Resour Crop Evol 48: 381–390.
17. Casler MD (1995) Patterns of variation in a collection of perennial ryegrass accessions. Crop Sci 35: 1169–1177.
18. Tohme J, Gonzalez DO, Beebe S, Duque MC (1996) AFLP analysis of gene pools of a wild bean core collection. Crop Sci 36: 1375–1384.
19. Bartish GI, Jeppsson N, Bartish IV, Nybom H (2000) Assessment of genetic diversity using RAPD analysis in a germplasm collection of sea buckthorn. Agric Food Sci 9: 279–289.
20. Fu YB, Peterson GW, Williams D, Richards KW, Fetch JM (2005) Patterns of AFLP variation in a core subset of cultivated hexaploid oat germplasm. Theor Appl Genet 111: 530–539.
21. Brown AHD, Spillane C (1999) Implementing core collections - principles, procedures, progress, problems and promise. Johnson RC, Hodgkin T, editors. Core collections for today and tomorrow. Rome, International Plant Genetic Resources Institute, 1–9.
22. Upadhyaya HD, Ortiz R (2001) A mini core subset for capturing diversity and promoting utilization of chickpea genetic resources in crop improvement. Theor Appl Genet 102: 1292–1298.
23. McKhann HI, Camilleri C, Bérard A, Bataillon T, David JL, et al. (2004) Nested core collections maximizing genetic diversity in *Arabidopsis thaliana*. Plant J 38: 193–202.
24. Furman BJ (2006) Methodology to establish a composite collection: case study in lentil. Plant Genet Resour 4: 2–12.
25. Polignano GB, Uggenti P, Scippa G (2001) Diversity analysis and core collection formation in Bari faba bean germplasm. Plant Genet Resour Newslett 125: 33–38.
26. Gepts P (2006) Plant genetic resources conservation and utilization: the accomplishments and future of a societal insurance policy. Crop Sci 46: 2278–2292.
27. Dwivedi SL, Upadhyaya HD, Stalker HT, Blair MW, Bertioli DJ, et al. (2008) Enhancing crop gene pools with beneficial traits using wild relatives. Plant Breed Rev 30: 180–230.
28. Pessoa-Filho M, Rangel PHN, Ferreira ME (2010) Extracting samples of high diversity from thematic collections of large gene banks using a genetic-distance based approach. BMC Plant Biol 10: 127. doi:10.1186/1471-2229-10-127
29. Xu Y (2010) Plant genetic resources: Management, evaluation and enhancement. In: Molecular plant breeding. Wallingford, UK, CAB International, 15–194.
30. Mackay M (1990) Strategic planning for effective evaluation of plant germplasm. In: Srivastava JP, Damania AB, editors. Wheat genetic resources: meeting diverse needs. John Wiley & Sons, Chichester, 21–25.
31. Mackay M (1995) One core collection or many? In: Hodgkin T, Brown AHD, Van Hintum TJL, Morales EAV, editors. Core collections of plant genetic resources. John Wiley & Sons Ltd., Chichester, 199–210.
32. Mackay M, Street K (2004) Focused identification of germplasm strategy – FIGS. In: Proceedings of the 54th Australian Cereal Chemistry Conference and the 11th Wheat Breeders' Assembly, , editors. Cereal Chemestry Division, Royal Australian Chemical Institute (RACI), Melbourne, Victoria, Australia. 138–141.
33. Mackay M, von Bothmer R, Skovmand B (2005) Conservation and utilization of plant genetic resources – future directions. Czech J Genet Plant Breed 41: 335–344.
34. Endresen DTF, Street K, Mackay M, Bari A, De Pauw E (2011) Predictive association between biotic stress traits and ecogeographic data for wheat and barley landraces. Crop Sci 51: 2036–2055.
35. Bari A, Street K, Mackay M, Endresen DTF, De Pauw E, et al. (2012) Focused identification of germplasm strategy (FIGS) detects wheat stem rust resistance linked to environmental variables. Genet Resour Crop Evol 59: 1465–1481.
36. Kaur N, Street K, Mackay M, Yahiaoui N, Keller B (2008) Molecular approaches for characterization and use of natural disease resistance in wheat. Eur J Plant Pathol 121: 387–397.
37. Bhullar NK, Street K, Mackay M, Yahiaoui N, Keller B (2009) Unlocking wheat genetic resources for the molecular identification of previously undescribed functional alleles at the *Pm3* resistance locus. Proc Natl Acad Sci USA 106: 9519–9524.
38. El Bouhssini M, Street K, Joubi A, Ibrahim Z, Rihawi F (2009) Sources of wheat resistance to sunn pest, *Eurygaster integriceps* Puton, in Syria. Genet Resour Crop Evol 56: 1065–1069.
39. El Bouhssini M, Street K, Amri A, Mackay M, Ogbonnaya FC, et al. (2011) Sources of resistance in bread wheat to Russian wheat aphid (*Diuraphis noxia*) in Syria identified using the Focused Identification of Germplasm Strategy (FIGS). Plant Breed 130: 96–97.
40. Endresen DTF (2010) Predictive association between trait data and ecogeographic data for Nordic barley landraces. Crop Sci 50: 2418–2430.
41. FAO (2010) The second report on the state of the world's plant genetic resources for food and agriculture. Rome, Italy, 398 p.
42. Hijmans RJ, Cameron SE, Parra JL, Jones PG, Jarvis A (2005) Very high resolution interpolated climate surfaces for global land areas. Int J Climatol 25: 1965–1978.
43. Wang H, Clarke JM (1993) Genotypic, intraplant, and environmental variation in stomatal frequency and size in wheat. Can J Plant Sci 73: 671–678.
44. Muratova V (1931) Common Beans (*Vicia faba*). Bull Appl Bot Genet Plant Breed 50: 1–298.
45. Cubero J (1974) On the evolution of *Vicia faba* L. Theor Appl Genet 45: 47–51.
46. Fischer RA, Rees D, Sayre KD, Lu Z-M, Condon AG, et al. (1998) Wheat yield progress associated with higher stomatal conductance and photosynthetic rate, and cooler canopies. Crop Sci 38: 1467–1475.
47. Barrs HD, Weatherley PE (1962) A re-examination of the relative turgidity technique for estimating water deficit in leaves. Aust J Biol Sci 15: 413–428.
48. R Development Core Team (2012) R: A language and environment for statistical computing. R Foundation for Statistical Computing, Vienna, Austria. ISBN 3-900051-07-0. Available: http://www.R-project.org.
49. Kuhn M (2008) Building predictive models in R using the caret package. J Stat Softw 28 (5).
50. Steinberg D (2009) CART: Classification and Regression Trees. Taylor & Francis Group, LLC.
51. Breiman L (2001) Random forests. Mach Learn 45: 5–32.
52. Cutler DR, Edwards Jr TC, Beard KH, Cutler A, Hess KT, et al. (2007) Random forests for classification in ecology. Ecology 88: 2783–2792.
53. Prasad AM, Iverson LR, Liaw A (2006) Newer classification and regression tree techniques: bagging and random forests for ecological prediction. Ecosystems 9: 181–199.
54. Dimitriadou E, Hornik K, Leisch F, Meyer D, Weingessel A (2010) R library (e1071). The R foundation for statistical computing. ISBN: 3-900051-07-0.
55. Karatzoglou A, Meyer D, Hornik K (2006) Support vector machines in R. J Stat Softw 15 (9).
56. Tirelli T, Pozzi L, Pessani D (2009) Use of different approaches to model presence/absence of Salmo marmoratus in Piedmont (Northwestern Italy). Ecol Inform 4: 234–242.

57. Drake JM, Randin C, Guisan A (2006) Modelling ecological niches with support vector machines. J Appl Ecol 43: 424–432.
58. Swets JA, Dawes RM, Monahan J (2000) Better decisions through science. Sci Am 283: 82–87.
59. Fawcett T (2006) An introduction to ROC analysis. Pattern Recognit Lett 27: 861–874.
60. Freeman EA, Moisen GG (2008) A comparison of the performance of threshold criteria for binary classification in terms of predicted prevalence and kappa. Ecol Modell 217: 48–58.
61. Landis JR, Koch GG (1977) The measurement of observer agreement for categorical data. Biometrics 33: 159–174.
62. Lin W, Bradshaw AD, Thurman DA (1975) The potential for evolution of heavy metal tolerance in plants. III. The rapid evolution of copper tolerance in *Agrostis stolonifera*. Heredity 34: 165–187.
63. Spieth PT (1979) Environmental heterogeneity: a problem of contradictory selection pressures, gene flow, and local polymorphism. Am Nat 113: 247–260.
64. Epperson BK (1990) Spatial autocorrelation of genotypes under directional selection. Genetics 124: 757–771.
65. Peleg Z, Fahima T, Abbo S, Krugman T, Nevo E, et al. (2005) Genetic diversity for drought resistance in wild emmer wheat and its ecogeographical associations. Plant Cell Environ 28: 176–191.
66. Bouchabke O, Chang F, Simon M, Voisin R, Pelletire G, et al. (2008) Natural variation in *Arabidopsis thaliana* as a tool for highlighting differential drought response. PLoS ONE 3: e1705. doi:10.1371/journal.pone.0001705
67. Vavilov NI (1926) Tzentry proiskhozhdeniya kulturnykh rastenii. (Studies on the origin of cultivated plants). Trudy Byuro prikl Bot [in Russian] 16: 139–248.
68. Mba C, Guimaraes EP, Ghosh K (2012) Re-orienting crop improvement for the changing climate conditions for the 21st century. Agric Food Secur 1: 7. doi:10.1186/2048-7010-1-7
69. Guerin GR, Wen H, Lowe AJ (2012) Leaf morphology shift linked to climate change. Biol Lett 8: 882–886.
70. Xu F, Guo W, Xu W, Wei Y, Wang R (2009) Leaf morphology correlates with water and light availability: what consequences for simple and compound leaves? Prog Nat Sci 19: 1789–1798.
71. Fan JW, Du YL, Turner NC, Li FM, He J (2012) Germination characteristics and seedling emergence of switchgrass with different agricultural practices under arid conditions in China. Crop Sci 52: 2341–2350.
72. Mut Z, Akay H (2010) Effect of seed size and drought stress on germination and seedling growth of naked oat (*Avena sativa* L.). Bulg J Agric Sci 16: 459–467.
73. Avola G, Cavallaro V, Patanè C, Riggi E (2008) Gas exchange and photosynthetic water use efficiency in response to light, CO_2 concentration and temperature in *Vicia faba*. J Plant Physiol 165: 796–804.
74. Parra-Quijano M, Iriondo JM, Torres E (2012) Ecogeographical land characterization maps as a tool for assessing plant adaptation and their implications in agrobiodiversity studies. Genet Resour Crop Evol 59: 205–217.
75. Blum A (2011) Plant breeding for water limited environments. Springer-Verlag, New York, 255 p.
76. Blum A (2009) Effective use of water (EUW) and not water-use efficiency (WUE) is the target of crop yield improvement under drought stress. Field Crops Res 112: 119–123.
77. Blum A (2011) Drought resistance – is it really a complex trait? Funct Plant Biol 38: 753–757.
78. Masi CEA, Maranville JW (1998) Evaluation of sorghum root branching using fractals. J Agric Sci 131: 259–265.
79. Therneau TM, Atkinson EJ (1997) An introduction to recursive partitioning using the rpart routine. Technical Report 61, Section of Biostatistics, Mayo Clinic, Rochester. Available: http://www.mayo.edu/hsr/techrpt/61.pdf.
80. Strobl C, Malley J, Tutz G (2009) An introduction to recursive partitioning: rationale, application, and characteristics of classification and regression trees, bagging, and random forests. Psychol Methods 14: 323–348.
81. Liknes GC, Woodall CW, Perry CH (2009) Predicting forest attributes from climate data using a recursive partitioning and regression tree algorithm. In: McWilliams W, Moisen G, Czaplewski R, editors. Forest Inventory and Analysis (FIA) Symposium 2008; Park City, UT. Proc. Fort Collins, CO: U.S. Department of Agriculture, Forest Service, Rocky Mountain Research Station, 7 p.
82. Hothorn T, Hornik K, Zeileis A (2006) Unbiased recursive partitioning: A conditional inference framework. J Comput Graph Stat 15: 651–674.

17

Favorable QTL Alleles for Yield and Its Components Identified by Association Mapping in Chinese Upland Cotton Cultivars

Hongxian Mei[1,2], Xiefei Zhu[1], Tianzhen Zhang[1]*

1 National Key Laboratory of Crop Genetics & Germplasm Enhancement, Cotton Research Institute, Nanjing Agricultural University, Nanjing, China, **2** Sesame Research Center, Henan Academy of Agricultural Sciences, Zhengzhou, China

Abstract

Linkage disequilibrium based association mapping is a powerful tool for dissecting the genetic basis underlying complex traits. In this study, an association mapping panel consisting of 356 representative Upland cotton cultivars was constructed, evaluated in three environments and genotyped using 381 SSRs to detect molecular markers associated with lint yield and its components. The results showed that abundant phenotypic and moderate genetic diversities existed within this germplasm panel. The population could be divided into two subpopulations, and weak relatedness was detected between pair-wise accessions. LD decayed to the background ($r^2=0.1182$, $P\leq0.01$), $r^2=0.1$ and $r^2=0.2$ level within 12–13 cM, 17–18 cM and 3–4 cM, respectively, providing the potential for association mapping of agronomically important traits in Chinese Upland cotton. A total of 55 marker-trait associations were detected between 26 SSRs and seven lint yield traits, based on a mixed linear model (MLM) and Bonferroni correction ($P\leq0.05/145$, $-\log_{10}P\geq3.46$). Of which 41 could be detected in more than one environment and 17 markers were simultaneously associated with two or more traits. Many associations were consistent with QTLs identified by linkage mapping in previous reports. Phenotypic values of alleles of each loci in 41 stably detected associations were compared, and 23 favorable alleles were identified. Population frequency of each favorable allele in historically released cultivar groups was also evaluated. The QTLs detected in this study will be helpful in further understanding the genetic basis of lint yield and its components, and the favorable alleles may facilitate future high-yield breeding by genomic selection in Upland cotton.

Editor: Tongming Yin, Nanjing Forestry University, China

Funding: The funders had no role in study design, data collection and analysis, decision to publish, or preparation of the manuscript.

* E-mail: cotton@njau.edu.cn

Introduction

Cotton is the most important natural textile fiber source globally. The worldwide economic impact of the cotton industry is estimated at approximately $500 billion per year with an annual utilization of about 27 million metric tons of cotton fiber. In recent years, demand for cotton fiber in the world market has dramatically increased, stock and use ratio dropped to 37% in 2010, compared to 55% in 2009. While cotton acreage has declined worldwide in the past few years, mainly due to strong competition from other crops as well as production costs (National Cotton Council, USA, http://www.cotton.org, 2012). The tetraploid species *Gossypium hirsutum* L. (n = 26, AD genome), commonly referred to as Upland cotton, accounts for 95% of the world's cotton production [1]. Thus, improving lint yield of Upland cotton cultivars will be critical for meeting worldwide demand, and maintaining profitability for cotton growers.

Lint yield is a complex trait in cotton, which is controlled by a large number of quantitative loci (QTLs). It is becoming progressively more difficult to improve lint yield using conventional breeding methods. Fortunately, the development in applied genomics research has provided alternative tools to improve efficiency in plant breeding programs. Molecular markers linked to the causal genes and QTLs can be used for marker-assisted selection (MAS) and/or genomic selection (GS) [2–3]. In the past two decades, a large number of QTLs for lint yield and fiber quality traits have been identified in Upland cotton [4–13]. However, approximately 80% of the previously reported QTLs could not be confirmed in subsequent studies, and few have actually been applied in breeding programs [14–16]. This may be because that most QTLs were population-specific, and the genetic variation detected in a unique bi-parental population might not be shared with other genetic populations, or shared but fixed in the parental lines. In addition, the limited genetic recombinations in most populations used for linkage mapping make it difficult to map QTLs with a high resolution, which severely limits their application in breeding programs. With the potential to exploit all recombination events that occurred in the evolutionary history of natural populations, linkage disequilibrium (LD) based association mapping (AM) has become a powerful approach for the dissection of complex traits and identification of causal variation with modest effects for target traits in many plant species [17–18] including cotton [19–24]. While the key constraints for the successful use of association mapping in plants are population structure and genetic relatedness, which can result in spurious marker-trait associations that may make it difficult to distinguish loci that truly affect the target traits [25–26]. Several statistical strategies have been developed to account for issues related to population structure and relatedness [27–29]. One powerful strategy is the unified mixed model approach (mixed linear model, MLM), which accounts for multiple levels of relatedness simultaneously, and can improve control of both type I and type II error rates [28]. In cotton, the first attempt of association mapping was reported by Kantartzi and Stewart in 2008. In that study, 30 marker and fiber trait associations were detected in 56 *Gossypium arboreum* accessions genotyped by 98 SSR markers [19]. Abdurakhmonov et al. performed an association mapping study, with the MLM model considering both kinship (K) and population structure (Q), of fiber quality traits by using a set of 95 core microsatellite markers in 285 exotic *Gossypium hirsutum* accessions and detected between 6% and 13% of SSR markers associated with the main fiber traits. Meanwhile, they indicated the genome-wide LD ($r^2 \geq 0.1$) declined at <10 cM in the landrace stocks and >30 cM in variety germplasm, but at $r^2 \geq 0.2$ which reduced to ~1–2 cM and ~6–8 cM, respectively [20]. Abdurakhmonov et al. performed another association mapping study of fiber quality traits using 202 microsatellite markers in a panel of 335 *G. hirsutum* varieties [21]. The result showed that the genome-wide LD extended up to 25 cM at $r^2 \geq 0.1$ and reduced to ~5–6 cM at $r^2 \geq 0.2$ and an average of ~20 SSR markers was associated with each main fiber quality trait in two environments. Zeng et al. carried out an association mapping study between 86 SSR markers and fiber traits using an exotic germplasm population of 260 lines derived from multiple crosses among *Gossypium* tetraploid species and found 59 markers were significantly associated with six fiber traits [22]. All the results mentioned above provided useful evidences of the potential for association mapping of agronomically important traits in cotton. But till now, association mapping study of lint yield traits has not been reported in cotton.

Although AM has been successfully used to detect the QTLs underlying quantitative traits in some crops, from a breeding standpoint, detecting associated loci is just the first step; analyzing the genetic effects of alleles and identifying favorable alleles will be more beneficial for target trait improvement. Breseghello & Sorrells identified several potentially beneficial alleles for kernel size and milling quality by comparing the average phenotypic value with specific alleles and null alleles in a soft winter wheat population [30]. Jia et al. identified some putative resistant alleles for Sheath Blight resistance in a rice panel composed of 217 accessions from the USDA core collection, and found that the number of putative resistant alleles presented in an entry was highly and significantly correlated with the decrease of ShB rating [31]. We performed a preliminary AM study in 81 Upland cotton cultivars and identified some elite alleles for yield and fiber quality traits [24]. China is the world's largest cotton-growing nation, but not a cotton domestication region. Most Upland cotton cultivars developed in China were derived from a few germplasm resources such as Deltapine (DPL), Stoneville (STV), Foster, King and Uganda, all of which were introduced from abroad [32]. Current and obsolete cultivars have been and continue to be the main resources for cotton breeding programs. Dissecting the genetic basis of lint yield and quality traits will be of great benefits to germplasm evaluation and future molecular breeding. In the present study, we aimed to detect QTLs underlying lint yield and its components, and to identify the favorable alleles in an AM panel composed of 356 accessions. We also analyzed genetic diversity, LD decay, population structure, genetic relatedness and favorable allele frequency in historically released cutivar groups. Our results should provide useful information for further understanding the genetic basis of lint yield and its components, and will facilitate future high-yield breeding by genomic selection in Upland cotton.

Materials and Methods

Association mapping panel construction

A total of 356 representative Upland cotton cultivars and breeding lines were selected from the cotton germplasm collections in our laboratory and the Cotton Research Institute, Chinese Academy of Agricultural Sciences (CRI-CAAS), and assembled to construct an AM panel. The population consisted of 348 cultivars developed in China, seven introduced from the U.S., including the genetic standard line TM-1, and one introduced from Uganda. According to their release year, the 348 Chinese cultivars could be divided into the following six groups: I (1930–1960, 26 lines); II (1961–1970, 26 lines); III (1971–1980, 39 lines), IV (1981–1990, 83 lines); V (1991–2000, 125 lines); and VI (2000–2005, 49 lines). The cultivars introduced from abroad, DPL 15, DPL 16, STV 2B, King, Foster 6 and Uganda 3 were used as a check group for genetic diversity and allele transmission evaluation, because they had been used as the main founder parents in China's Upland cotton breeding programs and are the progenitors of many cultivars [32]. All accessions have been self-pollinated for more than six generations and their detailed information are summarized in Table S1.

Trait phenotyping

All of the accessions were planted in the following three environments to evaluate phenotypic performance: (1) Jiangpu Breeding Station, Nanjing Agricultural University, Nanjing, in 2009 (designated as E1), (2) Dafeng Agronomy Farm, Yancheng, Jiangsu Province (E2), and (3) Zhengzhou Agricultural Research Institute, Zhengzhou, Henan in 2010 (E3). The first two locations were in the Yangtze River cotton-growing region, and the third was in the Yellow River region. A randomized complete block design with single row plot and two replications was used in all field trails. The sowing dates were from late March to early April in different years and locations, and seedlings having up to 3–4 leaves were transplanted from seedbeds to fields, with 20 plants per row, a 30 cm plant-to-plant spacing, and 80 cm between rows. For most of the accessions are non-BT cottons, chemical control were used for preventing from bollworm damage and field managements were adjusted to local practices.

Field planting has been approved by Nanjing Agric Univ.. No specific permissions were required for these locations/activities since they are pure-line cultivars and the field studies did not involve endangered or protected species.

Ten consecutive plants in the middle of each row were tagged for trait measurement. Yield traits evaluated included: lint yield (LY, g/plant), seed cotton yield (SY, g/plant), bolls per plant (BN), boll weight (BW, g), lint percentage (LP, %), lint index (LI, g/100 seeds) and seed index (SI, g/100 seeds).

SSR genotyping

Young leaves from each of the 356 accessions were collected and stored at −20°C. Total genomic DNA was extracted from the leaf samples as described by Guo et al. [33]. Based on the dense genetic linkage map constructed in our laboratory [33], 381 pairs of SSR primers that amplify loci evenly covering the tetraploid cotton genome (one marker per 10 cM, 186 on At and 195 on Dt subgenome with an average of 14.65 markers each chromosome) were selected to genotype the 356 accessions. The procedure for PCR-amplification and product analysis followed the published methods from our laboratory [34–35]. Since *G. hirsutum* is an allopolyploid species, SSR markers often yield complex band patterns and some of them had been located to more than one locus. To measure the complex band patterns of large scale genotypes, the band pattern in TM-1 (genetic standard line, one parent with which the reference linkage map was constructed) was treated as a check and the following criteria were used to assign the alleles to the corresponding loci: 1) when only one fragment was amplified in each accession, the fragments were regarded as alleles belonging to the single locus; 2) when multiple fragments were amplified in each line and the bands showed an obvious co-segregating relationship among different samples, they were regarded as alleles belonging to the same locus; and 3) when multiple bands produced in each line did not co-segregate among different accessions, the corresponding fragments in TM-1 that had been mapped to the reference map and co-segregated among different accessions were measured, other bands were discarded. According to the above criteria, the band pattern in TM-1 was designated as 1, the same patterns were also designated as 1, and the different ones were designated as 2, 3, 4, 5 and so on, thus the alleles from all accessions on each locus were measured. Markers with more than 10% missing data were not used in further analysis.

Genotypic data analysis

Summary statistics including the total number of alleles, the number of alleles per locus, and gene diversity values were calculated using the software PowerMarker 3.25 [36]. The Bayesian model-based program STRUCTURE 2.3 was used to infer the population structure using 66 unlinked or weakly linked SSR markers [37]. The length of the burn-in period and the number of Markov Chain Monte Carlo replications after burn-in were all assigned at 100,000 with an admixture and allele frequencies correlated model. Five independent run iterations were performed with the hypothetical number of subpopulations (k) ranging from 1 to 10. The correct estimation of k was provided by joining the log probability of data [LnP(D)] from the STRUCTURE output and an ad hoc statistic Δk [38]. Based on the correct k, each accession was assigned to a subpopulation for which the membership value (Q value) was >0.5 [39], and the population structure matrix (Q) was generated for further marker-trait association mapping. The software SPAGeDi was used to calculate the pair-wise relatedness coefficients (K, kinship matrix) in order to estimate the genetic relatedness among individuals, with the negative value of kinship set to zero [40]. To estimate LD pattern in Upland cotton genome, the weighted average of squared correlation coefficient r^2 of each pair of SSR loci was calculated using the software package TASSEL 2.1 with rare alleles (allele frequency less than 0.05) treated as missing data [41]. The r^2 was estimated for total, linked and unlinked markers both in the entire panel and each subpopulation, respectively. The 99th percentile of r^2 distribution for unlinked markers, which determined whether LD is due to physical linkage [42], was treated as the background LD level [43].The r^2 values of each pair of SSR loci were plotted against map distance (cM), and LD decay was estimated.

Phenotypic data analysis

Statistical analysis of all phenotypic data across three environments was performed with SAS 8.0 software (SAS Institute 1999). Analysis of variance (ANOVA) of all phenotypic data was calculated with PROC GLM, based on the trait means for each line across the three environments. Decomposition of variance components (genotype, environment, block, and the interactions among these factors) was evaluated using PROC VARCOMP, and the broad-sense heritability (h_B^2) of each trait was estimated with the variance components. Correlation coefficients between traits were calculated with PROC CORR.

Association mapping and favorable allele identification

Because the MLM model accounts for the effects both of population structure and genetic relatedness, and can significantly reduce spurious associations [28], the marker-trait AM was carried out with the MLM model as implemented in TASSEL software, and the P value and R^2 for each marker-trait association were determined [41]. Based on the results of AM, QTL alleles of loci significantly associated with the target traits were further analyzed. The phenotypic allele effect was estimated through comparison between the average phenotypic value over accessions with the specific allele and that of all accessions:

$$a_i = \sum x_{ij} \Big/ n_i - \sum N_k \Big/ n_k$$

where a_i is the phenotypic effect of the *ith* allele; x_{ij} is the phenotypic value over the *jth* accession with the *ith* allele; n_i is the number of accessions with the *ith* allele; N_k is the phenotypic value over all accessions; n_k is the number of accessions. If the value of $a_i>0$, the allele is considered to have a positive effect, if it is <0, it corresponds to a negative allele. The favorable alleles were then identified according to the breeding objective of each target trait [24].

Results

Genetic diversity, population structure and genetic relatedness

Of the 381 SSR markers selected, only 145 amplified polymorphism (67 of 186 in At and 78 of 195 in Dt subgenome) in the present panel and a total of 415 alleles were detected (Table S2). The allele number, gene diversity and polymorphism information content (PIC) value of the 145 loci averaged 2.86, 0.32 and 0.27, respectively; with ranges of 2–9, 0.01–0.73 and 0.27–0.68, respectively. Approximately 80% of the polymorphic loci (115 of 145) had only two or three alleles. Among the 415 alleles detected, population frequencies of 131 alleles were rare (less than 0.05) and 34 were unique (detected in only one accession). The total number of alleles and the number of alleles per locus detected in the six historically released cultivar groups were much greater than that in the six founder parents (Table S3).

The model-based evaluation of the population structure of the 356 Upland cotton cultivars showed that the LnP(D) value corresponding to each hypothetical k kept increasing with k value and did not show any peak. The Δk value showed a much higher likelihood at $k = 2$ than at $k = 3$–10 (**Figure 1**), suggesting that the total panel could be divided into two major subpopulations [38], designated as P1 and P2, respectively. The P1 group contained 115 accessions including 63 cultivars from Yellow River cotton growing region, 46 lines from North and Northwest China regions, and six cultivars from Yangtze River region. The P2 group consisted of 241 accessions including 116 lines from Yellow River cotton growing region, 107 lines from Yangtze River region, 10 lines from the North and Northwest China regions, and eight lines intrduced from abroad (Table S1). Then, the corresponding Q matrix at $k = 2$ was used for the following association analysis.

For the kinship coefficient values, 86.85% was less than 0.05, 8.56% had a range of 0.05–0.10, and the remaining 4.59% showed various degrees of genetic relatedness (data not shown).

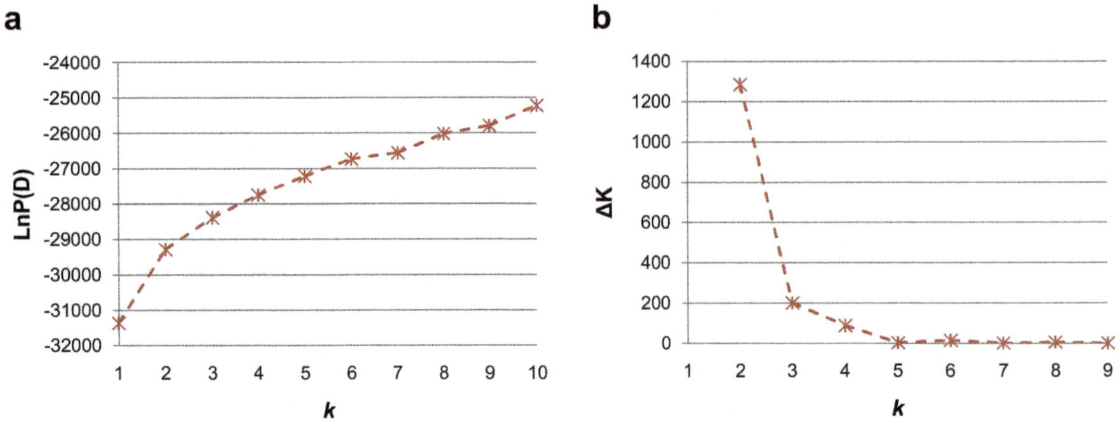

Figure 1. Estimated LnP(D) and ΔK over five repeats of STRUCTURE analysis. (a) LnP(D) for *k* from 1 to 10 for 356 accessions. LnP(D) value of each hypothetical *k* kept increasing with *k* value and did not show any peak. (b) ΔK for *k* from 2 to 9 for 356 accessions. The Δk value showed a much higher likelihood at $k=2$ than at $k=3$–10, suggesting that the total panel should be divided into two major subpopulations.

Based on the results of the relatedness analysis, a K matrix was constructed for association mapping.

Pair-wise linkage disequilibrium across the whole genome

The r^2 was calculated for total, linked and unlinked markers (**Table 1**), respectively, with SSR loci on the same chromosome considered as linked and those from different chromosomes as unlinked. In the entire panel, the average r^2 of locus pairs was 0.0103, and 18.29% were significant ($P \leq 0.01$). Moreover, 21.03% of the linked locus pairs and 18.18% of the unlinked pairs showed significant LD ($P \leq 0.01$) with the average r^2 of 0.0160 and 0.0101, respectively. In the subpopulation P1 and P2, the average r^2 of locus pairs was 0.0151 and 0.0104, respectively, and the proportion of significant LD ($P \leq 0.01$) was 5.10% and 10.78%, respectively. In the entire panel and subpopulations, both average r^2 and proportion of significant LD for linked loci were all higher than those for unlinked markers (**Table 1**).

The r^2 value and genetic distance of each pair of SSR loci was plotted into a scatter diagram, and then a curve was drawn to describe the trend of LD decay using the nonlinear regression model [43]. The curve exhibited a clear decay of LD with increase in genetic distance (**Figure 2**). In this study, the 99th percentile of r^2 distribution for unlinked markers, which determined the background level of LD, was 0.1182; and LD decayed to the background level within 12–13 cM. If the threshold of LD decay was set to $r^2 = 0.1$ and $r^2 = 0.2$, the genome-wide LD extended up to about 17–18 cM and 3–4 cM, respectively (**Figure 2**).

Variation of phenotypic traits

Seven traits for lint yield and its components were measured in 356 Upland cotton accessions across three different environments. Each trait varied widely (**Table 2**), and the ANOVA showed that the genotype (G) and the interactions between genotype and environmental factors (G×E) were both significant ($P \leq 0.01$) for all the seven traits. The mean coefficient of variance for FY, SY, BN, BW, LP, LI and SI was 29.09%, 23.19%, 19.36%, 9.11%, 9.33%, 12.48% and 9.02%, respectively, demonstrating that there was a high degree of diversity in lint yield traits of Chinese Upland cotton cultivars. The broad sense heritabilty (h_B^2) for the seven traits had a range of 27.34–75.77% in the reference population (**Table 2**). The highest h_B^2 value was for LP (75.77%), indicating that LP was less impacted by environmental factors than the other six traits.

Phenotypic correlation analysis showed that there were significant positive correlations between lint yield and its most components, while the negative correlation between lint yield and SI was also significant (**Table 3**). The correlation coefficients for LY with SY, BN, BW, LP, LI and SI were 0.963, 0.869, 0.461, 0.704, 0.671, and −0.257, respectively.

Table 1. LD in the entire panel and subpopulations.

Groups[a]	Total[b]		Linked[c]		Unlinked[d]	
	r^2	Sig. LD (%)[e]	r^2	Sig. LD (%)[e]	r^2	Sig. LD (%)[e]
P1	0.0151	5.10	0.0194	7.31	0.015	5.02
P2	0.0104	10.78	0.0172	11.11	0.0101	10.77
Entire panel	0.0103	18.29	0.0160	21.03	0.0101	18.18

[a]Groups P1 and P2 were classified based on the results of STRUCTURE analysis of the 356 Upland cotton accessions.
[b]The total set of locus pairs, including linked and unlinked loci.
[c]Pairs of loci on the same chromosome.
[d]Pairs of loci from different chromosomes.
[e]Significant threshold is set to $P \leq 0.01$.

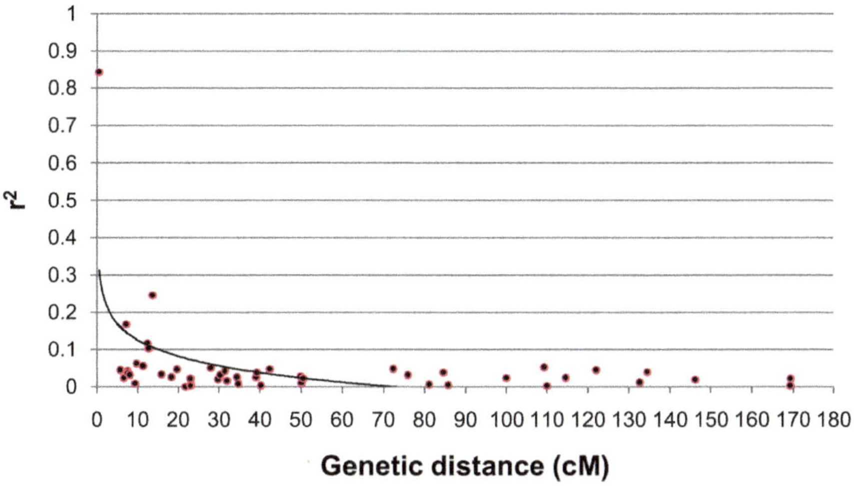

Figure 2. Scatter plots of r^2 values against genetic distance among linked loci (P≤0.01) in 356 *G. hirsutum* L. accessions. The trend line is a non-linear logarithmic regression curve of r^2 on genetic distance. LD decayed to the background (r^2 = 0.1182, P≤0.01), r^2 = 0.1 and r^2 = 0.2 level within about 12–13 cM, 17–18 cM and 3–4 cM, respectively.

Markers associated with lint yield and its components

The marker-trait AM was performed with the MLM model, considering both kinship (K) and population structure (Q), implemented in TASSEL software. At the $\alpha = 0.01$ ($-\log_{10}P = 2$) level, a total of 195 significant associations were detected between 82 SSR markers and seven lint yield traits (Table S4). Among these, most of the associations (125 of 195) were detected in only one environment, and the proportion of phenotypic variation explained by markers ranged from 0.0152 to 0.0940, with an average of 0.0370 (Table S4).

In this study, 145 markers were used for detecting association, so the same statistical test was performed 145 times at a significance level of 0.01, and the experimental type I error rate would be much higher than 0.01. To overcome this problem, the Bonferroni correction ($P \leq 0.05/145$, $-\log_{10}P \geq 3.46$) was used to obtain an appropriate significance threshold [44]. After Bonferroni correction, 55 associations were found to be significant between 26 SSR markers and seven lint yield traits, and the results are shown in **Table 4**. Most (41 associations between 23 SSR markers and seven lint yield traits) of the associations could be detected in more than one environment, and the proportion of phenotypic variation explained by markers ranged from 0.0163 to 0.0940, with an average of 0.0451. The number of SSR markers associated with LY, SY, BN, BW, LP, LI and SI were 9, 4, 6, 4, 14, 17 and 1, respectively. Seventeen loci were co-associated with two or more different traits (**Table 4**). For example, NAU3269 (Chr. 5) and NAU3100 (Chr. 23) were simultaneously associated with FY, SY, BN, LP, and LI, and most of the lint yield-associated loci were associated with at least one of its components.

Favorable QTL alleles and their transmission in Chinese Upland cotton cultivars

Phenotypic effects of each QTL allele for the 41 associated loci detected in more than one environment were measured according to the method mentioned above, and 5, 2, 3, 4, 12, 14 and 1 favorable alleles for FY, SY, BN, BW, LP, LI and SI were identified, respectively. Phenotypic effects and representative accessions for each favorable allele are shown in **Table 5**. Among the favorable alleles, NAU3100-2 had the most positive phenotypic effect for FY and SY, and increased FY and SY by 3.61 g and 7.27 g, respectively; NAU6584-2, NAU3398-2, NAU5166-2 and NAU3917-2 increased BN, BW, LP and LI by 0.89, 0.42 g, 4.93% and 0.94 g, respectively; while NAU493-1 deceased SI by 0.17 g.

Allele frequencies of the 23 favorable alleles in the CK group and the six Chinese historically released cultivar groups are summarized in **Table 6**. Based on allele frequencies across the different groups, these favorable alleles could be categorized into three classes. The alleles in the first class, such as JESPR135-1, BNL1404-1 and Gh508-1, presented in the founder cultivars and with high frequency in all populations, might have been passed down stably from the original parents and were almost fixed in modern cultivars by selection. Alleles in the second class, such as BNL3269-2, BNL1414-2, NAU3100-2 and JESPR208-2, presented in the founder cultivars and with moderate to low frequency in most populations, should have been underutilized in modern breeding programs. Those in the third class, such as NAU5166-2, NAU980-3, Gh369-3 and CIR246-3, not presented in the founder cultivars and presented at low frequency in modern cultivars, might be from other original parents or could have been generated by mutations and/or recombinations. Favorable alleles, especially of the latter two classes, should have a great potential in future Upland cotton genetic improvement.

Discussion

Genetic diversity and population structure of the association panel

A suitable association mapping panel should embrace as much phenotypic and genotypic diversity as can be reliably measured in common environments [45]. Most Upland cotton cultivars developed in China were derived from a few germplasm resources introduced from abroad and therefore the genetic base is narrow [46–47]. It is especially critical to select samples that encompass genetic diversity as much as possible. In this study, the 356 Upland cotton accessions, which can normally flower and ripen for target trait evaluation, were chosen from more than 1000 cultivars and breeding lines in CRI-CAAS and NAU germplasm collections. The phenotypic measurments in three different locations indicated that there was a high degree of diversity in lint yield and its component traits (**Table 2**). Of the 381 SSR markers, only 145

Table 2. Descriptive statistics, ANOVA and broad-sense heritability (h_B^2) for yield and its components across three different environments.

Traits[a]	Environments[b]	Mean	SD	Min	Max	CV(%)	G[c]	G×E[d]	h_B^2(%)
LY	E1	21.66	6.54	8.29	57.91	30.20	**[e]	**	69.10
	E2	27.32	10.24	2.84	68.22	37.48			
	E3	18.72	6.44	6.63	39.27	34.37			
	Mean	22.57	6.57	7.53	42.31	29.09			
SY	E1	54.68	14.02	18.87	133.46	25.65	**	**	55.00
	E2	81.32	25.12	10.88	161.04	30.90			
	E3	52.03	15.46	20.58	93.86	29.72			
	Mean	62.68	14.53	24.23	102.44	23.19			
BN	E1	16.01	3.13	7.00	28.20	19.55	**	**	50.87
	E2	23.06	6.32	4.40	44.30	27.39			
	E3	14.82	3.85	5.70	30.50	26.01			
	Mean	17.96	3.48	8.33	27.73	19.36			
BW	E1	4.58	0.44	3.22	6.04	9.66	**	**	60.96
	E2	4.87	0.59	3.16	6.66	12.03			
	E3	4.85	0.62	2.56	6.48	12.75			
	Mean	4.76	0.43	3.39	6.10	9.11			
LP	E1	39.31	3.57	25.92	50.33	9.08	**	**	75.77
	E2	33.03	3.60	20.03	43.75	10.91			
	E3	35.60	3.59	23.78	46.51	10.07			
	Mean	35.98	3.36	23.76	46.87	9.33			
LI	E1	6.72	0.88	3.74	10.06	13.16	**	**	71.06
	E2	5.92	0.83	3.06	8.70	14.03			
	E3	5.82	0.90	3.00	8.12	15.44			
	Mean	6.15	0.77	3.46	8.17	12.48			
SI	E1	10.35	0.91	8.08	14.71	8.74	**	**	61.09
	E2	12.01	1.28	9.15	15.33	10.67			
	E3	10.51	1.31	6.10	14.35	12.48			
	Mean	10.96	0.99	8.52	13.88	9.02			

[a]LY: lint yield (g/plant); SY: seed cotton yield (g/plant); BN: bolls per plant; BW: boll weight (g); LP: lint percentage (%); LI: lint index (g/100 seeds); SI: seed index (g/100 seeds).
[b]E1, E2, and E3 indicate Jiangpu in 2009, Dafeng in 2010 and Zhengzhou in 2010, respectively.
[c]Genotype across different environments.
[d]Genotype and environment interaction
[e]Significant at $P<0.01$ level.

Table 3. Phenotypic correlations among lint yield and its components based on trait means of 356 upland cotton accessions across three environments.

Traits[a]	LY	SY	BN	BW	LP	LI
SY	0.963***					
BN	0.869***	0.895***				
BW	0.461**	0.497***	0.144			
LP	0.740***	0.544***	0.523***	0.229***		
LI	0.671***	0.541***	0.384***	0.554***	0.796***	
SI	−0.257***	−0.119*	−0.311***	0.424***	−0.498***	0.121*

[a]See Table 2 for abbreviations.

were found to be polymorphic in the 356 Upland cotton accessions, indicating that intraspecific genetic diversity is far less than interspecific diversity; for the reference linkage map of allotetraploid cotton was constructed with a BC_1 mapping population derived from an interspecific cross (*G. hirsutum* TM-1×*G. barbadense* Hai7124) [33]. Eighty percent of the 145 polymorphic loci only generated two or three alleles, and the allele frequencies of 131 of the 415 alleles were <0.05 (Table S2), showing that the genetic diversity in this panel is relatively low, which might affect the QTL detection power of AM in Upland cotton.

Many crops have a long and complex history of domestication and breeding, such as Upland cotton, and complex population structures may confound AM [48]. It is important to consider the influence of population structure and relationships between individuals in the AM panel [27–28]. The model-based evaluation of the population structure of the 356 Upland cotton cultivars

Table 4. SSR marker loci significantly associated with lint yield traits and their explained proportion of phenotypic variation in three different environments.

Traits[a]	Marker loci	Chr.	Position	$-Log_{10}P$			R^2		
				E1[b]	E2[b]	E3[b]	E1[b]	E2[b]	E3[b]
LY	NAU3269[c]	A05(Chr.05)	182.215	3.43	4.42	3.33	0.0283	0.0460	0.0305
	NAU5166[c]	A10(Chr.10)	23.423	ns	5.49	ns		0.0540	
	NAU2935	A10(Chr.10)	51.592	ns	4.00	ns		0.0472	
	JESPR204[c]	A13(Chr.13)	59.714	4.12	3.00	ns	0.0460	0.0408	
	CIR246[c]	D02(Chr.14)	112.473	ns	ns	3.55			0.0444
	BNL3594[c]	D06(Chr.25)	7.66	ns	4.51	ns		0.0765	
	TMK19[c]	D06(Chr.25)	70.38	4.30	2.77	3.97	0.0421	0.0321	0.0428
	NAU3100[c]	D09(Chr.23)	25.262	3.67	5.38	4.31	0.0408	0.0708	0.0541
	NAU2776[c]	D10(Chr.20)	9.452	ns	3.81	2.54		0.0441	0.0276
SY	NAU3269	A05(Chr.05)	182.215	2.25	4.16	2.43	0.0185	0.0441	0.0219
	CIR246	D02(Chr.14)	112.473	ns	ns	4.23			0.0536
	BNL3594	D06(Chr.25)	7.66	ns	4.61	ns		0.0797	
	NAU3100	D09(Chr.23)	25.262	ns	4.03	2.82		0.0561	0.0384
BN	NAU6584	A03(Chr.03)	74.975	3.53	2.29	ns	0.0393	0.0291	
	NAU3269	A05(Chr.05)	182.215	2.08	3.52	3.45	0.0174	0.0364	0.0344
	BNL3594	D06(Chr.25)	7.66	ns	3.62	ns		0.0667	
	TMK19	D06(Chr.25)	70.38	3.58	ns	2.41	0.0371		0.0284
	NAU493	D07(Chr.16)	113.413	3.91	ns	ns	0.0338		
	NAU3100	D09(Chr.23)	25.262	ns	3.48	ns		0.0492	
BW	BNL1414	A09(Chr.09)	95.911	ns	3.26	3.72		0.0418	0.0445
	NAU4047	A12(Chr.12)	11.118	3.29	4.17	ns	0.0332	0.0461	
	NAU3398	A13(Chr.13)	3.311	3.47	3.46	ns	0.0565	0.0603	
	JESPR208	D09(Chr.23)	118.225	ns	4.01	3.69		0.0515	0.0445
LP	NAU3269	A05(Chr.05)	182.215	3.54	2.21	2.72	0.0318	0.0202	0.0259
	NAU5166	A10(Chr.10)	23.423	2.33	4.36	2.74	0.0180	0.0410	0.0244
	NAU2508	A10(Chr.10)	128.028	4.45	3.62	4.68	0.0523	0.0481	0.0609
	NAU980	A11(Chr.11)	0	3.40	ns	3.75	0.0565		0.0672
	JESPR135	A11(Chr.11)	55.787	3.79	6.64	3.80	0.0316	0.0645	0.0354
	NAU3398	A13(Chr.13)	3.311	3.50	ns	ns	0.0499		
	JESPR204	A13(Chr.13)	59.714	3.73	2.96	2.22	0.0441	0.0396	0.0311
	BNL3590	D03(Chr.17)	39.284	4.70	ns	2.89	0.0491		0.0338
	TMK19	D06(Chr.25)	70.38	5.97	4.58	3.67	0.0604	0.0518	0.0419
	NAU3100	D09(Chr.23)	25.262	4.50	3.35	2.89	0.0535	0.0458	0.0400
	NAU3917	D10(Chr.20)	31.125	ns	3.57	ns		0.0403	
	BNL1404	D11(Chr.21)	33.571	4.48	6.20	3.51	0.0381	0.0601	0.0323
	Gh508	D11(Chr.21)	54.48	2.15	3.37	3.81	0.0163	0.0306	0.0354
	NAU2361	D11(Chr.21)	101.215	4.05	4.30	4.03	0.0637	0.0734	0.0703
LI	NAU3269	A05(Chr.05)	182.215	4.98	2.77	2.38	0.0492	0.0263	0.0232
	NAU980	A11(Chr.11)	0	8.85	2.34	2.80	0.0896	0.0476	0.0572
	JESPR135	A11(Chr.11)	55.787	ns	5.43	2.34		0.0520	0.0209
	Gh369	A11(Chr.11)	84.701	3.77	2.09	ns	0.0479	0.0300	
	NAU1151	A12(Chr.12)	97.965	6.08	ns	ns	0.0563		
	NAU3398	A13(Chr.13)	3.311	6.01	2.49	ns	0.0819	0.0415	
	JESPR204	A13(Chr.13)	59.714	ns	3.68	ns		0.0481	
	CIR246	D02(Chr.14)	112.473	4.75	3.14	ns	0.0577	0.0417	
	BNL3590	D03(Chr.17)	39.284	3.30	2.70	3.47	0.0367	0.0313	0.0426
	NAU2233	D05(Chr.19)	171.278	4.86	2.15	ns	0.0596	0.0304	

Table 4. Cont.

Traits[a]	Marker loci	Chr.	Position	$-Log_{10}P$			R^2		
				E1[b]	E2[b]	E3[b]	E1[b]	E2[b]	E3[b]
	TMK19	D06(Chr.25)	70.38	4.83	4.02	ns	0.0519	0.0455	
	NAU3100	D09(Chr.23)	25.262	5.90	2.74	3.47	0.0719	0.0384	0.0493
	NAU2776	D10(Chr.20)	9.452	ns	ns	3.57			0.0430
	NAU3917	D10(Chr.20)	31.125	ns	3.66	2.03		0.0413	0.0245
	BNL1404	D11(Chr.21)	33.571	ns	5.85	2.72		0.0564	0.0250
	Gh508	D11(Chr.21)	54.48	ns	3.67	3.41		0.0338	0.0327
	NAU2361	D11(Chr.21)	101.215	6.27	5.16	4.58	0.0940	0.0845	0.0812
SI	NAU493	D07(Chr.16)	113.413	3.70	2.85	5.24	0.0385	0.0290	0.0575

[a]See Table 2 for abbreviations.
[b]E1: Jiangpu in 2009; E2: Dafeng in 2010; E3: Zhengzhou in 2010.
[c]Markers associated with lint yield simultaneously associated more than one its component.

showed that the population could be divided into two major subpopulations (**Figure 1**). Of the 115 accessions in P1 group, 63, 46 and 6 cultivars were from the Yellow River, North/Northwest China and Yangtze River cotton growing regions, respectively. Out of the 241 accessions in P2 group, 116 were from Yellow River, 107 from Yangtze River, 10 from North/Northwest China cotton growing regions and eight from abroad (Table S1). The North/Northwest China, Yellow River and Yangtze River region represented short-, middle- and long-growth-period cotton cultivation area in China, respectively. The P1 group contained almost all cultivars with early maturity and part of cultivars with middle maturity, while the P2 group contained almost all cultivars with late maturity and part of cultivars with middle maturity [32].

Linkage disequilibrium in Upland cotton

The extent of LD can provide information for the needed marker density and mapping resolution in AM study [25]. LD decay had been repeatedly estimated in many plant species [26,48], while that was limited in cotton. Abdurakhmonov et al. (2008) performed a pioneer estimation [20]. They reported that, in a panel contained 285 exotic *Gossypium hirsutum* accessions, the genome-wide LD ($r^2 \geq 0.1$) declined at <10 cM in the landrace stocks and >30 cM in variety germplasm, but at $r^2 \geq 0.2$ which reduced to about 1–2 cM and 6–8 cM, respectively. In another panel composed of 335 *G. hirsutum* variety germplasm, the genome-wide LD extended up to 25 cM at $r^2 \geq 0.1$ and reduced to about 5–6 cM at $r^2 \geq 0.2$ [21]. In the present panel, the average r^2 of locus pairs was 0.0103, and 18.29% were significant ($P \leq 0.01$) (**Table 1**), which is higher than that (13% siganificant at $P \leq 0.01$) reported by Abdurakhmonov et al. [21]. In our panel, the LD decayed to genome background level ($r^2 = 0.1182$) within 12–13 cM (Fig. 2). If the threshold of LD decay was set to $r^2 = 0.1$ and $r^2 = 0.2$, the genome-wide LD extended up to 17–18 cM and 3–4 cM (**Figure 2**), respectively, which is shorter than those previously reported [20–21].

In the entire panel and subpopulations, both average r^2 and proportion of significant LD for linked loci were all higher than those for unlinked markers (**Table 1**), demonstrating that physical linkage is predominant in determining LD compared with random forces in the present association panel [48]. Therefore this Upland cotton panel is suitable for association analysis and has the potential to identify QTLs in an interval equivalent to the distance of LD decay of 3–4 cM. Based on the LD decay in the panel of 335 *G. hirsutum* varieties, it is suggested that about 1,000 polymorphic markers be required for successful association mapping with LD extending to 5–6 cM [21]. In our panel, the LD decayed faster, suggesting that more markers are probably needed for genome wide association analysis (GWAS) of complex traits. As is often the case in self-pollinated crops [26], the level of LD in the Upland cotton genome was moderately high, suggesting that the mapping resolution gained from LD is likely to be limited. Given that genomic selection is less challenging than map-based cloning, the level of LD in the present population would guarantee that the identified SSR markers would facilitate breeding for high-yield in Upland cotton.

QTLs for lint yield identified by association mapping

Association mapping can be affected by many factors, such as population structure, relatedness among accessions, small sample size, and low frequency of specific alleles; these may increase the detection of false positive associations [25,28]. In this study, the AM was performed in a moderately large sized panel (356 accessions) with the optimal model of MLM, considering both population structure and relatedness, to detect SSR markers associated with lint yield and its components. A total of 195 significant associations were detected between 86 SSR markers and 7 lint yield and yield component traits at the $\alpha = 0.01$ ($-\log_{10}P = 2$) level (Table S4). It is very difficult to say which significance level is acceptable in a given association study. The use of stringent probability thresholds will reduce the danger of false positives, but meanwhile has the risk of rejecting true positives caused by setting the thresholds too high [49]. Since the present study aimed at mining favorable alleles of main QTL for lint yield, a relatively stringent significance threshold ($P \leq 0.05/145$, $-\log_{10}P \geq 3.46$) for the Bonferroni correction was adopted to reduce the experimental type I error rate induced by multiple tests [44]. After Bonferroni correction ($P \leq 0.05/145$, $-\log_{10}P \geq 3.46$), 55 associations remained significant and 74.55% (41 of 55) could be detected in more than one environment (**Table 4**). Population size had been considered as a factor that severely affects the QTL detection power in AM [18,26]. Many more associations were detected in our present panel than in another 81-accession panel between the same markers and target traits at the same significance level [24, unpublished data].

Although the markers used in different studies are different, and QTL mapping results are not easy to be compared, some of the

Table 5. Favorable QTL alleles, their phenotypic effects (a_i) and representative accessions.

Traits[a]	Favorable allele	a_i	Accessions	Representative accessions[b]
LY	NAU3269-2	0.27	133	Simian3, Zhongmiansuo9, Huakangmian1
	JESPR204-1	0.70	314	Simian3, Zhongmiansuo9, P164-2
	TMK19-2	1.02	240	Simian3, Zhongmiansuo9, P164-2
	NAU3100-2	3.61	87	Simian3, Zhongmiansuo9, Lumianyan16
	NAU2776-1	0.85	151	Zhongmiansuo9, P164-2, Lumianyan16
SY	NAU3269-2	0.42	133	Zhongmiansuo9, Zhongmiansuo19, Simian3
	NAU3100-2	7.27	87	Zhongmiansuo9, Han4849, Lumianyan16
BN	NAU6584-2	0.89	217	Lumianyan16, Zhongmiansuo44, Zhongmiansuo9
	NAU3269-2	0.08	133	Zhongmiansuo9, Wanmian73-10, Simian3
	TMK19-2	0.45	235	Zhongmiansuo44, Zhongmiansuo9, Wanmian17
BW	BNL1414-2	0.18	93	Zhongmiansuo18, Zhongmiansuo5, I40005
	NAU4047-2	0.03	221	Zhongmiansuo18, Zhongmiansuo5, I40005
	NAU3398-2	0.42	22	Zhongmiansuo5, I40005, Hua101
	JESPR208-2	0.20	86	Zhongmiansuo18, Zhongmiansuo5, I40005
LP	NAU3269-2	0.23	133	Simian3, Ekangmian9, Huakangmian1
	NAU5166-2	4.93	8	Simian3, Huakangmian1, Sumian4
	NAU2508-2	0.36	113	Nannongzao, 86-1, Yu668
	NAU980-3	2.79	9	Ekangmian6, Emian16, Ekangmian10
	JESPR135-1	0.13	343	XiangSC-24, Simian3, Ekangmian9
	JESPR204-1	0.36	309	XiangSC-24, Simian3, Ekangmian9
	BNL3590-1	0.26	327	XiangSC-24, Simian3, Ekangmian9
	TMK19-2	0.58	235	XiangSC-24, Simian3, Huakangmian1
	NAU3100-2	1.55	86	Simian3, Ekangmian9, Nannongzao
	BNL1404-1	0.13	343	XiangSC-24, Simian3, Ekangmian9
	Gh508-1	0.07	347	XiangSC-24, Simian3, Ekangmian9
	NAU2361-3	0.75	73	Ekangmian9, Yu668, Yumian21
LI	NAU3269-2	0.01	133	Huakangmian1, I40005, Ekangmian9
	NAU980-3	0.84	9	I40005, Zhongmiansuo5, Hua101
	JESPR135-1	0.02	343	Huakangmian1, Emian23, I40005
	Gh369-3	0.11	8	Emian16, Ekangmian8, Yumian20
	NAU3398-2	0.80	22	Huakangmian1, I40005, Zhongmiansuo5
	CIR246-3	0.63	16	Hua101, Zhongmiansuo41, Yumian9
	BNL3590-1	0.06	327	Huakangmian1, Emian23, I40005
	NAU2233-1	0.01	212	Huakangmian1, Emian23, I40005
	TMK19-2	0.10	235	Huakangmian1, Emian23, I40005
	NAU3100-2	0.38	86	I40005, Zhongmiansuo5, Hua101
	NAU3917-2	0.94	6	Huakangmian1, Simian4, Sumian9
	BNL1404-1	0.03	343	Huakangmian1, Emian23, I40005
	Gh508-1	0.01	347	Huakangmian1, Emian23, I40005
	NAU2361-3	0.28	73	Emian23, I40005, Zhongmiansuo5
SI	NAU493-1	−0.17	230	Chaoyangmian1, Xuzhou1818, XiangSC-24

[a]See Table 2 for abbreviations.
[b]Representative accessions are the top-3 entries for the target trait value of accessions with the corresponding favorable allele.

marker associations detected in this study were consistent with QTLs for lint yield and its components that had been mapped previously by conventional linkage mapping. The locus JESPR204 associated with LY on chromosome 13 (detected in 2 environments) was located in the same region as a QTL identified by Wu et al. [50]; TMK19 (Chr. 25, detected in 3 environments) was consistent with the results of our prevenient study [51]; NAU3100 (Chr. 23, detected in 2 environments) associated with SY was consistent with that found in the study of Wang et al. [8]; NAU980 (Chr. 11, detected in 2 environments), BNL3590 (Chr. 17, detected in 2 environments) and TMK19 (Chr. 25, detected in 3 environments) associated with LP were consistent with several

Table 6. Allele frequency for each favorable QTL allele in historically released Chinese Upland cotton cultivar groups.[a]

Favorable alleles	CK	I	II	III	IV	V	VI	Total
NAU6584-2	0.3333	0.5000	0.6923	0.7692	0.6747	0.4720	0.7755	0.6096
NAU3269-2	0.6667	0.5769	0.5385	0.6154	0.3976	0.3040	0.1020	0.3736
BNL1414-2	0.5000	0.3077	0.1923	0.4872	0.2410	0.2320	0.1633	0.2612
NAU5166-2	0.0000	0.0000	0.0000	0.0500	0.0000	0.0320	0.0417	0.0225
NAU2508-2	0.4000	0.1200	0.2400	0.2250	0.3012	0.3952	0.4694	0.3333
NAU980-3	0.0000	0.0200	0.0000	0.0256	0.0244	0.0369	0.0000	0.0233
JESPR135-1	1.0000	0.8462	1.0000	1.0000	0.9759	0.9840	1.0000	0.9775
Gh369-3	0.0000	0.0000	0.0000	0.0000	0.0241	0.0480	0.0000	0.0225
NAU4047-2	0.6667	0.5000	0.5769	0.6154	0.5663	0.6400	0.7551	0.6208
NAU3398-2	0.0000	0.0385	0.0000	0.1026	0.0843	0.0720	0.0204	0.0618
JESPR204-1	1.0000	0.8462	0.8077	0.7692	0.8675	0.9200	0.9388	0.8820
CIR246-3	0.0000	0.0000	0.0000	0.0256	0.0241	0.0800	0.0612	0.0449
BNL3590-1	0.5000	0.8077	0.8462	0.8462	0.9036	0.9440	0.8776	0.8904
NAU2233-1	0.6667	0.4231	0.5000	0.3590	0.5542	0.6560	0.8163	0.5955
TMK19-2	0.6667	0.5000	0.5769	0.6667	0.6024	0.7280	0.8163	0.6742
NAU493-1	0.6667	0.5000	0.7308	0.6923	0.6265	0.6800	0.5714	0.6461
NAU3100-2	0.3333	0.2308	0.0769	0.2308	0.2289	0.2400	0.3673	0.2444
JESPR208-2	0.5000	0.3077	0.1154	0.4359	0.2289	0.2240	0.1633	0.2444
NAU2776-1	0.1667	0.4231	0.3462	0.5385	0.4458	0.4480	0.3061	0.4242
NAU3917-2	0.0000	0.0000	0.0000	0.0513	0.0000	0.0240	0.0204	0.0169
BNL1404-1	1.0000	0.8462	1.0000	1.0000	0.9759	0.9840	1.0000	0.9775
Gh508-1	1.0000	0.9231	1.0000	0.9750	1.0000	0.9920	1.0000	0.9888
NAU2361-3	0.2143	0.0769	0.0417	0.1538	0.1125	0.2810	0.3673	0.2066

[a]CK, I, II, III, IV, V and VI indicates the founder parent group (CK), and the Chinese cultivars released in 1930–1960, 1961–1970, 1971–1980, 1981–1990, 1991–2000 and 2000–2005, respectively.

earlier reports [11,51,52]; CIR246 (Chr. 14, detected in 2 environments) and NAU3100 (Chr. 23, detected in 3 environments) associated with LI were also consistent with results from several recent QTL mapping studies [11,51,52].

Moreover, in our study, seventeen markers were co-associated with two or more different traits, and most of the lint yield-associated markers were associated with at least one of its components, which coincided with phenotypic correlations among these traits. This could result from pleiotropy of a single causal gene or tight linkage of multiple causal genes. We found that 10 of 14 markers associated with LP were detected in all three environments, which was consistent with the phenotypic statistical analysis that LP possessed the highest broad-sense heritability ($h_B^2 = 75.77\%$). The phenotype of complex traits often results from the combined actions of multiple genes and environmental factors, all these can easily lead to lost heritability [18]; only those traits with high heritability can be stably detected. The resulting stably associated markers should be useful for cotton breeding with broad adaptability to different environments.

Favorable alleles and their potential application in future cotton breeding programs

Since most Upland cotton cultivars developed in China were derived from limited founder parents, there is great challenge in genetic improvement and high risk of vulnerability to changing climate. New variations that have emerged and accumulated during the long breeding history in China should be fully exploited and additional diversity should be introduced into breeding programs to broaden the genetic basis of Chinese Upland cotton. By comparing the average phenotypic value of each allele for target traits in the 41 stably detected associations, we identified 5, 2, 3, 4, 12, 14 and 1 favorable alleles for FY, SY, BN, BW, LP, LI and SI, respectively (**Table 5**). We suggest that a multi-parent population should be constructed using cultivars that possess most of the favorable alleles, and in the meantime, a ranking system for MAS or genomic selection should be developed based on the results of AM. Favorable alleles that were passed down from the founder parents and have been almost fixed in modern cultivars formed the basis of lint yield of Chinese Upland cotton, and should be treated as fundamental elements in order to reject deleterious alleles at the corresponding loci. Alleles either absent in the founder cultivars or present at moderate to low frequencies in most cultivar groups have been underutilized in modern breeding programs, and should be regarded as essential elements for increasing lint yield potential.

Lint yield of cotton is the result of series components and their interactions, such as boll number, boll weight, lint percentage, lint index, and seed index. Developing potentially high-yielding cultivars thus relies to some extent on selecting the appropriate yield components. As some of the QTLs were associated with more than one yield component, favorable alleles must be treated with caution. Positively co-associated genetic loci could simultaneously improve multiple target traits, while negative linkages must be broken. In summary, the favorable alleles indentified in this study have great potential for developing high-yielding Upland cotton cultivars in future breeding programs.

Supporting Information

Table S1 Detailed information of the 356 accessions in the association panel.

Table S2 Detailed information of 145 polymorphic SSR markers.

Table S3 Genetic diversity in different historically released cultivar groups.

Table S4 SSR markers significantly ($P \leq 0.01$, $-\log_{10}P \geq 2.0$) associated with lint yield traits and their explained phenotypic variation across three different environments.

Acknowledgments

We thank Dr. XM Du of the Cotton Research Institute, Chinese Academy of Agriculture Science for providing part of Upland cotton germplasm resources used in the present study.

Author Contributions

Conceived and designed the experiment: TZZ, performed the experiments: HXM and XFZ, analyzed the data: HXM, and wrote the manuscript: HXM and TZZ.

References

1. Zhang HB, Li YN, Wang BH, Chee PW (2008) Recent advances in cotton genomics. Int J Plant Genomics ID 742304. doi:10.1155/2008/742304.
2. Xu Y, Crouch J (2008) Marker-assisted selection in plant breeding: from publications to practice. Crop Sci 48: 391–407.
3. Jannink JL, Lorenz AJ, Iwata H. (2010) Genomic selection in plant breeding: from theory to practice. Brief Funct Genomics 9: 166–177.
4. Shen X, Guo W, Zhu X, Yuan Y, Yu JZ, et al. (2005) Molecular mapping of QTLs for fiber qualities in three diverse lines in Upland cotton using SSR markers. Mol Breed 15: 169–181.
5. Zhang ZS, Xiao Y, Luo M, Li X, Luo X, et al. (2005) Construction of a genetic linkage map and QTL analysis of fiber-related traits in upland cotton (*Gossypium hirsutum* L.). Euphytica 144: 91–99.
6. Wang B, Guo W, Zhu X, Wu Y, Huang N, et al. (2006) QTL mapping of fiber quality in an elite hybrid derived-RIL population of upland cotton. Euphytica 152: 367–378.
7. Shen X, Guo W, Lu Q, Zhu X, Yuan Y, et al. (2007) Genetic mapping of quantitative trait loci for fiber quality and yield trait by RIL approach in Upland cotton. Euphytica 155: 371–380.
8. Wang B, Guo W, Zhu X, Wu Y, Huang N, et al. (2007) QTL mapping of yield and yield components for elite hybrid derived-RILs in upland cotton. J Genet Genomics 34: 35–45.
9. Qin H, Guo W, Zhang Y, Zhang T (2008) QTL mapping of yield and fiber traits based on a four-way cross population in *Gossypium hirsutum* L. Theor Appl Genet 117: 883–894.
10. Zhang ZS, Hu MC, Zhang J, Liu DJ, Zheng J, et al. (2009) Construction of a comprehensive PCR-based marker linkage map and QTL mapping for fiber quality traits in upland cotton (*Gossypium hirsutum* L.). Mol Breed 24: 49–61.
11. Liu R, Wang B, Guo W, Qin Y, Wang L, et al. (2012) Quantitative trait loci mapping for yield and its components by using two immortalized populations of a heterotic hybrid in *Gossypium hirsutum* L. Mol Breed 29: 297–311.
12. Sun FD, Zhang JH, Wang SF, Wang SF, Gong WK, et al. (2012) QTL mapping for fiber quality traits across multiple generations and environments in upland cotton. Mol Breed 30: 569–582.
13. Zhang K, Zhang J, Ma J, Tang S, Liu D, et al. (2012) Genetic mapping and quantitative trait locus analysis of fiber quality traits using a three-parent composite population in upland cotton (*Gossypium hirsutum* L.). Mol Breed 29: 335–348.
14. Lacape JM, Nguyen TB, Courtois B, Belot JL, Giband M, et al. (2005) QTL analysis of cotton fiber quality using multiple *Gossypium hirsutum*×*Gossypium barbadense* backcross generations. Crop Sci 45: 123–140.
15. Rong J, Feltus FA, Waghmare VN, Pierce GJ, Chee P, et al. (2007) Meta-analysis of polyploid cotton QTL shows unequal contributions of subgenomes to a complex network of genes and gene clusters implicated in lint fiber development. Genetics 176: 2577–2588.
16. Lacape JM, Llewellyn D, Jacobs J, Arioli T, Becker D, et al. (2010) Meta-analysis of cotton fiber quality QTLs across diverse environments in a *Gossypium hirsutum*×*G. barbadense* RIL population. BMC Plant Biol 10: 132.
17. Zhu C, Gore M, Buckler ES, Yu J (2008) Status and prospects of association mapping in plants. Plant Genome 1: 5–20.
18. Mackay TFC, Stone EA, Ayroles JF (2009) The genetics of quantitative traits: challenges and prospects. Nat Rev Genet 10: 565–577.
19. Kantartzi SK, Stewart JMcD (2008) Association analysis of fibre traits in *Gossypium arboreum* accessions. Plant Breed 127: 173–179.
20. Abdurakhmonov IY, Kohel RJ, Yu JZ, Pepper AE, Abdullaev AA, et al. (2008) Molecular diversity and association mapping of fiber quality traits in exotic *G. hirsutum* L. germplasm. Genomics 92: 478–487.
21. Abdurakhmonov IY, Saha S, Jenkins JN, Buriev ZT, Shermatov SE, et al. (2009) Linkage disequilibrium based association mapping of fiber quality traits in *G. hirsutum* L. variety germplasm. Genetica 136: 401–417.
22. Zeng L, Meredith Jr WR, Gutierrez OA, Boykin DL (2009) Identification of associations between SSR markers and fibre traits in an exotic germplasm derived from multiple cross among *Gossypium* tetraploid species. Theor Appl Genet 119: 93–103.
23. Rakshit A, Rakshit S, Singh J, Chopra SK, Balyan HS, et al. (2010) Association of AFLP and SSR markers with agronomic and fibre quality traits in *Gossypium hirsutum* L. J Genetics 89: 155–162.
24. Zhang TZ, Qian N, Zhu XF, Chen H, Wang S, et al. (2013) Variations and transmission of QTL alleles for yield and fiber qualities in Upland cotton cultivars developed in China. PLoS ONE 8: e57220.
25. Gupta PK, Rustgi S, Kulwal PL (2005) Linkage disequilibrium and association studies in higher plants: present status and future prospects. Plant Mol Biol 57: 461–485.
26. Myles S, Peiffer J, Brown PJ, Ersoz ES, Zhang Z, et al. (2009) Association mapping: critical consideration shift from genotype to experimental design. Plant Cell 21: 2194–2202.
27. Price A, Patterson N, Plenge R, Weinblatt ME, Shadick NA, et al. (2006) Principal components analysis corrects for stratification in genome-wide association studies. Nat Genet 38: 904–909.
28. Yu J, Pressoir G, Briggs WH, Bi IV, Yamasaki M, et al. (2006) A unified mixed-model method for association mapping that accounts for multiple levels of relatedness. Nat Genet 38: 203–208.
29. Yu J, Holland JB, McMullen MD, Buckler ES (2008) Genetic design and statistical power of nested association mapping in maize. Genetics 178: 539–551.
30. Breseghello F, Sorrells ME (2006) Association mapping of kernel size and milling quality in wheat (*Triticum aestivum* L.) cultivars. Genetics 172: 1165–1177.
31. Jia L, Yan W, Zhu C, Agrama HA, Jackson A, et al. (2012) Allelic analysis of sheath blight resistance with association mapping in rice. PLoS ONE 7: e32703.
32. Huang ZK (2007) The Cultivars and Their Pedigree of Cotton in China. China Agriculture Press, Beijing.
33. Guo WZ, Cai C, Wang C, Han Z, Song X, et al. (2007) A microsatellite-based, gene-rich linkage map reveals genome structure, function and evolution in *Gossypium*. Genetics 176: 527–541.
34. Zhang J, Guo W, Zhang T. (2002) Molecular linkage map of allotetraploid cotton (*Gossypium hirsutum* L.×*Gossypium barbadense* L.) with a haploid population. Theor Appl Genet 105: 1166–1174.
35. Zhao L, Cai CP, Mei HX, Guo WZ. (2012) Screening of microsatellite loci for identifying genome barcoding of cotton cultivars. ACTA Agronomica Sinica 38: 1810–1817.
36. Liu KJ, Muse SV (2005) PowerMarker: an integrated analysis environment for genetic marker analysis. Bioinfo Appl Note 9: 2128–2129.
37. Pritchard JK, Wen X, Falush D (2009) Documentation for structure software: Version 2.3 http://pritch.bsd.uchicago.edu/structure.html.
38. Evanno G, Regnaut S, Goudet J (2005) Detecting the number of clusters of individuals using the software STRUCTURE: a simulation study. Mol Ecol 14: 2611–2620.
39. Pritchard JK, Stephens M, Donnelly P (2000) Inference of population structure using multilocus genotype data. Genetics 155: 945–959.
40. Hardy O, Vekemans X (2002) SPAGeDi: a versatile computer program to analyse spatial genetic structure at the individual or population levels. Mol Ecol Notes 2: 618–620.
41. Bradbury PJ, Zhang Z, Kroon DE, Casstevens TM, Ramdoss Y, et al. (2007) TASSEL: software for association mapping of complex traits in diverse samples. Bioinformatics 23: 2633–2635.
42. Mather KA, Caicedo AL, Polato NR, Olsen KM, McCouch S, et al. (2007) The extent of linkage disequilibrium in rice (*Oryza sativa* L.). Genetics 177: 2223–2232.
43. Xiao Y, Cai D, Yang W, Ye W, Younas M, et al. (2012) Genetic structure and linkage disequilibrium pattern of a rapeseed (*Brassica napus* L.) association mapping panel revealed by microsatellites. Theor Appl Genet 125: 437–447.
44. Lander ES, Botstein D (1989) Mapping mendelian factors underlying quantitative traits using RFLP linkage maps. Genetics 121: 185–199.

45. Flint-Garcia SA, Thuillet AC, Yu J, Pressoir G, Romero SM, et al. (2005) Maize association population: a high-resolution platform for quantitative trait locus dissection. Plant J 44: 1054–1064.
46. Guo WZ, Zhang T, Pan J, Wang X (1997) A preliminary study on genetic diversity of Upland cotton cultivars in China. Acta Gossypii Sinica 9: 19–24.
47. Chen G, Du X (2006) Genetic diversity of source germplasm of Upland cotton in China as determined by SSR marker analysis. Acta Genetica Sinica 33: 733–745.
48. Flint-Garcia SA, Thornsberry JM, Bucker ES (2003) Structure of linkage disequilibrium in plants. Annu Rev Plant Biol 54: 357–374.
49. Yan J, Warburton M, Crouch J (2011) Association mapping for enhancing maize (*Zea mays* L.) genetic improvement. Crop Sci 51: 433–449.
50. Wu JX, Gutierrez OA, Jenkins JN, McCarty JC, Zhu J (2009) Quantitative analysis and QTL mapping for agronomic and fiber traits in an RI population of upland cotton. Euphytica 165: 231–245.
51. Qin YS, Liu RZ, Mei HX, Zhang TZ, Guo WZ (2009) QTL Mapping for yield traits in Upland cotton (*Gossypium hirsutum* L.). Acta Agronomica Sinica 35: 1812–1821.
52. Guo WZ, Ma GJ, Zhu YC, Yi CX, Zhang TZ (2006) Molecular tagging and mapping of quantitative trait loci for lint percentage and morphological marker genes in upland cotton. J Integr Plant Biol 48: 320–326.

18

Assessing Quantitative Resistance against *Leptosphaeria maculans* (Phoma Stem Canker) in *Brassica napus* (Oilseed Rape) in Young Plants

Yong-Ju Huang[1,2]*, Aiming Qi[1,2], Graham J. King[2,3], Bruce D. L. Fitt[1,2]

1 School of Life and Medical Sciences, University of Hertfordshire, Hatfield, Hertfordshire, United Kingdom, **2** Department of Plant Pathology and Microbiology, Rothamsted Research, Harpenden, Hertfordshire, United Kingdom, **3** Southern Cross Plant Science, Southern Cross University, Lismore, Australia

Abstract

Quantitative resistance against *Leptosphaeria maculans* in *Brassica napus* is difficult to assess in young plants due to the long period of symptomless growth of the pathogen from the appearance of leaf lesions to the appearance of canker symptoms on the stem. By using doubled haploid (DH) lines A30 (susceptible) and C119 (with quantitative resistance), quantitative resistance against *L. maculans* was assessed in young plants in controlled environments at two stages: stage 1, growth of the pathogen along leaf veins/petioles towards the stem by leaf lamina inoculation; stage 2, growth in stem tissues to produce stem canker symptoms by leaf petiole inoculation. Two types of inoculum (ascospores; conidia) and three assessment methods (extent of visible necrosis; symptomless pathogen growth visualised using the GFP reporter gene; amount of pathogen DNA quantified by PCR) were used. In stage 1 assessments, significant differences were observed between lines A30 and C119 in area of leaf lesions, distance grown along veins/petioles assessed by visible necrosis or by viewing GFP and amount of *L. maculans* DNA in leaf petioles. In stage 2 assessments, significant differences were observed between lines A30 and C119 in severity of stem canker and amount of *L. maculans* DNA in stem tissues. GFP-labelled *L. maculans* spread more quickly from the stem cortex to the stem pith in A30 than in C119. Stem canker symptoms were produced more rapidly by using ascospore inoculum than by using conidial inoculum. These results suggest that quantitative resistance against *L. maculans* in *B. napus* can be assessed in young plants in controlled conditions. Development of methods to phenotype quantitative resistance against plant pathogens in young plants in controlled environments will help identification of stable quantitative resistance for control of crop diseases.

Editor: Zhengyi Wang, Zhejiang University, China

Funding: This work was funded by the UK Biotechnology and Biological Sciences Research Council (BBSRC, BB/E001610/1, and BB/I017585/1), the Department for Environment, Food and Rural Affairs (Defra), HGCA, the British Council, DuPont, the Perry Foundation, the Felix Thornley Cobbold Agricultural Trust, and the Chadacre Agricultural Trust. The funders had no role in study design, data collection and analysis, decision to publish, or preparation of the manuscript.

Competing Interests: The authors have the following interests: This study was partly funded by DuPont. There are no patents, products in development or marketed products to declare.

* E-mail: y.huang8@herts.ac.uk

Introduction

With growing concern about world-wide food shortages and climate change [1], protecting food crops against pathogens that cause epidemic diseases is more important than ever. Using resistance against pathogens in crop cultivars is one of the most economical and environmentally friendly methods for control of crop diseases. Two types of cultivar resistance used are quantitative resistance and qualitative resistance [1–4]. Quantitative resistance, which is usually controlled by several genes, is often a 'partial' resistance that does not prevent pathogens from colonisation of plants but decreases symptom severity and/or epidemic progress over time [5–7]. By contrast, qualitative resistance is usually controlled by single, dominant resistance (*R*) genes and often effective in preventing pathogens from colonisation of plants [8–12]. Whilst *R* gene-mediated qualitative resistance is 'complete' resistance, which follows a gene-for-gene interaction, it is generally less durable than quantitative resistance because pathogen populations often rapidly evolve for virulence against the *R* genes [13–16].

Many arable crops rely on quantitative partial resistance that has been selected by breeding 'field' resistance into modern cultivars for control of their diseases. This reliance on resistance is essential for small-holder farmers in many regions of the world where use of fungicides is prohibitively expensive. In regions where crops are grown intensively on larger farms, the reliance on resistance for disease control will increase if use of certain effective fungicides is not permitted by legislation. However, it has been difficult to breed cultivars with quantitative resistance because of the difficulty in assessing it. By contrast, *R* gene-mediated qualitative resistance can be selected reliably by phenotyping in young plants (with distinct resistant and susceptible phenotypes after inoculation). Traditionally, selection for quantitative resistance has relied on field assessments of disease severity made towards the end of the cropping season before harvest [17–19]. Thus, quantitative resistance has also been referred to as 'adult plant resistance' [3] because it has been difficult to assess it in young plants. Furthermore, due to the influence of genotype by environment interactions [16,19], identification of stable quantitative resistance has required large replicated field plot experiments at different locations in different years, which is very costly and time-consuming. There is a need to develop new methods to select for quantitative resistance in arable crops that are rapid, cheap and reliable.

Phoma stem canker, caused by *Leptosphaeria maculans*, is an economically important disease on oilseed rape (*Brassica napus*) in Europe, North America and Australia, causing world-wide losses worth more than £1000 M (at a price of £300 t^{-1}) per cropping season, despite use of fungicides [20–21]. In Europe, these losses will increase if use of the most effective fungicides is no longer permitted by EU legislation (EU directive 91/414, http://ec.europa.eu/food/plant/protection/index_en.htm). Epidemics of phoma stem canker are initiated by air-borne *L. maculans* ascospores (sexual spores) [18,22], which produce germ tubes that infect the leaves of winter oilseed rape in autumn (October/November), causing leaf lesions [23–24]. Conidia (asexual spores), produced in pycnidia (asexual fruiting bodies) on leaf lesions and spread by rain-splash, can cause secondary leaf infections [25]. From leaf lesions, *L. maculans* grows symptomlessly along the leaf vein/petiole to reach the stem where, in spring/summer (April-July), it causes stem cankers that result in yield losses [9,18].

Both quantitative resistance and *R* gene-mediated qualitative resistance against *L. maculans* have been identified in *B. napus* [3,8,26–27]. *R* gene-mediated resistance against *L. maculans* operates in leaves to prevent leaf lesion development and thus prevent subsequent stem canker development [9,13–14]. Quantitative resistance against *L. maculans* is a partial resistance that does not prevent leaf lesion development but decreases the severity of stem cankers [6,28].

R gene-mediated resistance against *L. maculans* can be selected in young plants by scoring lesion phenotypes after inoculation of cotyledons or leaves [8,14]. However, *R* gene-mediated resistance against *L. maculans* usually loses its effectiveness within three cropping seasons of widespread use in commercial cultivars because of selection for virulence within the variable *L. maculans* populations. For example, the resistance genes *Rlm1* (introduced into several winter oilseed rape cultivars in France) and *LepR3* (introduced into the Australian cultivar Surpass 400 from *B. rapa* var. *sylvestris*) were rendered ineffective after 3 years of commercial use [13–14,29]. This is very costly for breeders since introduction of new resistance into elite commercial cultivars can take 10–15 years [20].

By contrast, quantitative resistance is considered to be more durable than *R* gene-mediated resistance against *L. maculans* [3,30]. However, it has been difficult to select for quantitative resistance against *L. maculans* due to the long period of symptomless growth after initial development of leaf lesions [6]. Furthermore, if effective *R* genes are present, it is difficult to assess quantitative resistance because *R* gene-mediated qualitative resistance operates in leaves to prevent the growth of the pathogen from leaf lesions to stems. Currently, selection of cultivars with quantitative resistance generally relies on winter oilseed rape field experiments in which stem canker severity is assessed just before harvest [18,30–31]. If quantitative resistance can be assessed in young plants, it will not only accelerate the process of breeding for resistance but also save money for the industry. A study of growth of *L. maculans* from the cotyledon to the hypocotyl showed that stem canker symptoms can be produced on hypocotyls after cotyledon inoculation in controlled conditions [28]. However, it was not clear whether resistance to growth of *L. maculans* to cause canker in hypocotyls is the same as that in stems. Recently, a petiole inoculation method has been used to compare pathogenicity of different *L. maculans* isolates and to screen different cultivars in terms of stem canker development at a young-plant stage in controlled conditions [32–33]. However, there is a need for further work to develop methods to test for quantitative resistance in young plants.

This paper reports work to investigate new methods for assessing quantitative resistance against *L. maculans* in controlled environment experiments with young plants of two oilseed rape doubled haploid (DH) lines differing in quantitative resistance, using different types of inoculum and different inoculation and assessment methods.

Materials and Methods

In this work, assessment of quantitative resistance against growth of *L. maculans* in oilseed rape was considered in two stages; stage 1, resistance against growth along leaf veins/petioles towards the stem; stage 2, resistance against growth in stem tissues to produce stem canker symptoms (Fig. 1). Inoculation methods, type of *L. maculans* inoculum, number of plants inoculated, design of experiments and assessment methods used in each of the controlled environment experiments are described in Table 1.

Preparation of plant material and pathogen inoculum

DH lines A30 and C119, selected from the mapping population developed from a cross between winter oilseed rape Darmor-*bzh* (with quantitative resistance against *L. maculans* and *R* gene *Rlm9*) and the spring oilseed rape Yudal (without quantitative resistance or *R* genes) [30], were used for all the experiments. These two DH lines were derived from microspore cultures of a single F_1 hybrid plant from the cross 'Darmor-*bzh* × Yudal'. Use of DH lines can decrease potential variation associated with genetic background. Since no effective major resistance gene segregates in this DH population, 'field' resistance provides an assessment of quantitative resistance against *L. maculans*. When 'field' resistance was assessed by scoring stem canker at the end of the cropping season in winter oilseed rape field experiments in France, A30 developed severe phoma stem cankers whilst C119 developed much less severe stem cankers [27,30]. Similar results were obtained when these two lines

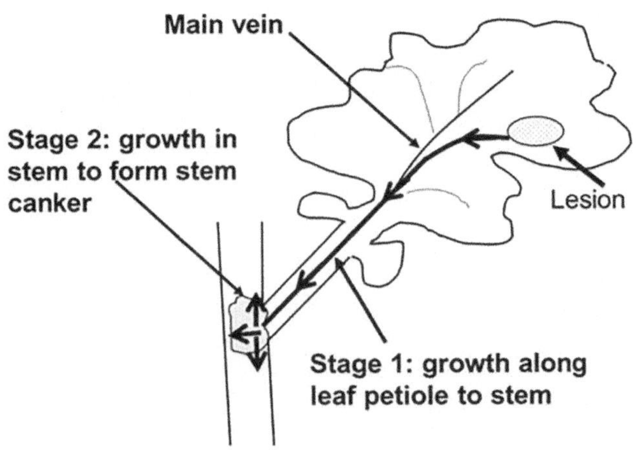

Figure 1. Stages of *Leptosphaeria maculans* growth in oilseed rape. Assessment of quantitative resistance against growth of *L. maculans* in oilseed rape was considered in two stages; stage 1, resistance against growth from phoma leaf lesion along the main leaf vein and petiole to the stem; stage 2, resistance against growth in stem to produce stem canker; both stages were investigated in controlled environment experiments.

Table 1. Inoculation methods, type of *Leptosphaeria maculans* inoculum, number of plants inoculated, design of experiments and assessment methods used in each of the controlled environment experiments with the doubled haploid (DH) lines A30 or C119.

Exp.	Inoculum[1]	Design	No. of plants inoculated[2]	Assessment method
L. maculans growth in leaf, leaf lamina inoculation				
LExpt 1	Conidia (GFP)	Randomised block	8×2 plants, 3 leaves per plant	Lesion area[3], distance grown (DG) along vein/petiole viewed by GFP
LExpt 2	Conidia (GFP)	Complete randomised	8 plants, 3 leaves per plant	Lesion area, DG viewed by GFP
LExpt 3	Ascospores	Randomised block	5×2 plants, 3 leaves per plant	DG assessed by extent of necrosis, *L. maculans* DNA in leaf petiole (qPCR)
LExpt 4	Ascospores	Complete randomised	12 plants, 3 leaves per plant	DG assessed by extent of necrosis, qPCR
LExpt 5	Ascospores	Complete randomised	8 plants, 2 leaves per plant	Lesion area, DG assessed by extent of necrosis, qPCR
L. maculans growth in stem, leaf petiole inoculation				
PExpt 1	Conidia (GFP)	Complete randomised	10 plants, 2 petioles per plant	Stem canker score, extent of *L. maculans* growth in stem viewed by GFP
PExpt 2	Ascospores	Complete randomised	5 plants, 3 petioles per plant	Stem canker score, *L. maculans* DNA in stem (qPCR)
PExpt 3	Ascospores	Complete randomised	12 plants, 2 petioles per plant	Stem canker score, qPCR
PExpt 4	Ascospores	Complete randomised	12 plants, 2 petioles per plant	Stem canker score, qPCR

[1]*L. maculans* ascospores obtained from naturally infected oilseed rape stem base debris collected in August 2007; conidia were produced by GFP-transformed isolate ME24/3.13.
[2]Plants were inoculated when they had three fully expanded leaves. For details of leaf lamina or leaf petiole inoculation, see Fig. 2. All experiments were done at 20°C with alternating 12 h light/12 h darkness.
[3]The lesion area was estimated by multiplying the lesion length by lesion width.

were assessed in field experiments in the UK (the mean disease score on a 0–4 scale was 1.9 for A30 and 0.6 for C119 in 2007/2008, and 2.7 for A30 and 1.3 for C119 in the 2008/2009 cropping season). Results of these field experiments show that A30 is susceptible and does not have quantitative resistance whereas C119 has good quantitative resistance against *L. maculans*. Furthermore, these two lines do not have the dwarfing gene (*bzh*) and do not differ greatly in terms of maturity date and plant height. Molecular markers in QTL regions for resistance to *L. maculans* showed that the difference in resistance between these two DH lines was mainly due to resistance inherited from the resistant parent (Personal communication, Dr Regine Delourme). Therefore, these two DH lines were chosen for this work.

Plants of lines A30 or C119 were grown in pots (9 cm diameter) containing a peat-based compost mixed with a soluble fertiliser (WE Hewitt & Sons Ltd, UK). Plants were initially grown in a

Figure 2. Leaf lamina and leaf petiole inoculation. Inoculation of oilseed rape leaf lamina (A) or leaf petiole (B) with conidia or ascospores of *Leptosphaeria maculans*. The arrows indicate the droplets of spore suspension (either conidia or ascospores) placed on the lamina (A) or petiole (B) of leaves of doubled haploid lines A30 (susceptible) or C119 (with quantitative resistance). Plants were inoculated at 21 days after sowing when they had three fully expanded leaves.

glasshouse (20–23°C) and thinned to one plant per pot 10 days after sowing. Three weeks after sowing, when each plant had three expanded leaves, the plants were transferred to a controlled environment cabinet (20°C day/20°C night, 12 h light/12 h darkness, light intensity 210 µmol $m^{-2}s^{-1}$) for 24 h before inoculation. After inoculation, plants were kept in the cabinet until the end of the experiments.

Both ascospores and conidia were used as inoculum. Although line C119 has resistance gene *Rlm9* (A30 has no *R* genes), the *L. maculans* populations throughout Europe are 100% virulent against *Rlm9* [8,34], so ascospores from natural populations were used. Both the *L. maculans* isolates used as conidial inoculum, ME24 (*AvrLm1, avrLm2, avrLm3, avrLm4, avrLm5, AvrLm6, AvrLm7, avrLm9*) and GFP-expressing isolate ME24/3.13 (GFP-transformed ME24; all work involving the use and storage of genetically modified *L. maculans* was done under Defra licence no. PHL 174E/5443/01/2007), are virulent against *Rlm9*. Conidial suspensions of isolates ME24 and ME24/3.13 were prepared from 12-day-old cultures on V8 agar [9]. The concentration of conidia was adjusted to 10^7 conidia mL^{-1} using a haemocytometer slide. Since phoma stem canker epidemics in crops are initiated by ascospores released from stem debris, *L. maculans* ascospores produced under natural conditions were also used as inoculum. An ascospore suspension of *L. maculans* was prepared from naturally infected UK oilseed rape stem base debris of the cultivar Courage collected after harvest in August 2007 (pieces of stem debris with mature pseudothecia were stored at −20°C until required), using the method described by Huang *et al.* [23]. The concentration of ascospores was adjusted to 10^4 ascospores mL^{-1} using a haemocytometer slide.

Growth of *L. maculans* in leaf petioles

In natural conditions in Europe, phoma stem cankers are usually initiated from phoma lesions on leaves of winter oilseed rape crops in the previous autumn [18,22]. Therefore, to investigate the growth of *L. maculans* from leaf lesions towards the stems, experiments were done by inoculation of leaf laminas. Before inoculation, each leaf lamina was rubbed gently using a wet tissue so that the droplet of inoculum would not run off. For inoculation with conidia, the rubbed sites were wounded using a sterile pin and a 10 μL droplet of conidial suspension was placed on each wounded site. For inoculation with ascospores, the rubbed sites were not wounded and a 15 μL droplet of ascospore suspension was placed on each site. Each leaf had two inoculation sites, one on each side of the main vein (Fig. 2A). The first two or three fully expanded leaves of each plant were inoculated.

Two experiments were done with conidia of GFP-expressing *L. maculans* isolate ME24/3.13 and three experiments were done with ascospores (Table 1). After inoculation, plants in trays were covered with tray covers to maintain high relative humidity for 48 h (after inoculation with ascospores) or 72 h (after inoculation with conidia). The time from inoculation to the appearance of phoma leaf lesions was recorded in each experiment. The maximum length and width of each lesion were measured at 17–18 days post inoculation (dpi), then the lesion area (mm^2) was estimated by multiplying the lesion length by lesion width.

For leaf inoculation with conidia of GFP-expressing *L. maculans*, the distance grown by *L. maculans* from the inoculation site along leaf the vein/petiole towards the stem was assessed by measuring extent of appearance of GFP fluorescence. At 22–25 dpi, the inoculated leaves were detached at the place where the petiole joins the stem (leaf scar). The leaves were viewed using a Leica MZ FLIII stereo-microscope. For observation of GFP fluorescence, filter GFP2 from Leica Microsystems (Milton Keynes, UK) was used.

For leaf inoculation with ascospores, the growth of *L. maculans* along the leaf vein/petiole towards the stem was measured by assessing the extent of visible necrosis and by quantification of the pathogen DNA in the tissue using quantitative PCR (qPCR). At 18–20 dpi, the inoculated leaves were detached at the leaf scar. The distance from the inoculation site on the leaf lamina to the furthest visible necrosis on the leaf petiole was measured on each leaf, then a piece of leaf petiole 8 cm long (measured from the inoculation site) was placed in a 15 mL tube to be freeze-dried for DNA extraction and qPCR.

Growth of *L. maculans* in stem tissues

To ensure that *L. maculans* reached the stem before the leaf abscission, the petioles of first two or three leaves of each plant were inoculated (Table 1). Before inoculation, the leaf petiole (1 cm from the stem) was gently rubbed using a wet tissue so that the droplet of inoculum would not run off. Then the leaf petiole was wounded using a sterile pin and either a 10 μL droplet of conidial suspension or a 12 μL droplet of ascospore suspension was placed over the wound (Fig. 2B). One experiment was done with conidia of GFP-expressing *L. maculans* isolate ME24/3.13 and three experiments were done with ascospores (Table 1). After inoculation, plants in trays were covered with tray covers to maintain high relative humidity for 48 h (after inoculation with ascospores) or 72 h (after inoculation with conidia). The time from inoculation to the appearance of first phoma stem canker symptoms at the leaf scar on the stem was recorded for each experiment.

For petiole inoculation with conidia of GFP-expressing isolate ME24/3.13, the growth of *L. maculans* in the stem was measured both by assessing stem canker severity and by observation of extent of GFP fluorescence within stem tissues. After inoculation, plants were regularly assessed for appearance of visible canker symptoms at the leaf scars of inoculated leaf petioles. At 87 dpi, the inoculated plants were sampled by cutting their stems at the soil surface and these stems were viewed using a Leica MZ FLIII stereo microscope for appearance of GFP fluorescence at the leaf scar of each inoculated leaf. The stems were then cut horizontally at the leaf scar of the inoculated leaf to assess the internal severity of stem canker on a 0–4 scale; where 0 = healthy; 1 = 1–25% stem cross-section necrotic; 2 = 26–50% stem cross-section necrotic; 3 = 51–75% stem cross-section necrotic; 4 = 76–100% stem cross-section necrotic. The scale was modified from that of Zhou *et al.* [35]. Then stems were cut vertically to assess internal vertical spread of GFP-expressiong *L. maculans* along the stem cortex or pith using a Leica MZ FLIII stereo-microscope.

For leaf petiole inoculation with ascospores, the growth of *L. maculans* in the stem was measured by assessing stem canker severity and by quantification of the pathogen DNA using qPCR. At 46 dpi, stems of the inoculated plants were sampled by cutting the stem at the soil surface. The stems were then cut horizontally at the leaf scar of the inoculated leaf to assess the internal severity of stem canker on the 0–4 scale. After stem canker assessment, a piece of each stem (5 cm long) was placed in a 50 mL tube to be freeze-dried for DNA extraction and qPCR.

Quantification of *L. maculans* DNA in leaf petioles and stems

Freeze-dried individual leaf petioles or stems were ground into powder using a mortar and pestle. DNA was extracted from each ground individual leaf petiole or stem using a DNA extraction kit (DNAMITE Plant Kit, Microzone Ltd, West Sussex, UK). The amount of *L. maculans* DNA in each leaf petiole or stem was quantified using a SYBR green quantitative PCR. The primers LmacF and LmacR, specific for amplification of the internal transcribed spacer (ITS) region of ribosomal DNA of *L. maculans*, that have been previously used for identification [36] and quantification [6] of *L. maculans* were used. All qPCR reactions were done in duplicate and each reaction volume was 20 μL, including 0.6 μL of primers at a final concentration of 300 nM, 10 μL of SYBR Green JumpStart Tag ReadyMix (Sigma, UK), 0.08 μL of ROX internal reference dye and the 50 ng DNA sample. Nuclease-free water (Sigma, UK) was used as the no-template control. All reactions were done in 96×0.2 mL PCR plates (ABgene) covered with cap strips, using a Stratagene Mx3000P quantitative PCR machine. The thermocycling profile consisted of an initial cycle of 95°C for 2 min, followed by 40 cycles of 95°C for 15 s, 60°C for 30 s, 72°C for 45 s and a read step at 83°C for 15 s, then a dissociation stage (thermal profile: 95°C 1 min, 60°C 1 min, 95°C 15 s). In each qPCR run, a standard dilution series consisting of 10000, 1000, 100, 10 and 1 pg of DNA of *L. maculans* isolate ME24 was included to produce a standard curve. A standard curve was generated by plotting the amount of *L. maculans* DNA against the threshold cycle (*Ct*) value for the series of dilutions. The amount of *L. maculans* DNA for each unknown sample was extrapolated from the *Ct* value and the value obtained from the standard curve using Stratagene MxPro-Mx3000 P v3.2 software.

Statistical analysis

For experiments to assess growth of *L. maculans* in leaf petioles, the data on leaf lesion area, distance grown by *L. maculans* along leaf petioles and amount of *L. maculans* DNA in leaf petioles were analysed by analysis of variance to assess the differences between DH lines A30 and C119 or between different experiments. For experiments to assess growth of *L. maculans* in stem tissues, the data on stem canker severity and amount of *L. maculans* DNA in stems were analysed by analysis of variance to assess the differences between DH lines A30 and C119 or between different experiments. To compare different methods used to assess quantitative resistance to *L. maculans* at growth stages 1 or 2, linear regressions were done for individual experiments; these regression lines were compared to assess whether they differed in intercept and/or slope for different experiments. The data on amount of *L. maculans* DNA in petioles or stems were $\log_{10}$-transformed before regression analysis. For stage 1 assessments, linear regressions were done for leaf lesion area against distance grown or amount of *L. maculans* DNA and for amount of *L. maculans* DNA against distance grown in the leaf vein/petiole. For stage 2 assessments, linear regressions were done for amount of *L. maculans* DNA in the stem against stem canker severity score. All the analyses were done using GENSTAT statistical software [37].

Results

Growth of *L. maculans* in leaf petioles

There were differences between conidial inoculum and ascospore inoculum in incubation period (time from inoculation to appearance of lesions) for phoma leaf lesion development, with the incubation period shorter for ascospores than for conidia. Phoma leaf lesions were observed at 7 dpi for plants inoculated with ascospores (Fig. 3B) but phoma leaf lesions were not observed on plants inoculated with conidia until 10 dpi (Fig. 3A). There was no difference between lines A30 and C119 in incubation period or visual appearance of phoma leaf spot symptoms, whether the plants were inoculated with conidia or ascospores.

In experiments with conidia of GFP-expressing isolate ME24/3/.13 (LExpt 1 & 2; Table 1), the leaf lesions on A30 were larger

Figure 3. Phoma leaf spot symptoms produced in controlled conditions. Phoma leaf spot symptoms on leaves of doubled haploid lines A30 (susceptible) or C119 (with quantitative resistance) inoculated with conidia or ascospores of *Leptosphaeria maculans*. Lesions on leaves of A30 at 15 days post inoculation (dpi) with conidia (A) or 10 dpi with ascospores (B); lesions on leaves of A30 (C) and C119 (D) at 22 dpi; symptomless growth of GFP-expressing *L. maculans* (isolate ME24/3.13) along a leaf vein towards the petiole of A30 viewed with brightfield illumination (E-1) or a GFP filter (E-2) at 18 dpi.

and the distances grown by *L. maculans* in petioles of A30 were greater than those on C119 (Fig. 3C, d). At 17 dpi, leaf lesion area (average of LExpt 1 & 2) on A30 (60.7 mm^2) was significantly larger than on C119 (27.4 mm^2) ($P<0.001$, 28 d.f., SED 7.4). GFP-expressing *L. maculans* (Fig. 3E) was observed in petioles of both A30 and C119; the distance grown by *L. maculans* (average of LExpt 1 & 2) was greater in petioles of A30 than in petioles of C119 (Fig. 4A) ($P<0.001$, 28 d.f., SED 1.6). There was a significant difference between LExpt 1 and LExpt 2 in distance grown by GFP *L. maculans* ($P<0.05$, 28 d.f., SED 1.6), since the distance grown by GFP *L. maculans* was assessed at 22 dpi in experiment LExpt 1 and 25 dpi in LExpt 2. However, there was no significant interaction between experiment and DH line in distance grown along the leaf petiole.

In experiments with ascospore inoculum (LExpt 3, 4 & 5; Table 1), the distance grown (average of LExpt 3, 4 & 5) by *L. maculans* (measured by the extent of necrosis) was greater in petioles of A30 than in petioles of C119 ($P<0.001$, 54 d.f., SED 1.3) (Fig. 4A). There was a significant difference between the three experiments in distance grown by *L. maculans* ($P<0.001$, 54 d.f., SED 1.4). However, there was no significant interaction between experiment and DH line. The difference between these three experiments may have occurred because the distance grown in leaf petioles was assessed at 18 dpi in LExpt 3 and at 20 dpi in LExpt 4 and LExpt 5, because there was no significant difference between LExpt 4 and LExpt 5. When symptomless growth of *L. maculans* in petioles was measured by qPCR, there was a significant difference between A30 and C119 in the amount of *L. maculans* DNA (average of LExpt 3, 4 & 5), with the amount of *L. maculans* DNA greater in leaf petioles of A30 than that in leaf petioles of C119 ($P<0.001$, 54 d.f., SED 2.9; Fig. 4B). There was no difference between experiments in amount of *L. maculans* DNA. Leaf lesion area was assessed in LExpt 5 (not assessed in LExpt 3 and LExpt 4); the leaf lesion area on A30 (327.5 mm^2) was larger than on C119 (137.6 mm^2) at 18 dpi ($P<0.01$, 14 d.f., SED 0.6).

For the experiments with conidial inoculum, there were significant linear relationships between distance grown (d) by *L. maculans* and leaf lesion area (f) in the leaf petiole in LExpt 1 ($d = 0.32f + 12.7$, $R^2 = 0.64$, $P<0.001$, 14 d.f.) (Fig. 5A) and LExpt 2 ($d = 0.31f + 14.1$, $R^2 = 0.62$, $P<0.001$, 13 d.f.). Comparison of these two regression lines showed that there was no significant difference in intercept ($P>0.85$) or slope ($P>0.71$) between them. A significant linear relationship was observed in experiment LExpt 5 with ascospores ($d = 3.90f + 10.7$, $R^2 = 0.40$, $P<0.01$, 14 d.f.). There was also a significant linear relationship between leaf lesion area (f) and *L. maculans* DNA (n) in the leaf petiole ($f = 130.8n - 163.0$, $R^2 = 0.59$, $P<0.001$, 13 d.f.) (Fig. 5B). For the three experiments with ascospore inoculum, there were significant linear relationships between *L. maculans* DNA (n) and distance grown (d) by *L. maculans* in experiments LExpt 3 ($n = 0.5d + 3.0$, $R^2 = 0.58$, $P<0.001$, 18 d.f.), LExpt 4 ($n = 0.7d + 1.6$, $R^2 = 0.78$, $P<0.001$, 22 d.f.; Fig. 5C) and LExpt 5 ($n = 0.7d + 1.7$, $R^2 = 0.54$, $P<0.005$, 13 d.f.). Comparison of these regression lines showed that the slope did not differ significantly ($P>0.36$) but the intercepts differed significantly ($P<0.001$) between them.

Growth of *L. maculans* in stem tissues

For leaf petioles inoculated with conidia of GFP-expressing isolate ME24/3.13, although no visible symptoms were observed at the leaf scars (Fig. 6A-1) at the time when the inoculated leaves abscised, growth of *L. maculans* in the leaf scar tissues was observed by using a GFP filter (Fig 6A-2) at 25 dpi. By 36 dpi, visible stem canker symptoms had started to develop at the leaf scars on some plants of line A30 (Fig 6B-1), as confirmed by observing GFP

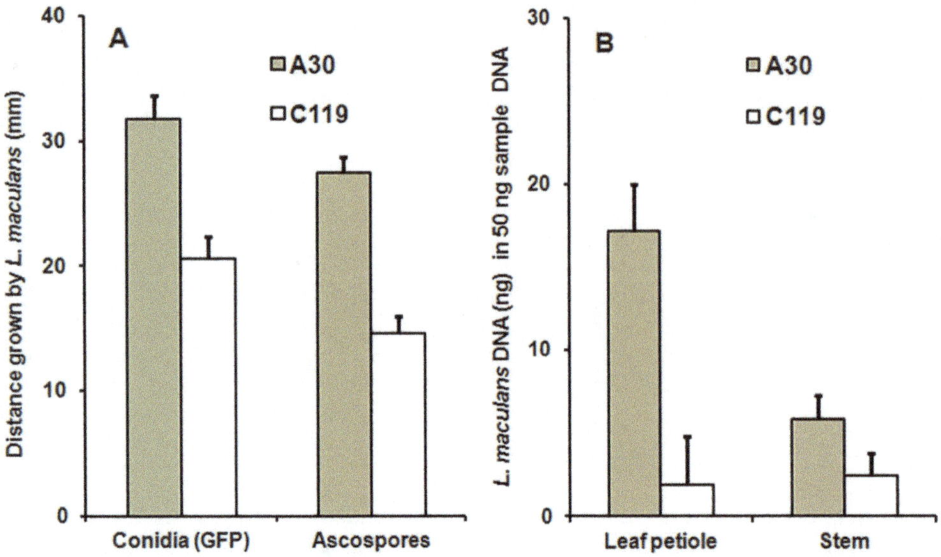

Figure 4. Distance grown along leaf petioles and *Leptosphaeria maculans* DNA in petioles and stems. Leaf laminas or leaf petioles of doubled haploid lines A30 (susceptible) or C119 (with quantitative resistance) were inoculated with conidia of GFP-expressing *L. maculans* isolate ME24/3.13 or ascospores from a natural population. The distance grown by *L. maculans* along the leaf vein/petiole towards the stem (leaf lamina inoculation, Fig. 2A) was measured by assessing extent of GFP fluorescence at 22–25 days post inoculation (dpi) for conidial inoculation (A) or by assessing extent of the visible necrosis (A) and by quantification of amount of *L. maculans* DNA (B) using qPCR at 18–20 dpi for ascospore inoculation. The growth of *L. maculans* in stem tissue (petiole inoculation, Fig. 2B) was assessed by quantification of amount of *L. maculans* DNA using qPCR (B) at 46 dpi inoculation with ascospores. Bars show standard errors (A, 28 d.f. for conidia and 54 d.f. for ascospores; B, 54 d.f. for leaf petiole and 50 d.f. for stem).

fluorescence (Fig 6B-2). Necrotic stem canker lesions were assessed at 87 dpi; there was a significant difference ($P<0.05$, 17 d.f., SED 0.96) between A30 and C119 (Fig. 7), with stem canker more severe on A30 (Fig. 6C) than on C119 (Fig. 6D). After it reached the stem, GFP-expressing *L. maculans* was observed to spread both vertically up/down the stem and horizontally towards the stem

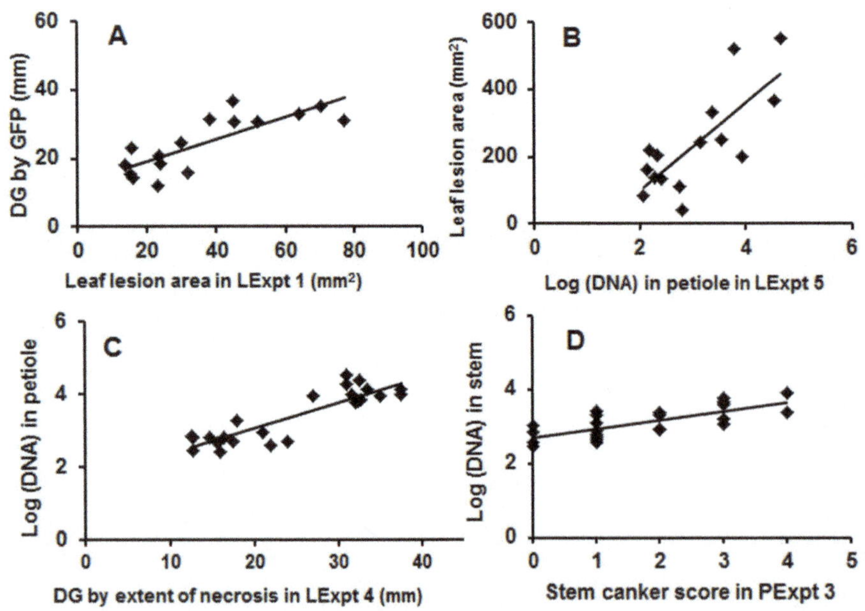

Figure 5. Relationships between different methods for assessment of quantitative resistance. Different methods were used to assess growth of *Leptosphaeria maculans* in leaf petioles or in stems of doubled haploid lines A30 (susceptible) or C119 (with quantitative resistance). Relationships between leaf lesion area and distance grown (DG) by *L. maculans* in the leaf petiole viewed by GFP in experiment LExpt 1 (A; $R^2=0.64$) or amount of *L. maculans* DNA in leaf petiole in experiment LExpt 5 (B; $R^2=0.59$); or between amount of *L. maculans* DNA and DG in leaf petiole viewed by extent of necrosis in experiment LExpt 4 (C; $R^2=0.78$); or between amount of *L. maculans* DNA in the stem and stem canker severity score in experiment PExpt 3 (D; $R^2=0.56$). The amount of *L. maculans* DNA in the petiole or stem was log_{10}-transformed. Details of these experiments are presented in Table 1.

pith. GFP-expressing *L. maculans* was observed in the pith of more A30 plants (80%) than C119 plants (56%) and spread further both up and down internal stem tissues of A30 (27.7 mm) than those of C119 (14.8 mm). After incubation of the stem cross-sections in darkness for 22 h, *L. maculans* grew more slowly in the cortex of C119 (Fig. 8B) than in that of A30 (Fig. 8A) but there was no difference between C119 (Fig. 8 D) and A30 (Fig. 8 C) in *L. maculans* growth after *L. maculans* had reached the stem pith.

There were differences between conidial inoculum and ascospore inoculum in incubation period for phoma stem canker development, with the incubation period shorter for ascospores than for conidia. Phoma stem cankers were observed at 25 dpi for plants inoculated with ascospores and at 36 dpi for plants inoculated with conidia. Stem canker developed more rapidly and more severely on plants inoculated with ascospores than on plants inoculated with conidia (Fig. 6C-F; Fig. 7).

In experiments with ascospore inoculum (PExpt 2, 3 & 4), 42% of inoculated leaves of A30 and 38% of those of C119 were senescent by 25 dpi. Stem canker developed more rapidly in A30 than in C119. By 32 dpi, 63% of inoculated leaves of A30 and 8% of inoculated leaves of C119 had developed stem canker symptoms at their leaf scars in PExpt 3; 60% of inoculated leaves of A30 and 30% of inoculated leaves of C119 had developed stem canker symptoms at their leaf scars by 33 dpi in PExpt 4. By 46 dpi, the internal stem canker severity score (average of PExpt 2, 3 & 4) was significantly greater in A30 than in C119 ($P<0.001$, 50 d.f., SED 0.27) (Fig. 7); the amount of *L. maculans* DNA was significantly greater in stems of A30 than in stems of C119 ($P<0.05$, 50 d.f., SED 1.48; Fig. 4B). However, there was a significant difference between the three experiments in stem canker severity ($P<0.01$, 50 d.f., SED 0.28) and in the amount of *L. maculans* DNA in stem tissues ($P<0.05$, 50 d.f., SED 1.59). There was no significant interaction between experiment and DH line in either stem canker severity or the amount of *L. maculans* DNA.

There were significant linear relationships between amount of *L. maculans* DNA (n) in stem tissues and stem canker severity score (s) in experiments PExpt 2 ($n = 0.87s - 1.61$, $R^2 = 0.86$, $P<0.001$, 7 d.f.) and PExpt 3 (s) ($n = 0.24s + 2.70$, $R^2 = 0.56$, $P<0.001$, 21 d.f.) (Fig. 5D). Differences between these two regression lines in both the slope ($P<0.001$) and the intercept ($P<0.001$) were significant. There was no significant linear relationship between amount of *L.*

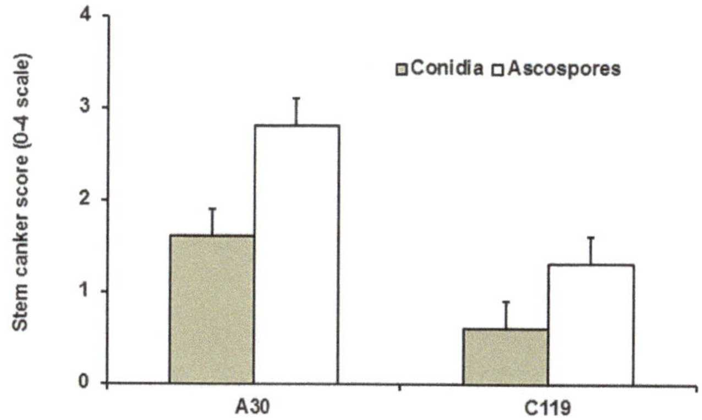

Figure 7. Severity of phoma stem canker symptoms. Leaf petioles of oilseed rape doubled haploid lines A30 (susceptible) or C119 (with quantitative resistance) were inoculated with conidia of isolate ME24/3.13 or ascospores of *Leptosphaeria maculans*; severity of stem canker was assessed on a 0–4 scale at 46 days post inoculation (dpi) (ascospore inoculum) or 87 dpi (conidial inoculum). Bars show standard errors (17 d.f. for conidial inoculation; 50 d.f. for ascospore inoculation).

Figure 6. Phoma stem canker symptoms produced in controlled conditions. Leaf petioles of oilseed rape doubled haploid lines A30 (susceptible) or C119 (with quantitative resistance) were inoculated with conidia of GFP-expressing *Leptosphaeria maculans* isolate ME24/3.13 (A, B, C) or ascospores produced on stem debris under natural conditions (E, F). No visible stem canker symptom at the leaf scar (A-1) of A30 after the inoculated leaf had abscised at 25 days post inoculation (dpi) but growth of the pathogen was observed using a GFP2 filter (A-2). Stem canker was visible at the leaf scar of A30 at 36 dpi (B-1) and growth of *L. maculans* was visualised by GFP fluorescence (B-2) (the selected region in B-1 was viewed using a GFP2 filter). Symptoms at leaf scars of A30 (C, E) or C119 (D, F) at 87 dpi with conidia (C, D) or at 31 dpi with ascospores (E, F) (arrows indicate the leaf scars of inoculated leaves).

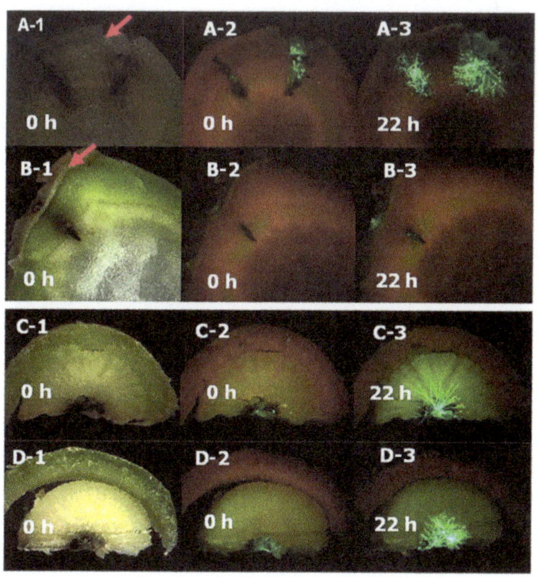

Figure 8. Growth of *Leptosphaeria maculans* in the stem cortex or pith. Leaf petioles of oilseed rape doubled haploid lines A30 (susceptible) or C119 (with quantitative resistance) were inoculated with conidia of GFP-expressing *L. maculans* isolate ME24/3.13, stems of A30 (A, C) and C119 (B, D) were cut horizontally at the leaf scar (A, B) of the inoculated leaf or at the hypocotyl (C, D) 2–3 cm below the leaf scar at 87 days post inoculation (dpi). Stem cross-sections were viewed with brightfield illumination (A-1, B-1, C-1, D-1) or a GFP2 filter (A-2-3, B-2-3, C-2-3, D-2-3) immediately after cutting (0 h) or after incubation for 22 hours in darkness at 20°C (22 h).

maculans DNA in stem tissues and stem canker severity score in experiment PExpt 4.

Discussion

Results of these experiments with DH lines C119 and A30 suggest that *B. napus* quantitative resistance against *L. maculans* can be detected in young plants in controlled conditions by leaf lamina or leaf petiole inoculation. These methods should be further evaluated on a wide range of cultivars or breeding lines with different levels of quantitative resistance to develop methods can be used reliably by breeders to select for quantitative resistance in young plants. This will not only save money but also accelerate the process of breeding for resistance. Traditionally, breeding for quantitative resistance against *L. maculans* in winter oilseed rape has relied on disease assessments before harvest 10–11 months after sowing of field experiments [3,18,30]. By contrast, an assessment of quantitative resistance in young plants in controlled conditions takes only 1–2 months by leaf lamina inoculation or 2–3 months by leaf petiole inoculation (Table 2). Furthermore, controlled environment assessment of young plants can test large numbers of lines. Such controlled environment assessment methods could be used for pre-selection of breeding material to decrease the number of lines required for final field testing and could thus save costs.

Differences between A30 and C119 in growth of *L. maculans* in leaf petioles and stems suggest that quantitative resistance against *L. maculans* decreases the severity of stem canker by impeding the growth of *L. maculans* in both petiole and stem tissues. The shorter distance grown by *L. maculans* in petioles of C119 than in those of A30 suggests that quantitative resistance in C119 impeded the growth of *L. maculans* along the petiole towards the stem. However, in previous work with commercial cultivars Darmor (with quantitative resistance) and Eurol (without quantitative resistance), there was no difference between the two cultivars in growth of *L. maculans* along the petiole although there was a difference between them in severity of stem canker [6]. The observation that growth of GFP-labelled *L. maculans* was slower in the stem cortex of C119 than in the cortex of A30 but did not differ between A30 and C119 after *L. maculans* reached the stem pith suggests that quantitative resistance in C119 mainly operated during growth of *L. maculans* from the stem cortex to the stem pith. This is consistent with results of previous work [6]. The development of more severe necrosis in hypocotyls of A30 than in those of C119 [28] suggests that quantitative resistance against *L. maculans* may also impede the growth of the pathogen from the cotyledon to the hypocotyl. There is a need to test a wide range of cultivars and DH lines under different conditions to investigate the relationship between quantitative resistance against *L. maculans* in leaf petiole and stem tissues identified in controlled environment experiments and quantitative 'field' resistance identified by end of cropping season disease assessments in field conditions.

Results of these experiments with C119 and A30 inoculated with either conidia or ascospores suggest that leaf lesion area might be used to assess quantitative resistance against *L. maculans* at stage 1 (Fig. 1). The good positive relationship between leaf lesion area and distance grown by *L. maculans* along the petiole towards the stem suggests that *L. maculans* may reach the stem more quickly and subsequently cause more severe stem canker before harvest if

Table 2. Comparison of different methods for assessing *Brassica napus* quantitative resistance against *Leptosphaeria maculans*.

Inoculum	Assessment method	Advantages	Disadvantages	Duration	Ref. [1]
Field expt., natural inoculation					
Ascospores	Score canker severity on stem before harvest	Test of field resistance under natural conditions, reliable results, no need to wound leaf	Variation between years and regions, costly to assess large numbers of lines, cannot distinguish quantitative from *R* gene resistance when effective *R* gene is present in the plant	10–11 months[2]	1, 2, 3, 4
CE expt., cotyledon inoculation – growth in hypocotyl					
Conidia	Score canker severity on hypocotyl	Easy to produce genetically homogeneous inoculum, can test large numbers of plants	Tests few isolates, less infective than ascospores, variation between expts, not easy to produce canker symptoms, hypocotyl infection does not occur in field in Europe	60–90 days	5
CE expt., leaf lamina inoculation – growth in leaf petiole					
Conidia	Extent (distance) of visible necrosis or GFP	Easy to produce homogeneous inoculum, rapid, can test large numbers of plants, visualize symptomless growth if GFP used	Tests few isolates, less infective than ascospores, need to wound leaf, may not distinguish lines with moderate resistance, restriction on use of GFP	40–45 days	3, 6
Ascospores	Extent (distance) of visible necrosis, qPCR	Easy to assess, tests natural population, can test large number of plants, no need to wound leaf, can distinguish quantitative from *R* gene resistance	Difficult to produce large amount of inoculum, inoculum not genetically homogeneous, may not distinguish moderate resistance lines	35–40 days	6
CE expt., leaf petiole inoculation – growth in stem tissue					
Conidia	Extent of visible necrosis or GFP	Easy to produce genetically homogeneous inoculum, visualize symptomless growth if GFP used	Tests few isolates, less infective than ascospores, restriction on use of GFP	70–90 days	6
Ascospores	Extent of visible necrosis, qPCR	Easy to assess, test natural population, results close to natural conditions, reliable results	Difficult to produce large amount of inoculum, inoculum not genetically homogeneous, cannot distinguish quantitative from *R* gene resistance when effective *R* gene is present	50–60 days	6

[1] 1,Delourme et al., 2006; 2, Fitt et al., 2006; 3, Huang et al., 2009; 4, Pilet et al., 1998; 5, Travadon et al., 2009; 6, this study.
[2] To obtain a reliable estimate, it is necessary to do several experiments at different sites in different seasons due to G × E interactions.

it is growing from large leaf lesions than if it is growing from small leaf lesions. This suggests that the measurement of areas of phoma leaf lesions might provide a more accurate assessment of quantitative 'field' resistance than assessing the numbers of phoma leaf lesions in field experiments in autumn since there is a poor relationship between the number of phoma leaf lesions in autumn and the subsequent severity of stem canker in summer [38]. Furthermore, it is quicker and more reliable to measure lesion area than to measure distance grown by *L. maculans* along leaf vein/petiole. For example, with conidial inoculation (e.g. LExpt 1 &2), lesion area can be measured at 17–18 dpi and no difference was observed between experiments. However, distance grown along leaf petiole cannot be measured until 22–25 dpi and differences between experiments were observed. Measurement of leaf lesion area is easy and simple. If area of leaf lesions could be used to assess quantitative resistance in field conditions, it would be a valuable method for breeders. Leaf lesion area has been used to assess quantitative resistance against other fungal pathogens [5,39–40]. However, there is a need to test a wide range of cultivars under different conditions to investigate the relationship between the area of phoma leaf lesions in autumn and the severity of phoma stem canker in the following summer.

Comparison of different inoculation methods for assessing *B. napus* quantitative resistance against *L. maculans* in young plants suggests that leaf petiole inoculation is better than other methods for use in controlled environments (Table 2). Although canker symptoms can be produced on hypocotyls after cotyledon inoculation [28], production of canker symptoms requires 4 weeks at 6°C followed by 6 weeks at 20°C and hypocotyl infection is not representative of what happens in field conditions in the UK. Furthermore, the structure of hypocotyl is different from that of stem [41]. Although the leaf lamina inoculation method is quicker than leaf petiole inoculation, the main damage caused by *L. maculans* is canker on stems, and results of leaf lamina inoculation experiments need to be further confirmed by assessing canker severity on stems. The consistency of the results of the four petiole inoculation experiments for stem canker score (i.e. less severe stem canker on C119 than on A30) suggests that quantitative resistance against *L. maculans* can be reliably assessed in young plants during colonisation of stem tissues (stage 2) after petiole inoculation. This conclusion is supported by previous work on cultivars Darmor and Eurol [6].

Results of these experiments with different types of inoculum suggest that ascospore inoculum of *L. maculans* is more effective than conidial inoculum in terms of leaf lesion and stem canker development. The incubation periods for leaf lesion and stem canker development may have been shorter for ascospore inoculum than for conidial inoculum because ascospores germinate more rapidly and contain more nutrients for the pathogen to colonise the host than conidia. Although phoma stem canker epidemics are initiated by ascospores released from crop debris, much experimental work with young plants to study resistance has used conidial inoculum [8,33,42]. Use of ascospores from natural populations as inoculum can simulate what happens in natural conditions and is more likely to identify 'quantitative resistance' expressed in field conditions. Recent work has shown that some components of quantitative resistance against *L. maculans* can be isolate-specific [43]; use of conidia of single *L. maculans* isolates may not accurately assess quantitative resistance while use of ascospore inoculum from natural populations is more likely to do so. Therefore, use of ascospore inoculum is more reliable than use of conidial inoculum for detecting and assessing quantitative resistance.

Comparison of different assessment methods used in controlled environments suggests that leaf lesion area could be a good measurement for use at stage 1 and stem canker severity score could be a good measurement for use at stage 2 (Table 2). At stage 1, measurement of lesion area was better than the other three methods used (GFP, qPCR and visible distance grown) and it might be used in field experiments whereas it is difficult to measure the spread of *L. maculans* from the leaf lesion along the leaf petiole in field conditions. In addition, assessment of leaf lesions (appearance and size) can distinguish between quantitative resistance (typical large grey lesion) and *R* gene-mediated resistance (small dark lesion) against *L. maculans* [6,9]. Although GFP and qPCR are valuable methods for assessing symptomless growth of *L. maculans* in leaf or stem tissues, GFP can be used only in controlled environments and is not practical for use with large numbers of lines. Whilst qPCR can be used with large numbers of lines, it is expensive and not always reliable. Because qPCR is very sensitive and *L. maculans* has a symptomless growth phase, *L. maculans* DNA was detected in stems (Fig. 5D) or hypocotyls [28] with no visible necrosis. By contrast, scoring stem canker severity in controlled environment experiments is more reliable at stage 2. More severe stem cankers developed on A30 than on C119 in all the four experiments with petiole inoculation. Results of stem canker assessment in young plants in controlled conditions (stage 2) could be used to predict how the lines or cultivars will operate in field conditions.

Development of methods to phenotype quantitative resistance against plant pathogens in young plants under controlled conditions will improve the process of breeding cultivars for control of crop diseases. Assessment of quantitative trait loci (QTL) for resistance is often inconsistent due to interactions with environment [4,16,31]. Detection of quantitative resistance in controlled conditions can avoid the influence of fluctuations in natural weather conditions. Furthermore, with the development of methods to assess quantitative resistance in controlled environments, it is possible to test effectiveness of the quantitative resistance under a range of different controlled conditions, which will help to optimise the use of quantitative resistance in different natural environments. Due to the incomplete nature of quantitative resistance and the difficulties in assessing it, mechanisms of operation of quantitative resistance are poorly understood [4]. Quantitative resistance has been thought to be expressed at later stages of plant development and it has been referred to as 'adult plant resistance'. There has been little work to investigate potential operation of quantitative resistance in young plants; results of this work suggest that quantitative resistance can be expressed at early stages of plant development. With the use of next generation sequencing techniques and reduced costs, more genome sequences of both pathogens and their hosts are available [44–47]. Identification of stable QTL in controlled conditions will improve our understanding of mechanisms of operation of quantitative resistance, which will help in the breeding of cultivars with durable resistance to contribute to food security.

Acknowledgments

We thank Regine Delourme and Hortense Brun for providing the seeds of DH lines A30 and C119 and for their advice throughout this work, including comments on this paper, Zhunyan Huang for help with DNA extraction, Jean Devonshire for help with GFP microscopy, Sandra Harvey for help with inoculation and data curating. All work involving the use and storage of genetically modified *L. maculans* was done under Defra licence no. PHL 174E/5443(01/2007).

Author Contributions

Conceived and designed the experiments: YJH GJK BDLF. Performed the experiments: YJH. Analyzed the data: YJH AQ. Contributed reagents/materials/analysis tools: YJH AQ GJK. Wrote the paper: YJH AQ BDLF.

References

1. Beddington J (2010) Food security: contributions from science to a new and greener revolution. Philosophical Transactions of the Royal Society B: Biological Sciences 365: 61–71.
2. Lindhout P (2002) The perspectives of polygenic resistance in breeding for durable disease resistance. Euphytica 124: 217–226.
3. Delourme R, Chèvre AM, Brun H, Rouxel T, Balesdent MH, et al. (2006) Major gene and polygenic resistance to *Leptosphaeria maculans* in oilseed rape (*Brassica napus*). European Journal of Plant Pathology 114: 41–52.
4. Poland JA, Balint-Kurti PJ, Wisser RJ, Pratt RC, Nelson RJ (2008) Shades of gray: the world of quantitative disease resistance. Trends in Plant Sciences 14: 21–29.
5. Chartrain L, Brading PA, Widdowson JP, Brown JKM (2004) Partial resistance to septoria tritici blotch (*Mycosphaerella graminicola*) in wheat cultivars Arina and Riband. Phytopathology 94: 497–504.
6. Huang YJ, Pirie EJ, Evans N, Delourme R, King GJ, et al. (2009) Quantitative resistance to symptomless growth of *Leptosphaeria maculans* (phoma stem canker) in *Brassica napus* (oilseed rape). Plant Pathology 58: 314–323.
7. Brun H, Chèvre AM, Fitt BDL, Powers F, Besnard AL, et al. (2010) Quantitative resistance increases the durability of qualitative resistance to *Leptosphaeria maculans* in *Brassica napus*. New Phytologist 185: 285–299.
8. Balesdent MH, Attard A, Ansan-Melayah D, Delourme R, Renard M, et al. (2001) Genetic control and host range of avirulence towards *Brassica napus* cvs Quinta and Jet Neuf in *Leptosphaeria maculans*. Phytopathology 91: 70–76.
9. Huang YJ, Evans N, Li ZQ, Eckert M, Chèvre AM, et al. (2006) Temperature and leaf wetness duration affect phenotypic expression of *Rlm6*-mediated resistance to *Leptosphaeria maculans* in *Brassica napus*. New Phytologist 170: 129–141.
10. Larkan NJ, Lydiate DJ, Parkin IAP, Nelson MN, Epp DJ, et al. (2013) The Brassica napus blackleg resistance gene *LepR3* encodes a receptor-like protein triggered by the *Leptosphaeria maculans* effector AVRLM1. New Phytologist 197: 595–605.
11. Hammond-Kosack KE, Kanyuka K (2007) Resistance genes (*R* genes) in plants. Encyclopedia of Life Sciences, John Wiley & Sons, Ltd. doi: 10.1002/9780470015902.a0020119.
12. Bent AF, Mackey D (2007) Elicitors, effectors, and *R* genes: the new paradigm and a lifetime supply of questions. Annual Review of Phytopathology 45: 399–436.
13. Rouxel T, Penaud A, Pinochet X, Brun H, Gout L, et al. (2003) A 10-year survey of populations of *Leptosphaeria maculans* in France indicates a rapid adaptation towards the *Rlm1* resistance gene of oilseed rape. European Journal of Plant Pathology 109: 871–881.
14. Sprague SJ, Balesdent MH, Brun H, Hayden HL, Marcroft SJ, et al. (2006) Major gene resistance in *Brassica napus* (oilseed rape) is overcome by changes in virulence population of *Leptosphaeria maculans* in France and Australia. European Journal of Plant Pathology 114: 33–40.
15. Soanes DM, Talbot NJ (2008) Moving targets: rapid evolution of oomycete effectors. Trends in Microbiology 16: 507–510.
16. McDonald B (2010) How can we achieve durable disease resistance in agricultural ecosystems? New Phytologist 185: 3–5.
17. Czembor PC, Arseniuk E, Czaplicki A, Song Q, Cregan PB, et al. (2003) QTL mapping of partial resistance in winter wheat to *Stagonospora nodorum* blotch. Genome 46: 546–554.
18. Fitt BDL, Brun H, Barbetti MJ, Rimmer SR (2006) World-wide importance of phoma stem canker (*Leptosphaeria maculans* and *L. biglobosa*) on oilseed rape (*Brassica napus*). European Journal of Plant Pathology 114: 3–15.
19. Oliver RP, Rybak K, Shankar M, Loughman R, Harry N, et al. (2008) Quantitative disease resistance assessment by real-time PCR using the *Stagonospora nodorum*-wheat pathosystem as a model. Plant Pathology 57: 527–532.
20. Fitt BDL, Fraaije BA, Chandramohan P, Shaw MW (2011) Impacts of changing air composition on severity of arable crop disease epidemics. Plant Pathology 60: 44–53.
21. Stonard JF, Latunde-Dada AO, Huang YJ, West JS, Evans N, et al. (2010) Geographic variation in severity of phoma stem canker and *Leptosphaeria maculans/L. biglobosa* populations on UK winter oilseed rape (*Brassica napus*). European Journal of Plant Pathology 126: 97–109.
22. Huang YJ, Fitt BDL, Jedryczka M, Dakowska S, West JS, et al. (2005) Patterns of ascospore release in relation to phoma stem canker epidemiology in England (*Leptosphaeria maculans*) and Poland (*L. biglobosa*). European Journal of Plant Pathology 111: 263–277.
23. Huang YJ, Toscano-Underwood C, Fitt BDL, Hu XJ, Hall AM (2003) Effects of temperature on ascospore germination and penetration of oilseed rape (*Brassica napus*) leaves by A-group or B-group *Leptosphaeria maculans* (phoma stem canker). Plant Pathology 52: 245–255.
24. Toscano-Underwood C, West JS, Fitt BDL, Todd AD, Jedryckzka M (2001) Development of phoma lesions on oilseed rape leaves inoculated with ascospores of A-group and B-group *Leptosphaeria maculans* (stem canker) at different temperatures and wetness durations. Plant Pathology 50: 28–41.
25. Travadon R, Bousset L, Saint-Jean S, Sache I (2007) Splash dispersal of *Leptosphaeria maculans* pycnidiospores and the spread of blackleg on oilseed rape. Plant Pathology 56: 595–603.
26. Rimmer SR (2006) Resistance genes to *Leptosphaeria maculans* in *Brassica napus*. Can J Plant Pathol 28: S288–S297.
27. Jestin C, Lode M, Vallee P, Domin C, Falentin C, et al. (2011) Association mapping of quantitative resistance for *Leptosphaeria maculans* in oilseed rape (*Brassica napus* L.). Molecular Breeding 27: 271–287.
28. Travadon R, Marquer B, Ribule A, Sache I, Masson JP, et al. (2009) Systemic growth of *Leptosphaeria maculans* from cotyledons to hypocotyls in oilseed rape: influence of number of infection sites, competitive growth and host polygenic resistance. Plant Pathology 58: 461–469.
29. Marcroft SJ, Van de Wouw, Salisbury PA, Potter TD, Howlett BJ (2012) Effect of rotation of canola (*Brassica napus*) cultivars with different complements of blackleg resistance genes on disease severity. Plant Pathology 61: 934–944.
30. Pilet ML, Delourme R, Foisset N, Renard M (1998) Identification of loci contributing to quantitative field resistance to blackleg disease, causal agent *Leptosphaeria maculans* (Desm.) Ces. et de Not., in winter rapeseed (*Brassica napus* L.). Theoretical and Applied Genetics 96: 23–30.
31. Delourme R, Brun H, Ermel M, Lucas MO, Vallee P, et al. (2008) Expression of resistance to *Leptosphaeria maculans* in *Brassica napus* double haploid lines in France and Australia is influenced by location. Annals of Applied Biology 153: 259–269.
32. Brun H, Ermel M, Mabon R, Miteul H, Balesdent MH, et al. (2011) A method to phenotype the aggressiveness of *Leptosphaeria maculans* isolates on stems of oilseed rape in controlled conditions. Abstract book of the 13th International Rapeseed Congress, Crop Protection Section: 507.
33. Delourme R, Miteul H, Ermel M, Brun H (2011) Assessment of quantitative resistance to *Leptosphaeria maculans* isolates in controlled conditions. Abstract book of the 13th International Rapeseed Congress, Crop Protection Section: 505.
34. Stachowiak A, Olechnowicz J, Jedryczka M, Rouxel T, Balesdent MH, et al. (2006) Frequency of avirulence alleles in field populations of *Leptosphaeria maculans* in Europe. European Journal of Plant Pathology 114: 67–75.
35. Zhou Y, Fitt BDL, Welham SJ, Gladders P, Sansford CE, et al. (1999) Effects of severity and timing of stem canker (*Leptosphaeria maculans*) symptoms on yield of winter oilseed rape (*Brassica napus*) in the UK. European Journal of Plant Pathology 105: 715–728.
36. Liu SY, Liu Z, Fitt BDL, Evans N, Foster SJ, et al. (2006) Resistance to *Leptosphaeria maculans* (phoma stem canker) in *Brassica napus* (oilseed rape) induced by *L. biglobosa* and chemical defence activators in field and controlled environments. Plant Pathology 55: 401–412.
37. Payne RW, Harding SA, Murray DA, Soutar DM, Baird DB, et al. (2011) *The Guide to GenStat Release 14, Part 2: Statistics*. 997 pp. Hemel Hempstead, UK: VSN International Ltd.
38. Powers SJ, Pirie EJ, Latunde-Dada AO, Fitt BDL (2010) Analysis of leaf appearance, leaf death and phoma leaf spot, caused by *Leptosphaeria maculans*, on oilseed rape (*Brassica napus*) cultivars. Annals of Applied Biology 157: 55–70.
39. Talukder ZI, Tharreau D, Price AH (2004) Quantitative trait loci analysis suggests that partial resistance to rice blast is mostly determined by race-specific interactions. New Phytologist 162: 197–209.
40. Silva MR, Martinelli JA, Federizzi LC, Chaves MS, Pacheco MT (2012) Lesion size as a criterion for screening oat genotypes for resistance to leaf spot. European Journal of Plant Pathology 137: 315–327.
41. Sprague SJ, Watt M, Kirkegaard JA, Howlett BJ (2007) Pathways of infection of *Brassica napus* roots by *Leptosphaeria maculans*. New Phytologist 176: 211–222.
42. Raman R, Taylor B, Marcroft S, Stiller J, Eckermann P, et al. (2012) Molecular mapping of qualitative and quantitative loci for resistance to *Leptosphaeria maculans* causing blackleg disease in canola (*Brassica napus* L.). Theoretical and Applied Genetics 125: 405–418.
43. Marcroft ST, Elliott VL, Cozijnsen AJ, Salisbury PA, Howlett BJ, et al. (2012) Identifying resistance genes to *Leptosphaeria maculans* in Australian *Brassica napus* cultivars based on reactions to isolates with known avirulence genotypes. Crop & Pasture Science 63: 338–350.
44. Rouxel T, Grandaubert J, Hane JK Hoede C, van de Wouw A (2011) Effector diversification within compartments of the *Leptosphaeria maculans* genome affected by Repeat-Induced Point mutations. Nature Communications 2: 202. (Doi: 10.1038/ncomms1189).
45. Rouxel T, Balesdent MH (2013) From model to crop plant–pathogen interactions: cloning of the first resistance gene to *Leptosphaeria maculans* in *Brassica napus*. New Phytologist 197: 356–358.
46. Wang X, Wang H, Wang J, Sun R, Wu J, et al. (2011) The genome of the mesopolyploid crop species *Brassica rapa*. Nature Genetics 43: 1035–1039.
47. Hayward A, McLanders J, Campbell E, Edwards D, Batley J (2012) Genomic advances will herald new insights into the Brassica: *Leptosphaeria maculans* pathosystem. Plant Biology 14 (Suppl. 1): 1–10.

Accelerating the Switchgrass (*Panicum virgatum* L.) Breeding Cycle Using Genomic Selection Approaches

Alexander E. Lipka[1¤a*], Fei Lu[1], Jerome H. Cherney[2], Edward S. Buckler[1,3,4], Michael D. Casler[5,6], Denise E. Costich[1,3¤b]

1 Institute for Genomic Diversity, Cornell University, Ithaca, New York, United States of America, **2** Department of Crop and Soil Sciences, Cornell University, Ithaca, New York, United States of America, **3** Agricultural Research Service, United States Department of Agriculture, Ithaca, New York, United States of America, **4** Department of Plant Breeding and Genetics, Cornell University, Ithaca, New York, United States of America, **5** Agricultural Research Service, United States Department of Agriculture, Madison, Wisconsin, United States of America, **6** Department of Agronomy, University of Wisconsin–Madison, Madison, Wisconsin, United States of America

Abstract

Switchgrass (*Panicum virgatum* L.) is a perennial grass undergoing development as a biofuel feedstock. One of the most important factors hindering breeding efforts in this species is the need for accurate measurement of biomass yield on a per-hectare basis. Genomic selection on simple-to-measure traits that approximate biomass yield has the potential to significantly speed up the breeding cycle. Recent advances in switchgrass genomic and phenotypic resources are now making it possible to evaluate the potential of genomic selection of such traits. We leveraged these resources to study the ability of three widely-used genomic selection models to predict phenotypic values of morphological and biomass quality traits in an association panel consisting of predominantly northern adapted upland germplasm. High prediction accuracies were obtained for most of the traits, with standability having the highest ten-fold cross validation prediction accuracy (0.52). Moreover, the morphological traits generally had higher prediction accuracies than the biomass quality traits. Nevertheless, our results suggest that the quality of current genomic and phenotypic resources available for switchgrass is sufficiently high for genomic selection to significantly impact breeding efforts for biomass yield.

Editor: David D. Fang, USDA-ARS-SRRC, United States of America

Funding: This work was supported by Department of Energy-United States Department of Agriculture Plant Feedstock Genomics for Bioenergy Program Project Number DE-A102-07ER64454 (http://genomicscience.energy.gov/research/DOEUSDA/) and United States Department of Agriculture—Agricultural Research Services (http://www.ars.usda.gov/main/main.htm). The funders had no role in study design, data collection and analysis, decision to publish, or preparation of the manuscript.

Competing Interests: The authors have declared that no competing interests exist.

* Email: alipka@illinois.edu

¤a Current address: Department of Crop Sciences, University of Illinois, Urbana, Illinois, United States of America
¤b Current address: International Maize and Wheat Improvement Center (CIMMYT), Texcoco, Mexico

Introduction

Switchgrass (*Panicum virgatum* L.) is undergoing development as a biofuel feedstock due to its high biomass yield, broad adaptation, perennial growth habit, and long-standing presence in the seed industry [1]. Once inhabiting prairie and savanna ecosystems from Canada to Mexico and east of the Rocky Mountains, native switchgrass is now confined to thousands of prairie and savanna remnants that range in size from a few plants to a few hundred hectares [2]. Driven largely by photoperiod and temperature, latitude is the principal source of adaptive phenotypic variability across a broad landscape [3,4].

Switchgrass contains three principal taxa: a tetraploid ($2n = 4x = 28$) lowland ecotype, a tetraploid upland ecotype, and an octoploid ($2n = 8x = 56$) upland ecotype [5]. Upland ecotypes originated from upland prairie and savanna habitats that were frequently exposed to drought, especially toward the western portion of the range [6]. Lowland ecotypes originated in low-lying riverine or lacustrine habitats that were exposed to seasonal wet periods [6]. Upland ecotypes tend to be more northern adapted, while lowland ecotypes tend to be more southern adapted, with a transition zone where both can be found, sometimes within a single prairie or savanna remnant [5]. Upland and lowland ecotypes are highly cross-fertile and significant gene flow has occurred between the ecotypes during glacial maxima of the past million years [7]. Ploidy is the secondary taxonomic division within the species, primarily within the upland ecotype; lowland plants at the octoploid level are rare [8]. Gene flow has occurred between tetraploid and octoploid levels, largely by 2n gametes (4x to 8x) or haploidy (8x to 4x), but at relatively low frequencies due to the role of ploidy as a hybridization barrier [9].

Since the establishment of switchgrass as the herbaceous model species for cellulosic biofuel feedstock development in 1992 [1], a total of 12 breeding programs have been developed in North America [5]. Due to phenotypic differences among the three principal taxa and to the magnitude of adaptive phenotypic variation for flowering time and temperature (cold and heat) tolerance, there is very little overlap or duplication among these breeding programs. Collectively, their target population of environments covers eastern North America, but their individual target regions are realistically broken down into a minimum of eight regional gene pools or cultivar deployment zones [5]. Because adaptive phenotypic variation is a strong driver of both adaptation and production traits, genotype × environment interactions are a dominant force and individual cultivars are rarely adapted to more than three hardiness zones, as defined by [10].

Due to the length of the breeding cycle and the need for frequent (perhaps constant) phenotypic assessment of adaptive traits, few cultivars have been developed with documented improvement in biomass production traits. 'Liberty' is the most notable example, demonstrating both an increase in biomass yield and broader adaptation into USDA hardiness zone 3 [11]. Recent advances in the development of genomic tools for measuring and quantifying DNA marker diversity and sophisticated statistical tools to associate marker variation with phenotypic variation have the potential to revolutionize switchgrass breeding methodology [12]. Switchgrass breeding is complicated by the perennial nature of the species and the need for accurate measurement of biomass yield on a per-hectare basis, the single trait that is most limiting for sustainable and economically viable biomass production [13]. Simple-to-measure surrogate traits are needed to speed up the breeding cycle. Genomic selection [14,15] offers such an opportunity by developing predictive equations that allow breeders to measure DNA markers on seedlings and to predict which seedlings will have the highest biomass yield potential as adult plants [12].

The potential of genomic selection for improving the effectiveness of breeding programs has been successfully demonstrated in livestock [16–18], annual crops [19–23], and forest trees [24–26]. In these species, genomic selection has been shown to increase selection accuracy, reduce evaluation cost per genotype, and reduce breeding cycle time compared to phenotypic selection. More specifically, a recent evaluation of genomic selection methods concluded that genomic selection for perennial biofuel crops, such as switchgrass, is most advantageous when biomass yield on a per-hectare basis is difficult or expensive to measure, when it is difficult or impossible to apply meaningful selection pressure on plants within families, and when cycle times are >5 years, which is typically the case with switchgrass [12].

The purpose of this study was to explore the potential for genomic selection to increase the breeding cycle in switchgrass, particularly for seven morphological traits and 13 biomass quality traits. For most of these traits, reasonably high prediction accuracies were obtained. Our analysis was conducted within an association panel of 515 genotypes defined as a random sample of switchgrass from the northern USA gene pools. The population was evaluated using a set of 16,669 single nucleotide polymorphisms (SNPs) obtained using genotyping by sequencing (GBS) techniques [27,28] that were subsequently mapped to the recently available *Panicum virgatum* genome sequence v1.1 reference genome [29].

Materials and Methods

Germplasm

We analyzed the switchgrass association panel described in [27]. Briefly, this panel included 66 diverse switchgrass populations derived from predominantly northern adapted upland germplasm. Both tetraploid and octoploid germplasm were included. This panel was grown from seed planted at the greenhouse in the USDA-ARS Dairy Forage Research Center in Madison, Wisconsin in 2007. Ten clones or genotypes from each population were vegetatively propagated, then planted in Ithaca, New York in 2008 in a randomized complete block design with two replicates. Subsequently, a total of 540 plants from the Ithaca location were used for genotypic and phenotypic evaluation.

Morphological traits

The association panel was evaluated for seven morphological traits in 482 of the plants grown in Ithaca, NY during the 2009, 2010, and 2011 field seasons. These traits included anthesis date, heading date, standability, leaf length, leaf width, plant height, and total plant height. Descriptions of how each of these traits was measured are presented in Table 1, and the tools used to obtain the measurements are described at http://www.maizegenetics.net/phenotyping-tools [30]. Prior to subsequent analysis, the heading and anthesis dates were converted to growing degree days (GDD) as follows:

1) The first day in which GDD was recorded occurs the day after the first five consecutive days where the average temperature is >32° F.
2) After this day, GDD for a single day is calculated as:

$$[(Adj.Min + Adj.Max)/2] - 32$$

where *Adj. Min* is the maximum of the minimum daily temperature and 32°F, and *Adj.Max* is the minimum of the of the maximum daily temperature and 86°F. Intuitively, *Adj. Min* and *Adj.Max* limits the recorded minimum and maximum daily temperatures to 32°F and 86°F, respectively.

3) For each day after the first day in which GDD is recorded, the cumulative GDD is also recorded. The cumulative GDD is used to record heading date and anthesis date.

Biomass Quality traits

Near-infrared reflectance spectroscopy (NIRS, described in [31]) was used to estimate 42 biomass quality traits for a total 515 genotypes grown during two field seasons at the Ithaca, NY location. Samples were ground in Ithaca, NY, shipped to Madison, WI, and scanned on an NIRS unit at the U.S. Dairy Forage Research Center, as described in [31]. A total of 42 biomass quality traits were predicted using equations developed by [31], but only 13 of those traits were analyzed in this study due to their direct relevance and practical value in a breeding program focused on improving conversion efficiency, and to minimize redundancy from collinear traits. Specifically, these traits include acid detergent lignin, minerals (total ash), carbon, high heating value, cell wall concentration, ethanol/g dry forage, etherified ferulates, in vitro dry matter digestibility, pentose sugars release/g dry forage, total soluble carbohydrates, starch, sucrose, and total sugar. No sample had an H-statistic >3.0, indicating that none of the samples could be classified as outliers.

Description of SNPs

The Universal Network-Enabled Analysis Kit (UNEAK) discovery pipeline [27] was used to generate 29,221 SNPs with a minimum call rate of 0.5 and minimum minor allele frequency (MAF) of 0.05 among the 540 plants grown at the Ithaca location. These SNPs were then aligned to the *Panicum virgatum* genome sequence v1.1 [29]. The resulting 16,669 uniquely aligned SNPs were used for subsequent analysis.

Table 1. Phenotyping protocol for seven morphology traits measured in three summer environments, in Ithaca, NY across three years.

Trait Name (units)	Trait Description	Measured in Following Years
Anthesis Date	50% of panicles have 50% open florets	2009–2011
Heading Date	at least 50% of stems are 50% emerged (panicle branches still upright, just starting to spread)	2009–2011
Standability	0 = prostrate	2010–2011
	10 = upright	
Leaf length (mm)	Leaf below flag; base to tip	2009–2011
Leaf width (mm)	Leaf below flag; widest part	2009–2011
Plant Height (cm)	Base of longest flowering stem to the node at the base of the panicle	2009–2011
Total Plant Height (cm)	Base of the longest flowering stem to the tip of the panicle	2009–2011

Phenotypic evaluation

A subset of the 540 plants that yielded sufficient biomass for at least one field season was evaluated for morphological and quality traits. Specifically, 482 plants were evaluated for seven morphological traits and 515 plants were evaluated for 13 quality traits. All 20 traits were examined for outliers using Studentized deleted residuals [32] from a mixed linear model including year, field, block, and population as random effects in SAS version 9.3 [33]. For each trait, best linear unbiased predictors (BLUPs) were obtained for each line across years and replicates, using a mixed linear model fitted in ASReml version 3.0 [34]. Details of the model fitting procedure have been described in [35]. The relationship between each of these BLUPs was then evaluated using the Pearson correlation coefficient (r). Variance component estimates from the model used to obtain BLUPs were also used to estimate repeatability on a clone mean basis ($\hat{h}^2_l$) [36,37]. These repeatability estimates are upper bounds of the heritabilities for each trait. The delta method was used to approximate the standard error of the repeatability estimates [36]. Finally, the Box-Cox procedure [38] was implemented to find the optimal transformation of the BLUPs, as described in [39].

Genomic Selection

Prior to evaluating the genomic selection models, missing allelic values among the 16,669 SNPs anchored to the *Panicum virgatum* genome sequence v1.1 reference genome were imputed using fastPhase version 1.4.0 [40]. The allele frequencies of these SNPs were calculated among the 482 plants evaluated for the morphology traits and again among the 515 plants evaluated for the quality traits. Within each of these two subsets, SNPs with MAF <0.05 were removed. Consequently, 11,857 SNPs were used in the genomic selection models for the morphology traits, and 12,180 SNPs were used in the models for the quality traits.

To assess the capability of our imputed markers to predict morphological and quality trait values, three genomic selection approaches were tested, namely ridge regression-best linear unbiased prediction (RR-BLUP) [14], least absolute shrinkage and selection operator (LASSO) [41], and elastic net [42]. Although these three approaches have been shown to produce similar results in practice (e.g., [21]), the performance of each approach could depend on the genetic architecture of the evaluated traits. Specifically, RR-BLUP should theoretically outperform LASSO for complex traits, while LASSO should be superior for simpler traits. The elastic net, whose penalty is a weighted average of the penalties of RR-BLUP and LASSO, is considered to be a compromise between the two approaches. In this study, the mixing parameter for the elastic net was set to $\alpha = 0.5$, meaning that the RR-BLUP and LASSO penalties were given equal weights. The RR-BLUP approach was conducted using the rrBLUP package [43] in the R programming language

Table 2. Means and ranges for best linear unbiased predictors (BLUPs) of seven morphological traits evaluated on a switchgrass association panel, and estimated repeatability on a clone-mean basis in three summer environments, in Ithaca, NY across three years.

Trait	No. Lines	BLUP Mean	BLUP SD[b]	BLUP Range	Repeatability	Repeatability SE[c]
Anthesis Date (GDD[a])	481	3840.53	450.21	2630.25–5272.48	0.93	0.01
Heading Date (GDD)	482	2870.47	343.81	2111.75–4547.04	0.91	0.01
Standability (0–10 scale)	481	5.36	1.60	1.47–8.59	0.88	0.01
Leaf Length (mm)	482	528.88	73.30	294.62–708.48	0.85	0.02
Leaf Width (mm)	482	13.32	1.91	6.56–20.75	0.82	0.02
Plant Height (cm)	482	88.78	16.45	44.75–146.16	0.75	0.03
Total Plant Height (cm)	482	162.22	20.24	105.43–222.81	0.85	0.02

[a]GDD, Growing degree dates
[b]SD, Standard deviation
[c]SE, Standard error

Table 3. Means and ranges for best linear unbiased predictors (BLUPs) of 13 quality traits evaluated on a switchgrass association panel, and estimated repeatability on a clone-mean basis in two summer environments, in Ithaca, NY, across two years.

Trait (µg/g)	No. Lines	BLUP Mean	BLUP SD[a]	BLUP Range	Repeatability	Repeatability SE[b]
Acid detergent lignin	514	75.62	5.45	61.47–90.02	0.81	0.02
Minerals (total ash)	514	69.20	4.60	54.47–83.62	0.67	0.03
Carbon	514	443.86	2.19	438.78–452.85	0.67	0.03
High Heating Value	514	4182.59	17.6	4136.14–4237.54	0.76	0.02
Cell wall concentration	514	673.11	47.31	564.12–832.92	0.87	0.01
Ethanol/g dry forage	514	82.73	7.46	60.83–106.78	0.78	0.02
Etherified ferulates	514	0.88	0.10	0.64–1.28	0.83	0.48
In vitro dry matter digestibility	514	410.54	35.16	311.86–494	0.82	0.01
Pentose sugars release/g dry forage	515	191.29	8.30	167.13–218.05	0.77	0.02
Total soluble carbohydrates	514	51.27	8.58	29.22–74.16	0.71	0.03
Starch	514	6.35	2.72	0.67–17.27	0.59	0.04
Sucrose	514	28.29	5.71	13.71–45.13	0.72	0.02
Total sugar	514	625.44	19.01	572.91–691.56	0.79	0.02

[a]SD, Standard deviation
[b]SE, Standard error

[44], while LASSO and elastic net were conducted using the glmnet R package [45].

For any genomic selection model, it is important to ensure that SNP effects arising from overall differences in population structure are factored out [17]. Given the genetic differences attributable to the observed ecotypes and ploidies in our association panel, it is crucial to account for such SNP effects prior to conducting our genomic selection study. Based on the results presented in [27], we hypothesized that the first two principal components (PCs) of a principal component analysis (PCA) of the 16,669 SNPs imputed with fastPhase would sufficiently account for these genetic differences. Accordingly, we fitted a model to each trait where the trait was the response variable and the first two PCs from the PCA of these SNPs were the explanatory variables. The residuals from each model were used for genomic selection.

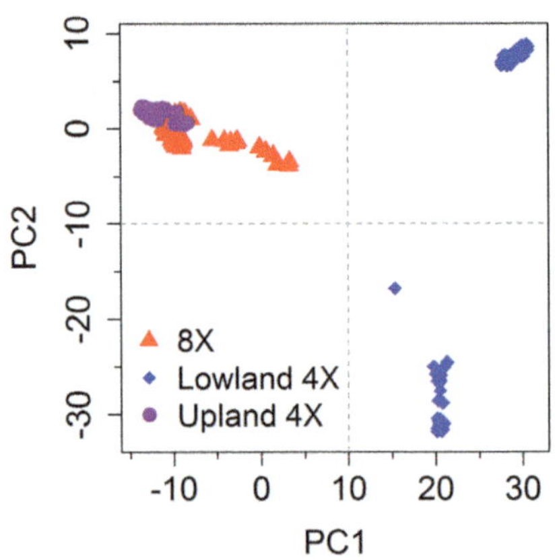

Figure 1. First two principal components of 16,669 single nucleotide polymorphisms separate plants by ploidy and ecotype. The first and second principal components (x- and y-axis, respectively) from a principal component analysis of 540 switchgrass closes separate octoploid (8X) and upland tetraploid (4X) accessions from lowland 4X accessions. The lowland accessions are also separated into two distinct clusters.

The performance of each model was assessed through ten-fold cross validation, as described in [46]. Briefly, the association panel was partitioned into ten equally-sized subgroups. Nine of the ten subgroups (i.e., the training set) were used to fit each prediction model while the remaining subgroup (the prediction set) was used to assess the correlation between the observed and predicted trait values. This process was repeated ten times, with each subgroup being the prediction set exactly once. For each trait, prediction accuracies were calculated by dividing the average Pearson's correlation coefficient across the ten folds by the square root of the repeatability [25]. To prevent inflated prediction accuracies arising from clones nested within populations, the data were partitioned for ten-fold cross validation so that none of the populations were in both the training and prediction sets. All phenotypic and genotypic data used to conduct this analysis are included in File S1.

Results

Extensive Phenotypic Variability among Clones

Substantial variation was observed for each of the seven morphological traits, with differences between minimum and maximum values of each trait ranging from 2-fold for anthesis date to 5.84-fold for standability (Table 2). In general, the majority of the morphological traits were highly correlated, with the strongest Pearson's correlation being between heading date and anthesis date ($r = 0.92$; Table S1). High correlations between leaf width, plant height, and total plant height were also observed (Pearson correlations ranging from $r = 0.53$ to $r = 0.88$). The average repeatability among the seven morphological traits was 0.86, with a range from 0.75 for plant height to 0.93 for anthesis date (Table 2). These high repeatabilities suggest that the majority of the phenotypic variation is attributable to genetic effects, and that genomic selection could be a useful breeding approach for morphological traits in switchgrass.

In comparison to the morphological traits, a greater range of fold differences between the minimum and maximum values of each trait was observed for the quality traits (Table 3). Although many of the correlations between the quality traits were generally

Table 4. Prediction accuracies of seven morphological traits in a switchgrass association panel.

Trait	Mean prediction accuracy	Prediction accuracy: RR-BLUP[a]	Prediction accuracy: Elastic net	Prediction accuracy: LASSO[b]
Anthesis Date	0.44	0.55 (0.21)	0.38 (0.23)	0.38 (0.23)
Heading Date	0.36	0.39 (0.14)	0.34 (0.20)	0.34 (0.19)
Standability	0.52	0.51 (0.27)	0.53 (0.19)	0.52 (0.19)
Leaf Length	0.40	0.55 (0.21)	0.34 (0.30)	0.32 (0.29)
Leaf Width	0.19	0.32 (0.29)	0.13 (0.24)	0.13 (0.24)
Plant Height	0.25	0.34 (0.18)	0.21 (0.26)	0.20 (0.26)
Total Plant Height	0.15	0.30 (0.28)	0.09 (0.19)	0.06 (0.19)

Standard errors of prediction accuracies are provided in parentheses.
Mean prediction accuracies were obtained by averaging results across ridge regression best linear unbiased prediction (RR-BLUP), least absolute shrinkage and selection operator (LASSO), and elastic net analysis.
[a]RR-BLUP, Ridge regression-best linear unbiased prediction
[b]LASSO, Least absolute shrinkage and selection operator

lower than those between the morphology traits, some individual quality traits were strongly correlated. For example, a Pearson correlation coefficient of $r = 0.95$ was observed between sucrose and total soluble carbohydrates (Table S2). Although lower than observed among the morphology traits, the estimated repeatabilities of the quality traits were sufficiently high enough to merit investigation into the utility of genomic selection.

First Two Principal Components of SNPs Sufficiently Account for Ploidy and Ecotype Differences

The first two PCs of the imputed GBS markers subdivided the plants used in this study into three genetically distinct subgroups (Figure 1). Specifically, the octoploid and upland tetraploid plants were clustered into one group, while the lowland tetraploid plants were subdivided into two distinct clusters. Collectively, these results suggest that the first two PCs of the SNPs capture a substantial amount of the major genetic differences between the ploidies and ecotypes of the plants included in our association panel. Moreover, these results justify our use of the first two PCs to factor out the SNP effects arising from overall population structure differences prior to conducting our genomic selection study.

Genomic Selection has Considerable Potential in Switchgrass

As expected, the observed prediction accuracies were similar across the three GS approaches (Tables 4–5). The predictive ability of the morphological traits were generally higher than those of the quality traits, with the highest prediction accuracies

Table 5. Prediction accuracies of 13 quality traits in a switchgrass association panel.

Trait	Mean prediction accuracy	Prediction accuracy: RR-BLUP[a]	Prediction accuracy: Elastic net	Prediction accuracy: LASSO[b]
Acid detergent lignin	0.34	0.41 (0.25)	0.31 (0.21)	0.30 (0.21)
Minerals (total ash)	−0.08	−0.09 (0.18)	−0.06 (0.13)	−0.10 (0.15)
Carbon	0.12	0.21 (0.25)	0.09 (0.27)	0.07 (0.27)
High Heating Value	0.22	0.26 (0.14)	0.21 (0.16)	0.20 (0.17)
Cell wall concentration	0.23	0.30 (0.23)	0.21 (0.19)	0.19 (0.18)
Ethanol/g dry forage	0.43	0.46 (0.20)	0.42 (0.20)	0.41 (0.21)
Etherified ferulates	0.22	0.27 (0.23)	0.20 (0.16)	0.19 (0.15)
In vitro dry matter digestibility	0.35	0.43 (0.27)	0.32 (0.25)	0.30 (0.25)
Pentose sugars release/g dry forage	0.06	0.15 (0.20)	0.03 (0.26)	0.01 (0.25)
Total soluable carbohydrates	0.30	0.39 (0.21)	0.26 (0.23)	0.25 (0.23)
Starch	0.08	0.19 (0.26)	0.03 (0.16)	0.03 (0.15)
Sucrose	0.32	0.44 (0.20)	0.26 (0.24)	0.25 (0.24)
Total sugar	0.04	0.16 (0.17)	0.00 (0.14)	−0.03 (0.17)

Standard errors of prediction accuracies are provided in parentheses.
Mean prediction accuracies were obtained by averaging results across ridge regression best linear unbiased prediction (RR-BLUP), least absolute shrinkage and selection operator (LASSO), and elastic net analysis.
[a]RR-BLUP, Ridge regression-best linear unbiased prediction
[b]LASSO, Least absolute shrinkage and selection operator

(averaged across all three tested GS models) obtained for standability (0.52), anthesis date (0.44), ethanol/g dry forage (0.43), leaf length (0.40), and heading date (0.36). We also obtained relatively strong positive Spearman's rank correlation coefficients between repeatabilities and unstandardized prediction accuracies for both the morphology ($r_{SP} = 0.61$) and the quality ($r_{SP} = 0.44$) traits. Consistent with the findings of previous studies (e.g., [25]), this result suggests that all three GS approaches successfully use the larger genetic contribution to phenotypic variability of the more heritable traits to obtain higher prediction accuracies.

Discussion

We evaluated the ability of three popular genomic selection approaches to predict the phenotypic values of seven morphological traits and 13 quality traits in a switchgrass association panel. Such a study is important because the successful application of genomic selection to switchgrass could significantly reduce the breeding cycle of this important biofuel feedstock. In general, our prediction accuracies are comparable to those reported in previous studies (e.g., [17] and [23]) that identified quantifiable advantages of genomic selection compared to traditional breeding programs. For perennial grasses such as switchgrass, one important quantity to consider is the expected genetic gain per unit of time. Because genomic selection does not require on-site phenotyping to identify accessions with superior trait values, multiple cycles of breeding could be completed with a genomic selection breeding program during the same amount of time required to achieve one cycle of breeding using traditional breeding programs [17,23]. For instance, it is demonstrated in [23] that it is possible for genomic selection breeding programs in maize and winter wheat to respectively achieve three cycles and two cycles of breeding during the same amount of time to complete one cycle of marker-assisted selection breeding. Moreover, the same study concluded that the expected genetic gain per year from a genomic selection breeding program will exceed that of a marker-assisted selection breeding program for traits with prediction accuracies as low as 0.20 in maize and 0.30 in winter wheat. Because many of our tested traits had prediction accuracies that exceed these thresholds, we believe that it is possible for similar advantages in expected genetic gain per unit of time to be achieved in switchgrass genomic selection breeding programs.

To our knowledge, the genetic architectures of the traits we evaluated are unknown in switchgrass. In particular, little is known about the number of genes underlying each trait. Therefore, we used three genomic selection models that have been hypothesized to perform differently under various genetic architectures. In general, we obtained similar prediction accuracies for all three genomic selection models. This result is especially apparent if we consider the standard errors of the prediction accuracies. Suppose we use the prediction accuracies and their standard errors from each genomic selection model (presented in Tables 4 and 5) to construct 95% confidence intervals. For each trait, the confidence intervals from the three genomic selection approaches overlap. This suggests that there are no discernible differences in prediction accuracies among the three genomic selection models. Indeed, this finding has been reported in other studies (e.g. [46]) and is theoretically justified in [47]. Nevertheless, we recommend repeating our study because we anticipate that the sampling, genotyping, and phenotypic resources available to the switchgrass community will continue to expand and improve, and it is imperative to confirm that these conclusions still hold given the new information we expect to obtain from these resources.

We observed a wide range of prediction accuracies across the traits. We suspect that this result was obtained because our markers provided incomplete coverage of the switchgrass genome, and it is likely that they tagged only a subset of the loci underlying the genetic sources of variation for each trait. It is therefore plausible that traits with higher prediction accuracies have causal loci that were in higher linkage disequilibrium with our markers compared to traits with lower prediction accuracies. Nonetheless, the prediction accuracies for many of the studied traits were suitably high enough to justify further investigation into the application of genomic selection to switchgrass breeding programs. Indeed one major factor contributing to our observed prediction accuracies was the availability of the *Panicum virgatum* genome sequence v1.1 reference genome. Because of this reference genome, we were able to use genotypic information from neighboring markers to impute missing genotypic data, and ultimately obtain substantial increases in the predictive abilities of our genomic selection models. Thus, we strongly recommend that switchgrass genomic selection breeding programs only use markers that are anchored to a reference genome. This will enable accurate imputation of missing data, and should ultimately result in genomic selection models with higher predictive abilities.

In general, lower prediction accuracies were obtained for the biomass quality traits relative to the morphology traits. We suspect that this result could have arisen from two different sources. In contrast to the morphological traits, the process of obtaining the quality traits was a lengthy procedure that was conducted in the laboratory. As such, it is possible that a greater amount of variability was introduced into the quality traits, which ultimately resulted in lower prediction accuracies. Factors such as spatial variability in the field, diurnal variation in biomass quality traits manifested by variation in sampling time, variation in grinding time and blade sharpness, and moisture content of the samples may all introduce variability to the measurement of biomass quality traits.

Our study suggests that the implementation of genomic selection approaches to switchgrass breeding programs will be highly beneficial. We believe that such an approach will revolutionize switchgrass breeding programs just as it has in at least four dairy cattle breeding programs [17]. Indeed, the large body of theoretical and empirical studies conducted in plant and animal species [15,22,25,46,48,49] suggests that genomic selection is a cost-effective approach that will substantially speed up breeding cycles, and we expect that these advantages will significantly benefit the development of switchgrass as a biofuel feedstock. As high as the prediction accuracies were in our study, we expect them to increase as more attention is focused on the characterization and exploitation of switchgrass phenotypic and genotypic resources. Specifically, we believe that increased prediction accuracies will arise from improvements to the switchgrass reference genome, improvements in phenotyping techniques, and the development of markers with higher levels of genomic coverage and density.

Supporting Information

Table S1 Correlation matrix for untransformed BLUPs of the seven morphological traits. Pearson correlation coefficients are presented in the upper triangle, and the P-values for the significance of associations are in the lower triangle.

Table S2 Correlation matrix for untransformed BLUPs of the 13 quality traits. Pearson correlation coefficients are presented in the upper triangle, and the P-values for the significance of associations are in the lower triangle.

File S1 Data files used to conduct analysis. All files used to conduct the genomic selection analysis are included in this file.

Acknowledgments

We would like to thank the propagators on our team, Ken Paddock (Cornell) and Nick Baker (USDA-ARS, Madison, Wisconsin), as well as the legions of undergraduate students who assisted in growing and phenotyping switchgrass, and preparing samples for both phenotypic and genotypic analyses. Nick Lepak (USDA-ARS, Ithaca, New York) provided valuable support at the Ithaca field site and contributed to the GDD calculations and selection of the appropriate formula. We also thank the Cornell University Genomics Core Facility and the staff at the Institute for Genomic Diversity (Cornell) for their support: they made the development of our GBS markers possible. AEL acknowledges the USDA-ARS for providing a Postdoctoral Fellowship to pursue this research. Prepublication data for the *Panicum virgatum* genome sequence v1.1 reference genome were provided by the Department of Energy Joint Genome Institute.

Author Contributions

Conceived and designed the experiments: ESB DEC JHC MDC. Performed the experiments: FL DEC JHC MDC. Analyzed the data: AEL FL. Contributed reagents/materials/analysis tools: FL DEC MDC JHC. Wrote the paper: AEL MDC DEC.

References

1. Sanderson MA, Adler PR, Boateng AA, Casler MD, Sarath G (2006) Switchgrass as a biofuels feedstock in the USA. Can J Plant Sci 86: 1315–1325.
2. Stubbendieck JL, Hatch SL, Butterfield CH (1992) North American range plants: U of Nebraska Press.
3. Casler M, Vogel K, Taliaferro C, Ehlke N, Berdahl J, et al. (2007) Latitudinal and longitudinal adaptation of switchgrass populations. Crop Sci 47: 2249–2260.
4. Casler M, Vogel K, Taliaferro C, Wynia R (2004) Latitudinal adaptation of switchgrass populations. Crop Sci 44: 293–303.
5. Casler MD (2012) Switchgrass breeding, genetics, and genomics. Switchgrass: Springer. pp. 29–53.
6. Porter CL Jr (1966) An analysis of variation between upland and lowland switchgrass, Panicum virgatum L., in central Oklahoma. Ecology: 980–992.
7. Zhang Y, Zalapa J, Jakubowski AR, Price DL, Acharya A, et al. (2011) Natural hybrids and gene flow between upland and lowland switchgrass. Crop Sci 51: 2626–2641.
8. Zhang Y, Zalapa JE, Jakubowski AR, Price DL, Acharya A, et al. (2011) Post-glacial evolution of Panicum virgatum: centers of diversity and gene pools revealed by SSR markers and cpDNA sequences. Genetica 139: 933–948.
9. Martinez-Reyna J, Vogel K (2002) Incompatibility systems in switchgrass. Crop Sci 42: 1800–1805.
10. Cathey HM (1990) USDA plant hardiness zone map.
11. Vogel KP, Mitchell R, Casler M, Sarath G (2014) Registration of 'Liberty' switchgrass. JPlant Regist.
12. Simeão Resende RM, Casler MD, Vilela de Resende MD (2014) Genomic selection in forage breeding: accuracy and methods. Crop Sci 54: 143–156.
13. Perrin R, Vogel K, Schmer M, Mitchell R (2008) Farm-scale production cost of switchgrass for biomass. Bioenerg Res 1: 91–97.
14. Meuwissen T, Hayes B, Goddard M (2001) Prediction of total genetic value using genome-wide dense marker maps. Genetics 157: 1819–1829.
15. de los Campos G, Hickey JM, Pong-Wong R, Daetwyler HD, Calus MP (2013) Whole-genome regression and prediction methods applied to plant and animal breeding. Genetics 193: 327–345.
16. Schaeffer L (2006) Strategy for applying genome-wide selection in dairy cattle. J Anim Breed Genet 123: 218–223.
17. Hayes B, Bowman P, Chamberlain A, Goddard M (2009) Invited review: Genomic selection in dairy cattle: Progress and challenges. J Dairy Sci 92: 433–443.
18. VanRaden P, Van Tassell C, Wiggans G, Sonstegard T, Schnabel R, et al. (2009) Invited Review: Reliability of genomic predictions for North American Holstein bulls. J Dairy Sci 92: 16–24.
19. Asoro FG, Newell MA, Beavis WD, Scott MP, Jannink JL (2011) Accuracy and training population design for genomic selection on quantitative traits in elite North American oats. Plant Gen 4: 132–144.
20. Zhao Y, Gowda M, Liu W, Würschum T, Maurer HP, et al. (2012) Accuracy of genomic selection in European maize elite breeding populations. Theor Appl Genet 124: 769–776.
21. Riedelsheimer C, Czedik-Eysenberg A, Grieder C, Lisec J, Technow F, et al. (2012) Genomic and metabolic prediction of complex heterotic traits in hybrid maize. Nat Genet 44: 217–220.
22. Heffner EL, Sorrells ME, Jannink JL (2009) Genomic selection for crop improvement. Crop Sci 49: 1–12.
23. Heffner EL, Lorenz AJ, Jannink JL, Sorrells ME (2010) Plant breeding with genomic selection: gain per unit time and cost. Crop Sci 50: 1681–1690.
24. Resende MD, Resende MF, Sansaloni CP, Petroli CD, Missiaggia AA, et al. (2012) Genomic selection for growth and wood quality in Eucalyptus: capturing the missing heritability and accelerating breeding for complex traits in forest trees. New Phytol 194: 116–128.
25. Resende MF, Muñoz P, Resende MD, Garrick DJ, Fernando RL, et al. (2012) Accuracy of genomic selection methods in a standard data set of loblolly pine (Pinus taeda L.). Genetics 190: 1503–1510.
26. Resende M, Munoz P, Acosta J, Peter G, Davis J, et al. (2012) Accelerating the domestication of trees using genomic selection: accuracy of prediction models across ages and environments. New Phytologist 193: 617–624.
27. Lu F, Lipka AE, Glaubitz J, Elshire R, Cherney JH, et al. (2013) Switchgrass genomic diversity, ploidy, and evolution: novel insights from a network-based SNP discovery protocol. PLoS Genet 9: e1003215.
28. Elshire RJ, Glaubitz JC, Sun Q, Poland JA, Kawamoto K, et al. (2011) A robust, simple genotyping-by-sequencing (GBS) approach for high diversity species. PLoS One 6: e19379.
29. Joint Genome Institute (2014) Phytozome 10. Panicum virgatum v1.1 DOE-JGI. Available: http://phytozome.jgi.doe.gov.
30. Buckler Lab for Maize Genetics and Diversity (nd) Buckler lab phenotyping tools. Available: http://www.maizegenetics.net/phenotyping-tools.
31. Vogel KP, Dien BS, Jung HG, Casler MD, Masterson SD, et al. (2011) Quantifying actual and theoretical ethanol yields for switchgrass strains using NIRS analyses. Bioenerg Res 4: 96–110.
32. Kutner MH, Nachtsheim CJ, Neter J, Li W (2004) Applied Linear Statistical Models. Boston, MA: McGraw-Hill.
33. SAS Institute (2012) The SAS system for Windows. Release 9.3. SAS Institute. Institute SAS: Sas Inst.
34. Gilmour AR, Gogel B, Cullis B, Thompson R, Butler D, et al. (2009) ASReml user guide release 3.0. VSN International Ltd, Hemel Hempstead, UK.
35. Chandler K, Lipka AE, Owens BF, Li H, Buckler ES, et al. (2013) Genetic Analysis of Visually Scored Orange Kernel Color in Maize. Crop Sci 53: 189–200.
36. Holland JB, Nyquist WE, Cervantes-Martínez CT (2003) Estimating and interpreting heritability for plant breeding: An update. Plant Breed Rev 22: 9–112.
37. Hung HY, Browne C, Guill K, Coles N, Eller M, et al. (2012) The relationship between parental genetic or phenotypic divergence and progeny variation in the maize nested association mapping population. Heredity (Edinb) 108: 490–499.
38. Box GEP, Cox DR (1964) An Analysis of Transformations. J Roy Stat Soc B Met 26: 211–252.
39. Lipka AE, Gore MA, Magallanes-Lundback M, Mesberg A, Lin H, et al. (2013) Genome-wide association study and pathway-level analysis of tocochromanol levels in maize grain. G3: Genes, Genomes, Genet 3: 1287–1299.
40. Scheet P, Stephens M (2006) A fast and flexible statistical model for large-scale population genotype data: applications to inferring missing genotypes and haplotypic phase. Am J Hum Genet 78: 629–644.
41. Tibshirani R (1996) Regression shrinkage and selection via the lasso. J Roy Stat Soc B Met: 267–288.
42. Zou H, Hastie T (2005) Regularization and variable selection via the elastic net. J Roy Stat Soc B Met 67: 301–320.
43. Endelman JB (2011) Ridge regression and other kernels for genomic selection with R package rrBLUP. Plant Gen 4: 250–255.
44. Team RC (2012) R: A language and environment for statistical computing.
45. Friedman J, Hastie T, Tibshirani R (2010) Regularization paths for generalized linear models via coordinate descent. J Stat Softw 33: 1.
46. Heslot N, Yang HP, Sorrells ME, Jannink JL (2012) Genomic selection in plant breeding: a comparison of models. Crop Sci 52: 146–160.
47. Gianola D (2013) Priors in whole-genome regression: the Bayesian alphabet returns. Genetics 194: 573–596.
48. Jannink J, Lorenz AJ, Iwata H (2010) Genomic selection in plant breeding: from theory to practice. Brief Funct Genomics 9.2 166–177.
49. Wolc A, Stricker C, Arango J, Settar P, Fulton JE, et al. (2011) Breeding value prediction for production traits in layer chickens using pedigree or genomic relationships in a reduced animal model. Genet Sel Evol 43.

20

Gene Discovery and Molecular Marker Development, Based on High-Throughput Transcript Sequencing of *Paspalum dilatatum* Poir

Andrea Giordano[1,2,3,4], Noel O. I. Cogan[1,2,3], Sukhjiwan Kaur[1], Michelle Drayton[1,2,3], Aidyn Mouradov[1,2,3¤], Stephen Panter[1,2,3], Gustavo E. Schrauf[5], John G. Mason[1,3,4], German C. Spangenberg[1,2,3,4*]

1 Department of Environment and Primary Industries, AgriBio, Centre for AgriBioscience, Bundoora, Victoria, Australia, **2** Molecular Plant Breeding Cooperative Research Centre, Bundoora, Victoria, Australia, **3** Dairy Futures Cooperative Research Centre, Bundoora, Victoria, Australia, **4** La Trobe University, Bundoora, Victoria, Australia, **5** Facultad de Agronomia, Universidad de Buenos Aires, Buenos Aires, Argentina

Abstract

Background: *Paspalum dilatatum* Poir. (common name dallisgrass) is a native grass species of South America, with special relevance to dairy and red meat production. *P. dilatatum* exhibits higher forage quality than other C4 forage grasses and is tolerant to frost and water stress. This species is predominantly cultivated in an apomictic monoculture, with an inherent high risk that biotic and abiotic stresses could potentially devastate productivity. Therefore, advanced breeding strategies that characterise and use available genetic diversity, or assess germplasm collections effectively are required to deliver advanced cultivars for production systems. However, there are limited genomic resources available for this forage grass species.

Results: Transcriptome sequencing using second-generation sequencing platforms has been employed using pooled RNA from different tissues (stems, roots, leaves and inflorescences) at the final reproductive stage of *P. dilatatum* cultivar Primo. A total of 324,695 sequence reads were obtained, corresponding to c. 102 Mbp. The sequences were assembled, generating 20,169 contigs of a combined length of 9,336,138 nucleotides. The contigs were BLAST analysed against the fully sequenced grass species of *Oryza sativa* subsp. *japonica*, *Brachypodium distachyon*, the closely related *Sorghum bicolor* and foxtail millet (*Setaria italica*) genomes as well as against the UniRef 90 protein database allowing a comprehensive gene ontology analysis to be performed. The contigs generated from the transcript sequencing were also analysed for the presence of simple sequence repeats (SSRs). A total of 2,339 SSR motifs were identified within 1,989 contigs and corresponding primer pairs were designed. Empirical validation of a cohort of 96 SSRs was performed, with 34% being polymorphic between sexual and apomictic biotypes.

Conclusions: The development of genetic and genomic resources for *P. dilatatum* will contribute to gene discovery and expression studies. Association of gene function with agronomic traits will significantly enable molecular breeding and advance germplasm enhancement.

Editor: Mark Gijzen, Agriculture and Agri-Food Canada, Canada

Funding: This work was supported by funding from the Victorian Department of Primary Industries, the Dairy Futures Cooperative Research Centre and the Molecular Plant Breeding Cooperative Research Centre. The funders had no role in study design, data collection and analysis, decision to publish, or preparation of the manuscript.

Competing Interests: The authors have declared that no competing interests exist.

* E-mail: german.spangenberg@depi.vic.gov.au

¤ Current address: Science, Engineering & Health, RMIT University, Bundoora, Victoria, Australia

Introduction

Paspalum dilatatum Poir. (common name dallisgrass) is a highly productive C4 grass with a wide distribution within temperate-warm regions and a growing season from late spring to late summer. *P. dilatatum* belongs to the Panicoideae subfamily within the PACMAD clade (Panicoideae, Arundinoideae, Chloridoideae, Micrairoideae, Aristidoideae and Danthonioideae), which contains all C4 grasses. It is believed that the PACMAD clade separated from the lineage that generated the more widely studied Pooideae and Bambusoideae grasses some 40–50 million years ago (Mya) [1]. Recently the C4 grasses have attracted considerable interest for bioenergy production and several of the key species under investigation (*Miscanthus giganteus*; *Panicum virgatum*; *Pennisetum purpuretum*) belong to the Paniceae clade and have probably diverged from the *P. dilatatum* clade by only c. 20 Mya. *P. dilatatum* is native to South America with special relevance to dairy and beef production [2,3] as its forage quality is higher than that of other C4 forage grasses [4,5,6], and it exhibits tolerance of frost [7,8] and water stress [9,10].

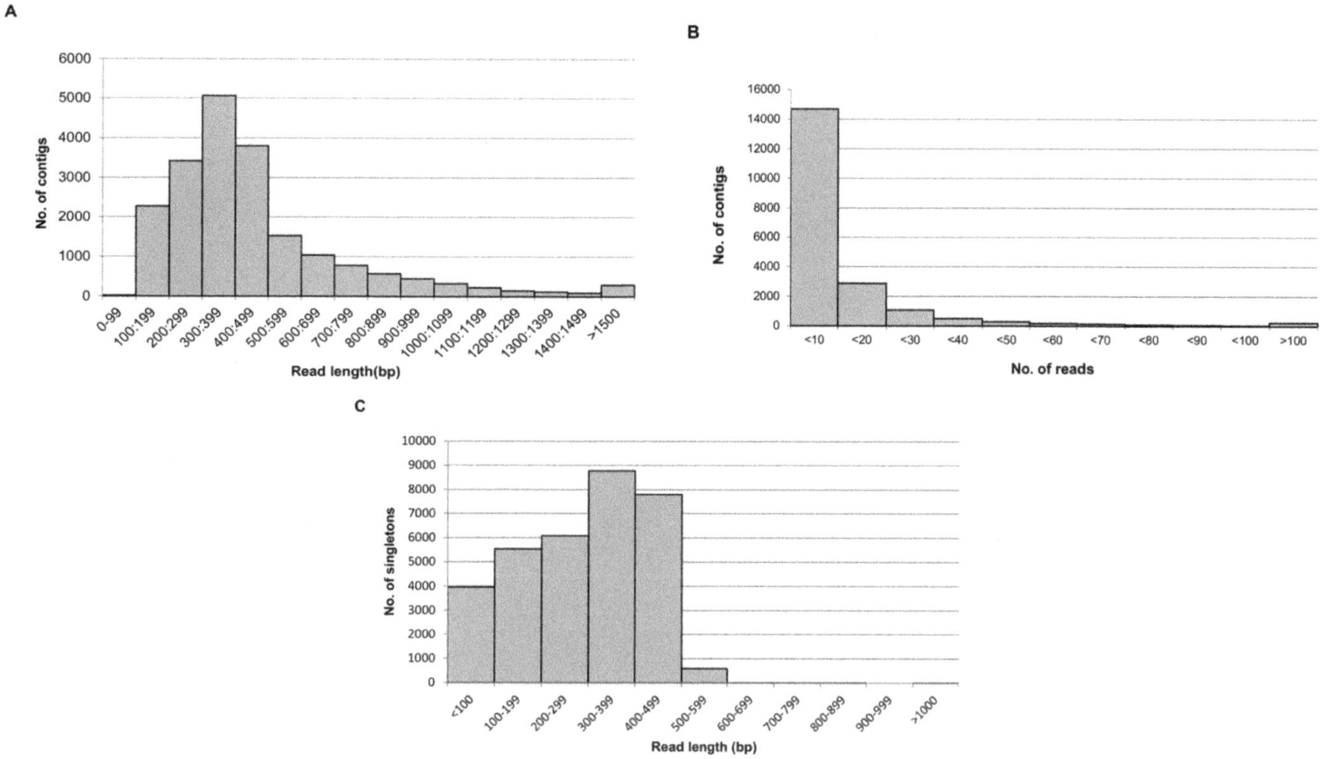

Figure 1. Frequency histograms showing the distribution of contigs compared to sequence read length (A), the distribution of the number of reads per contig (B), and distribution of singletons compared to read length (bp) (C).

There are several sexual and apomictic cytotypes (4×, 5×, 6×, 7×) and also biotypes of *P. dilatatum* that have been identified due to their morphological and distributional differences [11]. The predominant biotype found has purple anthers and apomictic reproduction (2n = 50), which is widespread in South America and naturalised in other parts of the world including China, USA, India, Japan, New Zealand and South Africa [12]. The sexual biotypes are typically characterised by yellow anthers. The genome constitution of the different ploidy levels has been described as IIJJ for tetraploids and IIJJX for the more common pentaploid biotype [13,14]. The I genome has been identified in the diploid *P. intermedium* [15], which is believed to be the progenitor of the original genome donor. Cytogenic and phylogenic studies suggest that the donor of the J genome is believed to be *P. juergensii* [16]. The donor of the X genome is unknown, though the apomictic character is thought to have originated from this genome. The genome size has been estimated in different biotypes of *P. dilatatum*, with the tetraploid and pentaploid genome contents being 2.43 pg/2C and 2.96 pg/2C per nucleus, respectively [11].

P. dilatatum is likely to have more widespread use in livestock grazing systems under predicted climate change scenarios due to its superior drought tolerance and water use efficiency, in comparison to commonly-used C3 grasses. However, conventional breeding methods for improvement of *P. dilatatum* are commonly limited by its asexual reproduction system [17]. In apomictic reproduction, seeds are produced without fertilisation, generating uniform progeny that are genetically identical to their parents, restricting the genetic variation available for conventional breeding programs. As a result of this apomictic character, large clonal monocultures are often maintained in production zones, potentially jeopardising the feed-base of the grazing industries if adverse biotic or abiotic pressures emerge. The common type of *P. dilatatum* (apomictic and pentaploid) has a documented susceptibility to *Claviceps paspali*, an ergot fungus, which is the causal agent of "*Paspalum* staggers" in cattle that graze infected pasture [18]. To address these issues, advanced novel breeding strategies that employ genetic diversity must be developed and enabled to ensure the production of advanced cultivars to generate options for production systems.

Previous efforts to improve dallisgrass through breeding include the use of somaclonal variation [19,20,21] and interspecific hybridisation [22,23]. Advanced molecular breeding techniques may be applied to the sexual cytotype or potentially to hybrids produced with the apomictic biotypes as pollen donors or with other closely related *Paspalum* species. Molecular markers are a valuable tool to identify and quantify genetic diversity within and between species, populations and available germplasm. These genetic resources can also link genotypes with economically important traits and assist breeders with germplasm selection and potential crossing schemes [24].

Genomic and transcriptomic resources for *P. dilatatum* are limited to only 77 nucleotide and 37 protein sequences. More broadly in the genus *Paspalum* the number of nucleotide sequences that are available is only 3,472 (NCBI 05/11/2013). Therefore, there is a significant need to increase the genomic resources for this species. Recently, second generation sequencing technology has been applied in a wide variety of areas including whole genome resequencing and *de novo* sequencing; as well as transcriptome analysis [25]. The Roche 454 Life Sciences GS FLX sequencing platform has been widely used for transcriptomic studies in non-model organisms (a full list of publications is directly available from http://454.com/publications/index.asp) [26,27,28]. Complete genome sequences have been generated for only a few grasses,

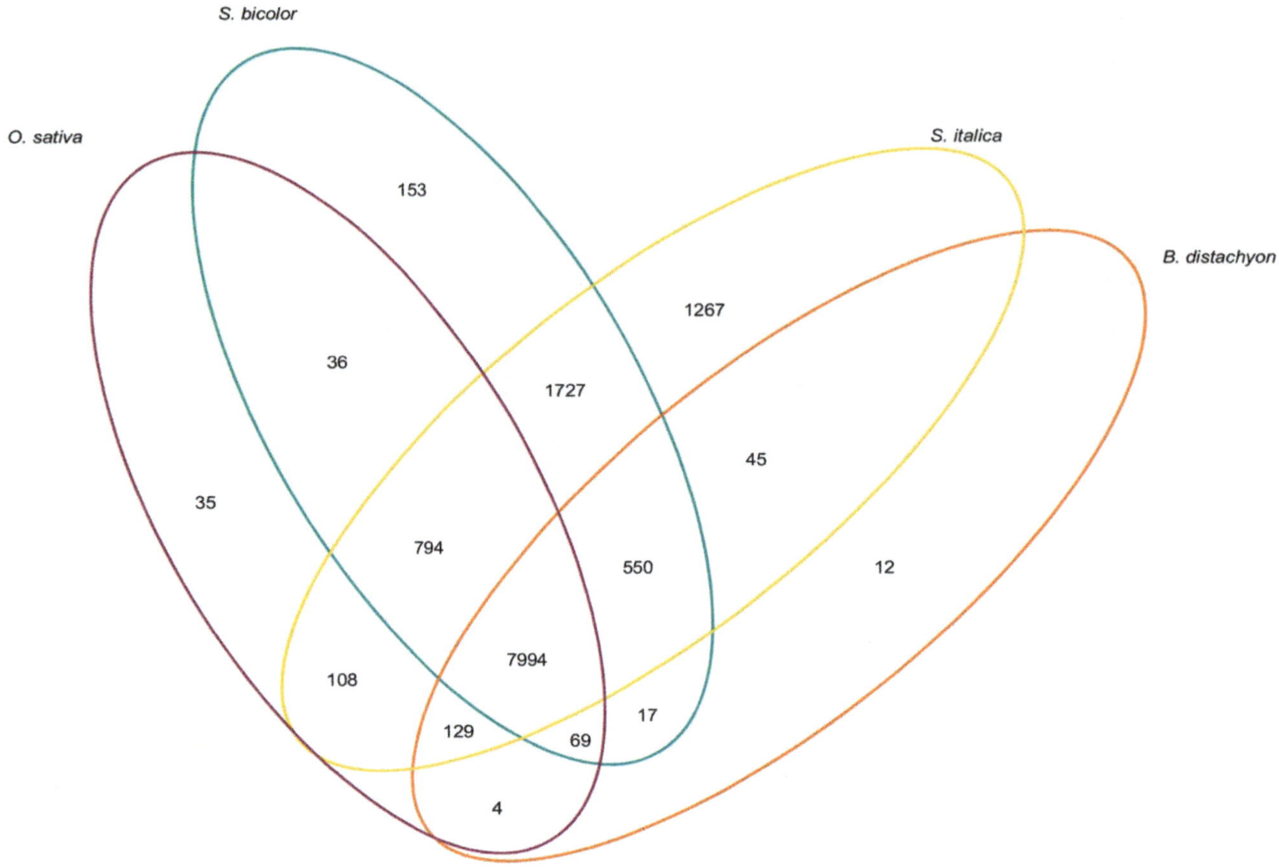

Figure 2. Sequence analysis of the generated *P. dilatatum* contigs compared to the reference grass genomes. The *P. dilatatum* contigs were compared using BLASTn with an e value threshold of 10^{-10} and significant matches were identified and totaled and presented as a Venn diagram.

as despite the advances made, undertaking to produce a whole genome sequence of a higher plant is still a significant task that requires considerable resources. The genome sequences of particular interest when studying *Paspalum* species are *Sorghum bicolor* L. [29] and foxtail millet, *Setaria italica* [30]. Both of these species are likely to have diverged from the *Paspalum* clade c. 10 Mya and represent two highly relevant genomes for enhancing any generated data. These two related reference genomes could assist in sequence annotation of *P. dilatatum*. Due to the complexities of undertaking complete sequencing of genomes from higher plants, transcriptomic sequencing of *P. dilatatum* would generate an initial resource of gene sequences as well as identifying valuable molecular markers such as EST-Simple Sequence Repeats (SSRs) for molecular breeding to improve the productivity and quality of this forage species.

The use of the second-generation sequencing technology is a cost-effective method to obtain large-scale EST sequences of non-model species. For the generation of a resource of largely full length sequences derived from transcriptomics, the longer read length of the GS FLX Titanium chemistry is ideally suited to generate such a data set. The GS FLX platform is the most widely used technology for the identification of SSR motifs and the development of molecular markers in plants [31]. EST-derived simple sequence repeat (SSR) molecular markers have the advantage that by definition they are genic and therefore can be potentially associated with functional regions, avoiding the non-coding and repetitive DNA in species with large genomes. EST-based molecular markers might also be linked to genes controlling important agronomic traits that could assist breeding programs and have been used to improve crop species [32,33,34]. Furthermore, due to their conserved location, EST-SSR can be transferable between related species [34]. A recent study has developed a cohort of 17 genomic SSR markers for *Paspalum atratum* Swallen and *Paspalum notatum* Flüggé. These SSRs were evaluated broadly across the Paspalum genus allowing species variability to be distinguished [35].

This study provides the first *de novo* assembly and annotation of a cDNA transcriptome dataset for *P. dilatatum*, derived from multiple tissues of a single plant at the reproductive stage of its life cycle. An extensive collection of EST-SSR molecular markers have subsequently been identified and a subset validated. This has generated a valuable resource of gene sequences and molecular markers for future breeding programs for this important forage species.

Materials and Methods

Plant material and growth conditions

P. dilatatum cv. Primo, a tetraploid sexually-reproductive cultivar and *P. dilatatum* cv. Relincho, a pentaploid apomictic cultivar, were kindly provided by Gustavo Schrauf (University of Buenos Aires, Argentina) and used as source plant material for all experiments. Individual plants were grown under glasshouse conditions (21°C, 14 h photoperiod/14°C, 10 h dark period). Plant tissues were harvested for RNA isolation from a single clone of *P. dilatatum* cv.

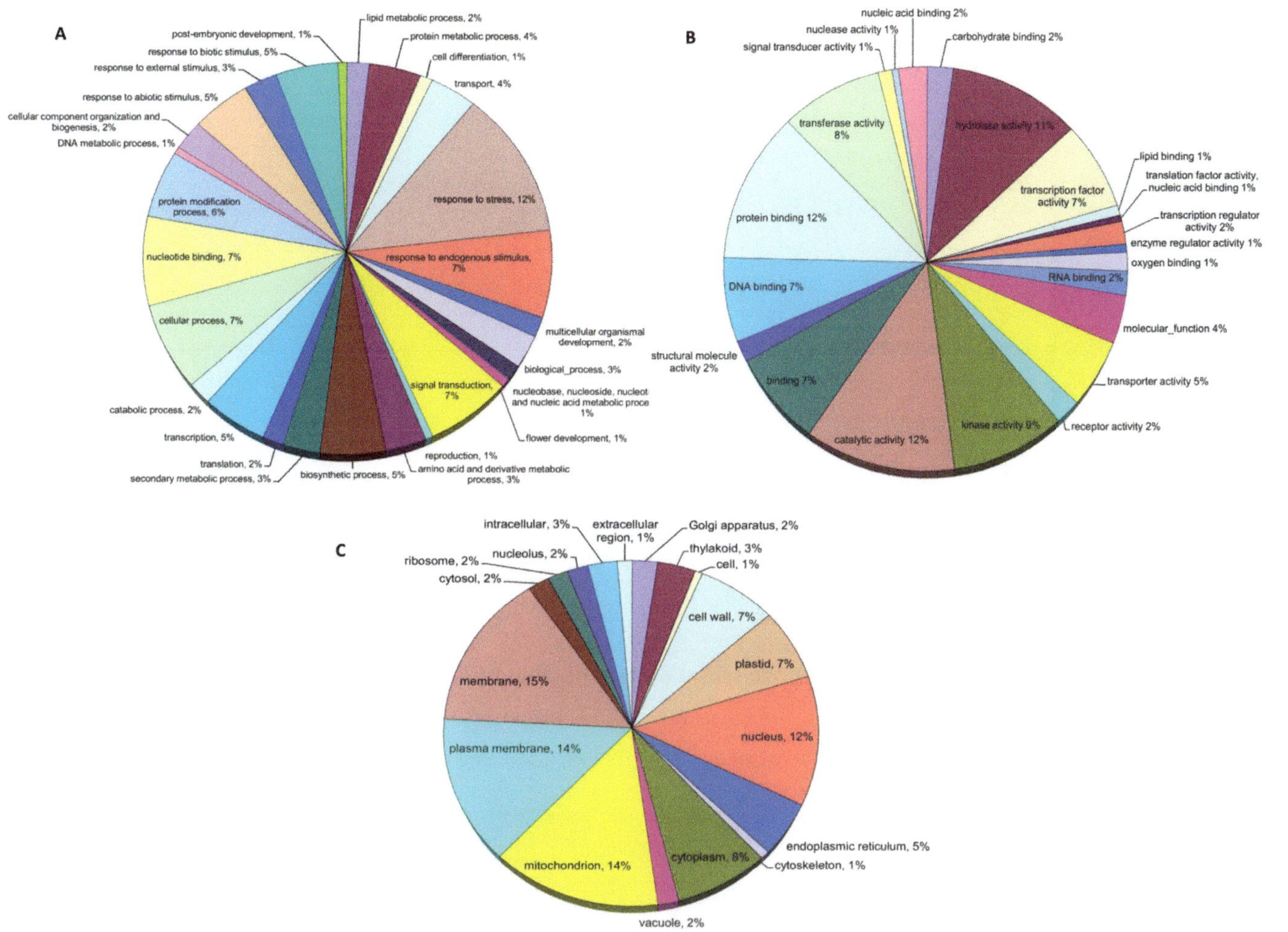

Figure 3. Assignment of putative function and gene ontology (GO) annotations to *P. dilatatum* sequence contigs based on Biological process classification (A), Molecular function (B) and Cellular components (C).

Primo at the late reproductive stage, from leaves, stems, roots and mature inflorescences.

RNA extraction and cDNA synthesis

Total RNA was isolated using RNeasy® Plant Mini Kit (Qiagen, Hilden, Germany) following manufacturer's instructions. RNA integrity was visually assessed on a 1% (w/v) agarose gel stained with SYBR® Safe DNA gel stain (Life Technologies, Carlsbad, USA), before proceeding with cDNA synthesis. A total of 250 ng of total RNA from each RNA extraction was combined to provide a total of 1 μg for use in cDNA synthesis using the SMART™ cDNA Synthesis kit (Clontech Laboratories, Mountain View, USA), according to the manufacturer's instructions. Messenger RNA (mRNA) was reverse-transcribed using SMARTScribe™ Reverse Transcriptase and SMART IV™ oligonucleotides. A modified poly-T primer sequence (5′-AAGCAGTGGTAT-CAACGCAGAGTCGCAGTCGGTACTTTTTTTCTTTTTTTV-3′) was used instead of the CDS-II primer to reduce sequencing problems associated with the poly-A tails [36].

EST sequence generation

Approximately 5 μg of the amplified cDNA library was sheared by nebulisation at 200 kPa for 2 min. The concentration and integrity of the nebulised cDNA sample was determined using a Bioanalyzer 2100, with a DNA 12000 Labchip (Agilent, Santa Clara, USA) according to manufacturer's instructions. The GS-FLX titanium shotgun sequence library was then prepared following the manufacturer's instructions (Roche Diagnostics, Basel, Switzerland). The quality of sequencing library was then assessed using a Bioanalyzer 2100, fitted with an RNA 6000 Pico Lab Chip (Agilent) according to the manufacturer's instructions. Quantification of the sequence library was performed using quantitative real-time PCR. Emulsions were prepared using the Large Volume Emulsion PCR Kit (Roche Diagnostics). Finally, enriched beads were loaded onto one half of a picotitre plate for sequencing following the manufacturer's protocol.

Assembly and annotation

Primary sequence output has been deposited in the sequence read archive of GenBank (accession number SRR1012849). Adapter and poly A sequences were trimmed from all sequence reads and the sequences were then *de novo* assembled, both procedures were performed using Newbler v 2.0.01.14 software (Roche Diagnostics). The coding sequences (CDS) from *Oryza sativa japonica* v6.1, *Sorghum bicolour* v1.0, *Brachypodium distachyon* v1.0 and *Setaria italica* v2.1 were downloaded from the GenBank and Phytozome websites (www.phytozome.net) and converted into a local custom BLAST database. The *P. dilatatum* sequence contigs

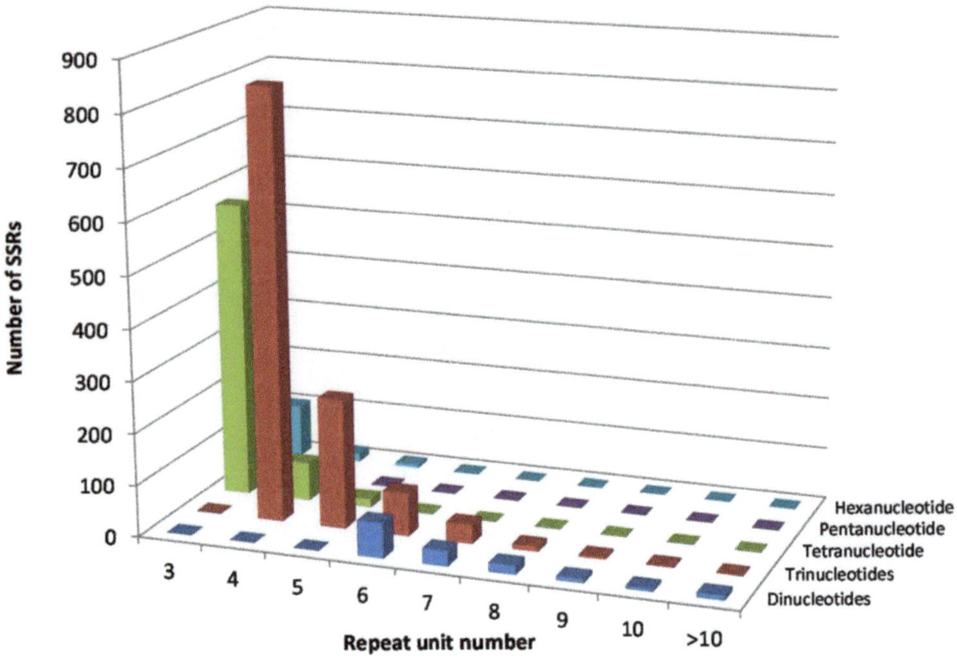

Figure 4. Distribution and frequency of different SSR repeat motifs identified from the *P. dilatatum* contig dataset.

were then used as the query sequences in a BLASTn analysis with a threshold E-value of 10^{-10}. Only the most significant match was recorded for each sequence contig.

The UniRef 90 (www.ebi.ac.uk/uniref/) protein database was downloaded and converted into a custom BLAST database. The *P. dilatatum* sequence contigs were used as queries using BLASTx analysis with a threshold E-value of 10^{-10} and the most significant matches were recorded. Gene Ontology (GO) classification was assigned for each contig that matched the *Oryza sativa* protein database.This was performed by selecting the rice gene IDs that were identified from the initial BLASTn analysis and extracting the GO terms for that subset of genes.

EST-SSR Identification and validation

Identification of di-, tri-, tetra-, penta- and hexanucleotide SSR loci, with minimum repeat numbers of 6, 4, 3, 3, and 3, respectively, and design of corresponding primers was performed using Batch Primer 3 software (http://probes.pw.usda.gov/batchprimer3/). The parameters for primer design were as follows: primer length range = 18 to 23 nucleotides (21 optimum); optimum annealing temperature = 55°C; GC content 30–70% (50% optimum).

A collection of 96 primer pairs were synthesized for empirical testing. Forward primers were synthesised with the addition of the M13 sequence to the 5′ end, to enable fluorescent tail addition through the PCR amplification process [37]. Genomic DNA was extracted from fresh leaf tissue from a collection of 49 genotypes from *P. diltatum* cv. Primo and one sample from *P. dilatatum* cv. Relincho using the DNeasy® 96 Plant Kit (QIAGEN) according to the manufacturers' instructions. PCR reactions were performed in a 12 μL volume containing Immolase PCR buffer, 1.5 mM $MgCl_2$, 200 μM dNTPs, 0.016 μM of M13 FAM universal primer, 0.041 μM of M13 tailed forward primer and 0.16 μM of reverse primer, 0.25 units of Immolase DNA polymerase (Bioline, London, UK) and approximately 10 ng of *P. dilatatum* genomic DNA. Cycling conditions were as follows: 95°C for 10 min, followed by 35 cycles of 95°C for 30 sec, 55°C for 30 sec and 72°C for 30 sec and a final elongation step of 72°C for 10 min, performed using an Applied Biosystems Geneamp 9700 thermal cycler (Life Technologies). Amplification products were diluted 40-fold with the addition of water, and 2 μL of product was then combined with 8.95 μL of Hi-Di™ formamide (Life Technologies) and 0.05 μL of GeneScan™ 500 LIZ® Size Standard (Life Technologies). PCR products were processed using an ABI3730xl DNA analyser and amplification product sizes were determined using GeneMapper® v 3.7 software (Life Technologies).

Results

EST sequencing and *de novo* assembly

After initial sequence base calling and quality assessment had been performed, a total of 324,695 sequence reads, generating a total of 102,539,862 nucleotides were obtained. The average and modal sequence lengths were 338 and 430 nucleotides, respectively.

After further quality filtering by removing adapter and primer sequences, reads were *de novo* assembled resulting in 20,169 contigs of a combined length of 9,336,138 bp and 32,751 singletons (Fasta Files S1, S2). The contig length varied from 104 to 7,877 bp. The majority of the contigs varied in length from 200–499 bp (60.9%) with the largest proportion being 300–399 bp (25%; Figure 1A). A total of 5,580 contigs (25%) were identified with a length greater than 500 nucleotides. The majority of the contigs (72.8%) were derived from less than 10 reads (Figure 1B).

The minimum accepted sequence length from the unassembled singletons was 50 bp. The average sequence length of the singletons was 374 bp and the largest proportion of the singletons (26.7%) were between 300–399 bp in length and only 25.6% were longer than 400 bp. There were a total of 17 singleton reads (1.8%) that were longer than 600 bp and represent sequences with long repeats of single nucleotides (Figure 1C).

Table 1. Details of SSR molecular markers from *P. dilatatum* validated as polymorphic within the current study.

SSR Id	Contig	Predicted Product size	SSR Motif	Motif Lenght	Forward Primer	Reverse Primer	Allele size range	Polymorphic in cv. Primo	Number of alleles in cv.Primo	Number of alleles in cv. Relincho
Pd0061	contig00359	128	AGGAGC	6	GCAGCTTCATCTCTATGTTTC	CTACATCAGGAGTAAGCGATG	139–149	No	1	4
Pd0228	contig01406	177	GGT	3	CTTCATTCCTTCAGATCCACT	TACACTGCATACACCGAAGA	153–205	No	4	5
Pd0240	contig01488	227	CCGA	4	AAACTGATGCTCCTCAACC	GTACAGCGGGTTGTTCTG	333–377	No	2	3
Pd0297	contig01864	157	CTGTGG	6	GTCCTCCTAGCTGTTGAAGAT	AGAAAAGACCTAAGCATGGAC	169–175	No	2	3
Pd0298	contig01883	126	CGC	3	CAAGATCCCTCCCTACTCC	ACATAGTCCTCCCCAACC	143–181	No	1	2
Pd0410	contig02637	165	TCG	3	GTCACCTCCTTATAGCCCTTA	CAAATCCAAAGACGATGC	180–187	No	1	2
Pd0630	contig04101	170	GCGA	4	AAGAGCTTCCCAACATCC	GATATTTCGGAGGAAGAAGAG	440–465	No	2	3
Pd0741	contig04909	181	TCT	3	GTGGATGGATGGATGGAT	TTCAGACACACTGCATCAAC	182–204	Yes	4	3
Pd0778	contig05150	184	GCGGAG	6	CTGAAGCAAGGGTAGGAGAT	ATCTCCTCCTGAGTTTGTACC	196–202	Yes	2	2
Pd0806	contig05407	157	GGA	3	AGGAGTGAGAGGAAGAACAAC	GATGACACAAGCAGTAGCAG	175–189	No	1	2
Pd1298	contig09623	154	GAG	3	TCCACCAAGAGCTCACAT	GCATCTGTCGTAGACTCGTAG	118–225	No	3	5
Pd1302	contig09667	183	CCTGTG	6	CAAGGTCATGTTCAACCTCTA	ACAGCACAAGAGCTAAAACC	175–236	No	2	3
Pd1357	contig10112	176	AGC	3	ATCATTTTTGCTGCTCTCTC	AGTATGGGAGGACGTTGTT	186–194	Yes	2	3
Pd1389	contig10385	159	CGG	3	CTCAGCTCCTAGTCGTCCTA	ACATTGCGGATAAATCTGAC	170–177	Yes	2	1
Pd1437	contig10830	150	GCA	3	CAACTTCACCAACAGCAAC	TGCTTATTCTTTCTGCGTTC	135–174	No	2	3
Pd1480	contig11038	167	CGG	3	ATCAGAAAGAGGGATTTTGC	AAATCAGGGCAAAGAGGTAG	182–185	No	1	2
Pd1488	contig11123	157	AAGAGG	6	CTCCTGGCTTCCTAGATTTAC	CAATTGAAACATGGTCTGC	173–189	Yes	2	2
Pd1505	contig11302	149	CGAGCA	6	ATCTTGCTCCATCACCATTA	CTTGTCTCCTACCTCACCAC	150–166	No	1	3
Pd1515	contig11456	160	CCG	3	AGTCCACTTTCACTTCCACTC	CTCTCCATCCTGCACCTC	171–179	No	2	4
Pd1533	contig11692	153	GGC	3	GGTCTTGAGGAGTGAGAGC	GAGAGAAAGCTAGGGTTTTGA	248–267	No	3	4
Pd1594	contig12224	160	GCA	3	TCAGTAGGACACACAACATCA	TAGTAGTTGGACGGATTTGAG	165–174	Yes	4	2
Pd1602	contig12343	122	GCG	3	ACTTCCGCTTCTTGTTGAT	ATCCAAGTCTAGGGTTCCAC	121–138	No	1	3
Pd1740	contig13647	148	AGCC	4	GGGATTATACAAATGGAGGAG	TGTTCGATTCTCATCAGTAGG	156–163	No	1	2
Pd1838	contig14705	162	TTTGTG	6	GAGATGTAATAGCGGTGTTTG	CTCAGCTCCACAGTCACAC	176–183	No	1	1
Pd1887	contig15149	140	TCA	3	GGACCAGCAGACACAAAC	CCACATCTTCACCTCTCTCTC	269–271	No	1	2
Pd1907	contig15377	139	CCG	3	ACAAGGCCAGAAGGGTTT	CAAAAGCAGTGGCAAGAG	150–153	No	1	1
Pd2162	contig18221	176	GCG	3	GTTGTACCTGATGGTGTTGTC	CCGATCAAGGGTTTCCTC	289–298	Yes	2	2
Pd2239	contig18901	148	CAT	3	GCTGCTGTTATTCTGTACCTG	CCTTCTCTTTCTCTCCAAGAC	160–166	Yes	2	2
Pd2251	contig19078	150	CGG	3	CTTCTGGTCGTAGGTGTAGTC	CTCCCCCTTCTTCACTCC	248–264	No	2	4
Pd2273	contig19311	155	GTCCCC	6	ACCGCCAAAAACCAAGAG	CTTCTTCTCCGCCTTCTC	166–177	No	1	4
Pd2337	contig20145	116	GCG	3	GGAAGGATTCTCTCCCTAGA	CTTCTTGTTGCCGAAGAAT	128–131	No	1	2
Pd2338	contig20158	153	GTG	3	GTCTGCGGCTCTTCTATTC	AAATGGACTAGAACCTCGAAC	168–203	No	2	4

Gene annotation

The *P. dilatatum* sequence contig dataset was reciprocally analysed, using the BLASTn algorithm against the CDS data sets of the relevant grass species with whole genome sequences. These data sets were *Oryza sativa* subsp. *japonica* (v6.1), *Brachypodium distachyon* (v1 assembly), *Sorghum bicolor* (v1.0 release, Sbi1.4 gene set) and foxtail millet (*Setaria italica* v2.1). This analysis revealed that 45.4% of the *P. dilatatum* cDNA contigs matched *O. sativa* genes; 43.7% matched *B. distachyon* genes, 56.2% matched *S. bicolor* genes and 62.5% matched CDS sequences from the foxtail millet assembly (Figure 2 and Table S1). A total of 7,994 (39.6%) of the *P. dilatatum* contigs had significant BLASTn matches to sequences from all of the reference genomes, while 7,229 (35.8%) failed to find a significant match sequences from any of the reference genomes.

The contigs were compared against the UniRef 90 protein database using the BLASTx algorithm with an E-value cutoff of 10^{-10} to assign putative functional roles and annotations to the *P. dilatatum* trascriptome data set (Table S1). A total of 11,417 contigs (56.6%) identified a protein from the uniref data base. Gene ontology (GO) terms were also assigned for the subset of contigs that matched the rice genome. Unigenes were classified into three GO terms: biological process, molecular function and cellular component. A total of 76,093 gene counts and 104,741 annotation counts were assigned. The majority of assignments belonged to the biological process ontology (41.4%) followed by molecular function (34.1%) and cellular components (24.5%). Among the biological process category, response to stress (12%) was the most highly represented category, followed by response to endogenous stimulus (7%), nucleotide binding (7%), cellular process (7%) and signal transduction (7%). Other functional classifications were represented at proportions less than 5% of the total (Figure 3A). In the molecular function classification class, catalytic activity (12%), protein binding (12%), hydrolase activity (11%), kinase activity (9%) and transferase activity (8%) constituted the major categories (Figure 3B). The mitochondrion (14%), plasma membrane (14%) and membrane (15%) categories of the cellular component ontology contributed to the largest proportion of annotations followed by nucleus (12%), cytoplasm (8%) and plastid (7%) (Figure 3C).

EST-SSR discovery and validation

The assembled contigs from the EST transcript data were chosen for SSR discovery and validation. Moderate length of sequence is required for SSR motif identification and reliable amplification following primer design. In addition to this, unique locus status for each SSR is desirable, which is best served through analysis of the assembled contigs to avoid partial sequences of ESTs. The 20,169 contigs were processed through batch Primer 3 software for motif identification and design. A total of 2,339 EST-simple sequence repeats (SSR) motifs (di-, tri-, tetra-, penta- or hexa-nucleotide) were identified from a total of 1,989 contigs (11.6%). The most frequent SSRs motifs in *P. dilatatum* transcriptome data set were tri-nucleotide repeats (Figure 4).

High quality primers were designed using BatchPrimer 3 software and a sub-set of 96 primer pairs were synthesised and used for empirical validation. All identified SSRs along with the designed primers are provided in table S2. Initial testing of the SSRs was performed on a single DNA sample from a distinct plant of the sexual tetraploid cultivar Primo and a DNA sample from the asexual apomictic cv. Relincho. Acceptable performance, defined as generation of distinct amplification products when resolved through capillary electrophoresis, was obtained for 64% of the SSRs that underwent initial validation. An additional collection of 48 plants from cv. Primo were screened with the selected SSR markers. The data from both rounds of SSR marker analysis was combined. A total of 32 SSR markers were able to identify variation between the sexual and apomictic cultivars, whilst the remaining markers only generated monomorphic marker profiles across all of the samples tested. Within these 32 SSR markers, a subset of eight identified polymorphic features within the samples from the cultivar Primo. The SSR markers typically generated fewer products from samples from cv. Primo than Relincho with median values of 2 and 3 respectively (Table 1).

Discussion

Efforts to sequence grass genomes have increased considerably in recent years. The first crop species with a complete genome sequence was rice [38], followed by several other species within the Poaceae: sorghum [29], maize [39], brachypodium [40], and most recently foxtail millet [30]. In addition, there are large collections of EST sequences available from species within the Poaceae family (7,180,207 nucleotides ESTs, 20/07/2013) along with collections of next generation RNA sequence based data (1,791 entries in the SRA database 20/07/2013). However, the majority of these studies in recent times have focused on maize (>1,000 of the RNA based SRA entries for the Poaceae). An EST sequencing approach can provide a valuable collection of cDNA sequences and has been widely used for gene discovery and development of molecular markers, as it is a rapid, cost effective approach that can avoid well known problems associated with repetitive regions and large genome sizes in many non-model organisms where no prior data is available. Even though complete genome sequences have now been generated for some grasses, there are >11,000 recognised species within the Poaceae [41] so it is not surprising that genomic resources have been limited for most grass species, including *P. dilatatum*, prior to this study.

Next generation sequencing has been widely used within many crop species due to its low cost and greater sequence yield [42]. The availability of next generation sequencing technology enables large sequence data sets to be generated at modest costs. However, *de novo* sequencing of complex, often repetitive genomes, which are common in the Poaceae family, is not trivial. The characterization of genes with low expression levels or genes induced by specific environmental cues, or with spatially-restricted expression patterns can be challenging. Sequencing technologies that generate short reads (35–150 bp) such as the Illumina HiSeq systems are ideally suited to model organisms where reads can be mapped to reference genomes or transcriptome sequences. However, as the GS FLX platform can generate reads of >400 bp, it has been used for *de novo* annotation in non-model plants such as olive [42,43], chestnut [44], ginseng [45],strawberry [46], bracken fern [47] and recently, in switchgrass [48].

This study reports the generation of genetic resources for *P. dilatatum* using the GS FLX Titanium platform. The EST-dataset generated in the current study provides a significant contribution towards the development of a resource in *P. dilatatum* that facilitates gene discovery and molecular marker development, for breeding programs. The study also provides a basis for future studies into gene expression under specific conditions with biotic or abiotic stresses. The GS FLX platform has a high error rate in homopolymeric regions (i.e., three or more consecutive identical DNA bases) caused by accumulated light intensity variance [49]. Therefore, a modified primer with an interrupted poly d(T) tail was used to avoid the limitations in sequence quality caused by the poly(A) tails of messenger RNA molecules. Similar approaches have been used in other transcriptome studies with significant benefits [45,50].

The data generated was of high quality as the majority of the generated sequence reads assembled into contigs (74%) with an average contig length of 463 bp; which is also comparable to the average contig length obtained in previous studies (454 bp [51], 334 bp [49], 440 bp [45]. There were still however a significant proportion of sequence reads that were not assembled into contigs and indicate that further transcriptomic studies could benefit the sequence assembly and provide a comprehensive catalogue of full length cDNA sequences, and also unequivocally identify low quality uninformative reads. A small proportion of singleton reads were longer than 600 bp, a length that is outside of the standard read length of the GS FLX 454 platform using Titanium chemistry. These sequences contain long repeats of single nucleotides, which increases the defined length of the read passed the expected output size. It is likely that these sequence reads are an artefact of the sequencing process and are not derived from *P. dilatatum*, even though they have passed through the standard pipeline approach to sequence quality filtering.

Results of comparative sequence analysis reflects taxonomic relationships, with a high proportion of *P. dilatatum* reads matching sequences from other members of the Panicoideae subfamily within the PACMAD clade, (*Sorghum bicolor* L. and *Setaria italica*). From the recent comprehensive and robust assessment of molecular phylogeny for the grass species [41], it would appear that the closest taxonomic relationship is between *P. dilatatum* and *Sorghum bicolor* L. However, this taxonomic relationship is not mirrored by the comparative sequence analysis performed in this study, where the highest number of matching sequence reads was obtained from comparisons to *Setaria italica*. This could indicate gene specific evolution that *S. bicolour* has undergone in isolation or potentially some differences in genome assembly and annotation between the two reference genomes. By undertaking the comparative genetic analysis within this study a significant description of the generated cDNA sequences has been attempted. Whilst this study has not focussed on single genes, it is anticipated that the resources generated, the gene ontology, protein as well as the reference genome analysis, will provide assistance to trait specific studies in the future.

Despite being described in plant species since 1992 [52], SSR markers are still widely used for many applications, including population genetics and marker assisted breeding due to their codominant and highly polymorphic nature [53]. EST-SSRs have previously been associated with agronomically important traits in many species including cotton and maize [54,55]. *De novo* development of a large cohort of SSRs was previously a costly and time-consuming process [56,57]. However, SSRs can now be rapidly identified in EST databases generated by sequencing of transcriptomes. EST-SSRs have an enhanced cross-species transferability and efficiency of amplification in comparison with SSRs located in non-transcribed regions [58,59], which could derive significant benefits for the c. 400 other *Paspalum* species [60]. A small cohort of genomic derived SSR markers were recently developed for *P. atratum* Swallen and *P. notatum* Flüggé and successfully transferred in 35 Paspalum species [35]. The use of SSRs across related species has been previously demonstrated in the Poaceae family: wheat EST-derived SSRs were transferable to eight related species [61] and to important crops including maize, barley and rice [62]. Barley (*Hordeum vulgare*) EST-derived SSRs were transferable to *H. chilense* lines and showed polymorphism [63]. The amplification of multiple products from tetraploid genotypes of *P. dilatatum* and amplification of even more products from pentaploid genotype in this study indicates that SSR markers are able to detect loci within the I, J and X genomes of *P. dilatatum*.

The frequency of EST- SSRs detected is typically influenced by the DNA template used, the criteria for defining SSRs, and the software and parameters used for identifying SSRs [64,65,66]. In this study, the frequency of trinucleotide repeat units was predominant, followed by tetra-, penta-, di- and hexanucleotide repeat units. The high frequency of trinucleotide repeat units is consistent with results previously reported in numerous plant species [67,68,69,70,71,72]. Polymorphism in trinucleotide repeats in coding regions of a transcript would be more tolerated as they would not cause shift in the amino acid translation [73].

Conventional plant breeding is a costly and time consuming approach that could be enhanced with molecular marker based system to facilitate the selection of desirable agronomic traits. In apomictic plants, novel approaches for interspecies hybridization must be employed to increase variation and effectively use available germplasm resources. Critically, the selection of germplasm for breeding would benefit from prior genomic characterisation to make informed decisions for cultivar development. As an example the cultivar Primo, used in this study, is characterized with high potential forage production, but was developed based on an interspecific hybridisation and backcross strategy to incorporate *C. paspali* resistance from *P. urvillei* (also a sexual tetraploid *Paspalum* species with a genomic constitution of IIJJ) into *P. dilatatum* [74]. The application and screening of a modest subset of the discovered SSR markers across the germplasm detailed in this study, has already demonstrated that these SSR are functional across both the sexual and apomictic biotypes of *P. dilatatum*. However, the low level of SSR marker variance that was identified in this study from multiple samples of Primo could potentially be accounted for by the breeding history of the cultivar, which suggests that the value of the markers may be more accurately assessed by screening a more diverse collection of *Paspalum* germplasm.

Conclusions

This study generated a substantial genomic EST-derived resource for the forage grass *Paspalum dilatatum* represented by 20,167 contigs, of which 12,940 have been sequence annotated. Furthermore, a collection of 2,339 SSR primer pairs have been designed and a subset validated for molecular breeding within the cultivated germplasm and can be assessed for efficiency across the genus *Paspalum*, dramatically advancing the available resources for this important forage grass species.

Author Contributions

Conceived and designed the experiments: AM SP JGM GCS GES. Performed the experiments: AG SK MD. Analyzed the data: AG NOIC SK. Wrote the paper: AG NOIC AM SP GES JGM GCS.

References

1. Christin PA, Samaritani E, Petitpierre B, Salamin N, Besnard G (2009) Evolutionary insights on C4 photosynthetic subtypes in grasses from genomics and phylogenetics. Genome Biol Evol 1: 221–230.
2. Clayton WD (1983) Tropical grasses. In: McIvor JG, Bray RA, editors. Genetic Resources of Forage Plants. Melbourne: CSIRO. pp 39–46.
3. Harlan JR (1983) The scope for collection and improvement if forage plants. In: McIvor JG, Bray RA, editors. Genetic Resources of Forage Plants. Melbourne: CSIRO.pp3–14.
4. Holt EC (1956) Dallisgrass. Texas Agricultural Experiment Station EUA Bulletin 829: 14.
5. Hutton EM, Nelson CJ (1968) Plant breeding and genetics. In: Commonwealth Agricultural Bureaux, editors. Some concepts and methods in sub-tropical pasture research. Brisbane: CSIRO. pp 79–92
6. Robinson DL, Wheat KG, Hubbert NL, Henderson M, Savoy HJ (1988) Dallisgrass yield, quality and nitrogen recovery responses to nitrogen and phosphorus fertilizers. Commun in soil Sci Plant Anal 19: 529–542.
7. Hacker JB, B.J F, Gow JM (1974) Simulated frosting of tropical grasses. Austr J of Agric Res 25: 45–57.
8. Davies LJ, Forde BJ (1991) Comparative responses of three subtropical grasses to combined frost and prolonged chilling treatments simulating a New Zealand winter. New Zealand J Agr Res 34: 249–256.
9. Blom CW, Voesenek LA (1996) Flooding the survival strategies of plants. Trends Ecol Evol11: 290–295.
10. Jackson MB, Armstrong W (1999) Formation of aerenchyma and the processes of plant ventilation in relation to soil flooding and submergence. Plant Biol 1: 274–287.
11. Vaio M, Mazzella C, Porro V, Speranza P, Lopez-Carro B, et al. (2007) Nuclear DNA content in allopolyploid species and synthetic hybrids in the grass genus *Paspalum*. Pl Syst Evol 265: 109–121.
12. Hayman DL (1956) Cytological evidence for apomixis in Australiam *Paspalum dilatatum*. Inst Agric Sci J. 22, 292–3.
13. Souza-Chies TT, Essi L, Rua GH, Valls JF, Miz RB (2006) A preliminary approach to the phylogeny of the genus *Paspalum* (Poaceae). Genetica 126: 15–32.
14. Miz BR, TT dS-C (2006) Genetic relationships and variation among biotypes of dallisgrass (*Paspalum dilatatum* Poir.) and related species using random amplified polymorphic DNA markers. Genet Resour Crop Evol 53: 541–552.
15. Burson BL (1978) Genome relations between *Paspalum conspersum* and two diploid *Paspalum* species. Can J Genet Cytol 20: 365–372.
16. Burson BL, Lee H, Benett HW (1973) Genome relations between tetraploid *Paspalum dilatatum* and four diploid *Paspalum* species. Crop Sci 13: 739–743.
17. Burton GW, Jackson JE (1962) Radiation breeding of apomictic postrate Dallisgrass, *Paspalum dilatatum* var. pauciciliatum. Crop Sci 2: 495–497.
18. Cole RJ, Dorner JW, Lansden JA, Cox RH, Pape C, et al. (1977) Paspalum staggers: isolation and identification of tremorgenic metabolites from sclerotia of *Claviceps paspali*. J Agric Food Chem 25: 1197–1201.
19. Davies LJ, Cohen D (1992) Phenotipic variation in somaclones of *Paspalum dilatatum* and their seedling offspring. Can J Plant Science 72: 773–784.
20. Akashi R, Adachi T (1992) Somatic embryogenesis and plant regeneration from cultured immature inflorescences of apomictic dallisgrass (*Paspalum dilatatum* Poir.). Plant Sci 82: 219–225.
21. Burson BL, Tischler CR (1993) Regeneration and somaclonal variation in apomictic *Paspalum dilatatum* Poir. Euphytica 67: 71–78.
22. Burton GW (1943) Interspecific hybrids in the genus *Paspalum*. J Hered 34: 14–23.
23. Benett HW, Bashaw EC (1966) Interspecific hybridization with *Paspalum* spp. Crop Sci 6: 52–54.
24. Gupta PK, Varshney RK (2000) The development and use of microsatellite markers for genetic analysis and plant breeding with emphasis on bread wheat. Euphytica 113: 163–185.
25. Zhou X, Ren L, Meng Q, Li Y, Yu Y, et al. (2010) The next-generation sequencing technology and application. Prot Cell 1: 520–536.
26. Vera JC, Wheat CW, Fescemyer HW, Frilander MJ, Crawford DL, et al. (2008) Rapid transcriptome characterization for a non model organism using 454 pyrosequencing. Mol Ecol 17: 1636–1647.
27. Garg R, Patel RK, Jhanwar S, Priya P, Bhattacharjee A, et al. (2011) Gene discovery and tissue-specific transcriptome analysis in chickpea with massively parallel pyrosequencing and web resource development. Plant Physiol 156: 1661–1678.
28. Kaur S, Pembleton LW, Cogan NO, Savin KW, Leonforte T, et al. (2012) Transcriptome sequencing of field pea and faba bean for discovery and validation of SSR genetic markers. BMC Genomics 13: 104.
29. Paterson AH, Bowers JE, Bruggmann R, Dubchak I, Grimwood J, et al. (2009) The Sorghum bicolor genome and the diversification of grasses. Nature 457: 551–556.
30. Zhang G, Liu X, Quan Z, Cheng S, Xu X, et al. (2012) Genome sequence of foxtail millet (*Setaria italica*) provides insights into grass evolution and biofuel potential. Nat Biotechnol 30: 549–554.
31. Zalapa JE, Cuevas H, Zhu H, Steffan S, Senalik D, et al. (2012) Using next-generation sequencing approaches to isolate simple sequence repeat (SSR) loci in the plant sciences. Am J Bot 99: 193–208.
32. Park YJ, Lee JK, Kim NS (2009) Simple sequence repeat polymorphisms (SSRPs) for evaluation of molecular diversity and germplasm classification of minor crops. Molecules 14: 4546–4569.
33. Hougaard BK, Madsen LH, Sandal N, de Carvalho Moretzsohn M, Fredslund J, et al. (2008) Legume anchor markers link syntenic regions between *Phaseolus vulgaris*, *Lotus japonicus*, *Medicago truncatula* and Arachis. Genetics 179: 2299–2312.
34. Zeng S, Xiao G, Guo J, Fei Z, Xu Y, et al. (2010) Development of a EST dataset and characterization of EST-SSRs in a traditional Chinese medicinal plant, *Epimedium sagittatum* (Sieb. Et Zucc.) Maxim. BMC Genomics 11: 94.
35. Cidade FW, Vigna BB, de Souza FH, Valls JF, Dall'agnol M, et al. (2013) Genetic variation in polyploid forage grass: Assessing the molecular genetic variability in the *Paspalum* genus. BMC Genet 14: 50.
36. Meyer E, Aglyamova GV, Wang S, Buchanan-Carter J, Abrego D, et al. (2009) Sequencing and de novo analysis of a coral larval transcriptome using 454 GSFlx. BMC Genomics 10: 219.
37. Schuelke M (2000) An economic method for the fluorescent labeling of PCR fragments. Nat Biotechnol 18: 233–234.
38. Goff SA, Ricke D, Lan TH, Presting G, Wang R, et al. (2002) A draft sequence of the rice genome (*Oryza sativa* L. ssp. japonica). Sci 296: 92–100.
39. Schnable PS, Ware D, Fulton RS, Stein JC, Wei F, et al. (2009) The B73 maize genome: complexity, diversity, and dynamics. Sci 326: 1112–1115.
40. Initiative IB (2010) Genome sequencing and analysis of the model grass *Brachypodium distachyon*. Nature 463: 763–768.
41. Grass Phylogeny Working Group II (2012) New grass phylogeny resolves deep evolutionary relationships and discovers C4 origins. New Phytol 193: 304–312.
42. Wall PK, Leebens-Mack J, Chanderbali AS, Barakat A, Wolcott E, et al. (2009) Comparison of next generation sequencing technologies for transcriptome characterization. BMC Genomics 10: 347.
43. Alagna F, D'Agostino N, Torchia L, Servili M, Rao R, et al. (2009) Comparative 454 pyrosequencing of transcripts from two olive genotypes during fruit development. BMC Genomics 10: 399.
44. Barakat A, DiLoreto DS, Zhang Y, Smith C, Baier K, et al. (2009) Comparison of the transcriptomes of American chestnut (*Castanea dentata*) and Chinese chestnut (*Castanea mollissima*) in response to the chestnut blight infection. BMC Plant Biol 9: 51.
45. Sun C, Li Y, Wu Q, Luo H, Sun Y, et al. (2010) *De novo* sequencing and analysis of the American ginseng root transcriptome using a GS FLX Titanium platform to discover putative genes involved in ginsenoside biosynthesis. BMC Genomics 11: 262.
46. Shulaev V, Sargent DJ, Crowhurst RN, Mockler TC, Folkerts O, et al. (2011) The genome of woodland strawberry (*Fragaria vesca*). Nat Genet 43: 109–116.
47. Der JP, Barker MS, Wickett NJ, dePamphilis CW, Wolf PG (2011) De novo characterization of the gametophyte transcriptome in bracken fern, *Pteridium aquilinum*. BMC Genomics 12: 99.
48. Wang Y, Zeng X, Iyer NJ, Bryant DW, Mockler TC, et al. (2012) Exploring the switchgrass transcriptome using second-generation sequencing technology. PLoS One 7: e34225. doi: 10.1371/journal.pone.0034225
49. Margulies M, Egholm M, Altman WE, Attiya S, Bader JS, et al. (2005) Genome sequencing in microfabricated high-density picolitre reactors. Nature 437: 376–380.
50. Beldade P, Rudd S, Gruber JD, Long AD (2006) A wing expressed sequence tag resource for *Bicyclus anynana* butterflies, an evo-devo model. BMC Genomics 7: 130.
51. Franssen SU, Shrestha RP, Brautigam A, Bornberg-Bauer E, Weber AP (2011) Comprehensive transcriptome analysis of the highly complex *Pisum sativum* genome using next generation sequencing. BMC Genomics 12: 227.
52. Akkaya MS, Bhagwat AA, Cregan PB (1992) Length polymorphisms of simple sequence repeat DNA in soybean. Genetics 132: 1131–1139.
53. Luikart G, England PR, Tallmon D, Jordan S, Taberlet P (2003) The power and promise of population genomics: from genotyping to genome typing. Nat Rev Genet 4: 981–994.
54. Abdurakhmonov IY, Abdullaev AA, Saha S, Buriev ZT, Arslanov D, et al. (2005) Simple sequence repeat marker associated with a natural leaf defoliation trait in tetraploid cotton. J Hered 96: 644–653.
55. Li X, Yang H, Buirchell B, Yan G (2011) Development of a DNA marker tightly linkedto low-alkaloid gene iucundus in narrow-leafed lupin (*Lupinus angustifolius L.*) for marker-assisted selection. Crop and Pasture Sci 62: 218–224.

56. Squirrell J, Hollingsworth PM, Woodhead M, Russell J, Lowe AJ, et al. (2003) How much effort is required to isolate nuclear microsatellites from plants? Mol Ecol 12: 1339–1348.
57. Zane L, Bargelloni L, Patarnello T (2002) Strategies for microsatellite isolation: a review. Mol Ecol 14: 1–16.
58. Barbara T, Palma-Silva C, Paggi GM, Bered F, Fay MF, et al. (2007) Cross-species transfer of nuclear microsatellite markers: potential and limitations. Mol Ecol 16: 3759–3767.
59. Kumpatla SP, Mukhopadhyay S (2005) Mining and survey of simple sequence repeats in expressed sequence tags of dicotyledonous species. Genome 48: 985–998.
60. Chase A (1929) The North American species of *Paspalum*. Contrib US Natl Herbarium 28:1–310.
61. Zhang LY, Bernard M, Leroy P, Feuillet C, Sourdille P (2005) High transferability of bread wheat EST-derived SSRs to other cereals. Theor Appl Genet 111: 677–687.
62. Tang J, Gao L, Cao Y, Jia J (2006) Homologous analysis of SSR-ESTs and transferability of wheat SSR-EST markers across barley, rice and maize. Euphytica 151: 87–93.
63. Castillo A, Budak H, Varshney RK, Dorado G, Graner A, et al. (2008) Transferability and polymorphism of barley EST-SSR markers used for phylogenetic analysis in *Hordeum chilense*. BMC Plant Biol 8: 97.
64. Parchman TL, Geist KS, Grahnen JA, Benkman CW, Buerkle CA (2010) Transcriptome sequencing in an ecologically important tree species: assembly, annotation, and marker discovery. BMC Genomics 11: 180.
65. Castoe TA, Poole AW, Gu W, Jason de Koning AP, Daza JM, et al. (2010) Rapid identification of thousands of copperhead snake (*Agkistrodon contortrix*) microsatellite loci from modest amounts of 454 shotgun genome sequence. Mol Ecol Resour 10: 341–347.
66. Tangphatsornruang S, Somta P, Uthaipaisanwong P, Chanprasert J, Sangsrakru D, et al. (2009) Characterization of microsatellites and gene contents from genome shotgun sequences of mungbean (*Vigna radiata* (L.) Wilczek). BMC Plant Biol 9: 137.
67. Eujayl I, Sledge MK, Wang L, May GD, Chekhovskiy K, et al. (2004) *Medicago truncatula* EST-SSRs reveal cross-species genetic markers for *Medicago* spp. Theor Appl Genet 108: 414–422.
68. Varshney RK, Thiel T, Stein N, Langridge P, Graner A (2002) In silico analysis on frequency and distribution of microsatellites in ESTs of some cereal species. Cell Mol Biol Lett 7: 537–546.
69. La Rota M, Kantety RV, Yu JK, Sorrells ME (2005) Non random distribution and frequencies of genomic and EST-derived microsatellite markers in rice, wheat, and barley. BMC Genomics 6: 23.
70. Gao L, Tang J, Li H, Jia J (2003) Analysis of microsatellites in major crops assessed by computational and experimental approaches. Mol Breeding 12: 245–261.
71. Thiel T, Michalek W, Varshney RK, Graner A (2003) Exploiting EST databases for the development and characterization of gene-derived SSR-markers in barley (*Hordeum vulgare* L.). Theor Appl Genet 106: 411–422.
72. Cardle L, Ramsay L, Milbourne D, Macaulay M, Marshall D, et al. (2000) Computational and experimental characterization of physically clustered simple sequence repeats in plants. Genetics 156: 847–854.
73. Metzgar D, Bytof J, Wills C (2000) Selection against frameshift mutations limits microsatellite expansion in coding DNA. Genome Res 10: 72–80.
74. Schrauf GE, Blanco MA, Cornaglia PS, Deregibus VA, Madia M, et al. (2003) Ergot resistance in plants of *Paspalum dilatatum* incorporated by hybridisation with *Paspalum urvillei*. Trop Grasslands 37: 182–186.

21

The Flavonoid Pathway Regulates the Petal Colors of Cotton Flower

Jiafu Tan, Maojun Wang, Lili Tu*, Yichun Nie, Yongjun Lin, Xianlong Zhang

National Key Laboratory of Crop Genetic Improvement, Huazhong Agricultural University, Wuhan, Hubei, China

Abstract

Although biochemists and geneticists have studied the cotton flower for more than one century, little is known about the molecular mechanisms underlying the dramatic color change that occurs during its short developmental life following blooming. Through the analysis of world cotton germplasms, we found that all of the flowers underwent color changes post-anthesis, but there is a diverse array of petal colors among cotton species, with cream, yellow and red colors dominating the color scheme. Genetic and biochemical analyses indicated that both the original cream and red colors and the color changes post-anthesis were related to flavonoid content. The anthocyanin content and the expression of biosynthesis genes were both increased from blooming to one day post-anthesis (DPA) when the flower was withering and undergoing abscission. Our results indicated that the color changes and flavonoid biosynthesis of cotton flowers were precisely controlled and genetically regulated. In addition, flavonol synthase (*FLS)* genes involved in flavonol biosynthesis showed specific expression at 11 am when the flowers were fully opened. The anthocyanidin reductase (*ANR)* genes, which are responsible for proanthocyanidins biosynthesis, showed the highest expression at 6 pm on 0 DPA, when the flowers were withered. Light showed primary, moderate and little effects on flavonol, anthocyanin and proanthocyanidin biosynthesis, respectively. Flavonol biosynthesis was in response to light exposure, while anthocyanin biosynthesis was involved in flower color changes. Further expression analysis of flavonoid genes in flowers of wild type and a flavanone 3-hydroxylase *(F3H)* silenced line showed that the development of cotton flower color was controlled by a complex interaction between genes and light. These results present novel information regarding flavonoids metabolism and flower development.

Editor: Jinfa Zhang, New Mexico State University, United States of America

Funding: This work was supported by a grant from the National High Technology Research and Development Program of China (863 Program, grant number 2012AA101108). The funders had no role in study design, data collection and analysis, decision to publish, or preparation of the manuscript.

Competing Interests: The authors have declared that no competing interests exist.

* E-mail: lilitu@mail.hzau.edu.cn

Introduction

The diversity of flower color is one of the most beautiful gifts of nature, and it serves as a versatile tool for plant biochemists, geneticists and ecologists. Different flower colors are constituted with various pigments and co-pigments. The betalains, carotenoids, and anthocyanins are the three major, well-characterized groups of pigments [1]. Flavonols, pH and ions also play important roles in determining flower color [2,3,4]. In addition to endogenous genetic effects, environmental factors such as light and temperature could also be responsible for the color of flowers [5,6,7,8]. In addition to its general phenotype, color also confers flowers with diverse biological functions, such as protection against UV-light and the attraction of pollinators [9,10]. Within a single flower, color changes during flower development are widespread throughout the angiosperms [11,12]. Multiple studies have indicated that, although it is important for pollinator interaction, the mechanism of flower color changes pre- and post- anthesis is not well understood [13,14,15].

Flavonoids, particularly anthocyanidin glycosides, are the major flower pigments [16]. The metabolism of the flavonoid pathway has been well analyzed in *Petunia*, *Arabidopsis* and *Antirrhinum* [17,18,19]. In most cases, anthocyanin accumulation was tightly linked with flower development and color changes [20]. The anthocyanin biosynthesis pathway has been mostly studied, and the biosynthesis of flavonol, which together with the biosynthesis of anthocyanin comprising the major biosynthesis pathways in flavonoid metabolism, has also been well studied in flower development due to its high antioxidation and UV protection [10]. Flavonol and anthocyanin metabolism can be induced by light [21,22]. Flavonoids are derived from phenylalanine, catalyzed by phenylalanine ammonia-lyase (PAL). Then it was mediated by a common step with chalcone synthase (CHS), flavanone 3-hydroxylase (F3H), and fluxed into anthocyanin biosynthesis by dihydroflavonol 4-reductase (DFR), anthocyanidin synthase (ANS) and UDP-glucose: flavonoid 3-Oglucosyltransferase (UFGT); fluxed into flavonol biosynthesis by flavonol synthase (FLS) following F3H; or fluxed into proanthocyandin biosynthesis through anthocyanidin reductase (ANR) following ANS [23]. Most flavonoid-related genes showed co-regulation and co-expression in plants [24,25,26]. The expression of these genes was mostly light dependent [22,27,28], while a light independent mechanism impacting flavonoids biosynthesis was also observed [29]. Various interactions were observed between different enzymes in *Arabidopsis*, which indicates that these enzymes function as a complex [30,31,32].

Cotton (*Gossypium spp.*) is composed of approximately 50 species, including approximately 5 allotetraploid species and 45 diploids

[33]. Many phenotypic polymorphisms exist among these species, of which the variety of flower color is one of the more obvious, and color variety had also been used in modern genetic study [34]. The biochemical study of cotton flower color has more than a century of history [35]. The color of the cotton flower is due to the effects of both flavonol and anthocyan [36]. Quercetin and cyanidin have been shown to be the major flavonol and anthocyan compounds, respectively, throughout the genus [35,37]. Varity, environment and fertilization could affect the composition of cotton flower pigments [35,38]. However, the pigment of the cotton flower is less affected by environmental factors than that of the leaf [39], and it could be stable inheritance [40]. The flavonoids in the cotton flower are kinds of biologically-active chemicals and show insect resistance and antibacterial properties [41,42]. The color of the cotton flower has also been employed as a tool for the genetic and taxonomic study of cotton [43,44,45]. Thousands of records of cotton flower color have been observed and presented in the international germplasm systems; however, these records all represent only one time-point phenotype. Cotton flowers suffer from a short life, as they bloom in the morning and begin to wither in the afternoon of the day of anthesis and abscise near noon of the next day under common field growth conditions; the colors dramatically change during flower development [35,38].

Here, we present data obtained through biological analysis showing that the color of the cotton flower is primarily composed of flavonoids. The change in color is associated with the expression of flavonoid genes, particularly those involved in anthocyanin biosynthesis. Light plays an important role in pigment accumulation and flavonoid biosynthesis, but an inherent factor also exists in the flavonoid pathway during flower development.

Materials and Methods

Plant materials

The materials were collected from the experimental field of Huazhong Agricultural University, Wuhan, China. No specific permissions were required for these locations or activities. It did not involve endangered or protected species. The cotton plants used in this study, *Gossypium hirsutum* YZ1, an F3H stable silence line of YZ1 (f3h) which was generated by our laboratory [46], brown cotton *G. hirsutum* T586, and *G. barbadense* 3-79, were cultivated with standard farming practices and management. Flowers were harvested on a typical sunny day in August at different time points ranging from the day before blooming to the day after blooming. The day of anthesis was noted as 0 DPA. Different time points were noted according to Beijing time (GMT +8). Flowers at six time points including 6, 11 and 12 am, 3 and 6 pm on 0 DPA and 6 am on 1 DPA, were collected for analysis. The anther and pistils were carefully removed to avoid contamination. Flowers were then cut flush with the margin of the sepal, weighed and either immediately ground into powder in liquid nitrogen or immersed in liquid nitrogen and stored at −70°C until use.

Cotton germplasm analysis

To collect information about the cotton flower, different Internet germplasm resources were consulted. Two of them, the Chinese Crop Germplasm Resources Information System (CGRIS) (http://icgr.caas.net.cn/#) and the National Plant Germplasm System (GRIN) (http://www.ars-grin.gov/npgs/searchgrin.html) containing abundant cotton petal data were referred to for analysis. Several representative pictures of cotton flower were obtained from the Internet (including http://www.learnnc.org/lp/multimedia/9447, http://commons.wikimedia.org/wiki/File:Darwin's-cotton-flower.jpg, http://www.photostuff.org/galapagos3.html, http://www.alohafriendsphotos.com/flowers2.html).

Quantification of anthocyanins

Petals were weighed and ground in liquid nitrogen, each flower was considered to be an individual sample. Anthocyanins were extracted from the flowers in acidic (1% HCl [w/v]) methanol for 48 h. The extraction was performed with a ratio of 10 ml buffer to 1 g sample. Anthocyanin was quantified by measuring the absorbance at 530 nm [47]. For each sample, more than five flowers were used for analysis.

Photography and image management

The flowers were collected from the experimental field and immediately photographed with a digital camera. Flowers that were harvested at the same time point were taken into one photo, and identical background and camera parameters were used at each time point. Photos of flowers taken at different time points were combined into one image with Adobe Photoshop 7.0.

Flower shading and emasculation treatment

The flowers were covered with a kraft bag before anthesis in the afternoon of -1 DPA to avoid light exposure; and the bags were large enough to permit flower opening. Flowers from the same plant and same position were taken as control and were studied from 11 am on 0 DPA to 6 am on 1 DPA. For flower emasculation treatment, the anther and pistil of each flower were carefully removed with scissors at approximately 8 am on 0 DPA, when the petals were not fully open. Flowers from the same plants and at the same position were taken as control and analyzed at 8 am on 1 DPA. For each treatment, more than five flowers were analyzed.

PH measurement

The pH of cotton flower was measured according to a previous report [4]. Three fresh flowers for each treatment were harvested from the field. The petals were collected, washed twice with deionized water within 15 min after removal from the plant, and then ground with a pestle and mortar for 1 min in 5 ml deionized water per flower. The pH of the homogenate was measured immediately with a pH electrode (Mettler-Toledo FE20, Shanghai).

Quantitative real-time PCR

Cotton flowers were harvested at different time points for flavonoid gene expression analysis. RNA extraction was performed following the procedure of the Spectrum Plant Total RNA Kit of Sigma-Aldrich (USA), cDNA synthesis and qRT-PCR were performed as previously described [48], with cotton ubiquitin gene *UBQ7* as the reference gene [49]. Flavonoid gene sequences were obtained from the public NCBI UniGene data bank and the cotton D genome sequence (http://www.phytozome.com/). The expression of all genes was normalized to *UBQ7*. Three biological replicates were performed. The error bars represent the standard deviations (SD). All primers are listed in Table S1. SuperScript® III reverse transcription kits were from Invitrogen (Carlsbad, USA). Reagents (iTaq SYBR Green supermix with ROX) for real-time PCR were from Bio-Rad (Foster City, USA).

Results

The cotton (*Gossyoium spp.*) flower contains a variety of colors

The flower color is one of the most variable traits of cotton. Based on an international survey of cotton germplasms, a total of 4664 germplasms containing flower color information were identified. Among them, 3557 germplasms are from the Chinese Crop Germplasm Resources Information System (CGRIS) (Figure 1A), and the remaining 1107 germplasms are from the National Plant Germplasm System (GRIN) (Figure 1B). In total, 14 types of colors are defined in CGRIS and 7 types are done in GRIN. These colors can be divided into three major colour schemes: cream is the most common color with composing more than 70% of the germplasms in both systems; yellow is the secondary color, composing nearly 16% of the germplasms from CGRIS and 3% from GRIN it; red is the third color, although it accounts for less than 5% in both systems. There are very few white flower germplasms listed in CGRIS. Different germplasm systems show different color patterns: most of the germplasms in both systems are *Gossypium hirsutum* with approximately 76% in CGRIS and 89% in GRIN having cream flowers. White and yellow flowers are also common in *G. hirsutum*. *G. barbadense* is the second-most abundant germplasm in both systems with the primary flower color of yellow (if well defined, because there was an ambiguous definition referred to 'segregating' within the GRIN system that included up to 81.6% of the *G. barbadense* in that system, although these flowers appeared to be yellow in most cases, we did not include them). There are several diploid A genome and wild cotton germplasms in CGRIS having yellow flowers. It also shows that most of the red flower germplasms are *G. hirsutum*, but only one germplasm flower of *G. barbadense* is red and defined as coceine in the CGRIS.

Cotton flower color changes dramatically after blooming

Three different flower color species were studied (Figure 2). YZ1 represents the most common cream-colored flower of *G. hirsutum*. T586 represents the red flower species, and 3-79 represents the yellow-colored flowers of *G. barbadense*. Flowers of these cotton species were collected from 11 am on the day of anthesis (0 DPA) to 6 am on the second day of anthesis (1 DPA) for analysis. All of the flowers changed the color during this period. Red pigment accumulated throughout the flower development (Figure 2A). The flower of YZ1 was cream at 11 am, and appeared a little bit red at 12 am, gradually accumulating to 6pm and then dramatically increasing color to 6 am on 1 DPA (Figure 2A). The flower of T586 was light red at 11 am, afterwards the red color deepened and the petal was dark red at 6 am on 1 DPA. Although the flower color of 3-79 was primarily yellow, red color was still emerged from the edge of the petal at 6 pm on 0 DPA and then increased further (Figure 2A). All of the flower color changes corresponded to an accumulation of anthocyanins (Figure 2B). T586 had about two times higher anthocyanin content than YZ1 and 3-79 during flower development in correspondence with its much intense color than the other species (Figure 2B). These results showed that anthocyanin biosynthesis was enhanced during flower development, conferring the petals with an intense red color. Dramatic anthocyanin accumulation occurred during the night of 0 DPA for YZ1 (Figure 2B).

Shade reduces accumulation of anthocyanins in cotton flowers

Light has been reported to be the main environmental factor that affects anthocyanin biosynthesis. The color of cotton flowers is also significantly affected by light. However, flowers subjected to shade treatment also underwent a mild color change (Figure 3A). With shade treatment, the accumulation of anthocyanin, which should kick in at 12 am under normal light, did not increase until 6 pm on 0 DPA in YZ1 (Figure 3A). The anthocyanin content of these flowers was significantly decreased during flower development when comparing with the flower under light (Figure 2B and 3B), while it still correlated with the flower color changes (Figure 3A and B). Although the concentration of anthocyanin was reduced in the shade, a gradual accumulation pattern still showed up in these flowers, which indicated the flower color change was controlled in a light independent pattern (Figure 3B). Fertilization and pH were considered as the key factors for flower color changes, but fertilized flowers showed little difference compared to unfertilized flowers in both phenotype and anthocyanin content of YZ1 (Figure S1A and B). The pH value did not significantly change during cotton flowers development, although it was changed at 6 am on 0 DPA, and then increased at 12 am, afterwards decreased again (Figure S2). The pattern of the pH changes was not associated with the color changes during flower development (Figure 2 and S2). These results indicated that light instead of fertilization or pH was responsible for the color changes during the development of the cotton flower. Although light could affect anthocyanin accumulation, the color of the flowers also changed under shaded conditions, which indicated that the color change of cotton flower was mainly under a genetic rather than light control and light acts as an enhancer to affect the concentration of pigment.

The expression levels of genes of the anthocyanin pathway are consistent with the color change

Phenotypic observation of cotton flower development suggested that anthocyanin biosynthesis is controlled by both environmental and genetic factors (Figure 3). Therefore, we further studied the transcription of genes involved in this pathway (Figure 4). The transcriptions of the flavonoid and anthocyanin biosynthesis-related genes *PAL*, *CHS*, *F3H*, *DFR*, *FLS*, *ANR*, *ANS* and *UFGT* were abundant in flowers and dramatically affected by light. The expression of *CHS*, *F3H*, *ANS* and *UFGT* showed a gradual increase pattern during flower development and lower levels in shading flowers compared to that in normal light (Figure 4B, C, E and F). These findings correlated with the accumulation of anthocyanin (Figure 2 and 3 B). Furthermore, the genes *CHS* and *UFGT* also showed a corresponding expression pattern with the accumulation of anthocyanin in 3-79 and T586 flowers (Figure S3A and B). These genes may play important roles in the accumulation of anthocyanin and changes in flower color.

The expression of *FLS1*, which is related to flavonol biosynthesis, peaked at 11 am on 0 DPA (when little red color was accumulated, Figure 2A) and then decreased to a very low level at 3 pm on 0 DPA (when red color had emerged, Figure 2A) and remained the base line for the rest time points of flower development (Figure 4 H); this gene was similarly expressed in the flowers of 3-79 and T586 (Figure S3C). The expression of *FLS1* was highly light-dependent, but very few transcripts were detected in shade-treated flowers (Figure 4 H). The expression of *PAL*, *DFR* and *ANR* reached the maximum at 6 pm on 0 DPA and then sharply declined (Figure 4A, D and G). *PAL* and *DFR* showed a light-induced expression pattern, but *ANR* was not strongly affected by light (Figure 4 A, D and G). Similar expression patterns of *ANR* were also observed in 3-79 and T586 flowers, although it was not dominant in T586 flowers (Figure S3D). The expression of most genes in the flavonoid pathway was light-induced, but a similar expression pattern for these genes, except for *FLS1*, was

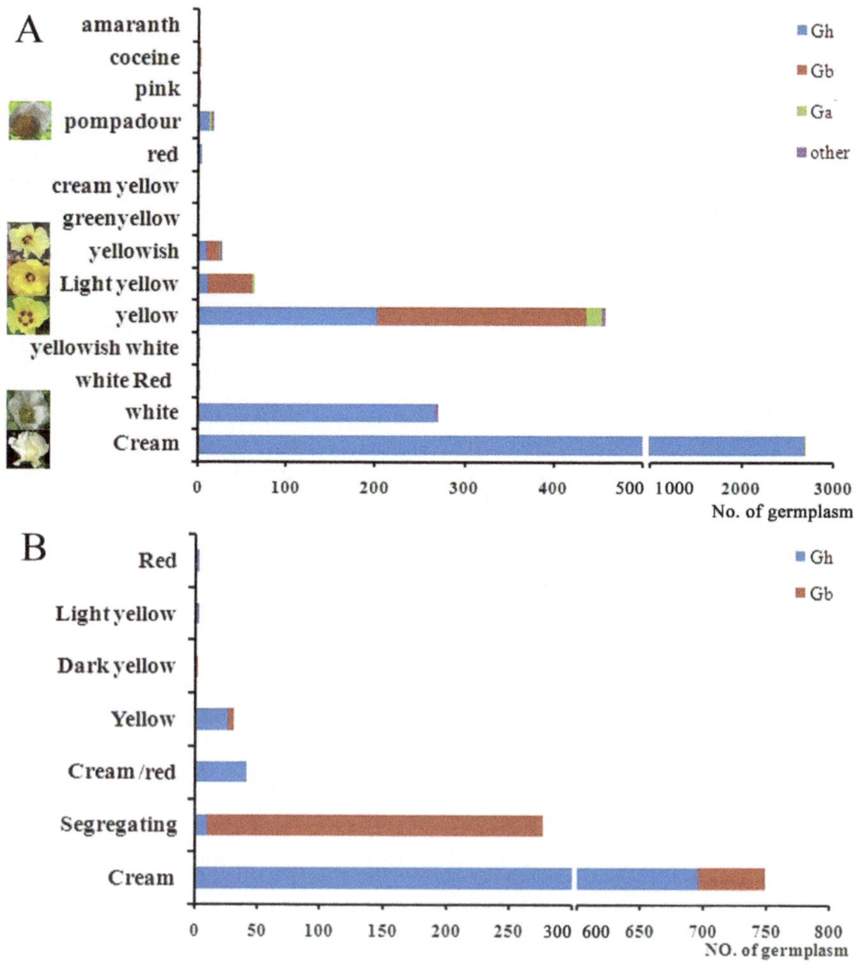

Figure 1. Variety of cotton flower colors obtained from the global germplasms collection systems. Germplasms from the Chinese Crop Germplasm Resources Information System (A) and the National Plant Germplasm System (B), with recorded petal colors, were analyzed. Colors were classified according to the record. *Gossypium hirsutum* (Gh), *G. barbadense* (Gb) and *G. arboreum* (Ga) were the three major cotton species in these systems. The remainders were classified as 'other' and mostly contained wild cotton.

observed during flower development regardless of light exposure. These genes expression profiles were correlated with the color changes of flower, which implies that genetic control is dominant for anthocyanin metabolism, while flavonol biosynthesis may be under a light-dependent control.

Flavonoid is the main pigment responsible for cotton flower color

A previous study showed that flavonol and anthocyanin were abundant in cotton flower [36], and the silencing of the flavanone 3-hydroxylase (F3H) gene caused an obvious change in the petal color of one day post anthesis (DPA) flowers in transgenic plants [46]. To determine whether flavonoids are the main pigment in cotton flowers, we additionally compared the petals of wild type *G. hirsutum* (YZ1) and *F3H* silence line (f3h) on 0 DPA and the abscission petals on 1 DPA (Figure 5). The color of YZ1 flowers was cream at 0 DPA, which represents the most common color of cotton species, while the flower color of f3h was white (Figure 5A), which indicated that lack of flavonoids caused color loss and the pigment in the cream petals was also mainly composed of flavonoids (Figure 5). This pigment could not be anthocyanin because the OD value of the extract of YZ1 flower on 0 DPA at 530 nm was lower than that of f3h (Figure 5B). However, the fresh red extract of the abscission 1 DPA petal and the corresponding increase in OD value at 530 nm indicated that the increase in red color of the developing flower was due to anthocyanin. *F3H* silencing line had the similar color changes as previously described for non-abscission 1 DPA petals [46]. These results confirmed that flavonoids are the main pigment responsible for cotton flower color including red, cream and white.

Genetic and environmental interactions control flavonoid metabolism in cotton flower

The flower color of the f3h line changed during its development. A red color appeared on 1 DPA in the f3h line which was significantly delayed compared to YZ1, which emerged at 12 am on 0 DPA (Figure 6A). The f3h line showed a slight decrease at 6 am on 1 DPA in anthocyanin accumulation following shade treatment compared to non-treatment (Figure 6B). Gene expression analysis showed that the expression pattern of these flavonoid genes was similar in the f3h and YZ1 lines during flower development, however, the expression of *PAL, DFR, UFGT* and *ANR* was suppressed by the silence of *F3H* (Figure 4 and 7). Shade also significantly affected gene expression in the f3h line, while

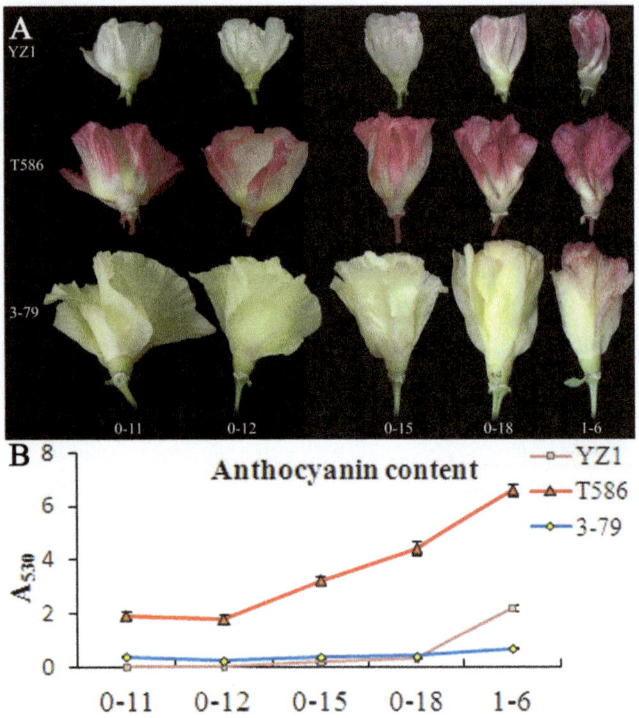

Figure 2. Petal colors change during flower development. A, flowers from *G. hirsutum* YZ1, T586 and *G. barbadense* 3-79 were collected from the field at 11 am, 12 am, 3 pm, and 6 pm on 0 DPA (referred as 0-11, 0-12, 0-15, and 0-18, respectively) and 6 am on 1 DPA (referred as 1-6). B, the anthocyanin contents of these flowers were measured at A_{530}. Three repeats with more than five flowers for each repeat were analyzed. Error bars represent SD.

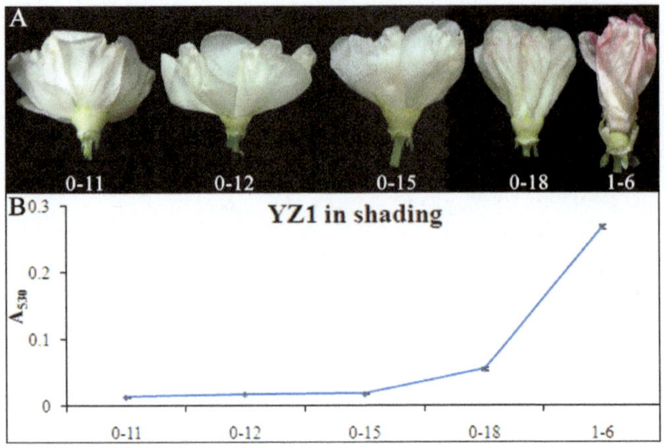

Figure 3. Flower color is affected by shade treatment. A, Phenotypes of YZ1flowers with shade treatment were analyzed from 11 am on 0 DPA (0-11) to 6 am on 1 DPA (1-6). B, the anthocyanin contents for corresponding timepoint flowers were measured at A_{530}. Error bars represent SD with three repeats performed. For each repeat, more than five flowers were analyzed.

PAL, F3H, DFR and *ANR* were all expressed at higher levels after 3 pm on 0 DPA when grown under the shade treatment compared to normal light, at that time the anthocyanin was dramatically accumulated in the YZ1 flower (Figure 7A, C, D and G). All of these flavonoid genes showed higher expression levels at 6 am on 1 DPA in the f3h line compared to YZ1 with shade treatment (Figure 4 and 7), at that time point anthocyanin was abundantly accumulated in the YZ1 flower (Figure 2). These results showed that there was a complicated interaction between gene expression and light exposure, especially for *PAL, DFR* and *ANR,* which were indirectly related with anthocyanin biosynthesis. These genes were all reduced by either shade treatment or silence of *F3H,* but they showed similar expression levels in the shade treated f3h line as YZ1 under normal light treatment.

Additionally, the expression of *FLS1* was strictly light dependent in f3h flowers (Figure 7H). Silence of *F3H* caused a maximum expression level at 12 am on 0 DPA, although it showed a similar expression level at 11 am as that of YZ1 under normal light (Figure 4H). Afterwards, the expression of *FLS1* decreased but still maintained relatively higher expression levels compared to YZ1 (Figure 7H). We further investigated two additional FLS genes, identified as *FLS2* and *FLS3,* from the published cotton D genome. *FLS3* showed a more dominant expression in flowers, similar to *FLS1,* compared with *FLS2* (Figure S4A and B). *FLS2* and *FLS3* also showed a peak expression at 11 am and then decreased in YZ1, 3-79 and T586 (Figure S4A and B). Light and silence of *F3H* showed less effect on the expression of *FLS2,* although the expression of *FLS2* was also enhanced in the flower of f3h silencing lines (Figure S4A). On the other side, the expression of *FLS3* showed a similar expression pattern as that of *FLS1,* which was significantly affected by light and showed a higher and delayed maximum expression at 12 am on 0 DPA in flowers f3h lines compared to YZ1 (Figure S4B). The significant effect of light on the expression of the dominantly expressed *FLS1* and *FLS3* indicates that the flavonol pathway was mainly light dependent in cotton flowers, and silence of *F3H* could increase the expression of *FLS* under light exposure. These results imply that flavonol is not associated with flower color changes but has a significant role under light response.

Discussion

The cotton flowers have been studied for more than a century. Although many different flower colors have been observed, it has been shown that the pigments of cotton flower mostly contain flavonols [35]. Yellow flowers of *Gossypium herbaceum* and *G. barbadense* contain gossypitrin, isoquercitrin quercimeritrin and herbacitrin, and the red flowers of *G. arboreum* also contain isoquercitrin [35,50]. Throughout many years of introduction and breeding, the major cultivated and collected cotton germplasms were consisted of *G. hirsutum,* and the major color of cotton flower was cream (Figure 1). However, the pigment responsible for the cream color is also composed of flavonoids, and likely also flavonols, because silence of *F3H* in cotton results in white flowers (Figure 5A). The extract of red cotton flowers showed a high OD value at 530 nm, which indicates high anthocyanin content (Figure 2 and 5). All of these results confirmed that the color of cotton flowers is mainly related to flavonoids content.

The color of flavonoids is not stable and changes as the result of differences in pH and ion concentration [3,4]. The cotton flowers undergo dramatic color changes during their short lifespan (Figure 2). It has been hypothesized that these color changes could correlate with changes in the environment. Our results indicated that the color change was not primarily associated with pH variation but showed a strong relationship with the expression of anthocyanin biosynthesis genes (Figure 4 and S2). The results also showed that the accumulation of anthocyanin is responsible for the color changes (Figure 2 and 3). The color change is an endogenous genetic process rather than solely the effect of environment, which was in correspondence with the previous findings showing that the

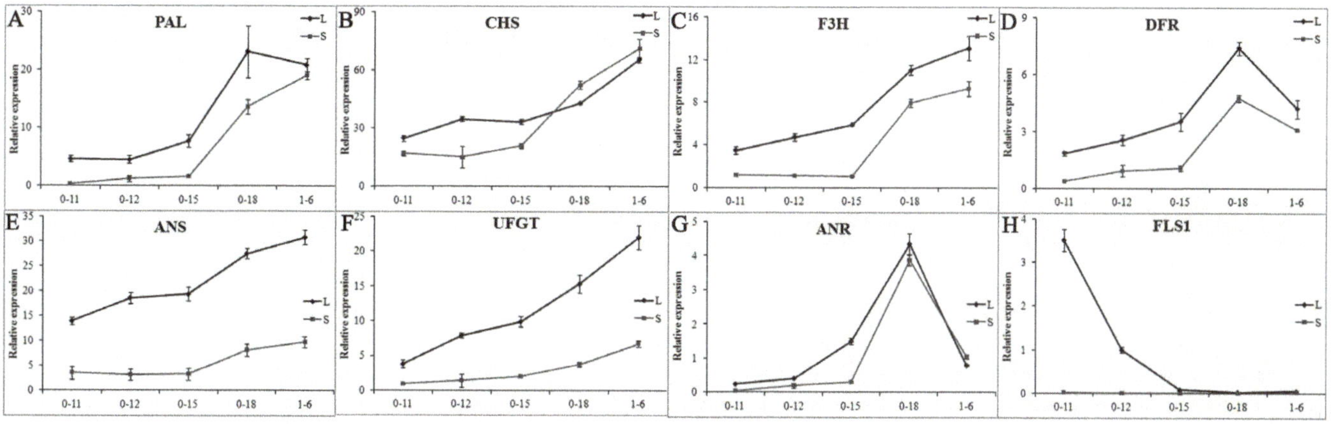

Figure 4. Gene expression analysis of the flavonoids biosynthesis pathway. Transcripts of *PAL* (A), *CHS* (B), *F3H* (C), *DFR* (D), *ANS* (E), *UFGT* (F), *ANR* (G) and *FLS1* (H) were analyzed by qPCR. Flowers of YZ1 with normal light (L) or in shade treatment (S) were collected from the field at five time points (from 11 am on 0 DPA (0-11) to 6 am on 1 DPA (1-6)). Transcripts were normalized with the expression of *UBQ7*. Three repeats were performed. Error bars represent SD.

constituents of cotton flower petal appear to be less variable as a result of environmental effects [39]. Light also play an important role in color changes of cotton flower, and the flower under shade treatment showed decreased and delayed anthocyanin accumulation (Figure 3). However, light only had limited effects on anthocyanin accumulation, as the blooming flowers under shade treatment showed the same cream color as those under normal light (Figure 3A), while the f3h lines showed white petals in both normal light and shading treatment during flower development (Figure 6A). Additionally, all flowers including F3H silencing lines with or without light exposure showed the same patterns of anthocyanin accumulation (Figure 3B and 6B). Gene expression analysis also showed the same expression pattern between flowers under light and shading treatment (Figure 4). These results further confirmed that the color change of cotton flowers is genetically controlled (Figure 3). The transcription of *ANR* genes that respond

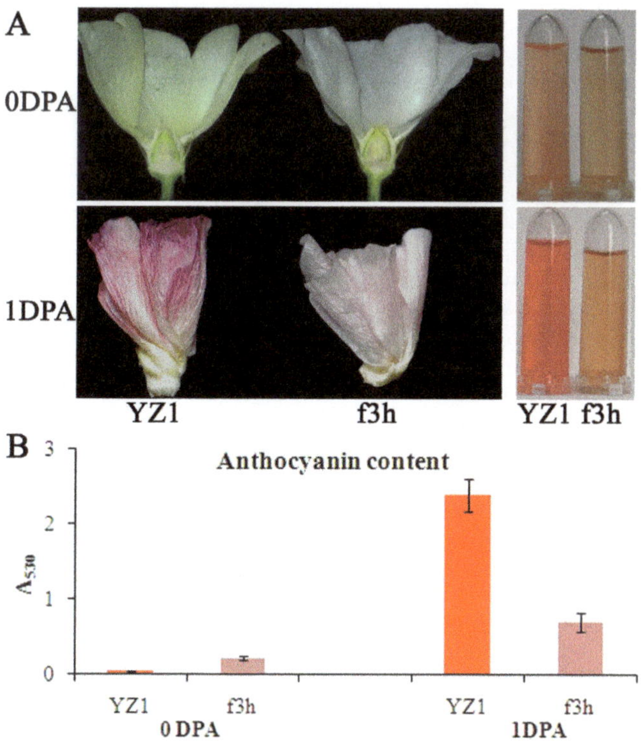

Figure 5. Anthocyanin analysis of YZ1 and f3h petals. A, petals of YZ1 and f3h were collected from the field at 0 and 1 DPA. The methanol extract is shown on the right side. Significant color change was observed. B, anthocyanin contents were measured at A_{530}. Error bars represent SD, and three repeats were performed. For each repeat, more than five flowers were analyzed.

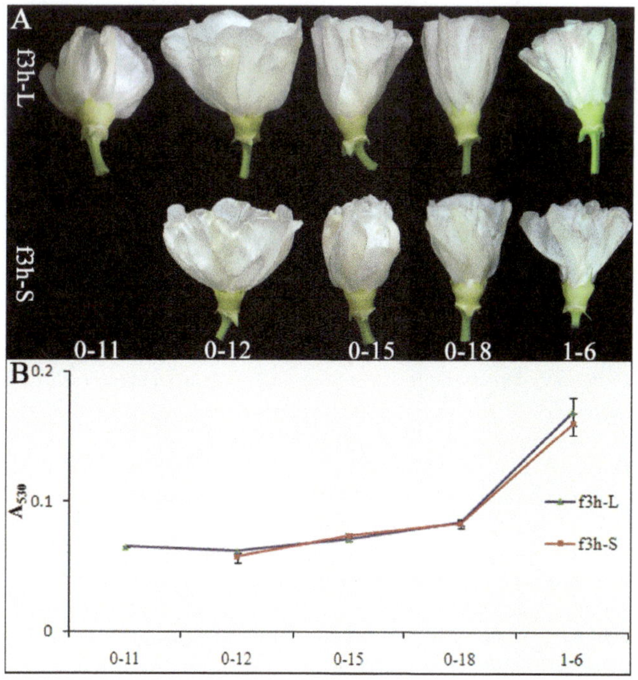

Figure 6. Flower color is affected by light and gene interaction. A, flowers of f3h lines were analyzed from 11 am on 0 DPA (0-11) to 6 am on 1 DPA (1-6). Flowers that were treated with normal light or shade are indicated by '-L' and '-S', respectively. B, anthocyanin contents for corresponding flowers were measured at A_{530}. Error bars represent SD, and three repeats were performed. For each repeat, more than five flowers were analyzed.

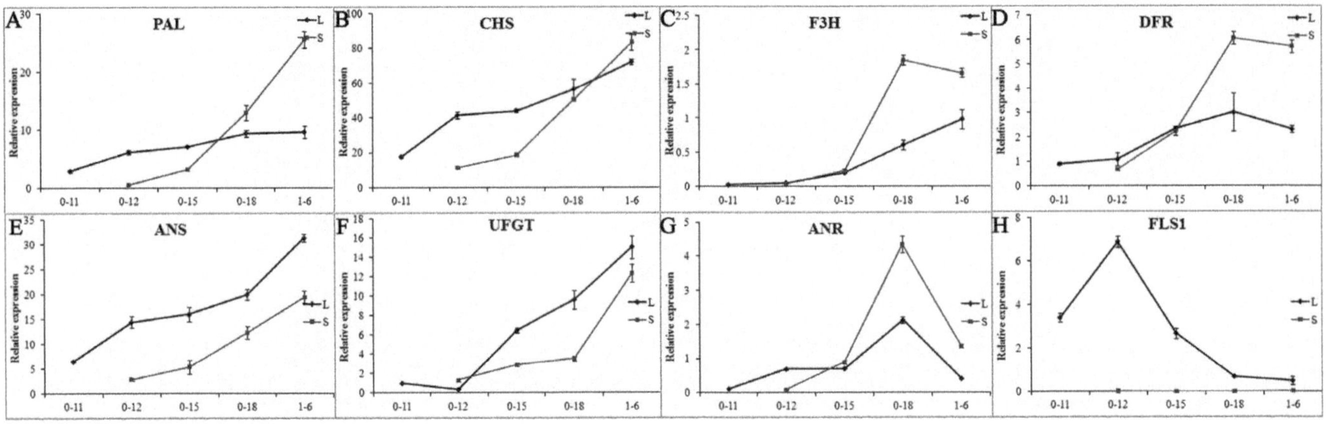

Figure 7. Gene expression of the flavonoids biosynthesis pathway was affected by both shade and *F3H* silence. Transcripts of *PAL* (A), *CHS* (B), *F3H* (C), *DFR* (D), *ANS* (E), *UFGT* (F), *ANR* (G) and *FLS1* (H) were analyzed by qPCR. Flowers of f3h with normal light (L) or in shade treatment (S) were collected from the field at five time points (from 11 am on 0 DPA (0-11) to 6 am on 1 DPA (1-6)). Transcripts were normalized with the expression of *UBQ7*, and three repeats were performed. Error bars represent SD.

to proanthocyanidin biosynthesis was showed to be specially time-point regulated, with a gradual increase from 11 am to a peak expression at 6 pm on 0 DPA; however, *ANR* showed less light response than the other genes (Figure 4G and S3D) and may not respond to the flower color change. Silence of *F3H* could significantly affect the expression of flavonoid genes, especially *PAL*, *DFR*, and *ANR*, which showed co-suppression with *F3H* (Figure 4 and 7). All of these genes showed higher expression levels in f3h lines following shade treatment compared to normal light exposure after 3 pm (Figure 7), when anthocyanin was beginning to accumulate in normal flower. The results indicated that there was a feedback control for these genes under normal conditions. The excess accumulation of flavonoids upstream of F3H or the deficiency of chemicals downstream of F3H could suppress the expression of *PAL*, *DFR* and *ANR* under light exposure. However, in shade treatment, the levels of excess flavonoids and deficient chemicals in the f3h lines should be less different with that of YZ1 comparing with that under light exposure. These results imply that there is a complex interaction between light and endogenous genetic factors to affect the flower color biosynthesis. It was different from that in *Viola cornuta* L. (Violaceae) flowers whose color change was totally light dependent [12].

Flavonol and anthocyanin were the main flavonoids found in the cotton flower. The accumulation of anthocyanin and the expression of anthocyanin biosynthesisrelated genes were gradually increased from 12 am on 0 DPA (Figure 2, 3 and 4). For flavonols, the expression of *FLS1* also showed light exposure controlled (Figure 4H and 7H). The highest transcriptional level of *FLS1* occurred at 11 am in all the three cotton species and was then dramatically decreased to a very low level by 3 pm (Figure S3). Silence of *F3H* not only increased the expression of *FLS* genes but also delayed the highest expression level at 12 am (Figure 7H and S4). These findings indicate that there is a light driven flavonols biosynthesis procedure during flower blooming in the morning of 0 DPA. Once this procedure was activated and suitable FLS proteins and flavonols were produced, the expression of *FLS* genes would be decreased. If the level of flavonols was insufficient under light exposure, such as what occurred in the *F3H* silenced flowers, then there was a feedback procedure to increase the expression of *FLS* genes to produce higher levels of flavonols. Additionally, the feedback procedure was strongest at 12 am and medially lasted to 6 pm of 0 DPA (Figure 7H). The two dominantly expressed *FLS* genes, *FLS1* and *FLS3*, were showed to be at a higher expression level at 12 am in the f3h lines and maintained a moderate level of expression afterwards, when low transcripts level of *FLS* genes were detected in YZ1 flowers (Figure 4H, 7H and S4B). According to the expression levels of flavonoid genes, a special flavonoid flux mechanism was present during cotton flower development. The previous high expression levels of *FLS* genes at 11 am made the flavonoid metabolism switch to flavonol biosynthesis in response to light exposure. However, it is still not clear about the mechanisms of the metabolic flux into proanthocyandin biosynthesis at 6 pm on 0 DPA and the final metabolic flux into anthocyanin biosynthesis at the end of the flower life in 1DPA.

Cotton is not a model plant and its genome is complex, so its genetic manipulation is difficult, even though one cotton genome has been sequenced [51,52]. However, the study of the short developmental life span and special genetic regulation of flavonoid biosynthesis in cotton flower would benefit to understand the flower development. Although fiber is the main product of cotton, the cotton flower could be engineered to improve the economic value of cotton. The yield of cotton flowers is far greater than that of cotton fiber; therefore, there is a great potential for cotton flowers to yield by-products, including abundant flavonoids and sugars, which would be beneficial for the chemical industry and agriculture. Furthermore, the cotton flower is the primary target of the cotton bollworm, which accounts for substantial yield losses in the cotton field [53]. Traditional genetically modified cotton cannot guarantee a persistent level of toxic BT proteins in the cotton flower, which would increase the survival of bollworm larvae [54]. Thus the cotton flower could be manipulated to protect against invasion by cotton bollworms. Our findings on the factors and regulations of the flavonoid pathways during the development of cotton flower color would provide a starting point for such studies.

Supporting Information

Figure S1 Effect of fertilization on the accumulation of anthocyanin in cotton flowers. A, normally fertilized (Y-F) and emasculated YZ1 flowers (Y-U) were collected from the field at 8 am on 1 DPA. B, anthocyanins of the Y-F and Y-U flowers were measured at A_{530}. Three repeats were performed. Error bars represent SD.

Figure S2 pH values of the developing cotton flowers. YZ1 and f3h flowers taken from five time points between 6 pm on -1 DPA (-1-18) and 6 am on 1 DPA (1-6) were collected and analyzed. Three repeats were performed. Error bars represent SD.

Figure S3 Expression pattern of flavonoid genes in developmental flowers of 3-79 and T586. The transcripts of *CHS* (A), *UFGT* (B), *FLS1* (C) and *ANR* (D) were analyzed at five time points (from 11 am, 12 am, 3 pm and 6 pm of 0 DPA (0-11, 0-12, 0-15 and 0-18) to 6 am on 1 DPA (1-6)). Transcripts were normalized with the expression of *UBQ7*. Three repeats were performed. Error bars represent SD.

Figure S4 Expression analysis of FLS2 and FLS3 in cotton flowers. Expression analysis of *FLS2* (A) and *FLS3* (B) in flowers of YZ1-L, shade-treated YZ1 (YZ1-S), f3h, 3-79 and T586 was performed with qPCR. Flowers at 11 am (0-11), 12 am (0-12) and 3 pm (0-15) on 0 DPA were collected for analysis. Transcripts were normalized with the expression of *UBQ7*. Three repeats were performed. Error bars represent SD.

Author Contributions

Conceived and designed the experiments: JT MW LT YN XZ. Performed the experiments: JT MW. Analyzed the data: JT MW LT XZ. Contributed reagents/materials/analysis tools: JT MW LT YN YL. Wrote the paper: JT MW LT YL XZ.

References

1. Grotewold E (2006) The genetics and biochemistry of floral pigments. Annu Rev Plant Biol 57: 761–780.
2. Thompson WR, Meinwald J, Aneshansley D, Eisner T (1972) Flavonols: pigments responsible for ultraviolet absorption in nectar guide of flower. Science (New York, NY) 177: 528–530.
3. Maekawa S, Inagaki N, Terabun M (1983) Effect of aluminum ions on the bluing of petal color in cut chinese bellflower, *Platycodon grandiflorum*. Plant Cell Physiol 24: 759–764.
4. Vlaming P, Schram AW, Wiering H (1983) Genes affecting flower colour and pH of flower limb homogenates in *Petunia hybrida*. Theor Appl Genet 66: 271–278.
5. Asen S, Stewart RN, Norris KH (1975) Anthocyanin, flavonol copigments, and pH responsible for larkspur flower color. Phytochemistry 14: 2677–2682.
6. Griesbach RJ (1992) Correlation of pH and light intensity on flower color in potted *Eustoma grandiflorum* grise. Hortscience 27: 817–818.
7. Irani N, Grotewold E (2005) Light-induced morphological alteration in anthocyanin-accumulating vacuoles of maize cells. BMC Plant Biol 5: 7 doi:10.1186/1471-2229-5-7.
8. Takos AM, Jaffé FW, Jacob SR, Bogs J, Robinson SP, et al. (2006) Light-induced expression of a MYB gene regulates anthocyanin biosynthesis in red apples. Plant Physiol 142: 1216–1232.
9. Delph LF, Lively CM (1989) The evolution of floral color change: pollinator attraction versus physiological constraints in *Fuchsia excorticata*. Evolution 43: 1252–1262.
10. Li J, Ou-Lee TM, Raba R, Amundson RG, Last RL (1993) Arabidopsis flavonoid mutants are hypersensitive to UV-B irradiation. Plant Cell 5: 171–179.
11. Weiss MR (1995) Floral color change: a widespread functional convergence. Am J Bot: 167–185.
12. Farzad M, Griesbach R, Weiss MR (2002) Floral color change in *Viola cornuta* L. (Violaceae): a model system to study regulation of anthocyanin production. Plant Science 162: 225–231.
13. Ne'eman G, Nesher R (1995) Pollination ecology and the significance of floral color change in *Lupinus pilosus* L.(Fabaceae). Isr J Plant Sci 43: 135–145.
14. Niesenbaum RA, Patselas MG, Weiner SD (1999) Does flower color change in *Aster vimineus* cue pollinators? Am Midl Nat 141: 59–68.
15. Ida TY, Kudo G (2003) Floral color change in *Weigela middendorffiana* (Caprifoliaceae): reduction of geitonogamous pollination by bumble bees. Am J Bot 90: 1751–1757.
16. Martin C, Gerats T (1993) Control of pigment biosynthesis genes during petal development. Plant Cell 5: 1253–1364.
17. Martin C, Prescott A, Mackay S, Bartlett J, Vrijlandt E (1991) Control of anthocyanin biosynthesis in flowers of *Antirrhinum majus*. Plant J 1: 37–49.
18. Mol J, Grotewold E, Koes R (1998) How genes paint flowers and seeds. Trends Plant Sci 3: 212–217.
19. Koes R, Verweij W, Quattrocchio F (2005) Flavonoids: a colorful model for the regulation and evolution of biochemical pathways. Trends Plant Sci 10: 236–242.
20. Weiss D (2000) Regulation of flower pigmentation and growth: Multiple signaling pathways control anthocyanin synthesis in expanding petals. Physiol Plant 110: 152–157.
21. Vogt T, Ibdah M, Schmidt J, Wray V, Nimtz M, et al. (1999) Light-induced betacyanin and flavonol accumulation in bladder cells of *Mesembryanthemum crystallinum*. Phytochemistry 52: 583–592.
22. Albert NW, Lewis DH, Zhang H, Irving LJ, Jameson PE, et al. (2009) Light-induced vegetative anthocyanin pigmentation in *Petunia*. J Exp Bot 60: 2191–2202.
23. Buer CS, Imin N, Djordjevic MA (2010) Flavonoids: new roles for old molecules. J Integr Plant Biol 52: 98–111.
24. Gonzalez A, Zhao M, Leavitt JM, Lloyd AM (2008) Regulation of the anthocyanin biosynthetic pathway by the TTG1/bHLH/Myb transcriptional complex in Arabidopsis seedlings. Plant J 53: 814–827.
25. Dixon RA, Xie DY, Sharma SB (2005) Proanthocyanidins – a final frontier in flavonoid research? New Phytol 165: 9–28.
26. Pelletier MK, Shirley BW (1996) Analysis of flavanone 3-hydroxylase in Arabidopsis seedlings. Coordinate regulation with chalcone synthase and chalcone isomerase. Plant Physiol 111: 339–345.
27. van Tunen AJ, Koes RE, Spelt CE, van der Krol AR, Stuitje AR, et al. (1988) Cloning of the two chalcone flavanone isomerase genes from *Petunia hybrida*: coordinate, light-regulated and differential expression of flavonoid genes. EMBO J 7: 1257–1263.
28. Koes R, Spelt C, Mol JM (1989) The chalcone synthase multigene family of *Petunia hybrida* (V30): differential, light-regulated expression during flower development and UV light induction. Plant Mol Biol 12: 213–225.
29. Kubasek W, Ausubel F, Shirley B (1998) A light-independent developmental mechanism potentiates flavonoid gene expression in Arabidopsis seedlings. Plant Mol Biol 37: 217–223.
30. Burbulis IE, Winkel-Shirley B (1999) Interactions among enzymes of the Arabidopsis flavonoid biosynthetic pathway. Proc Natl Acad Sci U S A 96: 12929–12934.
31. Winkel BSJ (2004) Metabolic channeling in plants. Annu Rev Plant Biol 55: 85–107.
32. Owens DK, Alerding AB, Crosby KC, Bandara AB, Westwood JH, et al. (2008) Functional Analysis of a Predicted Flavonol Synthase Gene Family in Arabidopsis. Plant Physiol 147: 1046–1061.
33. Fryxell P (1992) A revised taxonomic interpretation of *Gossypium* L.(Malvaceae). Rheedea 2: 108–165.
34. Stephens S (1948) A biochemical basis for the pseudo-allelic anthocyanin series in *Gossypium*. Genetics 33: 191–214.
35. Neelakantam K, Ramachandra Rao RH, Seshadri TR (1935) Pigments of cotton flowers. Part I. Cambodia (*G. hirsutum*). Proc Indian Acad Sci - Section A 1: 887–890.
36. Stephens S (1954) Interspecific homologies between gene loci in *Gossypium*. II. Corolla color. Genetics 39: 712–723.
37. Parks CR (1965) Floral pigmentation studies in the genus *Gossypium*. I. Species specific pigmentation patterns. Am J Bot: 309–316.
38. Chhabra N, Singh O (1978) Factors affecting anthocyanin synthesis in cotton flowers. Physiology of sexual reproduction in flowering plants, Kalyani Publishers, New Delhi: 136–141.
39. Parks C, Sandhu S, Montgomery K (1972) Floral pigmentation studies in the genus *Gossypium*. IV. Effects of different growing environments on flavonoid pigmentation. Am J Bot: 158–164.
40. Parks CR (1965) Floral Pigmentation Studies in the Genus *Gossypium*. II. Chemotaxonomic Analysis of Diploid Gossypium Species. Am J Bot: 849–856.
41. Hedin PA, Jenkins JN, Collum DH, White WH, Parrott WL, et al. (1983) Cyanidin-3-β-glucoside, a newly recognized basis for resistance in cotton to the tobacco budworm *heliothis virescens* (Fab.) (lepidoptera: Noctuidae). Experientia 39: 799–801.
42. Waage SK, Hedin PA (1984) Biologically-active flavonoids from *Gossypium arboreum*. Phytochemistry 23: 2509–2511.
43. Fryxell PA (1971) Phenetic analysis and the phylogeny of the diploid species of *Gossypium* L.(Malvaceae). Evolution: 554–562.
44. Parks C, Ezell W, Williams D, Dreyer D (1975) The application of flavonoid distribution to taxonomic problems in the genus *Gossypium*. Bulletin of the Torrey Botanical Club: 350–361.

45. Liang Z, Jiang R, Zhong W (1997) New red flower germplasm lines of cotton selected from hybrid of *Gossypium hirsutum* X *G. bickii*. Science in China Series C: Life Sciences 40: 284–292.
46. Tan J, Tu L, Deng F, Hu H, Nie Y, et al. (2013) A genetic and metabolic analysis revealed that cotton fiber cell development was retarded by flavonoid naringenin. Plant Physiol 162: 86–95.
47. Mancinelli AL, Schwartz OM (1984) The Photoregulation of Anthocyanin Synthesis IX. The photosensitivity of the response in dark and light-grown tomato seedlings. Plant Cell Physiol 25: 93–105.
48. Tu LL, Zhang XL, Liang SG, Liu DQ, Zhu LF, et al. (2007) Genes expression analyses of sea-island cotton (*Gossypium barbadense* L.) during fiber development. Plant Cell Rep 26: 1309–1320.
49. Tan J, Tu L, Deng F, Wu R, Zhang X (2012) Exogenous jasmonic acid inhibits cotton fiber elongation. J Plant Growth Regul 31: 599–605.
50. Neelakantam K, Seshadri TR (1937) Pigments of cotton flowers. PartIV. Constitution of herbacitrin and herbacetin-new glucoside and aglucone (flavonol). Proc Indian Acad Sci - Section A 5: 357–364.
51. Paterson AH, Wendel JF, Gundlach H, Guo H, Jenkins J, et al. (2012) Repeated polyploidization of *Gossypium* genomes and the evolution of spinnable cotton fibres. Nature 492: 423–427.
52. Wang K, Wang Z, Li F, Ye W, Wang J, et al. (2012) The draft genome of a diploid cotton *Gossypium raimondii*. Nat Genet 44: 1098 – 1103.
53. Casimero V, Tsukuda R, Nakasuji F, Fujisaki K (2000) Effect of larval diets on the survival and development of larvae in the cotton bollworm, *Helicoverpa armigera* Hübner (Lepidoptera: Noctuidae). Appl Entomol Zool 35: 69–74.
54. Yang Y, Johnson ML, Zalucki MP (2008) Possible effect of genetically modified cotton on foraging habits of early instar *Helicoverpa armigera* (Hübner) (Lepidoptera: Noctuidae) larvae. Aust J Entomol 47: 137–141.

22

Resveratrols in Grape Berry Skins and Leaves in *Vitis* Germplasm

Lijun Wang[1], Man Xu[1,2], Chunyan Liu[1], Junfang Wang[1,2], Huifen Xi[1,2], Benhong Wu[1], Wayne Loescher[3], Wei Duan[1], Peige Fan[1], Shaohua Li[1,4]*

1 Key Laboratory of Plant Resources and Beijing Key Laboratory of Grape Science and Enology, Institute of Botany, the Chinese Academy of Sciences, Beijing, China, **2** University of Chinese Academy of Sciences, Beijing, China, **3** College of Agriculture and Natural Resources, Michigan State University, East Lansing, Michigan, United State of America, **4** Key Laboratory of Plant Germplasm Enhancement and Specialty Agriculture, Wuhan Botanical Garden, the Chinese Academy of Sciences, Wuhan, China

Abstract

Background: Resveratrol is an important stilbene that benefits human health. However, it is only distributed in a few species including grape and is very expensive. At present, grape has been an important source resveratrol. However, the details are scarce on resveratrol distribution in different *Vitis* species or cultivars.

Methodology/Principal Finding: The composition and content of resveratrols were investigated by HPLC for assessing genotypic variation in berry skins and leaves of 75 grape cultivars, belonging to 3 species and 7 interspecific hybrids. *Trans*-resveratrol, *cis*-piceid and *trans*-piceid were detected in berry skins and leaves, but *cis*-resveratrol was not. Resveratrol content largely varied with genetic background as well as usage. In most cultivars, total resveratrol including the above three compounds was higher in berry skins than leaves. In berry skins of most cultivars and leaves of almost all cultivars, *cis*-piceid was the most abundant resveratrol; *trans*-resveratrol and *trans*-piceid were minor components. Some specific cultivars were found with extremely high levels of *trans*-resveratrol, *cis*- piceid, *trans*-piceid or total resveratrols in berry skins or leaves. In skins and leaves, rootstock cultivars had a higher content of total resveratrols, and the cultivated European type cultivars and their hybrids with *V. labrusca* had relatively low totals. There were no significant correlations of the amounts of total resveratrols or any individual resveratrol between berry skins and leaves. All 75 cultivars can be divided into four groups based on the composition of resveratrols and their concentration by principal component analysis.

Conclusion: Resveratrol content of grape berries and leaves varied largely with their genetic background and usage. Rootstock cultivars had a higher content of total resveratrols than the other germplasm. Total resveratrols were lower in leaves than berry skins in most cultivars. *Cis*-piceid was the most abundant resveratrol in most cultivars, and *trans*-res and *trans*-pd were minor components.

Editor: M. Lucrecia Alvarez, TGen, United States of America

Funding: This work is supported by the grants from National Natural Science Foundation of China (Grant number 31171918). No additional external funding was received for this study. The funders had no role in study design, data collection and analysis, decision to publish, or preparation of the manuscript.

Competing Interests: The authors have declared that no competing interests exist.

* E-mail: shhli@wbgcas.cn

Introduction

The polyphenol *trans*-resveratrol (*trans*-3,5,4′-trihydroxystilbene), a member of the stilbene family, was first isolated from white hellebore (*Veratrum grandiflorum O.*) [1]. It has become the focus of a number of studies in medicine and plant physiology, and has also emerged as a molecule that may affect human health. In leaves and berries of grape (*Vitis*), *trans*-resveratrol is a phytoalexin that is produced in response to stresses, such as wounding or pathogen attack [2]. Due to its daily consumption in the form of red wine [3–4], *trans*-resveratrol has been associated with the "French paradox" for having beneficial effects [5], particularly protection against coronary diseases [6]. In addition, *trans*-resveratrol may have anticancer properties, as suggested by its ability to suppress proliferation of a wide variety of tumor cells, including lymphoid and myeloid cancers; multiple myeloma; cancers of the breast, prostate, stomach, colon, pancreas, and thyroid; melanoma; head and neck squamous cell carcinoma; ovarian carcinoma and cervical carcinoma [7–8]. Nowadays, the main market for *trans*-resveratrol is in nutraceuticals, and an increasing awareness of the possible health benefits of the chemical is likely to further increase consumer demand for the product. Consequently, it may be necessary to develop new sources for obtaining the chemical.

The various resveratrols include *trans*- or *cis*-resveratrol and the corresponding piceids. *Trans*-resveratrol is formed via the phenylalanine/polymalonate route, where stilbene synthase plays a critical step. *Trans*-resveratrol may also be transformed into *cis*-resveratrol. *Trans*- and *cis*-resveratrols may also be transformed into *trans*- and *cis*-piceids by glycosyltransferases. Glycosylation of polyphenolic compounds commonly occurs in plants to protect the plant cell from their potential toxic effects. At the same time, glycosylation may protect resveratrol from oxidation and enzymatic degradation, thus enhancing its stability [9]. Piceids are especially interesting for human health because they have also been reported to inhibit DNA synthesis in Lewis lung carcinoma cells and the formation of capillary-like tube networks (angiogenesis) of human umbilical vein endothelial cells [10]. They also inhibit thromboxane B2 in the human body [9].

Grape (*Vitis*) is one of the most important fruit crops in the world. Viticulture and enology play an important role in the economy of many developed and emerging countries. Gross world production of grape berries was estimated at more than 68.35 million metric tons in 2010 (FAO STAT Database at www.fao.org). About 27% of the grapes are consumed as fresh fruit (table grapes) and 2% as dried fruit, whereas 71% of the crop is processed, especially for winemaking.

The grapes cultivated throughout the world today mainly belong to three types, the European type (*V. vinifera* L.), the American bunch type (*V. labrusca* L.), and their derivatives, especially the hybrids obtained from *V. labrusca* and *V. vinifera*. Given the potential importance of the resveratrols, it is important to evaluate the content of resveratrols in grape germplasm in order to better utilize germplasm resources directly or in breeding programs to obtain new grape cultivars rich in resveratrols in berries and other tissues. Previous evaluations have only focused on *trans*-resveratrol in berries [11–14], and none have evaluated all forms of resveratrol, i.e., *trans*-resveratrol, *cis*-resveratrol, *trans*-piceid and *cis*-piceid. Although grape leaves are also rich in resveratrols [7], there are no evaluations of foliar resveratrols in grape cultivars. The objective of the present study was to evaluate resveratrol distribution in cultivars, hybrids, and species, focusing on content of all four forms in berries and leaves in order to provide information for direct use in the wine and resveratrol industries and to assist in the choice of parents in breeding programs.

Materials and Methods

Plant material

Seventy-five cultivars were evaluated in this study in 2009 (Table 1). These cultivars included: 34 table grapes from hybrids between *V. vinifera* and *V. labrusca*; 15 table grapes of *V. vinifera*; 15 wine grapes of *V. vinifera*; 3 interspecific juice grapes: Honey Juice (*V. labrusca* × *V. vinifera*), Beixiang (*V. thunbergii* × *V. vinifera*), Russian Concord (*V. labrusca* × *V. amurensis*) and 1 *V. labrusca* juice grape; 4 interspecific rootstock cultivars: Beta (*V. labrusca* × *V. riparia*), Shanhe 2 (*V. amurensis* × *V. riparia*), Zhi 168 (*V. monticola* × *V. riparia*) and Dog Ridge (*V. rupestris* × *V. berlandieri*), and one *V. berlandieri* rootstock cultivar, respectively. All samples were collected from the experimental vineyard of the grape germplasm repository in the Institute of Botany, the Chinese Academy of Sciences located in Beijing. All of these cultivars were planted in the spring of 1993. The grapevines were trained to cordons, spaced 1.5 m apart within the row and 2.5 m apart between rows with a north-south row orientation. They were subjected to the same management practices, such as irrigation, fertilization, soil management, pruning, and disease control.

Grape berries were sampled at ripening according to the former years' ripening date and as judged from seed color change to dark brown without senescence of berry tissue. Berries were sampled from three clusters, randomly chosen in three vines of each cultivar as three replicates. The leaf samples were taken from the seventh to ninth leaves of a new shoot on each grapevine with three replicates. The skins were separated by hand. All the samples of fruit skins and leaves were immediately frozen in liquid nitrogen, and then stored at −40°C.

Extraction of resveratrols from berry skins and leaves

The frozen materials were ground to a powder in liquid nitrogen using a grinding machine (A11-b-s25, IKA, China). Each sample was extracted for 24 h in methanol and ethyl acetate (1/1, v/v) (1000 mg per 10 mL of organic solvent) at 25°C in darkness according to Liu et al. [15]. The suspension was centrifuged at 10,000× g for 10 min. The supernatant liquid was separated for collection. The remaining residue was extracted with 3 mL of methanol and ethyl acetate (1/1, v/v) and centrifuged. The two supernatants were combined and vacuum-dried with a rotary evaporator (N-1001D-WD, EYELA, Tokyo Rikakikai, Japan) at 40°C. The dried samples were then re-dissolved in 2 mL of pure methanol and stored at −40°C for resveratrol analysis.

Measurement of resveratrols

Each liquid sample was filtered through a 0.22 µm PTFE membrane filter, then the analyses were carried out on a Dionex P680 HPLC system (Dionex Corporation, CA, USA) with a Dionex PDA-100 detector. Separation was achieved using a reverse-phase C18 column (Atlantis® T3; 4.6 mm×250 mm, 5.0 µm particle size, Waters, USA) and a guard column (Atlantis T3, 4.6 mm×20 mm, 5.0 µm cartridge, Waters, USA) maintained at 30°C with a Dionex TCC-100 thermostated column compartment, and an injection volume was 10 µL. Separation was performed at a flow rate of 1.0 mL·min^{-1} with the mobile phase consisting of H_2O (A) and acetonitrile (B). The solvent gradient was as follows: 0–5 min from 10% to 17% solvent B; 5–12 min from 17% to 18% solvent B; 12–22 min from 18% to 22% solvent B; 22–30 min from 22% to 33% solvent B; 30–45 min from 33% to 38% solvent B; 45–50 min from 38% to 80% solvent B; 50–53 min from 80% to 10% solvent B; and, 53–60 min 10% solvent B. For fluorimetric detection, the maximum absorption wavelength of the two *trans*-isomers (*trans*-resveratrol and *trans*-piceid) was 306 nm, and the two *cis*-isomers (*cis*-resveratrol and *cis*-piceid) was 288 nm. Three replicates of each sample were also scanned from 240 to 600 nm. *Trans*-resveratrol and *trans*-piceid standards were purchased from Sigma-Aldrich (St. Louis, MO, USA) and the Chinese Standards Research Institute, respectively. The mixed solution of the *trans*-resveratrol (*trans*-res) and *trans*-piceid (*trans*-pd) standards were partly converted to the *cis*-resveratrol (*cis*-res) and *cis*-piceid (*cis*-pd) after UV-C irradiation at 6 W m^{-2} and a distance of 30 cm for 30 min. Conversion coefficients were computed from the two *trans*-isomers, respectively, and standard curves of the four isomers were made. The total resveratrol content was obtained from the sum of the four isomers.

Graphs and data analysis

The variations of extractable amounts of the different forms of resveratrol in berry skins and leaves were analyzed, as was the distribution of the resveratrols in the samples, by S-Plus 2000 (MathSoft Inc., Cambridge, MA). Boxplots were used to display range, median, and distribution density of variables in the samples [16]. The lower and upper quartile values were indicated by the height of the box, or the interquartile range (IQR). Whiskers indicated the range of the data and were represented as vertical lines ending in a small horizontal line. Extreme values outside these whiskers were indicated by asterisks. In all the Tables and Figures, experimental data were subjected to analysis of variance using the SPSS 13.0 program (SPSS, USA.). Means were separated by Student-Newman-Keuls multiple range tests at $P<0.05$.

Results and Discussion

Resveratrol in grape berry skins

The frequency distribution and median of total resveratrols in grape berry skins are shown in Fig. 1. Among all the genotypes,

Table 1. Grape cultivars used in this study.

Germplasm groups	Cultivar number	Cultivars
Table grape of LV[a] (T-LV)	34	White Olimpia (1), Triumph (2), Dabao (3), Takasumi (4),Takatsuma (5), Guixiangyi (6), Himrod (7), Black Olimpia (8), Benni Fuji (9), Red Queen (10), Yigawa 1011(11), Beniyamabiko (12), Beniizu (13), Hanazawa Kyoho (14), Mars Seedless (15), Jiangshang (16), Venus Seedless (17), Yigawa 1055 (18), Yigawa 11015 (19), Yigawa 11060 (20), Kyoho (21), Queenora Seedless (22), Honey Red (23), Jasmine (24), Pondicherry (25), Beni Sajku (26), Fujiminori (27), Tano Red (28), Venus (29), Izunishiki (30), Zhi 180 (31), Zhiyuan 540 (32), Ziguang (33), Shigyoku (34)
Table grape of V[b] (T-V)	15	Gros Colman (35), Misket Dounvaski (36), Flame Tokay (37), Fenniu (38), 'Fenghuang 51' (39), Hiro Hamburg (40), Guibao (41), Jingdajing (42), Jingfeng (43), Jingzaojing (44), Jingzhijing (45), Suffolk Seedless (46),Su 44 (47), Superior Seedless (48), Zhengzhouzaohong (49)
Wine grape of V (W-V)	15	Baiyu (50), Ugni Blanc (51), Bujiesuli (52), Meichun (53), Merlot (54), Cabernet Franc (55), Semillon (56), Suntory (57), Cabernet Gernischet (58), Lion Riesling (59), Wuyuezi (60), Chardonnay (61), Yan 73 (62), Italian Riesling (63), Zexiang (64)
Juice grape (J)	11	Honey juice (*V. labrusca* × *V. vinifera*, 65), Beixiang (*V. thunbergii* × *V. vinifera*, 66), Russian Concord (*V. labrusca* × *V. amurensis*, 67), Concord (*V. labrusca*, 68)
Wine grape of VA[c] (W-VA)	2	Beihong (69), Beimei (70)
Rootstock grape (R)	5	Beta (*V. labrusca* × *V. riparia*, 71), Shanhe 2 (*V. amurensis* × *V. riparia*, 72), Zhi 168 (*V. monticola* × *V. riparia*, 73), Berlandier Resseguier 2 (*V. Berlandieri*,74); Dog Ridge (*V. Berlandieri*,75)

The numbers in parentheses following the cultivar indicates the accession number.
[a]LV, hybrids between *V. labrusca* and *V. vinifera;*
[b]V, V. vinifera;
[c]VA, hybrids between *V. vinifera* and *V. amurensis*.

rootstock cultivars had the highest amounts of resveratrols in skins, ranging from 20.28 to 365.98 μg g^{-1} fresh weight (FW) with a median of 235.32 μg g^{-1} FW. However, total resveratrols in berry skins of the rootstocks varied significantly with their genetic backgrounds. The highest amounts in skin were found in 'Berlandier Resseguier No.2' (74) at 365.98 μg g^{-1} FW, followed by 'Zhi168' (73) with 344.68 μg g^{-1} FW, and 'Shanhe 2' (72) with 235.32 μg g^{-1} FW. Among the rootstock cultivars, 'Dog Ridge' (75) had the lowest amount of resveratrols with only 20.28 μg g^{-1} FW.

Juice grapes and two interspecific hybrid wine grapes from *V. vinifera* and *V. amurensis* had relatively high resveratrol content. Their medians were higher than those of table grapes. In the above juice and wine groups, 'Concord' (68) had very high amounts of resveratrols with 146.60 μg g^{-1} FW, while resveratrols in 'Russian Concord'(67) were very low with only 0.5 μg g^{-1} FW.

Among most of the grape germplasm groups studied, berry skin resveratrol content was lower in the wine and table grapes of *V. vinifera*, and the interspecific hybrids between *V. labrusca* and *V. vinifera*, as compared with the previous groups. Moreover, the distribution of their totals largely skewed towards a low content with a median much lower than the average (Fig. 1). However, high resveratrol content was found in berry skins of several cultivars: 'Venus' (154.77 μg g^{-1} FW, 29), 'Queenors Seedless'(102.19 μg g^{-1} FW, 22) and 'Takasumi'(91.09 μg g^{-1} FW, 4) in hybrid table grapes between *V. labrusca* and *V. vinifera*; 'Gros Colman' (123.12 μg g^{-1} FW, 35) for table grapes of *V. vinifera*; 'Cabernet Franc'(100.44 μg g^{-1} FW, 55) and Yan 73 (88.70 μg g^{-1} FW, 62) for wine grapes of *V. vinifera*.

Resveratrols in grape leaves

The content of total resveratrols were much lower in leaves than in fruit skins (Fig. 2). Similar to fruit skins, the rootstock cultivars had the highest resveratrol content in leaves among all the grape genotypes studied, and juice and hybrid wine grapes from *V. vinifera* and *V. amurensis* had a relatively high content. All cultivars of *V. vinifera* and the hybrid cultivars of *V. labrusca* and *V. vinifera* had low total resveratrol content in the leaves.

As regards rootstocks, 'Dog Ridge' (75) had the highest leaf content, with 71.50 μg g^{-1} FW, and the four other cultivars had a much lower leaf resveratrol content than 'Dog Ridge', ranging from 10 to 30 μg g^{-1} FW. Several germplasm groups had a relatively high leaf content of total resveratrols: the hybrid winemaking cultivar 'Beihong' (69) with 32.63 μg g^{-1} FW, three table grape cultivars from hybrids between *V. labrusca* and *V. vinifera*, 'Queenora Seedless' (22), 'Fujiminori' (25) and 'Ziguang' (33) (with 51.03, 25.08 μg g^{-1} FW, and 23.26 μg g^{-1} FW, respectively), a wine grape of *V. vinifera*, 'Zexiang' (64) with 58.78 μg g^{-1} FW, two

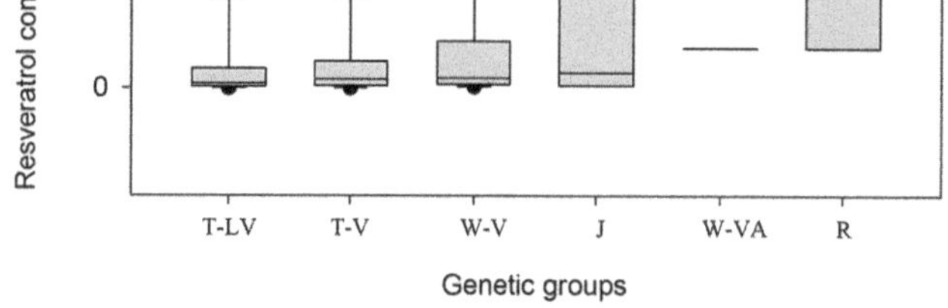

Figure 1. Range and distribution of total resveratrols in grape skins. The horizontal lines in the interior of the box are the median values. The height of a box is equal to the interquartile distance, indicating the distribution for 50% of the data. Approximately 99% of the data falls inside the whiskers (the lines extending from the top and bottom of the box). The data outside these whiskers are indicated by solid spots. T-LV, T-V, W-V, J, W-VA and R represent the germplasm groups as in Table 1.

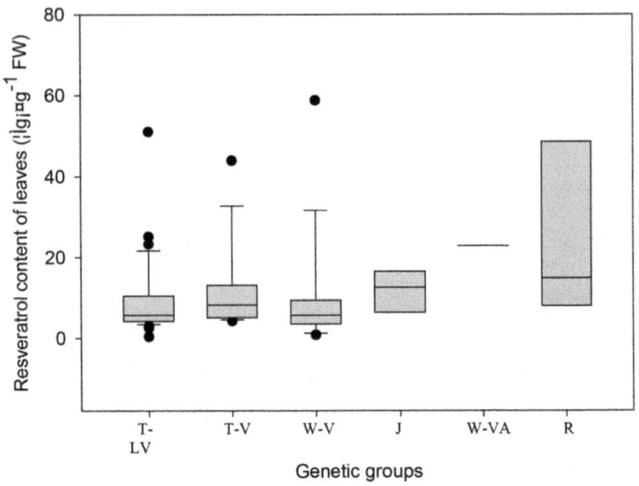

Figure 2. Range and distribution of total resveratrols in grape leaves. The notes about the figure and grape genotypes are the same as in Fig. 1.

table grapes of *V. vinifera*, 'Fenghuang 51' (39) and 'Gros Colman' (35) (with 43.93 and 25.30 μg g^{-1} FW, respectively).

Composition of resveratrols in berry skin and leaves

In this study, *cis*-res was not detected in any germplasm, and was thus omitted in Table 2. As shown in Table 2, the composition of berry skin resveratrols in the germplasm differed from that in leaves. In the berry skins, there are 10 cultivars which had only *trans*-res and *cis*-pd, 13 cultivars with only *trans*-res and *trans*-pd, and 5 cultivars with only *trans*-res. But, in leaves, almost all cultivars had three resveratrols, except 'Himrod (7)' and 'Bujiesuli (52)'.

The content of the different resveratrols varied with the genetic background. In berry skins, *trans*-res, *trans*-pd and *cis*-pd contents ranged from 0.01 to 84.63 μg g^{-1}FW, from 0 to 24.49 μg g^{-1}FW and from 0 to 360.12 μg g^{-1}FW, respectively.

There were five cultivars whose *cis*-pd contents in berry skins were more than 100 μg g^{-1}FW, especially 'Zhi168' (73), 'Berlandier Resseguier No. 2' (74) and 'Dog Ridge' (75) with more than 200 μg g^{-1}FW. The content of *trans*-res was also high in 'Venus' (29) with 84.63 μg g^{-1}FW. However, the content ranges of *trans*-res, *trans*-pd and *cis*-pd were 0.02–4.58 μg g^{-1}FW, 0.03–25.99 μg g^{-1}FW and 0–40.93 μg g^{-1}FW, respectively, in leaves.

As regard the percentages of the different resveratrol components, *cis*-pd accounted for more than 90% of the total resveratrols in berry skins of rootstocks while less than 56% in the other germplasm groups (Table 2). Moreover, both *cis*-pd and *trans*-pd were not detected in 'White Olympia' (1), 'Red Queen' (10), 'Ziguang' (33), 'Jingzaojing' (44) and 'Merlot' (54). Interestingly, the total resveratrol content of rootstock berry skin were much higher than that in the other groups, while 5 cultivars without *trans*-pd and *cis*-pd had very low total resveratrol. Moreover, the correlation among different resveratrol types (*trans*-res, *trans*-pd and *cis*-pd) was very low in berry skins and leaves; the only high correlation coefficients (0.973) were between total resveratrols and *cis*-pd in skins (Table 3). Those results signified that the total resveratrols in berry skins were mainly from the accumulation of *cis*-pd. This was likely due to the stability of *cis*-pd in the tissues [17]. It also suggested that glycosylation, especially glycosylation of *cis*-pd, was a critical step in accumulation of total resveratrols in berry skins. In addition, *cis*-res was not detected in berry skins even though *cis*-pd was a main component, indicating that *cis*-res was rapidly glycosylated to *cis*-pd after *cis*-res was converted from *trans*-res.

The percentage of *cis*-pd was about 60% of total resveratrols in leaves of rootstock grapevines, much lower than that in berry skins (Table 2), and was even lower than that in leaves from the other groups. Moreover, almost all germplasm had all three resveratrols in leaves at low levels. There was no correlation in resveratrol content between leaves and berry skins (Table 3, diagonal). In contrast to berry skin, there were significant correlations between all three resveratrols in leaves (Table 3). From the differences in content and the proportion of each resveratrol, we suggest that the mechanism regulating the accumulation of resveratrols in leaves differs from that in berry skins. As in skins, *trans*- and *cis*-pd were also the main contributors to the total resveratrols in leaves. Gatto et al. (2008) also showed that the grape cultivars with higher resveratrol accumulate resveratrol preferentially in the glycosylated forms, *trans*- and *cis*-pd [18]. Three possible roles for resveratrol glucosides in berry skins and leaves include storage of resveratrol, transport from cytoplasm to apoplast, and protection of *trans*-resveratrol from peroxidative degradation [17].

Principal component analysis (PCA)

PCA was used to analyze the data for *trans*-res, *trans*-pd and *cis*-pd from berry skins and leaves of 75 grape cultivars. Fig. 3 shows the relationships between the cultivars (scores) and their resveratrols (loadings). The two PCs carried a large amount of important information and accounted for 73.6% and 95.3% of the total variance for resveratrols in berry skins and in leaves, respectively. The PCA scatter plots of berry skins and leaves are shown in Fig. 3A and Fig. 3B, respectively, and the corresponding loading plots, which establish the relative importance of the variables, are shown in Fig. 3E and Fig. 3F, respectively.

With regard to resveratrols in berry skin, the grape cultivars can be divided into four groups based on the positions in the PCA scatter plots (Fig. 3A). Group 1: The group situated in the low right part of the third quadrant was comprised of three rootstock cultivars, 'Shanhe 2' (72), 'Zhi168' (73), 'Berlandier Resseguier No.2' (74); and one wine-making cultivar 'Cabernet Franc' of *V. vinifera* (55). This group was characterized by a very high content of *cis*-pd. Group 2: The group was also situated in the left part of the third quadrant and was comprised of three table grape cultivars from hybrids between *V. labrusca* and *V. vinifera* ['Takasumi' (4); 'Kyoho'(21); 'Izunishiki'(30)],four table grape cultivars of *V. vinifera* ['Gros Colman'(35); 'Jingfeng'(43); 'Jingzhijing'(45) and 'Zhengzhouzaohong'(49)], one wine-making cultivar of *V. vinifera* ['Bujiesuli' (52)], and one juice cultivar from *V. labrusca* ['Concord' (68)]. This group was characterized by a relatively higher content of *trans*-pd or *cis*-pd. Group 3: This group was situated mainly in the fourth quadrant and consisted of three table grape cultivars from hybrids between *V. labrusca* and *V. vinifera* ['Venus Seedless'(17); 'Queenora Seedless'(22); 'Venus'(29)], two wine-making cultivars of *V. vinifera* ['Cabernet Gernischet'(58); 'Yan 73',(62)], and two wine-making cultivars from hybrids between *V. vinifera* and *V. amurensis* ['Beihong'(69); 'Beimei'(70)]. This group was characterized by a very high content of *trans*-res. 'Venus' (29), which is a hybrid between *V. vinifera* and *V. labrusca*, was in the top right quadrant which indicates that it is associated with an extremely high positive value for PC1 and PC2. This cultivar was characterized by an extremely high *trans*-res. Group 4: The other *Vitis* cultivars were found near the PC1 axis. Their *trans*-res, *trans*-pd and *trans*-pd contents were very low.

As shown in Fig. 3B for leaves, the grape cultivars can be divided into four groups based on positions in the scores scatter

Table 2. Content (C) of different forms of resveratrols and their percentage (P) accounting for the total resveratrols in berry skins and leaves of 75 grape cultivars.

Germ-plasm group	Cultivar number	Berry skins							Leaves						
		Trans-res		*Trans*-pd		*Cis*-pd		Total	*Trans*-res		*Trans*-pd		*Cis*-pd		Total
		C[a]	P[b]	C	P	C	P	C	C	P	C	R	C	R	C
T-LV[c]	1[d]	0.02	100.00	-[e]	0.00	-	0.00	0.02	0.16	1.16	1.47	10.64	12.19	88.21	13.82
	2	1.23	25.26	-	0.00	3.64	74.74	4.87	0.05	1.66	0.55	18.21	2.42	80.13	3.02
	3	1.12	84.21	0.21	15.79	-	0.00	1.33	0.82	13.36	0.77	12.54	4.55	74.10	6.14
	4	21.99	24.14	24.49	26.89	44.60	48.96	91.09	0.05	1.06	0.37	7.86	4.29	91.08	4.71
	5	0.03	0.37	3.45	42.28	4.68	57.35	8.16	0.02	0.53	0.15	3.99	3.59	95.48	3.76
	6	6.71	77.30	0.13	1.50	1.85	21.31	8.68	0.04	1.00	0.14	3.49	3.83	95.51	4.01
	7	0.16	51.61	0.15	48.39	-	0.00	0.31	0.11	29.73	0.27	72.97	-	0.00	0.37
	8	0.33	1.70	4.45	22.93	14.63	75.37	19.41	0.36	5.15	0.20	2.86	6.43	91.99	6.99
	9	0.82	13.02	1.43	22.70	4.06	64.44	6.30	0.08	0.85	0.25	2.67	9.05	96.48	9.38
	10	0.15	100.00	-	0.00	-	0.00	0.15	0.14	2.55	0.14	2.55	5.20	94.89	5.48
	11	0.2	48.78	0.21	51.22	-	0.00	0.41	0.11	0.69	4.37	27.54	11.38	71.71	15.87
	12	3.32	26.95	0.09	0.73	8.90	72.24	12.32	0.06	0.39	0.13	0.84	15.20	98.70	15.40
	13	2.27	54.83	-	0.00	1.87	45.17	4.14	0.49	9.82	0.92	18.44	3.58	71.74	4.99
	14	0.36	11.39	-	0.00	2.80	88.61	3.16	0.08	1.59	0.10	1.99	4.84	96.41	5.02
	15	0.1	8.47	-	0.00	1.07	90.68	1.18	0.50	4.07	2.14	17.41	9.65	78.52	12.29
	16	0.79	23.44	0.94	27.89	1.64	48.66	3.37	0.07	1.75	0.20	5.00	3.73	93.25	4.00
	17	16.66	55.40	0.61	2.03	12.80	42.57	30.07	0.12	3.06	0.34	8.67	3.47	88.52	3.92
	18	1.31	51.78	-	0.00	1.21	47.83	2.53	0.05	0.68	0.31	4.20	7.01	94.99	7.38
	19	0.09	7.50	-	0.00	1.10	91.67	1.20	0.02	0.34	0.45	7.73	5.35	91.92	5.82
	20	0.35	21.34	-	0.00	1.29	78.66	1.64	0.10	1.02	0.16	1.64	9.52	97.34	9.78
	21	0.06	0.19	12.88	40.10	19.18	59.71	32.12	0.39	10.18	0.15	3.92	3.29	85.90	3.83
	22	14.55	14.24	0.39	0.38	87.25	85.38	102.1	0.18	0.35	13.38	26.22	37.48	73.45	51.03
	23	0.26	14.86	0.16	9.14	1.32	75.43	1.75	0.10	2.37	0.63	14.93	3.49	82.70	4.22
	24	0.67	87.01	0.10	12.99	-	0.00	0.77	1.05	13.32	0.47	5.96	6.35	80.58	7.88
	25	0.74	87.06	0.12	14.12	-	0.00	0.85	2.43	12.15	0.15	0.75	17.43	87.15	20.00
	26	0.37	25.17	-	0.00	1.10	74.83	1.47	2.20	36.73	0.34	5.68	3.45	57.60	5.99
	27	0.01	9.09	0.10	90.91	-	0.00	0.11	0.63	2.51	6.10	24.32	18.35	73.17	25.08
	28	0.46	10.48	0.45	10.25	3.48	79.27	4.39	-	0.00	0.28	6.62	3.95	93.38	4.23
	29	84.63	54.68	3.36	2.17	66.78	43.15	154.8	0.10	2.19	0.34	7.46	4.13	90.57	4.56
	30	0.07	0.12	14.53	24.41	44.94	75.49	59.53	0.30	6.47	0.13	2.80	4.21	90.73	4.64
	31	2.57	31.93	4.01	49.81	1.48	18.39	8.05	0.05	1.99	0.10	3.98	2.36	94.02	2.51
	32	12.91	40.12	1.10	3.42	18.17	56.46	32.18	0.02	0.21	0.09	0.93	9.58	98.86	9.69
	33	2.29	100.00	-	0.00	-	0.00	2.29	1.57	6.75	4.88	20.98	16.82	72.31	23.26
	34	7.27	48.15	0.38	2.52	7.45	49.34	15.10	0.07	1.62	0.65	15.01	3.61	83.37	4.33
	Average	5.44	38.55	3.21	15.37	14.29	46.05	18.12	0.38	5.21	1.21	10.91	7.87	83.96	9.22
T-V	35	1.03	0.84	3.56	2.89	118.53	96.27	123.1	0.02	0.08	9.82	38.81	15.46	61.11	25.30
	36	1.14	55.34	0.92	44.66	-	0.00	2.06	1.60	19.32	0.65	7.85	6.02	72.71	8.28
	37	0.89	21.71	3.21	78.29	-	0.00	4.10	1.46	20.74	1.53	21.73	4.04	57.39	7.04
	38	6.65	59.80	4.11	36.96	0.36	3.24	11.12	1.82	23.73	0.50	6.52	5.35	69.75	7.67
	39	5.12	49.66	3.14	30.46	2.06	19.98	10.31	3.24	7.38	15.87	36.13	24.83	56.52	43.93
	40	4.81	64.56	2.64	35.44	-	0.00	7.45	0.12	1.60	1.77	23.57	5.62	74.83	7.51
	41	1.16	31.35	1.33	35.95	1.20	32.43	3.70	0.08	1.58	0.78	15.42	4.20	83.00	5.06
	42	0.04	0.55	2.37	32.47	4.89	66.99	7.30	0.18	3.56	0.49	9.68	4.39	86.76	5.06
	43	4.68	16.73	12.73	45.50	10.57	37.78	27.98	0.10	0.94	3.05	28.77	7.45	70.28	10.60
	44	0.02	100.00	-	0.00	-	0.00	0.02	0.11	1.34	0.80	9.77	7.28	88.89	8.19
	45	1.25	2.44	10.79	21.07	39.18	76.49	51.22	0.58	6.21	0.79	8.46	7.97	85.33	9.34
	46	0.11	24.44	-	0.00	0.34	75.56	0.45	0.44	3.35	1.78	13.57	10.90	83.08	13.12
	47	0.18	21.43	0.66	78.57	-	0.00	0.84	0.02	0.47	0.03	0.71	4.16	98.58	4.22
	48	0.48	25.13	0.08	4.19	1.36	71.20	1.91	0.03	0.63	0.23	4.86	4.47	94.50	4.73

Table 2. Cont.

Germ-plasm group	Cultivar number	Berry skins							Leaves						
		Trans-res		*Trans*-pd		*Cis*-pd		Total	*Trans*-res		*Trans*-pd		*Cis*-pd		Total
		C[a]	P[b]	C	P	C	P	C	C	P	C	R	C	R	C
	49	0.62	2.78	11.64	52.27	10.01	44.95	22.27	1.02	7.25	4.41	31.34	8.64	61.41	14.07
	Average	1.88	31.78	4.40	33.25	18.85	34.99	18.26	0.72	6.55	2.83	17.15	8.05	76.28	11.61
W-V	50	0.82	27.89	0.45	15.31	1.67	56.80	2.94	0.19	12.67	0.09	6.00	1.22	81.33	1.50
	51	0.88	24.44	0.86	23.89	1.86	51.67	3.60	0.12	5.08	0.10	4.24	2.14	90.68	2.36
	52	1.81	4.30	15.45	36.73	24.80	58.96	42.06	0.06	7.59	0.74	93.67	-	0.00	0.79
	53	7.51	64.57	0.05	0.43	4.07	35.00	11.63	0.26	1.92	0.05	0.37	13.24	97.71	13.55
	54	8.44	100.00	-	0.00	-	0.00	8.44	0.12	3.46	0.07	2.02	3.28	94.52	3.47
	55	5.28	5.26	22.02	21.92	73.14	72.82	100.4	0.38	4.06	1.87	19.96	7.12	75.99	9.37
	56	0.36	33.33	0.44	40.74	0.28	25.93	1.08	0.05	0.88	0.22	3.89	5.39	95.23	5.66
	57	0.26	2.73	0.69	7.25	8.57	90.02	9.52	0.29	2.61	0.19	1.71	10.62	95.68	11.10
	58	38.69	99.59	0.16	0.41	-	0.00	38.85	0.68	8.72	0.10	1.28	7.02	90.00	7.80
	59	1.14	35.74	0.80	25.08	1.26	39.50	3.19	0.12	2.69	0.44	9.87	3.90	87.44	4.46
	60	0.36	7.39	0.34	6.98	4.17	85.63	4.87	0.15	2.55	0.33	5.60	5.41	91.85	5.89
	61	3.71	45.35	1.34	16.38	3.13	38.26	8.18	0.04	0.90	0.91	20.54	3.48	78.56	4.43
	62	25.05	28.24	8.43	9.50	55.22	62.25	88.70	0.82	10.68	0.30	3.91	6.56	85.42	7.68
	63	0.33	16.75	0.82	41.62	0.81	41.12	1.97	0.23	4.87	0.06	1.27	4.43	93.86	4.72
	64	1	75.76	0.32	24.24	-	0.00	1.32	0.83	1.41	25.34	43.11	32.60	55.46	58.78
	Average	6.38	38.09	3.73	18.03	14.92	43.86	21.79	0.29	4.67	2.05	14.50	7.60	80.91	9.44
J	65	6.43	98.92	0.06	0.92	-	0.00	6.50	0.15	3.02	0.03	0.60	4.79	96.38	4.97
	66	12.22	64.49	1.53	8.07	5.21	27.49	18.95	1.01	6.85	3.14	21.29	10.61	71.93	14.75
	67	0.01	2.00	-	0.00	0.48	96.00	0.50	0.07	0.67	0.08	0.77	10.26	98.56	10.41
	68	0.41	0.28	0.68	0.46	145.51	99.26	146.6	0.23	1.34	0.37	2.16	16.53	96.50	17.13
	Average	4.77	41.42	0.76	2.37	50.40	55.69	43.14	0.37	2.97	0.91	6.21	10.55	90.84	11.82
W-VA	69	21.17	56.85	0.21	0.56	15.86	42.59	37.24	0.54	1.65	8.34	25.56	23.74	72.76	32.63
	70	15.09	50.86	5.90	19.89	8.67	29.22	29.67	0.10	0.77	2.63	20.23	10.27	79.00	13.00
	Average	18.13	53.85	3.06	10.22	12.27	35.91	33.46	0.32	1.21	5.49	22.90	17.01	75.88	22.82
R	71	1.59	3.46	2.30	5.00	42.13	91.55	46.02	4.13	16.07	9.28	36.11	12.29	47.82	25.70
	72	7.05	3.00	12.78	5.43	215.49	91.57	235.3	0.18	5.39	0.50	14.97	2.66	79.64	3.34
	73	6.21	1.80	2.21	0.64	336.25	97.55	344.7	2.10	14.12	4.54	30.53	8.23	55.35	14.87
	74	4.04	1.10	1.82	0.50	360.12	98.40	365.9	1.07	8.53	3.23	25.74	8.25	65.74	12.55
	75	3.87	19.08	-	0.00	16.41	80.92	20.28	4.58	6.41	25.99	36.35	40.93	57.24	71.50
	Average	4.55	5.69	4.78	2.31	194.08	92.00	202.5	2.41	10.10	8.71	28.74	14.47	61.16	25.59

Notes:
[a]Content (C) unit is μg g^{-1}FW;
[b]Individual resveratrol as percentage of the total resveratrol pool;
[c]T-LV, T-V, W-V, J, W-VA and R represent the germplasm groups as in Table 1.
[d]Cultivar number corresponds to the accession number in Table 1.;
[e]Indicates not detected.

Table 3. Correlation of resveratrols between and in grape berry skins and leaves.

	Trans-res	*Trans*-pd	*Cis*-pd	total
Trans-res	**−0.043**	0.565**[y]	0.473**[y]	0.578**[y]
Trans-pd	0.118[z]	**−0.078**	0.852**[y]	0.946**[y]
Cis-pd	0.141[z]	0.196[z]	**0.057**	0.973**[y]
total	0.293*[z]	0.286*[z]	0.983**[z]	**0.077**

The data on the diagonal line from top left corner to lower right corner in the table shows the correlation coefficient of the contents of individual resveratrol and total resveratrols between berry skins and leaves in all germplasms.
[z]Berry skin data and.
[y]leaf data shows the correlation coefficients among different individual resveratrols from the tissues.
* and ** indicate a significant correlation at $P<0.05$ and $P<0.01$, respectively.

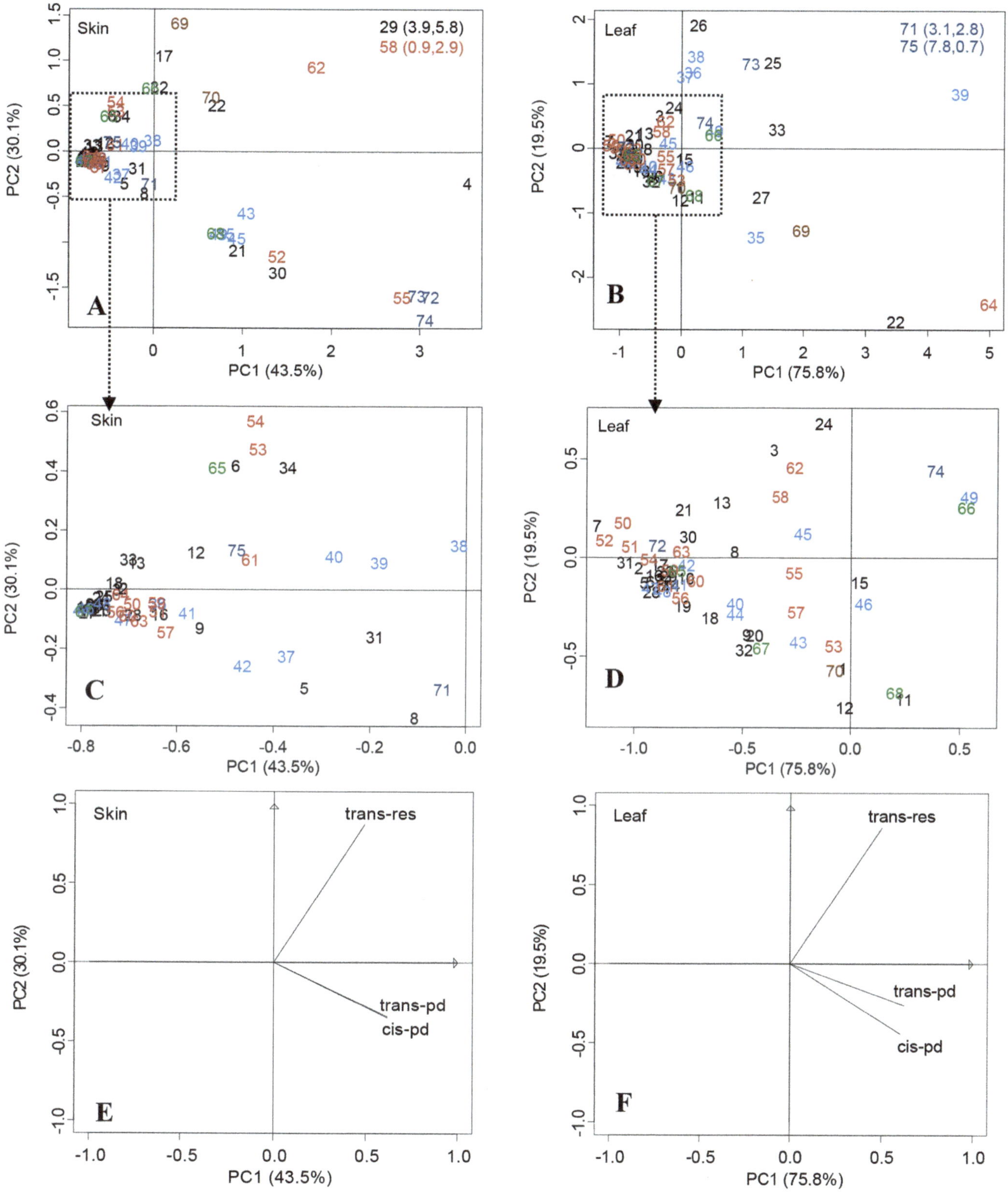

Figure 3. Principal Component Analysis (PCA) of seventy five grape berry skins and leaves. Fig. 3A and Fig. 3C: Scores scatter plot of PCA in berry skins; Fig. 3E: Loadings plot of PCA in berry skins; Fig. 3B and Fig. 3D: Scores scatter plot of PCA in leaves; Fig. 3F: Loadings plot of PCA in leaves. Percentages in parentheses represent the variance of each component. The numbers in the figure represent the sample numbers, which correspond to those in Table 1. Cultivars 1 to 34 are table grapes from hybrids between *V. labrusca* and *V. vinifera*; 35 to 49 are table grapes of *V. vinifera*; 50 to 64 are table grapes of *V. vinifera*; 65 to 68 are juice grapes; 69 and 70 are wine grapes of *V. vinifera* × *V. amurensis*; and, 71 to 75 are rootstock cultivars.

plot of PCA. Group 1: This group was situated mainly in middle and lower right part of the third quadrant and consisted of two table grape cultivars from hybrids between *V. labrusca* and *V. vinifera* ['Queenora Seedless'(22); 'Fujiminori'(27)], one table grape cultivar of *V. vinifera* ['Gros Colman' (35)], one wine-making cultivar of *V. vinifera* ['Zexiang' (64)], and one wine-making cultivar from a hybrid between *V. vinifera* and *V. amurensis* ['Beihong' (69)]. This group was characterized by higher contents of *trans*-pd and *cis*-pd. Group 2: This group was situated primarily in the left and right parts of the four quadrants and was comprised of three table grape cultivars from hybrids between *V. labrusca* and *V. vinifera* ['Pondicherry'(25); 'Beni Sajku'(26); 'Ziguang'(33)], four table grape cultivars of *V. vinifera* [('Misket Dounvaski'(36); 'Flame Tokay'(37); 'Fenniu'(38); 'Fenghuang 51'(39)], and three rootstock cultivars ['Beta'(71); 'Zhi168'(73)]. This group was characterized by a higher *trans*-res content. Group 3: This group was situated in the right side of the fourth quadrant and was comprised of the rootstock grape *V. Berlandieri* 'Dog Ridge'(75) with higher *trans*-res, *trans*-pd and *cis*-pd. Group 4: The other *Vitis* cultivars were found near the PC1 axis. Their *trans*-res, *trans*-pd and *cis*-pd contents were very low.

Conclusions

Resveratrol content varied significantly with genetic background as well as their usage. *Trans*-res, *cis*-pd and *trans*-pd were detected in berry skin and leaves, but *cis*-res was not. Generally, resveratrol totals were lower in leaves than berry skins in most cultivars. Rootstock cultivars had a higher content of total resveratrols than the other cultivars. *Cis*-piceid was the most abundant resveratrol in most cultivars, while *trans*-res and *trans*-pd were minor components. Some cultivars were found to have extremely high or low *trans*-res, *cis*-pd, *trans*-pd or total resveratrols in berry skins or leaves. There were no significant correlations of total resveratrols between berry skins and leaves.

Author Contributions

Conceived and designed the experiments: SL LW. Performed the experiments: LW MX CL JW HX BW. Analyzed the data: LW MX CL. Contributed reagents/materials/analysis tools: WD PF. Wrote the paper: LW SL MX CL WL.

References

1. Takaoka MJ (1940) The phenolic substances of white hellebore (*Veratrum grandiflorum* Loes.). J Faculty Sci 3: 1–16.
2. Langcake P, Pryce RJ (1976) The production of resveratrol by *Vitis vinifera* and other members of the Vitaceae as a response to infection or injury. Physiol Plant Pathol 9: 77–86.
3. Jeandet P (2002) Phytoalexins from the Vitaceae: biosynthesis, phytoalexin gene expression in transgenic plants, antifungal activity, and metabolism. J Agr Food Chem 50: 2731–2741.
4. Stervbo U (2007) A review of the content of the putative chemopreventive phytoalexine resveratrol in red wine. Food Chem 101: 449–457.
5. Wang J (2006) Moderate consumption of Cabernet Sauvignon attenuates Abeta neuropathology in a mouse model of Alzheimer's disease. FASEB J 20: 2313–2320
6. Grønbæk M (2000) Type of alcohol consumed and mortality from all causes, coronary heart disease and cancer. Ann Intern Med 133: 411–419.
7. Kalantari H, Das DK (2010) Physiological effects of resveratrol. Biofactors 36: 401–406
8. Jeandet P (1995) Effect of enological practices on the resveratrol isomer content of wine. J Agr Food Chem 43: 313–316.
9. Chong J, Poutaraud A, Hugueney P (2009) Metabolism and roles of stilbenes in plants. Plant Sci 177: 143–155.
10. Kimura Y, Okuda H (2000) Effects of naturally occurring stilbene glucosides from medicinal plants and wine, on tumour growth and lung metastasis in Lewis lung carcinoma-bearing mice. J Pharm Pharmacol 52: 1287–1295.
11. Li XD, Wang L, Wu B, Li S (2006) *Trans*-resveratrol concentrations in seed and berry skin in *Vitis* evaluated at the germplasm level. J Agr Food Chem 54: 8804–8811
12. Gerogiannaki-Christopoulou M, Athanasopoulos P, Kyriadis N, Gerogiannaki IA, Spanos M (2006) *Trans*-Resveratrol in wines from the major Greek red and white grape varieties. Food Control 17, 700–706
13. Moreno A, Castro M, Falqué E (2008) Evolution of *trans*- and *cis* resveratrol content in red grapes (*Vitis vinifera* L. cv Mencía, Albarello and Merenzao) during ripening. Eur Food Res Technol 227: 667–674
14. Park HJ, Ko JM, An N, Kim YS, Cha HC (2009). Contents of trans-resveratrol in Korean Grape Cultivars including 'Kyoho'. J Plant Biol 52: 319–324
15. Liu CY, Wang LJ, Wang JF, Wu BH, Liu W, et al. (2013) Resveratrols in *Vitis* berry skins and leaves: their extraction and analysis by HPLC. Food Chem 136,643–649
16. Becker RA, Chambers JM, Wilks AR (1998) Advance graphics. InThe New S Language, A Programming environment for data analysis and graphics; Wadsworth and Brooks/Cole Advanced Books and Software: Pacific Grove, CA, pp 307–336.
17. Morales M, Bru R, Garcia-Carmona F, Barcelo AR, Pedreno MA (1998) Effect of dimethyl-beta-cyclodextrins on resveratrol metabolism in Gamay grapevine cell cultures before and after inoculation with *Xylophilus ampelinus*. Plant Cell Tiss Org 53: 179–187.
18. Gatto P, Vrhovsek U, Muth J, Segala C, Romualdi C, et al. (2008) Ripening and genotype control stilbene accumulation in healthy grapes. J Agr Food Chem 56: 11773–11785.

Permissions

All chapters in this book were first published in PLOS ONE, by The Public Library of Science; hereby published with permission under the Creative Commons Attribution License or equivalent. Every chapter published in this book has been scrutinized by our experts. Their significance has been extensively debated. The topics covered herein carry significant findings which will fuel the growth of the discipline. They may even be implemented as practical applications or may be referred to as a beginning point for another development.

The contributors of this book come from diverse backgrounds, making this book a truly international effort. This book will bring forth new frontiers with its revolutionizing research information and detailed analysis of the nascent developments around the world.

We would like to thank all the contributing authors for lending their expertise to make the book truly unique. They have played a crucial role in the development of this book. Without their invaluable contributions this book wouldn't have been possible. They have made vital efforts to compile up to date information on the varied aspects of this subject to make this book a valuable addition to the collection of many professionals and students.

This book was conceptualized with the vision of imparting up-to-date information and advanced data in this field. To ensure the same, a matchless editorial board was set up. Every individual on the board went through rigorous rounds of assessment to prove their worth. After which they invested a large part of their time researching and compiling the most relevant data for our readers.

The editorial board has been involved in producing this book since its inception. They have spent rigorous hours researching and exploring the diverse topics which have resulted in the successful publishing of this book. They have passed on their knowledge of decades through this book. To expedite this challenging task, the publisher supported the team at every step. A small team of assistant editors was also appointed to further simplify the editing procedure and attain best results for the readers.

Apart from the editorial board, the designing team has also invested a significant amount of their time in understanding the subject and creating the most relevant covers. They scrutinized every image to scout for the most suitable representation of the subject and create an appropriate cover for the book.

The publishing team has been an ardent support to the editorial, designing and production team. Their endless efforts to recruit the best for this project, has resulted in the accomplishment of this book. They are a veteran in the field of academics and their pool of knowledge is as vast as their experience in printing. Their expertise and guidance has proved useful at every step. Their uncompromising quality standards have made this book an exceptional effort. Their encouragement from time to time has been an inspiration for everyone.

The publisher and the editorial board hope that this book will prove to be a valuable piece of knowledge for researchers, students, practitioners and scholars across the globe.

List of Contributors

Erica N. C. Renaud and Edith T. Lammerts van Bueren
Wageningen UR Plant Breeding, Plant Sciences Group, Wageningen University, Wageningen, The Netherlands

James R. Myers
Department of Horticulture, Oregon State University, Corvallis, Oregon, United States of America

Maria João Paulo and Fred A. van Eeuwijk
Biometris, Plant Sciences Group, Wageningen University, Wageningen, The Netherlands

Ning Zhu and John A. Juvik
Department of Crop Sciences, University of Illinois, Urbana, Illinois, United States of America

Giovanni Laidò, Luigi Cattivelli, Daniela Marone, Anna M. Mastrangelo, Roberto Papa and Pasquale De Vita
Consiglio per la Ricerca e la sperimentazione in Agricoltura, Cereal Research Centre, Foggia, Italy

Giacomo Mangini, Francesca Taranto, Agata Gadaleta and Antonio Blanco
Department of Soil, Plant, and Food Sciences, Section of Genetics and Plant Breeding, University of Bari, Via Amendola, Bari, Italy

Xiling Zou, Chengwei Hu, Liu Zeng, Yong Cheng, Mingyue Xu and Xuekun Zhang
Key Laboratory of Oil Crop Biology and Genetic Improvement of the Ministry of Agriculture, Oil Crops Research Institute of the Chinese Academy of Agricultural Sciences, Wuhan, Hubei, China

Gangaprasad Choudhary., Nagireddy Ranjitkumar., Malathi Surapaneni., Dondapati Annekitty Deborah, Abhilash Vipparla, Ghanta Anuradha, Ebrahimali Abubacker Siddiq and Lakshminarayana Reddy Vemireddy
Institute of Biotechnology, Acharya NG Ranga Agricultural University, Rajendranagar, Hyderabad, Andhra Pradesh, India

Dadong Zhang
Department of Agronomy, Kansas State University, Manhattan, Kansas, United States of America

Robert L. Bowden
USDA-ARS Hard Winter Wheat Genetics Research Unit, Manhattan, Kansas, United States of America

Jianming Yu
Department of Agronomy, Iowa State University, Ames, Iowa, United States of America

Brett F. Carver
Plant Science Department, Oklahoma State University, Stillwater, Oklahoma, United States of America

Guihua Bai
Department of Agronomy, Kansas State University, Manhattan, Kansas, United States of America
USDA-ARS Hard Winter Wheat Genetics Research Unit, Manhattan, Kansas, United States of America

Elisa Vendramin, Maria Teresa Dettori, Laura Gazza and Ignazio Verde
Consiglio per la Ricerca e la Sperimentazione in Agricoltura - Centro di Ricerca per la Frutticoltura (CRA-FRU), Rome, Italy

Giorgio Pea and Francesco Strozzi
Parco Tecnologico Padano, Via Einstein, Loc. C.na Codazza, Lodi, Italy

Luca Dondini and Stefano Tartarini
Universitá di Bologna, DipSA, Bologna, Italy

Igor Pacheco and Daniele Bassi
Universitá degli Studi di Milano, DiSAA, Milan, Italy

Simone Scalabrin
IGA Technology Services, Udine, Italy

Laura Rossini
Parco Tecnologico Padano, Via Einstein, Loc. C.na Codazza, Lodi, Italy
Universitá degli Studi di Milano, DiSAA, Milan, Italy

Eva Konečná and Pavel Hanáček
Department of Plant Biology, Mendel University in Brno, Brno, Czech Republic
CEITEC MENDELU, Mendel University in Brno, Brno, Czech Republic

Dana Šafářová and Milan Navrátil
Department of Cell Biology and Genetics, Palacky University in Olomouc, Olomouc, Czech Republic

Clarice Coyne
Western Regional Plant Introduction Station - USDA, Pullman, Washington, United States of America

Andrew Flavell
Division of Plant Sciences, University of Dundee at James Hutton Institute, Invergowrie, United Kingdom

Margarita Vishnyakova
Vavilov Institute of Plant Industries, Saint Petersburg, Russian Federation

Mike Ambrose
John Innes Centre, Norwich, United Kingdom

Robert Redden
Australian Grains Genebank, Horsham, Victoria, Australia

Petr Smýkal
Department of Botany, Palacky University in Olomouc, Olomouc, Czech Republic

Jeffrey C. Glaubitz, Terry M. Casstevens, Fei Lu, James Harriman and Robert J. Elshire
Institute for Genomic Diversity, Cornell University, Ithaca, New York, United States of America

Qi Sun
Biotechnology Resource Center Bioinformatics Facility, Cornell University, Ithaca, New York, United States of America

Edward S. Buckler
Institute for Genomic Diversity, Cornell University, Ithaca, New York, United States of America
USDA Agricultural Research Service, Ithaca, New York, United States of America

Jun Niu, Baoqing Zhu, Jian Cai, Zhixiang Zhang and Shanzhi Lin
College of Biological Sciences and Biotechnology, College of Nature Conservation, National Engineering Laboratory for Tree Breeding, Key Laboratory of Genetics and Breeding in Forest Trees and Ornamental Plants, Ministry of Education, Beijing Forestry University, Beijing, China

Peixue Li Huitang Dai, Lin Qiu, Denglong Ha and Haiyan Zhao
Jigongshan National Nature Reserve, Xingyang, China

Libing Wang and Haiyan Yu
Research Institute of Forestry, Chinese Academy of Forestry, Beijing, China

Nailin Xing, Chuchuan Fan and Yongming Zhou
National Key Laboratory of Crop Genetic Improvement, Huazhong Agricultural University, Wuhan, China

Xuehui Li, Yuanhong Han, Yanling Wei, Ananta Acharya, Maria J. Monteros and E. Charles Brummer
Forage Improvement Division, The Samuel Roberts Noble Foundation, Ardmore, Oklahoma, United States of America

Andrew D. Farmer
National Center for Genome Resources, Santa Fe, New Mexico, United States of America

Julie Ho
Forage Genetics International, Davis, California, United States of America

Luca Bianco, Alessandro Cestaro, Daniel James Sargent, Elisa Banchi, Riccardo Velasco, Michela Troggio and Roberto Viola
Research and Innovation Centre, Fondazione Edmund Mach, San Michele all'Adige, Trento, Italy

Eric van de Weg
Wageningen UR Plant Breeding, Wageningen University and Research Centre, Wageningen, The Netherlands

Sophia Derdak and Ivo Gut
CNAG – Centro Nacional de Análisis Genómico, Parc Científic de Barcelona, Barcelona, Spain

Mario Di Guardo
Research and Innovation Centre, Fondazione Edmund Mach, San Michele all'Adige, Trento, Italy
Wageningen UR Plant Breeding, Wageningen University and Research Centre, Wageningen, The Netherlands

Silvio Salvi
DipSA - University of Bologna, Bologna, Italy

Johannes Jansen
Biometris, Wageningen University and Research Centre, Wageningen, The Netherlands

Francois Laurens
INRA, UMR1345 Institut de Recherche en Horticulture and Semences, Beaucouzé, France

David Chagné
Plant & Food Research, Palmerston North Research Centre, Palmerston North, New Zealand

Tong Zhou, Linlin Du, Hui Feng, Lijiao Wang, Ying Lan, Feng Sun, Lihui Wei, Yongjian Fan and Yijun Zhou
Institute of Plant Protection, Jiangsu Academy of Agricultural Sciences, Nanjing, China

Wenbiao Shen
College of Life Science, Nanjing Agricultural University, Nanjing, China

Cunyi Gao
Institute of Plant Protection, Jiangsu Academy of Agricultural Sciences, Nanjing, China
College of Life Science, Nanjing Agricultural University, Nanjing, China

Chun Li, Hongmei Miao, Libin Wei, Tide Zhang, Xiuhua Han and Haiyang Zhang
Henan Sesame Research Center, Henan Academy of Agricultural Sciences, Zhengzhou, Henan, People's Republic of China

Marzena Nowakowska, Marcin Nowicki, Urszula Kłosińska and Elżbieta U. Kozik
Department of Genetics, Breeding, and Biotechnology of Vegetable Crops, Research Institute of Horticulture, Skierniewice, Poland

Robert Maciorowski
Unit of Economics and Statistics, Research Institute of Horticulture, Skierniewice, Poland

Hamid Khazaei and Frederick L. Stoddard
Department of Agricultural Sciences, University of Helsinki, Helsinki, Finland

Kenneth Street and Abdallah Bari
International Centre for Agricultural Research in the Dry Areas, Aleppo, Syria

Michael Mackay
Queensland Alliance for Agricultural and Food Innovation, The University of Queensland, St Lucia, Queensland, Australia

Hongxian Mei
National Key Laboratory of Crop Genetics & Germplasm Enhancement, Cotton Research Institute, Nanjing Agricultural University, Nanjing, China
Sesame Research Center, Henan Academy of Agricultural Sciences, Zhengzhou, China

Xiefei Zhu and Tianzhen Zhang
National Key Laboratory of Crop Genetics & Germplasm Enhancement, Cotton Research Institute, Nanjing Agricultural University, Nanjing, China

Yong-Ju Huang, Aiming Qi and Bruce D. L. Fitt
School of Life and Medical Sciences, University of Hertfordshire, Hatfield, Hertfordshire, United Kingdom
Department of Plant Pathology and Microbiology, Rothamsted Research, Harpenden, Hertfordshire, United Kingdom

Graham J. King
Department of Plant Pathology and Microbiology, Rothamsted Research, Harpenden, Hertfordshire, United Kingdom
Southern Cross Plant Science, Southern Cross University, Lismore, Australia

Alexander E. Lipka and Fei Lu
Institute for Genomic Diversity, Cornell University, Ithaca, New York, United States of America

Jerome H. Cherney
Department of Crop and Soil Sciences, Cornell University, Ithaca, New York, United States of America

Edward S. Buckler
Institute for Genomic Diversity, Cornell University, Ithaca, New York, United States of America
Agricultural Research Service, United States Department of Agriculture, Ithaca, New York, United States of America
Department of Plant Breeding and Genetics, Cornell University, Ithaca, New York, United States of America

Michael D. Casler
Agricultural Research Service, United States Department of Agriculture, Madison, Wisconsin, United States of America
Department of Agronomy, University of Wisconsin-Madison, Madison, Wisconsin, United States of America

Denise E. Costich
Institute for Genomic Diversity, Cornell University, Ithaca, New York, United States of America
Agricultural Research Service, United States Department of Agriculture, Ithaca, New York, United States of America

Andrea Giordano and German C. Spangenberg
Department of Environment and Primary Industries, AgriBio, Centre for AgriBioscience, Bundoora, Victoria, Australia
Molecular Plant Breeding Cooperative Research Centre, Bundoora, Victoria, Australia
Dairy Futures Cooperative Research Centre, Bundoora, Victoria, Australia
La Trobe University, Bundoora, Victoria, Australia

Noel O. I. Cogan, Michelle Drayton, Aidyn Mouradov and Stephen Panter
Department of Environment and Primary Industries, AgriBio, Centre for AgriBioscience, Bundoora, Victoria, Australia
Molecular Plant Breeding Cooperative Research Centre, Bundoora, Victoria, Australia

Dairy Futures Cooperative Research Centre, Bundoora, Victoria, Australia

Sukhjiwan Kaur
Department of Environment and Primary Industries, AgriBio, Centre for AgriBioscience, Bundoora, Victoria, Australia

Gustavo E. Schrauf
Facultad de Agronomia, Universidad de Buenos Aires, Buenos Aires, Argentina

John G. Mason
Department of Environment and Primary Industries, AgriBio, Centre for AgriBioscience, Bundoora, Victoria, Australia
Dairy Futures Cooperative Research Centre, Bundoora, Victoria, Australia
La Trobe University, Bundoora, Victoria, Australia

Jiafu Tan, Maojun Wang, Lili Tu, Yichun Nie, Yongjun Lin and Xianlong Zhang
National Key Laboratory of Crop Genetic Improvement, Huazhong Agricultural University, Wuhan, Hubei, China

Lijun Wang, Chunyan Liu, Benhong Wu, Wei Duan and Peige Fan
Key Laboratory of Plant Resources and Beijing Key Laboratory of Grape Science and Enology, Institute of Botany, the Chinese Academy of Sciences, Beijing, China

Man Xu, Junfang Wang and Huifen Xi
Key Laboratory of Plant Resources and Beijing Key Laboratory of Grape Science and Enology, Institute of Botany, the Chinese Academy of Sciences, Beijing, China
University of Chinese Academy of Sciences, Beijing, China

Wayne Loescher
College of Agriculture and Natural Resources, Michigan State University, East Lansing, Michigan, United State of America

Shaohua Li
Key Laboratory of Plant Resources and Beijing Key Laboratory of Grape Science and Enology, Institute of Botany, the Chinese Academy of Sciences, Beijing, China
Key Laboratory of Plant Germplasm Enhancement and Specialty Agriculture, Wuhan Botanical Garden, the Chinese Academy of Sciences, Wuhan, China

Index